Lecture Notes in Artificial Intelligence 8041

Subseries of Lecture Notes in Computer Science

LNAI Series Editors

Randy Goebel
University of Alberta, Edmonton, Canada
Yuzuru Tanaka
Hokkaido University, Sapporo, Japan
Wolfgang Wahlster
DFKI and Saarland University, Saarbrücken, Germany

LNAI Founding Series Editor

Joerg Siekmann
DFKI and Saarland University, Saarbrücken, Germany

Mingzheng Wang (Ed.)

Knowledge Science, Engineering and Management

6th International Conference, KSEM 2013
Dalian, China, August 10-12, 2013
Proceedings

 Springer

Volume Editor

Mingzheng Wang
Dalian University of Technology
Faculty of Management and Economics
Dalian 116023, China
E-mail: mzhwang@dlut.edu.cn

ISSN 0302-9743   e-ISSN 1611-3349
ISBN 978-3-642-39786-8   e-ISBN 978-3-642-39787-5
DOI 10.1007/978-3-642-39787-5
Springer Heidelberg Dordrecht London New York

Library of Congress Control Number: 2013943405

CR Subject Classification (1998): I.2.4, H.3, I.2, H.4, J.1, H.2.8

LNCS Sublibrary: SL 7 – Artificial Intelligence

Typesetting: Camera-ready by author, data conversion by Scientific Publishing Services, Chennai, India

Printed on acid-free paper

Springer is part of Springer Science+Business Media (www.springer.com)

Preface

Building on the success of previous events in Guilin, China (KSEM 2006), Melbourne, Australia (KSEM 2007), Vienna, Austria (KSEM 2009), Belfast, UK (KSEM 2010), and Irvine, USA (KSEM 2011), the 2013 International Conference on Knowledge Science, Engineering and Management (KSEM 2013) was held during August 10–12, 2013, in Dalian, China. The conference was hosted by Dalian University of Technology of China, which provided a leading international forum for sharing views, ideas, and original work among researchers, practitioners, and developers to offer new insights into KSEM-related areas. In addition, this conference was also combined with a Workshop on Chinese Practices of Innovation and Creativity Industry. This workshop attracted many famous experts and senior government officials who are interested in Chinese practices in the area of innovation in the creativity industry.

KSEM 2013 attracted numerous high-quality, state-of-the-art research papers from various countries. All submitted manuscripts were peer-reviewed by at least two PC members as well as external reviewers. Based on the referee reports, 33 papers were accepted as regular presentations and 18 papers for short presentations. Besides, the conference also featured keynote speakers and invited talks.

The success of KSEM 2013 was assured by the team efforts of a large number of people. We would like to acknowledge the contributions of the Organizing Committee and PC members for this conference. Thanks to Prof. Zhongtuo Wang, Prof. Ruqian Lu, Prof. Chengqi Zhang, and Prof. Zhi Jin for their valuable advice and suggestions. Our sincere gratitude goes to the participants and all the authors.

On behalf of the Organizing Committee of KSEM 2013, we thank Dalain University of Technology, the Natural Science Foundation of China (NSFC), and the Dalian Municipal Science and Technology Bureau for their sponsorship.

We wish to express our gratitude to the Springer team directed by Alfred Hofmann for their help and coordination.

Yanzhong Dang
Mingzheng Wang

Organization

KSEM 2013 was hosted and organized by the School of Management Science and Engineering, Dalian University of Technology, China. The conference was held on August 10–12, 2013, at Dalian City, Liaoning Province, China.

Organizing Committee

General Co-chairs

Yanzhong Dang   Dalian University of Technology, China
Maria Gini   University of Minnesota - Twin Cities, USA

Program Co-chairs

Geoff Webb   Monash University, Australia
     Data Mining and Knowledge Discovery, Editor-in-Chief
Jiangning Wu   Dalian University of Technology, China

Publicity Chairs

Jiangning Wu   Dalian University of Technology, China
Shuigeng Zhou   Fudan University, China

Publication Chair

Mingzheng Wang   Dalian University of Technology, China

Local Arrangements Chairs

Chonghui Guo   Dalian University of Technology, China
Haoxiang Xia   Dalian University of Technology, China

Web Chairs

Guangfei Yang   Dalian University of Technology, China
Donghua Pan   Dalian University of Technology, China

Steering Committee

David Bell   Queen's University, UK
Cungen Cao   Chinese Academy of Sciences, China
Dimitris Karagiannis   University of Vienna, Austria
Zhi Jin   Peking University, China
Jérome Lang   University Paul Sabatier, France
Yoshiteru Nakamori   JAIST, Japan
Jorg Siekmann   DFKI, Germany
Eric Tsui   The Hong Kong Polytechnic University,
                                      Hong Kong, China
Zhongtuo Wang   Dalian University of Technology, China
Kwok Kee Wei   City University of Hong Kong, Hong Kong,
                                      China
Mingsheng Ying   Tsinghua University, China
Zili Zhang   Southwest University, China
Yaxin Bi   Ulster University, UK
Ruqian Lu (Honorary Chair)   Chinese Academy of Sciences, China
Chengqi Zhang (Chair)   University of Technology, Sydney, Australia

Program Committee

Barry O'Sullivan   University College Cork, Ireland
Chonghui Guo   Dalian University of Technology, China
Chunlai Zhou   Renmin University of China, China
Chunxia Zhang   Beijing Institute of Technology, China
Carl Vogel   Trinity College Dublin, Ireland
Dan Oleary   University of Southern California, USA
Editor I.Czarnowski   Gdynia Maritime University, Poland
Enhong Chen   University of Science and Technology of China
Elsa Negre   Université Paris-Dauphine, LAMSADE, France
Eric Tsui   The Hong Kong Polytechnic University, SAR
                                      China
Gabriele Kern-Isberner   Technische Universität
                                      Dortmund, Germany
Haoxiang Xia   Dalian University of Technology, China
Hui Xiong   Rutgers University, USA
John-Jules Meyer   Utrecht University, The Netherlands
Juan J. Rodriguez   University of Burgos, Spain
Josep Domenech   Universitat Politècnica de València Spain
Josep Domingo-Ferrer   Rovira i Virgili University Spain
Jun Hong   Queen's University Belfast, UK
Jia-Huai You   University of Alberta, Canada
Jie Wang   ASU
Jun Hong   Queen's University Belfast, UK

Table of Contents

Short Papers

Co-anomaly Event Detection
in Multiple Temperature Series

Xue Bai[1], Yun Xiong[1], Yangyong Zhu[1], Qi Liu[2], and Zhiyuan Chen[3]

[1] Fudan University
[2] University of Science and Technology of China
[3] University of Maryland Baltimore County
{xuebai,yunx,yyzhu}@fudan.edu.cn, feiniaol@mail.ustc.edu.cn,
zhchen@umbc.edu

Abstract. Co-anomaly event is one of the most significant climate phenomena characterized by the co-occurrent similar abnormal patterns appearing in different temperature series. Indeed, these co-anomaly events play an important role in understanding the abnormal behaviors and natural disasters in climate research. However, to the best of our knowledge the problem of automatically detecting co-anomaly events in climate is still under-addressed due to the unique characteristics of temperature series data. To that end, in this paper we propose a novel framework *Sevent* for automatic detection of co-anomaly climate events in multiple temperature series. Specifically, we propose to first map the original temperature series to symbolic representations. Then, we detect the co-anomaly patterns by statistical tests and finally generate the co-anomaly events that span different sub-dimensions and subsequences of multiple temperature series. We evaluate our detection framework on a real-world data set which contains rich temperature series collected by 97 weather stations over 11 years in Hunan province, China. The experimental results clearly demonstrate the effectiveness of *Sevent*.

Keywords: Event Mining, Co-anomaly Event, Time Series.

1 Introduction

Since climate events reveal seasonal or interannual variations in climate change from periodic weather behaviors, mining them from climate data recorded in temperature series has attracted much attention in the literature [1,2]. Among climate events, the co-anomaly event, which represents the co-occurrence of similar abnormal behaviors in different temperature series, is one of the most important events in climate research for understanding climate variability and analyzing the process of abnormal events.

For better understanding co-anomaly event, Fig. 1 illustrates the subsequences of six daily temperature series from Dec.13 to Dec.24, 1998 in Hunan, China, where Fig. 1 (a) presents three stations with normal temperature behaviors and Fig. 1 (b) presents three temperature series that have unusual higher values than expected. Though the six weather stations in Fig. 1 are similar to each other in magnitudes, we can see the ones in Fig. 1 (b) are suffering from a co-anomaly event represented by unusual warm in the middle of winter. As the abnormal temperature behaviors in one co-anomaly event (e.g., that in Fig 1 (b)) are much likely to be caused by the similar climatic factors, mining

M. Wang (Ed.): KSEM 2013, LNAI 8041, pp. 1–14, 2013.

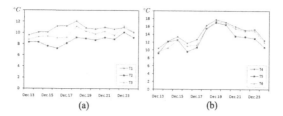

Fig. 1. Subsequences of six temperature series ($^\circ C$) from different weather stations. (a) T1, T2, and T3 are normal in winter. (b) T4, T5, and T6 are in a co-anomaly event.

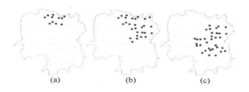

Fig. 2. A cold wave event moving from north to south of Hunan in adjacent three days

and identifying such co-anomaly events can provide a detailed exploration on these climate phenomena. For instance, it helps experts quickly identify whether a co-occurrent unusual phenomenon occurred by chance or not and the value of further analysis. Thus, capturing this co-occurrence of similar abnormal behaviors (co-anomaly event) is of growing interests in many real-world applications [3].

However, there are many technical and domain challenges inherent in detecting co-anomaly climate events in temperature series. First, temperature series in climate are relatively smooth curves, e.g., much smoother than stock price time series and vehicle sensor time series. In other words, the values of temperature series usually do not deviate far from the average. Thus, some co-anomaly events taking place at a limited number of cities are not that obvious with respect to the average temperature series or each single series, and this raises difficulties for traditional methods. Secondly, different from traditional anomaly events, the similar abnormal behaviors should co-occur in a number of series (i.e., a sub-dimension of the entire series set) before we can claim this is a co-anomaly event. However, we can not simply use the frequency as an evaluation to find interesting patterns because the frequent patterns in climate usually represent well-known normal phenomena. Thirdly, co-anomaly climate events often evolve with time, thus the associated groups of temperature series are changing too, i.e., they usually correlate with different sub-dimensional subsequences of multiple temperature series. For instance, Fig. 2 shows a cold wave moved from north to south during three days in spring, where we can see that the affected cities on the first day (blue points in Fig. 2 (a)) were quite different from the ones on the third day (blue points in Fig. 2 (c)). Since identifying co-anomaly events from temperature series data is not technically straightforward, the researchers and experts usually have to search and analyze these events

Table 1. Mathematical notations

Notation	Description
$D = T_1, T_2, ... T_m$	The set of temperature series data
$T = t_1, t_2, ..., t_n$	A temperature series
$S = t_p, ..., t_{p+k-1}$	A subsequence of a temperature series T
$\overline{S} = \overline{s}_1, ..., \overline{s}_w$	A Piecewise Aggregate Approximation of a subsequence S
$\hat{S} = \hat{s}_1, ..., \hat{s}_w$	A symbol representation (word) of a subsequence S
$E = \hat{e}_1, ..., \hat{e}_u$	A co-anomaly event E
$B = \beta_1, ..., \beta_{\Psi-1}$	Breakpoints
w	The number of PAA elements
Ψ	Alphabet size. The total number of different symbols
ϕ	The number of common temperature series between words

manually. However, the volume and complexity of the data preclude the use of manual visualization to identify these co-occurrent patterns.

To address the above challenges, in this paper we propose a novel co-anomaly event detection framework called *Sevent* which includes three phases: first, map multiple temperature series to symbolic representations based on data distributions; Second, apply statistical tests to extract interesting co-anomaly patterns; Third, the co-anomaly patterns are connected into co-anomaly events by their correlations. Thus, the co-anomaly events are finally generated and ranked. Our main contributions can be summarized as:

To the best of our knowledge, we are the first to solve the co-anomaly event mining problem in multiple temperature series. Meanwhile, we propose a symbolic representation framework that can differentiate group behaviors. Though we describe the work in a domain-depended way, worth noting that similar idea is generally applicable to mine co-anomaly events from other types of series data. We carry out extensive experiments on real-world data set [4] of temperature series collected from 97 weather stations over 11 years in Hunan province, China. The results show that the proposed *Sevent* can successfully detect co-anomaly underlying events interested in meteorology.

2   Problem Statement and Data Description

In this paper, we focus on dealing with the problem of detecting and ranking significant co-anomaly climate events from a given set of temperature series, and meanwhile, identifying the corresponding cities(or sub-dimensions) and time-spans (or subsequence) affected. As have said our solutions can be generally applied to pattern mining problems for multiple time series, including but not limited to detecting climate co-anomaly events from temperature series.

We exploit a real-world temperature dataset [4] collected from 97 weather stations over a period of 11 years in Hunan province, China. Thus, each station stands for one temperature series, and each temperature series records the daily average temperatures of the corresponding weather station. In all, there are 365×11 data points for each temperature series to represent the temperature behaviors over time. The timestamps at February 29th are directly removed from the data set for simplicity. Here, we choose the temperature data because temperature is a well accepted important climate variable and many of the well known climate indices are based upon it. At last, worth noting that the spatial distances of stations, although important, are not taken into consideration for two reasons: First, the weather stations are not far from each other in our data set (all locating in one province); Second, in this work, we focus on the problem of detecting

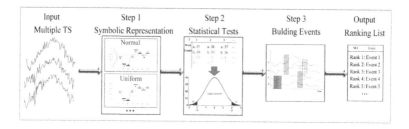

Fig. 3. The flowchart of Sevent

co-anomaly events from multiple temperature series and we would like to leave the detection of co-anomaly events from Geo-referenced time series as a future work.

3 Mining Co-anomaly Events

In this section, we present our framework $Sevent$ in detail. To facilitate understanding, the related important mathematical notations are illustrated in Table 1. To be specific, we define a ***temperature series*** $T = t_1, t_2, ..., t_n$, which records the temperature values over time, as an ordered set of n real-valued variables, where data points $t_1, t_2, ..., t_n$ are temporally ordered and spaced at equal time intervals (e.g., one day). Second, we define a ***temperature series data set*** D as a set of m temperature series. Moreover, a ***subsequence*** S of a temperature series $T = t_1, t_2, ..., t_n$ is a sampling of length $k \leq n$ of contiguous position from T, i.e., $S = t_p, ..., t_{p+k-1}$ for $1 \leq p \leq n - k + 1$.

Generally, to detect co-anomaly events from multiple temperature series, we need to identify sub-group behaviors among these temperature series. However, a simple clustering-like method (e.g., based on Euclidean distance) would not be appropriate for this task: First, it is very time consuming to search for all possible subsequences in all sub-dimensions; Second, we focus on group abnormal behaviors mining rather than the abnormal behaviors of one temperature series, i.e., the subsequences in an co-anomaly event do not need to be abnormal if we only look into each single time series. Thus, for detecting co-anomaly events from multiple temperature series effectively and efficiently, we propose $Sevent$, a novel framework with three major steps. First, we represent temperature series by symbolic characters. In this way, behaviors of each temperature series can be easily represented by combinations of characters. Then, we apply statistical tests to identify co-anomaly patterns. Finally, correlated patterns are connected into co-anomaly events from different time-spans. The overall flowchart is illustrated in Fig. 3, and each step of $Sevent$ is introduced in the following subsections.

3.1 Symbolic Representation

Symbolic representation is a popular way for time series representation with the benefits of reducing the volume of data points and preserving the evolving trends of time series simultaneously. A general framework for symbolic representation usually includes three steps: First, apply Piecewise Aggregate Approximation (PAA) [5] to reduce the dimensions of temperature series; Second, determine a list of breakpoints, which are usually drawn from a pre-defined distribution (e.g., Uniform). Third, transform PAA results to symbolic characters by comparing their positions with breakpoints.

In the first step, the PAA representation of a temperature series T can be denoted by a vector $\overline{T} = \bar{t}_1, ..., \bar{t}_w$ (or $\overline{S} = \bar{s}_1, ..., \bar{s}_w$ for S). Specifically, the ith element of $\overline{T}$ is calculated by the following equation,

$$\bar{t}_i = \frac{w}{n} \sum_{j=\frac{n}{w}(i-1)+1}^{\frac{n}{w}i} t_j.$$

In the second step, the breakpoints can be drawn from the distribution of the specific data that we need to analysis. Here an alphabet size Ψ is required to be predefined, which is leveraged for determining breakpoints of symbolic representation. Since most of the time series can be approximately fitted by normal distribution or uniform distribution, in the following we take these two distributions as an example to illustrate the way to generate breakpoints. Specifically, the breakpoints for normal distribution can be defined as follows, which is the same with that used in Symbolic Aggregate Approximation (SAX) [6].

Definition 1 (Breakpoints of N(0,1) Distribution). *Breakpoints are a sorted list of numbers $B = \beta_1, ..., \beta_{\Psi-1}$ such that the area under a $N(0,1)$ normal curve from β_i to β_{i+1} is equal to $\frac{1}{\Psi}$ (β_0 and β_Ψ are defined as $-\infty$ and ∞, respectively).*

Similarly, if the data is drawn from uniform distribution, and the corresponding breakpoints can be defined as follows.

Definition 2 (Breakpoints of Uniform Distribution). *Breakpoints are a sorted list of numbers $B = \beta_1, ..., \beta_{\Psi-1}$ such that $\beta_{i+1} - \beta_i = \frac{\beta_\Psi - \beta_0}{\Psi}$ (β_0 and β_Ψ are defined as minimum and maximum value of the temperature series, respectively).*

Noting that the breakpoints for other data distributions can be defined in the same way. After we obtain a list of breakpoints (B), a subsequence can be mapped into symbolic representation which is defined as a **word** [6].

Definition 3 (Word). *A subsequence S of length k can be represented as a word $\hat{S} = \hat{s}_1, ..., \hat{s}_w$. Let α_i denote the ith element of the alphabet, e.g., $\alpha_1 = a$ and $\alpha_2 = b$. Then the mapping from a PAA approximation $\overline{S}$ to a word $\hat{S}$ is obtained as follows,*

$$\hat{s}_i = \alpha_i, \quad iff. \quad \beta_{j-1} < \bar{s}_i \leq \beta_j. \tag{1}$$

For example, the data points whose value locates between the first two breakpoints ($[\beta_0, \beta_1)$) are mapped to "a", and the ones within $[\beta_1, \beta_2)$ are mapped to " b".

3.2 Detecting Co-anomaly Patterns

After transforming temperature series to words, we can calculate the number of each word at every timestamp. Words of different expressions represent different behaviors, e.g., the word *abcd* stands for a behavior of a rising temperature. Then by counting the number of the behaviors(words) at the same timestamp, we can find the frequent patterns(words) which are representations of group behaviors. However, frequency does not guarantee that pattern is interesting, and statistical tests are widely used to evaluate the importance of patterns. Specifically, co-anomaly patterns can be defined as follows.

Definition 4 (Co-anomaly pattern). *A word $\hat{S} = \hat{s}_1, ..., \hat{s}_w$ is a co-anomaly pattern if its count is statistically significant.*

The "co-anomaly pattern" is different from "anomaly pattern". The behavior of a co-anomaly pattern may not be abnormal if we only look into one single temperature series. It is abnormal and statistically significant in history only when we consider a group of consistent behaviors as a whole. For instance, in every year, there are always several cities experience extremely cold temperatures in winter. However, if dozens of the cities all have the same severe cold temperatures in one year's winter, it can be a co-anomaly event that may be caused by the same cold wave. For finding these co-anomaly patterns, a null hypothesis is defined and statistical hypothesis tests are used to calculate the P-value of each observed word.

Definition 5. *For a given word $\hat{S}$ and a timestamp t we define hypotheses H_0 and H_1:*
  H_0: $\hat{S}$ is uninteresting at t.
  H_1: $\hat{S}$ has a frequency that is significantly greater than the expected count at t.

Here, the expected count of each word at timestamp t are learned from the historic data, and it is used as the baseline of each concurrent behavior. The probability of $\hat{S}$ is,

$$\mu^t(\hat{S}) = \frac{N(\hat{S})^t}{nN_y}, \tag{2}$$

where $N(\hat{S})^t$ is the count of $\hat{S}$ at timestamp t for all years in history, N_y is the year number, and n is the number of temperature series. The expected count of $\hat{S}$ is,

$$\hat{N}(\hat{S})^t = n\mu^t(\hat{S}). \tag{3}$$

Then, for a word frequency x, we use the normal approximation to calculate its P-value, i.e., $N(\hat{S})^t$ follows the normal distribution $N(\hat{S})^t \sim \mathcal{N}(n\mu, n\mu(1-\mu))$.

$$\mathbb{P}(\mathcal{N}(\mu, \sigma^2) \geq N^{obs}(\hat{S})^t) = 1 - \frac{1}{2}\left[1 + \mathrm{erf}\left(\frac{x-\mu}{\sigma\sqrt{2}}\right)\right], \tag{4}$$

where $\mathrm{erf}(x)$ is the Normal Error Function and the formula is as follows,

$$\mathrm{erf}(x) = \frac{2}{\sqrt{\pi}} \int_0^x e^{-t^2} dt. \tag{5}$$

The P-value is then compared to a predefined critical value α. If $P < \alpha$, the null hypothesis H_0 is rejected and the word is accepted as a co-anomaly pattern. Noting that there are some other statistical methods for computing P-values, e.g., the Binomial trails, and Poisson approximation. Any of them can be used to test whether a word is statistically significant, while a detailed analysis of the pros and cons of these methods is beyond the scope of this paper.

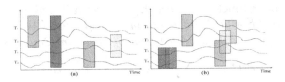

Fig. 4. Different ways for correlated significant words to form co-anomaly events

3.3   Building Co-anomaly Events

The co-anomaly patterns, which are adjacent in timestamps and correlated with the similar group of temperature series, are much likely to be involved in one co-anomaly event. Thus, the time-span of a co-anomaly event is not limited by the length of sliding windows. Here, a threshold ϕ is pre-defined, and two adjacent co-anomaly patterns will be connected if they have more than ϕ temperature series in common. Therefore, the co-anomaly events are able to have different durations(time-spans). Finally, we propose a ranking function $Pscore$ to evaluate a co-anomaly event. Generally speaking, one co-anomaly event with a higher $Pscore$ are likely to be formed by patterns with lower P-values and more affected temperature series. The co-anomaly event is defined as,

Definition 6 (Co-anomaly event). $E = \hat{e}_1, ..., \hat{e}_u$ is a co-anomaly event if $\forall \hat{e}_i \in E$ is a co-anomaly pattern, and $|\hat{e}_i \cap \hat{e}_{i+1}| > \phi$.

Under this definition, co-anomaly events are clusters of correlated significant words with various time-length, and those significant words that are connected by timestamps and temperature series should be put into one event. However, the temperature series associated with each word at different timestamps are not necessarily to be the same. For instance, Fig. 4 shows several possible ways for words to form events. In Fig. 4 (a), four separate words form four independent events. Fig. 4 (b) shows three events formed by different number of associated words. Specifically, the left one represents an event covering T_3 and T_4 for two timestamps. The right one displays an event moving from T_4 and T_3 to T_3 and T_2, and finally to T_2 and T_1, which could be a cold front or a typhoon flowing from west to east. Noting that threshold ϕ is defined to be the minimum of temperature series in common for adjacent words, i.e., if two words adjacent by timestamps and have more than ϕ common temperature series, then they can also be connected as a candidate event. Thus there may be multiple words at t_{k+1} qualified to be connected. It is natural in real word phenomenon for some events to change with time because they may have different kinds of evolutions. In this connection process, we are able to deal with this scenario by capturing every kind of the evolution record.

Due to the differences in word expressions and time-span, it is difficult to establish comprehensive evaluations for events. Generally, the anomaly of events are associated with the rareness of each behavior and the range of its influence: The lower the probability, the rarer the behavior, and the bigger the coverage, the more serious the behavior. As the P-value is the probability of each observed word count ranging from 0 to 1, we propose to compute $-\log$ of the P-value such that a rarer word can have a bigger positive value. In this way, we design a ranking function to evaluate each event according to the P-value of each word and the number of affected temperature series.

Definition 7 (Ranking function: $Pscore(E)$). *The overall Ranking value* $Pscore(E)$ *of a co-anomaly event* E *with regard to a set of relevant words* $RW(E)$, *the P-value of each word* $Pvalue(\hat{S})$, *and the observed count of each word* $Count(\hat{S})$ *is defined as,*

$$Pscore(E) = \sum_{\hat{S} \in RW(E)} -Count(\hat{S}) \cdot \log Pvalue(\hat{S}) \qquad (6)$$

In summary, the connecting phase includes three steps: First, connect the words adjacent in timestamps if they share over ϕ common temperature series; Second, repeat the first step until no more new connections are formed; At last, rank events by $Pscore$.

The pseudocode of the proposed co-anomaly climate event detection framework $Sevent$ is shown in Algorithm 1. Specifically, the procedures in line 1, line 2 to line 5, and line 6 to line 15 can be mapped into Symbolic Representation(Step 1), Statistical Tests(Step 2) and Building Events(Step 3) of the flowchart in Fig. 3, respectively. The runtime complexity for symbolic representation is $O(m \cdot n)$, where m is the number of temperature series, and n is the length of one temperature series. Let Ψ denote the alphabet size, and w be the number of PAA elements . Then, the runtime complexity for calculating P-values and connecting words is $O(w \cdot \Psi) + O(w \cdot \Psi)$. Thus, the total runtime complexity for $Sevent$ is $O(m \cdot n) + O(w \cdot \Psi)$.

Algorithm 1. Sevent(D, Ψ, w, α, ϕ)

Input: D: the m-dimensional temperature series dataset; Ψ: the alphabet size; w: PAA length; α:significance level; ϕ: the minimum number of common temperature series.
Output: The event set E .
 1: Mapping D into symbolic words using PAA and breakpoints;
 2: **for** each timestamp $T_i \in D$ **do**
 3:    Calculate the P-value of each words;
 4:    Delete the words with the P-value bigger than α;
 5: **end for**
 6: **for** each timestamp $T_i \in D$ **do**
 7:    **for** each word $\hat{S}_j \in T_i$ **do**
 8:       **if** $\hat{S}_j$ can not connect with adjacent words **then**
 9:          $E = E \cup \hat{S}_j$;
10:       **else**
11:          Connect $\hat{S}_j$ with associated words;
12:       **end if**
13:    **end for**
14: **end for**
15: Sort E by $Pscore$;
16: **return** E;

---

4 Experiments

In this section, we evaluate the proposed $Sevent$ on the real-world data set from Meteorology(Section 2). Specifically, we demonstrate: (1)The results of the co-anomaly events detection;(2)Two case studies of the detected co-anomaly events;(3) In-depth analysis on the generation of breakpoints(i.e., normal distribution and uniform distribution). We implemented our approaches in java, and all experiments were run on a personal computer with 2.0GB RAM and 2.26GHz CPU. In the following we fix the parameter settings as: $\Psi = 8$; $w = 122$ (122 timestamps in a year, i.e., 3 daily temperatures are combined to one timestamp) ; $\alpha = 0.01$; $\phi = 10$.

4.1 Co-anomaly Event Detection

One of our primary goals is to automatically detect the time-spans of co-anomaly events in multiple temperature series. The ground truth comes from The Climate Reports of Hunan Province [7] with the date, the magnitude of influences and other descriptions of some co-anomaly events.

Fig. 5 shows the time-length of co-anomaly events mining results of year 2001-2002. The reason of choosing year 2001 to 2002 for test is that there are plenty of detailed descriptions of the beginning and ending time in 2001-2002, while in other years, the ground truth of time-spans are not that clearly recorded in the report [7]. We demonstrate the time-spans of detected co-anomaly events based on the *Sevent* with uniform distribution for generating breakpoints (Fig. 5 (b)) and that with normal distribution(Fig. 5 (c)), and the baseline [1] (Fig. 5 (d)), respectively. Meanwhile, Fig. 5 (a) shows the average temperature, and (e) shows the real time-spans of events recorded in Climate Reports (Ground truth). Specifically, there are eight recorded co-abnormally events from 2001 to 2002, and the detailed descriptions are listed in Table 2. From Fig. 5 we can see that the detected time-spans of normal distribution are the most similar ones to the ground truth. The *Sevent* with uniform distribution detected most of the time-spans, however, it just found the beginning of event *b* (the warm spring in March), but the whole time durations. In contrast, the results of baseline are composed by lots of small fragments of timestamps, which has the least overlaps with the ground truth. Correspondingly, Table 3 lists the detailed information of top ten co-anomaly events detected by *Sevent* along with the uniform distribution, and table 4 lists the results along with the normal distribution. From the results we can find that the two top ten rankings have approximately 9 events in common (Please also refer to Fig. 5), and most of the detected events can be found in annual report (i.e., Table 2). For example, the first event in Table 4 (the 7*th* event in Table 3, and **f** in Table 2 and Fig. 5) resulted in disasters for crops production, and the economic losses were 1.23 billion (RMB).

Then, we compare the recall of results on the whole data set. For simplicity, we only select the following events as the ground truth for the comparison: (1) January 12 to February 8, 2008. The most severe snow storm disaster since 1949, with the direct economic losses of over 680 billion RMB. (2) the abnormal spring in 1998, (3) the warm winter in early 1999, (4) late spring coldness and hailstorm in 1999, (5-8) the 4 events in 2002 (Fig. 5 (e) *f* − *i*), (9) the late spring coldness in 2006,(10) the warm winter in 2007. One reason for choosing these events is that they are all important climate events, and most of them cause significant economic losses. The other reason

---

[1] Which is calculated by comparing the difference between the average temperature of multiple temperature series of each year and the total average temperature series, and then the timestamps that have a gap greater than 3 are chosen as candidates.

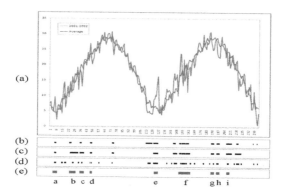

Fig. 5. Co-anomaly events mining results in 2001-2002. (a) The average temperatures. (b) The time-spans results when applying uniform distribution, (c) The time-spans results when applying normal distribution. (d) The baseline. (e) The time-spans recorded in the climate reports.

is that there are precise date records for each event and thus easy for us to compare. To test the effectiveness of $Sevent$, we only consider top 50 detected events ranked by $Pscore$, i.e., if these important events are not in top 50, the recall value will be low. The final results are illustrated in Table 5, where we can see that $Sevent$ based on both normal distribution and uniform distribution performs much better than the baseline. Specifically, we find that the snow storm in 2008 is detected as the most rare event(the same as the evaluation in the climate reports) by uniform distribution with the total $Pscore$ 6002.16. However it is not reported in the top 50 of normal distribution. That is because in normal distribution, the partitions in the lowest(or the highest) range are much coarser than that of in the middle range. Thus, although the snow storm in 2008 is very severe, it is not significant under the normal distribution. In contrast, all of the important events are successfully detected by the method with uniform distribution.

Table 2. The co-anomaly events in 2001-2002

Label	Duration	Description
a	Jan. and Feb.	Warm winter
b	Early Mar.	Warm spring
c	Late Apr.	Early summer
d	Late May.	Early heat waves
e	Jan.	Warm winter.
f	Apr. 1 - May. 10	Hailstone, coldness
g	Jul. 18 - Jul. 27	Low temperature
h	Aug. 6 - Aug. 15	Low temperature
i	Sep. 14 - Sep. 16	Cold dew wind

Table 3. Top ten co-anomaly events (uniform)

NO.	Durations	Score
1	Aug. 8 - Aug. 17, 2002	2430.80
2	Jan.1 - Jan. 12, 2002	2361.28
3	Oct. 16 - Oct. 22, 2002	2194.96
4	Apr. 8 - Apr. 17, 2002	1804.53
5	Mar.27 - Apr. 3, 2002	1775.56
6	Apr. 19 - Apr. 22 , 2001	1531.90
7	Apr. 23 - May. 4, 2002	1504.58
8	Dec. 6 - Dec. 18, 2001	1466.07
9	May. 2 - May. 11, 2002	1321.16
10	Dec. 24 - Dec. 27, 2002	1230.57

Table 4. Top ten co-anomaly events (normal)

NO.	Durations	Score
1	Apr. 23- May. 7, 2002	3453.85
2	Oct. 13 - Oct. 30, 2002	3244.40
3	Aug. 8 - Aug. 17, 2002	3228.31
4	Jan.1 - Jan. 12, 2002	3089.23
5	Sept. 13 - Oct. 7, 2002	2972.07
6	Apr. 8 - Apr. 17, 2002	2575.32
7	Mar.27 - Apr. 3, 2002	1708.21
8	Mar.8 - Mar. 23, 2002	1533.34
9	Apr.19 - Apr. 28, 2001	1378.92
10	Nov. 3 - Nov. 12, 2001	1232.93

Table 5. The Recall result

Alg.	Normal	Uniform	Baseline
Recall	0.9	1.0	0.6

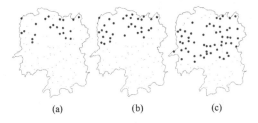

(a)   (b)   (c)

Fig. 6. A cold-wave event started from north and then expanded to the middle of Hunan province in late June, 1999. Only sub-dimensions of temperature series are involved in this events.

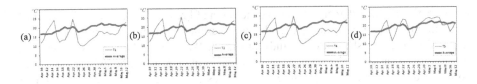

Fig. 7. Detail subsequences of four temperatures series. (a) - (c) are temperature series of the same event. (d) does not have continuous low temperature patterns.

4.2 Case Studies

To further explore the extracted co-anomaly events, we present two case studies to show how *Sevent* can capture the evolutions of co-anomaly events, and trace the involved sub-dimensions as well. First, we present the evolution of one cold-wave event captured in parts of the stations(i.e., the sub-dimensions of the temperature series in June, 1999). Fig. 6 (a)-(c) shows the evolution of this event in adjacent timestamps. Here we can see that *Sevent* successfully detected and traced the evolution of sub-dimensional events. In Fig. 6 (a), only 22 stations suffer from this cold-wave. Then the event expands to 32 stations (Fig. 6 (b)). Finally the event spreads to the middle of the province and 56 stations are affected (Fig. 6 (c)). From this result we can easily trace a cold wave movement, which is from north to the middle of Hunan province, and then blocked by the mountains in the middle and the south. Thus the evolutions of co-anomaly events can well support the further research of cold-wave abnormal activities for domain experts.

Then, we show a more detailed example of involved sub-dimensions in Fig. 7, where the original temperature series are from 10th April to 12th May in 2002. We find the corresponding event description from the climate report: "late spring coldness and low temperature in May". As shown in Fig. 7, station (a) (b) (c) all have similar behaviors during this period, while station (d) does not have "low temperature in May". This co-anomaly event does not span full dimensions or full durations (station (d) only joined the first half of the duration). From these two case studies, we can conclude that various co-anomaly events of sub-dimension and sub-durations can be detected by our *Sevent*.

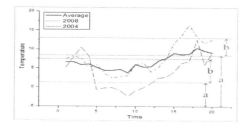

Fig. 8. The breakpoints under uniform and normal distribution

4.3   Normal Distribution *vs* Uniform Distribution

Here we give a detailed analysis on the generation of breakpoints, i.e., based on normal distribution or uniform distribution, and it provides more insights for future application. Along this line, we show a case study in Fig. 8. We choose the year 2008, when a severe snow disaster stroke many cities in Hunan province, and another year 2004 with correspondingly normal temperature as examples (the time is from Jan 1 to Feb 28 of each year). Fig. 8 shows the breakpoints under different distributions, uniform and normal, respectively. The purple lines are breakpoints under uniform distribution, where the gaps between lines are of the same. The blue lines are breakpoints under normal distribution, where the gaps in the middle (medium temperatures, e.g., the temperatures of springs and autumns) of the whole temperature series are much smaller than the up and low parts (high and low temperatures, e.g., the temperatures of summers and winters). Thus when mapping the subsequences of the winter temperatures in Fig. 8, the breakpoints under normal distribution has a much coarser differentiations than that of uniform distribution, where most of the temperature points in winters, no matter severe coldness or not, are mapped to the symbol "a". In this way, the severe cold disaster is not that significant when we adopt normal distribution. In contrast, the uniform distribution maps most of the normal temperature points to the symbol "b", and when it comes to severe cold event in 2008, a significant number of "a" co-occurrent together and lasting a long period of time forms a severe co-anomaly event. Based on the discussion, we can capture the pros and cons of each distribution, and they can guide us design more effective co-anomaly event detection framework.

5   Related Work

In the past several decades, there has been numerous work on finding abnormal patterns, change-points, and events in time series data. Due to the space limit, we just present a brief survey on major research directions most relevant to ours.

Event Detection in Time Series have been proposed in [2,8,9,10,11]. For instance, Guralnik and Srivastava [8] proposed an iterative algorithm that used a likelihood criterion for time series segmentation. Preston et al. [10] proposed a method to search for subintervals that are statistically significantly different from the underlying noise. Ihler et al. [11]proposed a time varying Poisson process to model periodic count data that can detect bursty events embedded in time series. Cho et al. [9] proposed a framework based on episode rules for event prediction over event streams. Minaei et al. [2] explored the correlation between time series streams and events stream, where event streams are logs from domain experts. Anomaly detection in time series is also similar to the problem of searching for events in time series. E.g., Keogh et al. [12] presented an anomaly detection method that searched for subsequences that differ most from the rest of its subsequences in one time series.

Finding Patterns in Multiple Time Series are explored in [1,13,14,15]. McGovern et al. [1] introduced a multi-dimensional motif mining approach to predict severe weather. They used labeled time series data to build the trie structure [12] to find subsequences that are relevant to severe weather, and then grow motifs into longer patterns. Minnen et al. [13] formulated multivariate motifs as regions with high estimated density via k-nearest neighbor search. An expected linear-time algorithm [14] was proposed to detect subdimensional motifs in multivariate time series. Tanaka et al. [15] used principal component analysis to project the multi-dimensional time series into one dimension signal. However, the frequent patterns are not necessarily the most interesting ones. To find significant patterns, many work on significant motif mining [10,11,16] have been proposed, while most of them do not detect significant subdimensional motifs. Xiong et al. [3] detected peculiar groups in day-by-day behavior datasets that are similar to the co-anomaly events problem in our work. However, they assume that most objects are dissimilar with each other, which was difficult to satisfy in real-word datasets.

Although these approaches are related to our work, they are fundamentally different and are not particularly well suited for our application. In summary, our work differs from them in four aspects: 1) We focus on detecting group abnormal behaviors, rather than single abnormal behaviors; 2) We consider periodic calendar time constrains for multiple temperature series modeling; 3) We propose a connection method based on correlation between objects and timestamp adjacency to form events with different time-span and even evolving with time; 4) We propose an abnormal ranking function based on statistical significance to evaluate the abnormal degree of events.

6 Conclusion

In this paper, we provided a focused study of exploiting multiple temperature series data for co-anomaly climate event detection. Specifically, we first map the original temperature series to symbolic representations based on data distributions. Then, we detect the co-anomaly patterns by statistical tests and finally generate the co-anomaly events that span different sub-dimensions and subsequences of multiple temperature series. Meanwhile, this proposed detection framework $Sevent$ also captures the evolutions of the co-anomaly events in multiple temperature series. The experimental results on real-world data of temperature series demonstrate that our $Sevent$ can successfully detect co-anomaly events interested in meteorology. In the future, we plan to apply and evaluate our framework in the co-anomaly event detection from other types of series data.

Acknowledgments. This work is supported in part by the National Science Foundation Project of China (Grant No. 61170096) and the National High Technology Research and Development Program ("863" Program) of China (Grant No. 2011AA100701).

References

1. McGovern, A., Rosendahl, D.H., Brown, R.A., Droegemeier, K.K.: Identifying predictive multi-dimensional time series motifs: an application to severe weather prediction. Data Mining and Knowledge Discovery 22(1), 232–258 (2011)
2. Minaei-Bidgoli, B., Lajevardi, S.B.: Correlation mining between time series stream and event stream. In: NCM 2008 (2008)
3. Xiong, Y., Zhu, Y.: Mining peculiarity groups in day-by-day behavioral datasets. In: ICDM 2009 (2009)
4. Liao, Y., Wang, K., Zhao, F., Bai, S.: Modern agro-climatic zoning of Hunan Province. Hunan University Press, Changsha (2010)
5. Keogh, E., Chakrabarti, K., Pazzani, M., Mehrotra, S.: Dimensionality reduction for fast similarity search in large time series databases. Knowledge and Information Systems 3(3), 263–286 (2001)
6. Lin, J., Keogh, E., Patel, P., Lonardi, S.: Finding motifs in time series. In: the 2nd Workshop on Temporal Data Mining (July 2002)
7. The climate reports of hunan province, http://www.hnqx.gov.cn
8. Guralnik, V., Srivastava, J.: Event detection from time series data. In: KDD 1999 (1999)
9. Cho, C.-W., Wu, Y.-H., Yen, S.-J., Zheng, Y., Chen, A.: On-line rule matching for event prediction. The VLDB Journal 20, 303–334 (2011)
10. Preston, D., Protopapas, P., Brodley, C.: Event discovery in time series. Arxiv preprint arXiv:0901.3329 (2009)
11. Ihler, A., Hutchins, J., Smyth, P.: Adaptive event detection with time-varying poisson processes. In: KDD 2006 (2006)
12. Keogh, E., Lin, J.: Hot sax: Efficiently finding the most unusual time series subsequence. In: ICDM 2005 (2005)
13. Minnen, D., Essa, I., Isbell, C.: Discovering multivariate motifs using subsequence density estimation. In: AAAI Conf. on Artificial Intelligence (2007)
14. Minnen, D., Essa, I., Isbell, C.L., Starner, T.: Detecting subdimensional motifs: An efficient algorithm for generalized multivariate pattern discovery. In: ICDM 2007 (2007)
15. Tanaka, Y., Iwamoto, K., Uehara, K.: Discovery of time-series motif from multidimensional data based on mdl principle. Machine Learning 58(2-3), 269–300 (2005)
16. Castro, N., Azevedo, P.J.: Time series motifs statistical significance. In: SDM 2011 (2011)

Frequent Patterns Based Word Network: What Can We Obtain from the Tourism Blogs?

Hua Yuan[1,*], Lei Guo[1], Hualin Xu[1], and Yong Xiang[2]

[1] School of Management and Economics, University of Electronic Science and Technology of China, 610054 Chengdu, China
yuanhua@uestc.edu.cn, {uestcgl,bruce123.xu}@gmail.com
[2] Public Security Department of Sichuan Province, 610041 Chengdu, China
redarmy74@hotmail.com

Abstract. In this work, we present a method to extract interesting information for a specific reader from massive tourism blog data. To this end, we first introduce the web crawler tool to obtain blog contents from the web and divide them into semantic *word segments*. Then, we use the frequent pattern mining method to discover the useful frequent 1- and 2-itemset between words after necessary data cleaning. Third, we visualize all the word correlations with a *word network*. Finally, we propose a local information search method based on the *max-confidence* measurement that enables the blog readers to specify an interesting topic word to find the relevant contents. We illustrate the benefits of this approach by applying it to a Chinese online tourism blog dataset.

Keywords: blog mining, frequent pattern, word segmentation, word network, max-confidence.

1 Introduction

With the development of Web 2.0 technology, many Internet users start recording their outdoor movements and publishing them on popular blog systems for many reasons, such as route tracing [1], travel experience sharing [2] and multimedia content (photo/video) management [3], etc. By obtaining reference knowledge from others' online speeches, individuals are able to visualize and manage their own travel plans. Meanwhile, some readers are more likely to enjoy a high quality travel experience from others' blogs. For instance, a person is able to find some places that attract him from other people's travel routes, and schedule an efficient and convenient (even economic) path to reach these places.

However, from the perspective of the recommender role, the potential blog readers may be confronted with too much data of *information overload* when a large number of blog contents are offered, which means that, it is impossible for average readers to find out the interesting information from such a huge mass of data. So, data mining (association analysis, clustering etc.) has been introduced

---

* Corresponding author.

M. Wang (Ed.): KSEM 2013, LNAI 8041, pp. 15–26, 2013.

as a powerful tool to extract blog information efficiently [4]. While the common data mining algorithm would generate a lot of patterns and most correlations hidden in these patterns are hard to explain, one of the most powerful methods for information extraction is blog topics co-expression analysis. Here, the correlation in expression pattern between pairs of blog words is measured, and those exhibiting strong correlations are "joined" in a graphical representation to create a network, which can be visualized with graph network viewers [5]. However, there are few impressive researches on how to visualize the extracted knowledge as a whole and how to provide useful or sufficient knowledge from the massive blogs contents to readers who are really interested in.

In this work, we employ the frequent patterns mining method to obtain useful frequent 1- and 2-itemsets and then visualize all the correlations with a *word network* with which people can obtain all the mined knowledge from blogs as a whole. In addition, we present a local information search method based on the *max-confidence* measurement that enables the blog readers to specify an interesting topic word to find relevant blog content. The remainder of this paper is organized as follows. Section 2 presents the related work. Section 3 sketches out the approaches about blog extraction and word network construction. Section 4 discusses how to extract information from a word network in detail. Section 5 shows the experimental results. This paper is concluded in Section 6.

2 Related Work

In this section, we briefly review the important studies related to this research.

Firstly, we note that the basic technology used in online text contents processing is text-mining [6,7] which is used to derive insights from user-generated contents and primarily originated in the computer science literature [8,9]. Thus the previous text-mining approaches focused on automatically extracting the opinions of online contents [10,11]. This method used in blog mining not only involves reducing a larger corpus of multiple documents into a short paragraph conveying the meaning of the text, but also is interested in features or objects on which customers have opinions. In [12], the authors presented a method to extract such temporal discussions, or stories, occurring within blogger communities, based on some query keywords. Paper [13] presented the opinion search engine for the TREC 2006 Blog Task which is based on an enhanced index on words denoting. Paper [4] proposed a shallow summarization method for blogs as a preprocessing step for blog mining which benefits from specific characteristics of the blogs. Bai et. al. proposes a system for extracting knowledge hidden in Chinese blogs [14]. In their work, blogs are clustered into categories before extraction and then knowledge are extracted basing on some domain ontologies.

In the application of blog mining, some researches have been done on sentiment analysis which is to judge whether an online contents expresses a positive, neutral or negative opinion [15]. The typical work is the methods presented by Pang and Lee of sentiment classification on the document level [9]. O'Leary has reviewed some of the literature aimed at gathering opinion, sentiment and information from blogs in [16]. Some research have studied the problem of mining sentiment information from online resources and investigate ways to use such information to predict the performance of product sales [17,18].

However, there are few impressive researches on providing useful knowledge to blog readers from massive content that they are individually interested in.

3   Frequent Patterns Based Word Network

In general, this part of work primarily consists of three aspects: content extraction, frequent patterns mining and word network construction.

3.1   Blog Contents Extraction

In this work, the *web crawling technology* [19,20] is used to obtain large-scale and high-quality users generated contents from the website, such as tourism-related BBS, blog systems and independent reviewing sites.

First, the blogs in a targeting website are crawled into data set $\mathbb{B}$ with software; Then, for each piece of blog $B_i \in \mathbb{B}$ (proposed by one user at one time), the contents are split into n_i semantic *word segment* of b_j (where $j = 1, ..., n_i$) based on some user-defined rules. Third, all these word segments are used to compose of a vector $b_i, i = 1, ..., |\mathbb{B}|$, (the symbol $|.|$ is used to denote the number of elements in a set). Finally, some useless words are cleared and the remained part of b_i can be seen as a record in transactional data set T. Algorithm 1 shows the presented method as a whole.

Algorithm 1. Blog Data Extraction and Processing Algorithm

1: **Input**: Blog data set $\mathbb{B}$;
2: **Output**: A transactional data set T;
3: $T = \phi$;
4: **for** $i = 1$ to $|\mathbb{B}|$ **do**
5:     Split $B_i \in \mathbb{B}$ into n_i word segments $\{b_j\}, j = 1, ..., n_i$;
6:     **for** $j = 1$ to n_i **do**
7:         **if** b_j is useless **then** $b_i = b_i \setminus \{b_j\}$;
8:         **end if**
9:     **end for**
10:    **if** $b_i \neq \phi$ **then** $T \leftarrow b_i$;
11:    **end if**
12: **end for**
13: **return** T;

Actually, there is an important work on how to judge the usefulness of a *word segment* generated by the contents. This work is in relation to the data cleaning process, in which we remove all the noise phrases, stop words, meaningless symbols and finally only nouns, meaningful verbs, adjectives and semantic phrases are kept in T. Note that, the data set T is transactional whose items are the valuable *word segment* from each blog contents.

3.2 Frequent 2-Itemset Mining

Frequent patterns (FPs) are itemsets that appear in a data set with frequency no less than a user-specified threshold. In classic data mining tasks, frequent itemsets play an essential role in finding interesting patterns from databases, such as association rules, correlations, sequences, episodes, classifiers, and clusters. The objective of FP mining in this work is as follows: given a database T with a collection of items (word segments) and a user predefined minimum support of $mini\_supp$, all the patterns that are frequent with respect to $mini\_supp$ are discovered.

Note that, the FP mining technology is not the main concern in this work, thus any feasible FP algorithms can be selected as the potential algorithm depending on contents and data formation of T. And, the *absolute support* [21] of an itemset X, denoted by $supp(X)$, is used to specify the number of transactions in which X is contained. Mathematically, the *absolute support* for an itemset X can be stated as follows:

$$supp(X) = |\{\boldsymbol{b}_i | X \subseteq \boldsymbol{b}_i, \boldsymbol{b}_i \in T\}|. \tag{1}$$

In addition, only frequent 1- and 2-itemset patterns would be generated for the following two reasons:

- All the frequent n-itemset (pattern) can be generated by frequent $(n-1)$-itemset with other frequent items according to Apriori principle [21];
- Any network $G = (V, E)$ constructed by nodes $v \in V$ and connections $(u, v)_{u \in V, v \in V} \in E$ represents only 2-pair relations of u and v directly.

The 2-item frequent patterns generated from data set T is denoted by FP_T. Each frequent pattern in FP_T can be transformed into a text set of fp, in which are only one or two elements of *word segments*.

3.3 Frequent Patterns Based Word Network

From a semantic perspective, the vector $\boldsymbol{b}_i \subset T$ shows the hidden correlations between words in it, and the total relations of all the valuable words from a blog website, $\mathbb{B}$, thus lie in T. Apparently, the frequent patterns set FP_T indicate the common and frequent concerns of the bloggers. In the following, we will reveal these common concerns (relations) with the help of *word network*.

From a network perspective, each item in fp can be seen as an entity (node). If two items b_u and b_v are shown in the same fp, i.e. $fp = \{b_u, b_v\}$, we can expect that they have some potential semantic relations, thus b_u and b_v can be linked with edge (b_u, b_v). To construct a *word network*, the set of vertices is defined as

$$V = \bigcup_{i=1}^{|FP_T|} fp_i, \tag{2}$$

where $fp_i \subset FP_T$. And edges set is defined as

$$E = \{(b_u, b_v) | b_u \in fp_i, b_v \in fp_i, i = 1, ..., |FP_T|\}. \tag{3}$$

So, we obtain a *word network* of $G = (V, E)$, in which, the nodes are *word segments* and edges are word correlations. In order to facilitate the calculation, we label here each node b with $supp(\{b\})$ and weight the edge (b_u, b_v) with $w_{(b_u, b_v)} = supp(\{b_u, b_v\})$. Fig.1 is an construction example for such a *word network* with $mini\_supp = 2$. Obviously, for any edge (b_u, b_v), $w_{(b_u, b_v)} \geq mini\_supp$ holds true. The process of word network generation is shown in algorithm 2.

Algorithm 2. *Word Network* generation Algorithm

1: **Input**: Frequent patterns set FP_T (minimum support threshold $mini_supp$);
2: **Output**: Word network: $G = (V, E)$;
3: $V = \phi; E = \phi$;
4: **for** $i = 1$ to $|FP_T|$ **do**   /*Obtain network nodes.*/
5:     Get frequent 1-itemset fp_i from FP_T;
6:     $V \leftarrow fp_i$;
7:     $FP_T = FP_T \setminus \{fp_i\}$;
8: **end for**
9: **for** $i = 1$ to $|FP_T|$ **do**   /*Obtain network edges.*/
10:     Get frequent 2-itemset fp_i from FP_T;
11:     Obtain b_u and b_v from fp_i;
12:     $E \leftarrow (b_u, b_v)$;
13: **end for**
14: **return** G;

TID	Items
1	Bread, Milk
2	Bread, Milk, Beer, Eggs
3	Milk, Diaper, Coke
4	Bread, Diaper
5	Bread, Milk, Diaper, Coke

(a) Sample transactions.

Itemset	Count
{Bread, Milk }	3
{Bread, Diaper }	2
{Diaper, Coke }	2
{Milk, Coke }	2
{Milk, Diaper }	2

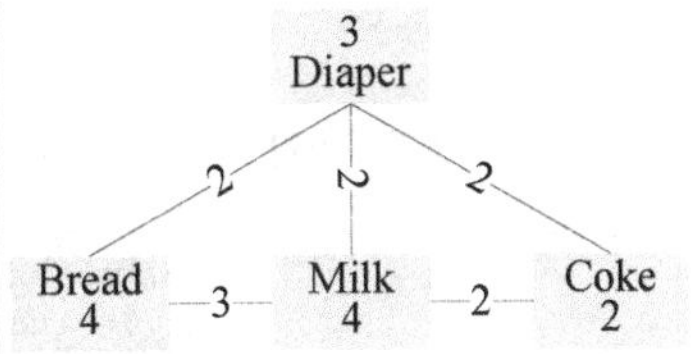

(b) Frequent patterns based word network.

Fig. 1. Example for word network construction

4   Information Extraction from Word Network

In this section, we present a method to extract information from the weighted *word network*.

4.1   Max-confidence

In real-world applications, many transaction data sets have inherently skewed support distributions which often lead to the so-called "cross-support patterns" [22]. The "cross-support patterns" typically represent spurious associations among items with substantially different support levels.

Since we know that the support of reviewed *word segments* in blogs are really askew distributed because some aspects are common (popular) topics for bloggers while some others are not [23], then, these "cross-support patterns" may reflect a master-slave relationship between itemsets which can be used to reveal some semantic information in online information retrieval. For example, when bloggers record a tourist city (b_1), some of them tend to review a famous scenic spot (b_2) in b_1. Under such a case, the appearances of scenic spot $(supp(\{b_2\}))$ are totally dependent on the blogs that reviewed both the city and the scenic spot $(supp(\{b_1, b_2\}))$. That is to say, the scenic spot has a strong relation (dependency) with the city, if somebody inquire the information about the city, then the scenic spot as an necessary composition of the city should be provided.

Here, we identify the dependent relation between b_1 and b_2 in the 2-itemset $X = \{b_1, b_2\}$ with a *max-confidence*, which is defined as [24]:

$$\theta_X = \frac{supp(X)}{\min\{supp(\{b_1\}), supp(\{b_2\})\}} = \max\{Pr(b_2|b_1), Pr(b_1|b_2)\}. \qquad (4)$$

In this work, the *max-confidence* is used to measure the impact of common parts on the item which has the minimum support.

Let $j^* = \arg\min_j\{supp(\{b_j\})\}$, we call the item b_{j^*} as the *dependent node* and item $X \setminus \{b_{j^*}\}$ as the *master node* of b_{j^*}. Relation $\theta \to 1.0$ indicates that the dependent intensity of b_{j^*} on item $X \setminus \{b_{j^*}\}$ is almost 100%. See Fig. 2(b), the common parts of $supp(\{b_1, b_2\})$ is almost equal to $supp(b_2)$, therefor b_1 has a heavy impact on b_2, and almost all the appearances of b_2 are dependent on b_1.

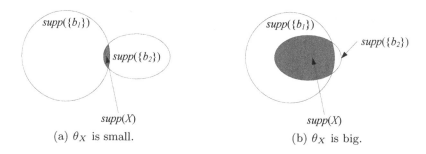

(a) θ_X is small.   (b) θ_X is big.

Fig. 2. Example for θ_X $(supp(\{b_2\})$ is the smaller one)

4.2   Information Extraction

In the following, we would try to solve the problem of extracting useful information from the network generated by segment words. To simplify the calculation, we assume that:

- There is a specific topic that the blog readers are interested in and would like to collect its information from the published blogs. This topic can be represented by a *word segment* and presented as a focal node b_k in *word network*.
- For the interested topic, the reader want to know: how is this topic reviewed online and what are the correlation topics?

As to the first problem, we can solve it with calculating support of the topic (high support means hot topic) and checking the sentiment words (adjective) frequently connected with the topic. For the last problem, we need to check the correlation topics (usually represented by nouns in *word network*) frequently connected with it. So, all the information about a topic which a reader is interested in can be collected from a well-constructed *word network*. Now the useful information extraction problem is, given a focal node, how to extract correlation information from a *word network*.

Given an itemset of $X = \{b_1, b_2\}$ satisfy that $supp(X) \geq mini\_supp$, a bigger max-confidence of X means the smaller-support item has strong dependency with the bigger-support one, and then it is a necessary supplement information for the item that has bigger support. Thus, the smaller-support item should be deemed as in a same clique with the bigger-support item. So, when we explore information about a focal node b_k, all the dependent nodes should be extracted. In addition, the master nodes of b_k are important information, then they should also be extracted. Now, the basic extraction principle is that: all the node $b_j \in V, j = 1, ..., |V|$ should be extracted if the following relations are satisfied:

$$\theta_{\{b_k, b_j\}} = \max\{Pr(b_k|b_j), Pr(b_j|b_k)\} \geq \theta_0, \tag{5}$$

where θ_0 is a predefined threshold.

More over, if b_j is dependent on b_k, i.e. $\theta_{\{b_k, b_j\}} = Pr(b_k|b_j) \geq \theta_0$, then all the nodes that depend on b_j can be extracted as the additional information for focal node b_k:

$$\theta_{\{b_j, b_s\}} = Pr(b_j|b_s) \geq \theta_0, \tag{6}$$

where b_j satisfies relation (5).

The following codes show the skeleton of our information extraction processes. As illustrated in algorithm 3, it goes through the two phases: first we extract the direct dependent relations of the focal nodes b_k, second we obtain the nodes which depend on b_k's neighbors.

5 Experimental Results

In this section, we present a case study to demonstrate the proposed method.

5.1 Experiment Setup

Altogether, 617 blogs posted from 05-31-2006 to 12-31-2012 in "Emei-mountain tourism" channel of mafengwo.com, one of the most famous tourism web-blog site in China, were collected with the blog extraction tool LocoySpider which is

---

Algorithm 3. Information Extraction Algorithm

1: **Input**: Word network $G = (V, E)$, focal node $b_k \in V$ and max-conference threshold $0 < \theta_0 \leq 1.0$;
2: **Output**: A sub network for b_k: $G_{b_k} = (V_{b_k}, E_{b_k})$;
3: $V_{b_k} = \{b_k\}$; $V_{dependent} = \phi$; $E_{b_k} = \phi$;
4: **for** $j = 1$ to $|V|$ **do**   /*Extract nodes that depend on b_k or b_k depends on.*/
5:     **if** $\theta_{\{b_k, b_j\}} \geq \theta_0$ **then** $\{V_{b_k} \leftarrow b_j; V_{dependent} \leftarrow b_j; E_{b_k} \leftarrow (b_k, b_j)\}$;
6:     **end if**
7: **end for**
8: **repeat**
9:     Select a node b_j from $V_{dependent}$;   /*Extract nodes that depend on b_j.*/
10:     **for** $s = 1$ to $|V \setminus V_{b_k}|$ **do**
11:         $V_{temp} = \phi$;
12:         **if** $Pr(b_j|b_s) \geq \theta_0$ **then** $\{V_{temp} \leftarrow b_s; V_{dependent} \leftarrow b_s; E_{b_k} \leftarrow (b_j, b_s)\}$;
13:         **end if**
14:     **end for**
15:     $V_{b_k} = V_{b_k} \cup V_{temp}$;
16:     $V_{dependent} = V_{dependent} \setminus \{b_j\}$;
17: **until** $V_{dependent} = \phi$;
18: **return** G_{b_k};

---

available freely at `http://www.locoy.com/`. The blogs with empty content were removed (some collected blogs are full of pictures only) and 575 useful travel blogs were remained for experiments.

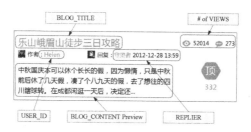

Fig. 3. An example of a travel blog on mafengwo.com

The extracted blogging attributes include BLOG_TITLE, BLOG_CONTENT, USER_ID, PUBLISH_TIME, REPLY_COTENTS, REPLIER, and PAGEURL. Fig.3 is an example of such a blog. Note that, as the reviews were mostly written in Chinese. In the following, we do experiments with data in Chinese character and report the results in English.

5.2   Word Segmentation and Frequent Patterns Mining

The frequencies of all the *words segment* in "Emei-mountain tourism" blogs (obtained with $mini\_supp \rightarrow 0$) are sorted in Fig.4. Compared with those contents

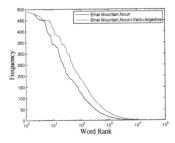

Fig. 4. Frequencies of word segments

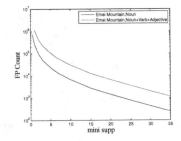

Fig. 5. FP Counts of 1- & 2-itemset

extracted from online reviewing systems having a shorter text, such as BBS and micro-blogs [25], the *words segment* frequencies to their corresponding ranks does not follow the common power-law distribution (See Fig.4) This result indicates that: 1) the potential blog readers need not to read all the blogs that they are interested in while the topics mentioned in blogs are numerous, and most of the common topics would be covered by most bloggers; 2) the blogging behaviors of bloggers are independent from each other, but they try to sketch out the travel experiences in detail to obtain the praise of readers. Thus the most likely reason is that the document length of blogs is much longer than that of the contents in BBS or microblogs. In other words, if people are to obtain relevant information from the BBS or micro-blog system, they should try to read all of the published contents while the topics in such a system are various and only few extreme popular topics would be mentioned commonly.

In frequent pattern mining, the important parameter of minimum support should be specified properly. Too low support may introduce noisy patterns, and too high support discards even useful patterns. The different *mini_supp* to the total number of generated frequent patterns (FP Counts) are shown in Fig.5. With the increase of the mining threshold, the proportion of frequent adjectives and verbs is growing. This illustrates that, the adjectives and verbs used by independent bolggers are similar to each other and are somehow homogenous. Accordingly, we conduct the frequent pattern mining experiments in this work with *mini_supp* = 57 (10%) and only the nouns are remained in the final data set for the following computation.

5.3 The Word Network for "Emei-Mountain Tourism" Blogs

With the blog data, 267 of word segments and 6552 word-pair correlations (edges) are generated. The whole network is as follows (see Fig.6) and the averaged degree is 49.1 and clustering coefficient is 0.86.

We use the famous scenic spots of "Leidongping" as the examples for information extraction. The extraction results lie in Fig.7 and the important measurements for these sub-graphs are listed in Table 1.

As we can see, the degree centrality of the selected focal node decreases while the max-confidence threshold increases, which makes the dependent relations

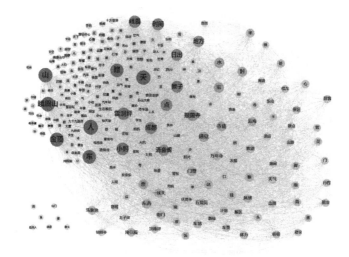

Fig. 6. The word network for "Emei-mountain tourism" channel

Table 1. Some measurements for the extracted sub-graphs

θ_0	Averaged degree	Degree centrality (focal node)	Clustering coefficient
0.5	15.21	201	0.398
0.7	6.41	186	0.375
0.9	1.37	24	0.025

around the focal words are more clearly. In Fig.7(a)-7(b), the dependent relationship around "Leidongping" is really complex, from which, we know only that the focal node is a famous scenic spot in Emei-mountain and a lots of interesting things are connected with it. Whereas the same situation in Fig.7(c) is relative

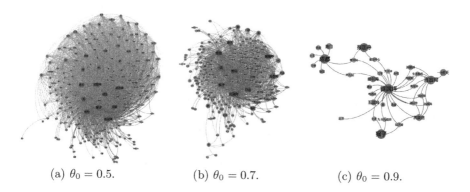

(a) $\theta_0 = 0.5$.   (b) $\theta_0 = 0.7$.   (c) $\theta_0 = 0.9$.

Fig. 7. Effects of max-confidence thresholds (focal node is "Leidongping")

simple, we can see that "Leidongping" is dependent highly on "Jinding" which means that the most (bigger than 90%) people who are interested in "Leidong-ping" are also interested in "Jinding". In addition, the aspects depended mainly on the focal node are transportation("Bus station", "Speed", "Slopes"), nearby attractions ("Jieyindian", "Taiziping", "Leiyinsi", "Hongchunping" "Yuxiansi", "Wuxiangang") visiting tips ("Ticket price", "Electric blanket", "Coat", "Bam-boo rod") and related hot scenic spots ("Qingyinge", "Baoguo temple", "Monkey area" "Xixiangchi"). These are key information for visiting "Leidongping".

6 Conclusion

In this work, we present a frequent patterns visualization method and a max-confidence based correlation analysis method to extract interesting information for a specific reader from massive blog data. We first introduce the web crawler tool to obtain note contents from web-blogs and divide them into semantic *word segments*. Then, we use the frequent pattern mining method to discover the useful frequent 1- and 2-itemset between words after necessary data cleaning. Third, we visualize all the word correlations with a *word network*. Finally, we present a local information search method based on the *max-confidence* measurement that enables the blog readers to specify an interesting topic word to find the relevant contents. We illustrate the benefits of this approach by applying it to a Chinese online tourism blog dataset.

Future work is about the algorithm complexity analysis and presents an op-timization method to extract more precise correlations for a focal blog word.

Acknowledgments. The work was partly supported by the National Natural Science Foundation of China (71271044/U1233118) and the Specialized Research Fund for the Doctoral Program of Higher Education (20100185120024).

References

1. Zheng, Y., Zhang, L., Xie, X., Ma, W.-Y.: Mining interesting locations and travel sequences from GPS trajectories. In: Proceedings of WWW 2009, pp. 791–800 (2009)
2. Pan, B., MacLaurin, T., Crotts, J.: Travel blogs and the implications for destination marketing. Journal of Travel Research 46, 35–45 (2007)
3. Sigurbjörnsson, B., van Zwol, R.: Flickr tag recommendation based on collective knowledge. In: Proceedings of the WWW 2008, pp. 327–336 (2008)
4. Asbagh, M.J., Sayyadi, M., Abolhassani, H.: Blog Summarization for Blog Min-ing. In: Lee, R., Ishii, N. (eds.) Software Engineering, Artificial Intelligence. SCI, vol. 209, pp. 157–167. Springer, Heidelberg (2009)
5. Provart, N.: Correlation networks visualization. Frontiers in Plant Science 3(artical 240), 1–6 (2012)
6. Cao, Q., Duan, W., Gan, Q.: Exploring determinants of voting for the "helpfulness" of online user reviews: A text mining approach. Decision Support Systems 50(2), 511–521 (2011)

7. Ghose, A., Ipeirotis, P.G.: Estimating the helpfulness and economic impact of product reviews: Mining text and reviewer characteristics. IEEE Transactions on Knowledge and Data Engineering 23, 1498–1512 (2011)
8. Hu, M., Liu, B.: Mining and summarizing customer reviews. In: Proceedings of the KDD 2004, pp. 168–177 (2004)
9. Pang, B., Lee, L.: Opinion mining and sentiment analysis. Foundations and Trends in Information Retrieval 2(1-2), 1–135 (2008)
10. Dave, K., Lawrence, S., Pennock, D.M.: Mining the peanut gallery: Opinion extraction and semantic classification of product reviews. In: Proceedings of the WWW 2003, pp. 519–528 (2003)
11. Cui, H., Mittal, V., Datar, M.: Comparative experiments on sentiment classification for online product reviews. In: Proceedings of the AAAI 2006, pp. 1265–1270 (2006)
12. Qamra, A., Tseng, B., Chang, E.Y.: Mining blog stories using community-based and temporal clustering. In: Proceedings of the CIKM 2006, pp. 58–67 (2006)
13. Attardi, G., Simi, M.: Blog Mining Through Opinionated Words. In: Proceedings of the Fifteenth Text REtrieval Conference (TREC 2006), pp. 14–17 (2006)
14. Bai, X., Sun, J., Che, H., Wang, J.: Towards Knowledge Extraction from Weblogs and Rule-Based Semantic Querying. In: Paschke, A., Biletskiy, Y. (eds.) RuleML 2007. LNCS, vol. 4824, pp. 215–223. Springer, Heidelberg (2007)
15. Liu, B.: Sentiment analysis and subjectivity, 2nd edn. Handbook of Natural Language Processing (2010)
16. O'Leary, D.E.: Blog mining-review and extensions: "From each according to his opinion". Decision Support Systems 51(4), 821–830 (2011)
17. Liu, Y., Yu, X., Huang, X., An, A.: Blog Data Mining: The Predictive Power of Sentiments. In: Data Mining for Business Applications, pp. 183–195 (2009)
18. Wang, F., Wu, Y.: Mining Market Trend from Blog Titles Based on Lexical Semantic Similarity. In: Gelbukh, A. (ed.) CICLing 2012, Part II. LNCS, vol. 7182, pp. 261–273. Springer, Heidelberg (2012)
19. Kobayashi, M., Takeda, K.: Information retrieval on the web. ACM Computing Surveys 32(2), 144–173 (2000)
20. Raghavan, S., Garcia-Molina, H.: Crawling the Hidden Web. In: Proceedings of the VLDB 2001, pp. 129–138 (2001)
21. Han, J., Kamber, M.: Data Mining: Concepts and Techniques, 2nd edn. Morgan Kaufmann (2005)
22. Xiong, H., Tan, P.-N., Kumar, V.: Hyperclique pattern discovery. Data Mining and Knowledge Discovery Journal 13(2), 219–242 (2006)
23. Jo, Y., Oh, A.H.: Aspect and sentiment unification model for online review analysis. In: Proceedings of the WSDM 2011, pp. 815–824 (2011)
24. Wu, T., Chen, Y., Han, J.: Re-examination of interestingness measures in pattern mining: a unified framework. Data Min. Knowl. Discov. 21(3), 371–397 (2010)
25. Java, A., Song, X., Finin, T., Tseng, B.: Why we twitter: understanding microblogging usage and communities. In: Proceedings of the WebKDD/SNA-KDD 2007, pp. 56–65. ACM, New York (2007)

Vague Sets and Closure Systems

Sylvia Encheva

Stord/Haugesund University College, Bjørnsonsg. 45, 5528 Haugesund, Norway
sbe@hsh.no

Abstract. Just in time learning is a very important issue in any rapidly changing society. Initially well defined projects are often forced to introduce changes requiring current team members to learn new skills or alternatively find new people ready to step in. Such decisions have to be seriously considered before taking a risk that current team members might not manage to master those new skills sufficiently enough in order to complete a particular task. The other option is to invest time looking for new employees that might need quite some time to learn about the project and then begin with a real contribution. Aiming at providing automated support for such decision making processes we propose applications of vague sets and closure systems. Their functions are to formalise establishment of correlations between basic and complicated skills needed to complete a particular task.

Keywords: Optimization, skills, learning.

1 Introduction

Today's project leaders face a number of challenges related to balansing between rapidly changing tasks requirements and work force qualifications. Thus the need of tools providing automated advises in the process figuring out what is available in terms of skills a team posses, how much time team members need to learn specific new skills required by the project's undertakings and should additional consultancy be involved.

In this work we are looking at ways to employ decision support systems for extracting correlations between sets of basic skills and sets of more complicated skills that have the basic ones as prerequisites. In practice all descriptions of skills are somewhat imprecise. To handle such imprecise data in a formal way we suggest application of vague sets. They are characterized by a truth-membership function and a false-membership function, [5] which allows further tuning of rules in a decision support system.

Another interesting problem arising while solving skills' issues is which basic skills are needed in addition to the available ones for a successful project outcome. In that sense we refer to closure systems [1] because they are providing us with structures where addition of new elements is well defined.

The rest of this article is presented as follows. Section 2 contains basic terms and statements. Our approach is discussed in Section 3. Conclusion remarks can be found in Section 4.

M. Wang (Ed.): KSEM 2013, LNAI 8041, pp. 27–35, 2013.

2   Preliminaries

Knowledge management has been of interest to many authors, see for example [8], [9], [11], and [12]. Fuzzy multi-criteria decision making is discussed in [2]. Vague sets have been exploited in relation to decision making problems in [3] and [10]. Information obtained via fuzzy equivalence relations can be used to determine a limit for the degree of precision in which inputs should be measured, [7].

2.1   Lattices

A lattice is a partially ordered set, closed under least upper and greatest lower bounds. The least upper bound of x and y is called the join of x and y, and is sometimes written as $x + y$; the greatest lower bound is called the meet and is sometimes written as $x\dot{y}$, [4]. X is a sublattice of Y if Y is a lattice, X is a subset of Y and X is a lattice with the same join and meet operations as Y. A lattice L is meet-distributive if for each $y \in L$, if $x \in L$ is the meet of (all the) elements covered by y, then the interval $[x; y]$ is a boolean algebra, [4].

Let L be a lattice.

i) L is non-modular if and only if L has a sublattice isomorphic to N_5.

ii) L is non-distributive if and only if L has a sublattice isomorphic to N_5 or M_3.

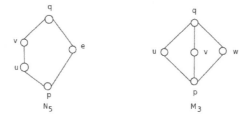

Fig. 1. The two lattices N_5 and M_3

2.2   Closure System

Closure systems can be defined by a set of implications (a basis), and in such form they appear as Horn formulas in logic programming, dependencies in relational data bases, CNF representations of Horn Boolean functions and directed hypergraphs in discrete optimization, [1].

A closure system on a finite set M is a set F of subsets of M such that $M \in F$ and $C, C' \in F \Rightarrow C \cap C' \in F$.

A closure system on set M is convex geometry if it satisfies the following properties - the empty set is closed and for every closed set $M_1 \neq M$ there exists $m \notin M_1$ such that $M_1 + m$ is a closed set.

Convex geometries are closure systems which satisfy anti-exchange property (which is an antipodal concept to the Steinitz MacLane exchange property of matroids), and they are known as dual of antimatroids. Antimatroids can be seen as a combinatorial abstraction of convexity, while matroids can be seen as a combinatorial abstraction of linear independence. Antimatroids are related to matroids in that both can be defined by a apparently similar axioms.

The set of closed sets of a convex geometry, form a lattice when ordered by set inclusion. Such lattices are precisely the meet-distributive lattices. A lattice is said to be infinitely meet-distributive if all existing meets in it are distributive.

2.3   Vague Sets

Notations in this subsection are as in [6]. Let U be the universe of discourse, $U = \{u_1, u_2, ..., u_n\}$, with a generic element of U denoted by u_i. A vague set A in U is characterized by a truth-membership function t_A and a false-membership function f_A,

$$t_A : U \to [0,1],$$

$$f_A : U \to [0,1],$$

where $t_A(u_i)$ is a lower bound on the grade of membership of u_i derived from the evidence for u_i, $f_A(u_i)$ is a lower bound on the negation of u_i derived from the evidence against u_i, and $t_A(u_i) + f_A(u_i) \leq 1$. The grade of membership of u_i in the vague set A is bounded to a subinterval $[t_A(u_i), 1 - f_A(u_i)]$ of $[0,1]$. The vague value $[t_A(u_i), 1 - f_A(u_i)]$ indicates that the exact grade of membership $\mu(u_i)$ of u_i may be unknown. But it is bounded by

$$t_A(u_i) \leq \mu(u_i) \leq 1 - f_A(u_i),$$

where

$$t_A(u_i) + f_A(u_i) \leq 1.$$

When the universe of discourse U is continuous, a vague set A can be written as

$$A = \int_J [t_A(u_i), 1 - f_A(u_i)]/u_i$$

3   Combining Basic Skills

In this section we first consider single skills hereafter referred to as *basic* skills and other skills that are combinations of two or more of the basic skills, hereafter referred to as complicated skills. The idea is to draw a clear picture of which skill is necessary to master first, in order to go on with more complicated ones and if an employ is believed to master a complicated one, is that person in a possession of all the required basic skills.

In a particular project where prerequisites are clear one can attach a parameter for addressing time necessary to master a new skill, provided sufficient level of proficiency of the rest of the needed skills. A simple way to calculate time for learning is to use the following function

$$T = \sum t_{a_i}$$

where T is the total time needed for a person to learn certain skills, t_{a_i} is the needed to learn a skill a_i, $1 \le i \le n$, where n is the number of skills needed for the task. A cost can be estimated by

$$T = \sum t_{a_i} \times c_{a_i}$$

where c_{a_i} is the amount of expenses needed to secure mastering of skill a_i by a person. This might make the model a bit more demanding but will facilitate the process of making fast decisions on optimal solutions like for example should one send team members to a course of involve new members who can just do the job.

A basic skill and the related complicated skills requiring the first one are shown in Fig. 2. In this case if an employ posses only one skill (say a) and she is assigned to

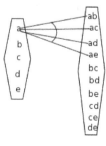

Fig. 2. A basic kill and four consecutive skills

work on tasks requiring any of the other complicated skills (say *ab, ac, ad, ae*) additional time has to be anticipated for mastering for example one extra skill (say *b*) and mastering usage of the required complicated still *ab*.

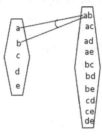

Fig. 3. Two basic questions and one related to them question

In Fig. 3 we illustrate the reverse connection. A complicated skill (*ab*) is needed and all basic skills (*a, b*) that are prerequisites for it are shown. Correlations in Fig. 2 and Fig. 3 have one thing in common, in both cases there is exactly one element either as a premise or as a consequence.

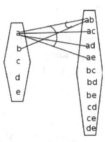

Fig. 4. Two basic skills related to four complicated skills

In Fig. 4 we observe a different setting, two basic skills are related to four complicated skills. One of the basic skills is related to four complicated skills while the other basic skill is related to one complicated skill.

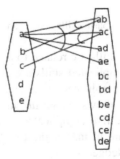

Fig. 5. Three basic skills related to four complicated skills

Correlations in Fig. 5 are more entangled. Three basic skills are related to four complicated skills where one of the basic skill is related to all of the complicated skills and the other two basic skills are related to two different complicated skills.

Four basic skills are related to four complicated skills in Fig. 6. One of the basic skills is related to all complicated skills supported by it while the other three basic skills are related to each one of three of the complicated skills.Three of the complicated skills are related to the supporting basic skills and one of complicated skills is related to one of the supporting basic skills.

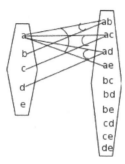

Fig. 6. Four basic skills are related to four complicated skills

Two basic skills are related to seven complicated skills in Fig. 7. Each of the two basic skills is related to the corresponding complicated skills. Only one of the complicated skills is related to the two supporting basic skills.

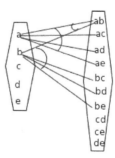

Fig. 7. Two basic skills and seven complicated skills

Finally three of the basic skills are related to all of the involved complicated skills Fig. 8. Only three of the complicated skills are related to the supporting them basic skills. The rest of the five complicated skills are related to a single basic skill only. This can be interpreted as follows. Skills *ab, ac* and *bc* are mastered by the team members, for the rest of the five complicated skills some consideration have to be done. Extra time for the team to master the two basic skills or new team members possesing the required knowledge.

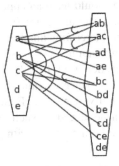

Fig. 8. Three basic skills and nine complicated skills

The process of mastering an additional skill and applying in combination with other skills is not necessarily linear in terms of time and efforts. In Fig. 9 we see that on each level of the lattice the amount of skills is increased by one going upwards.

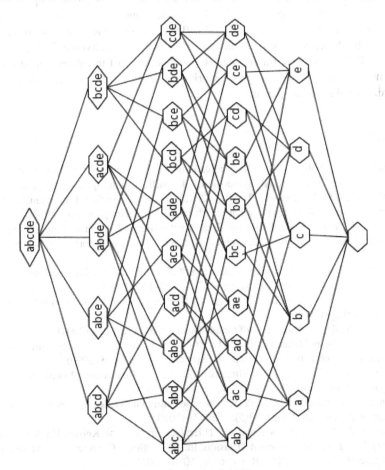

Fig. 9. Five basic skills

Under a project planning one should have an approximate idea about the time needed to master some new skills. Difficulties might occur due to the nature of some particular skills, due to their increasing number as well as due to the necessity of applying several skills in combination.

Five basic skills are considered in Fig. 9. In this case only relations among complicated skills are shown where again the difference between two nodes connected by an edge is exactly one skill. Following such a lattice illustrates all possible paths for reaching a node and as such can be used to calculate amount of time and additional cost for gathering a team with predifined qualifications.

4   Conclusion

In this article we discuss relationships between skills with different level of complexity. Complicated tasks require a possession of various basic or elementary skills as well as abilities to apply a combination of them. A decision support system can be used for connecting people, skills they possess and current projects needs. The presented approach can be used in several different ways some of which are relating what should be done, what is available and what has to be delivered from outside in terms of skillfull workforce. Based on such conclusions staff members can be sent for just in time learning courses when they work on a project or for longer training when the organizational schedule permits.

References

1. Adaricheva, K., Nation, J.B.: On implicational bases of closure systems with unique critical sets. In: International Symposium of Artificial Intelligence and Mathematics (ISAIM 2012), Ft. Lauderdale, FL, USA (2012), Results are included into plenary talk on conference Universal Algebra and Lattice Theory, Szeged, Hungary (June 2012)
2. Chang, T.H., Wang, T.C.: Using the fuzzy multi-criteria decision making approach for measuring the possibility of successful knowledge management. Information Sciences 179, 355–370 (2009)
3. Chen, S.M., Tan, J.M.: Handling multicriteria fuzzy decision-making problems based on vague set theory. Fuzzy Sets and Systems 67(2), 163–172 (1994)
4. Davey, B.A., Priestley, H.A.: Introduction to lattices and order. Cambridge University Press, Cambridge (2005)
5. Gross, J.L., Yellen, J.: Handbook of Graph Theory. CRC Press Inc. (2004)
6. Hong, D.H., Chang-Hwan Choi, C.-H.: Multicriteria fuzzy decision-making problems based on vague set theory. Fuzzy Sets and Systems 114, 103–113 (2000)
7. Klawonna, F., Castrob, J.L.: Similarity in Fuzzy Reasoning. Mathware & Soft Computing 2, 197–228 (1995)
8. Percin, S.: Use of analytic network process in selecting knowledge management strategies. Management Research Review 33(5), 452–471 (2010)
9. Pourdarab, S., Nadali, A., Nosratabadi, H.E.: Determining the Knowledge Management Strategy Using Vague Set Group Decision. In: International Conference on Management and Artificial Intelligence IPEDR, vol. 6, pp. 60–64 (2011)

10. Verma, M., Kumar, A., Singh, Y.: Vague modelling for risk and reliability analysis of compressor system. Concurrent Engineering 20, 177–184 (2012)
11. Ye, J.: Improved method of multicriteria fuzzy decision-making based on vague sets. Computer-Aided Design 39, 164–169 (2007)
12. Wu, W.W.: Choosing knowledge management strategies by using a combined ANP and DEMATEL approach. Expert Systems with Applications 35, 828–835 (2008)
13. Zhang, D., Zhang, J., Lai, K.K., Lu, Y.: An novel approach to supplier selection based on vague sets group decision. Expert Systems with Applications 36(5), 9557–9563 (2009)

Building Enhanced Link Context by Logical Sitemap

Qing Yang[2], Zhendong Niu[2], Chunxia Zhang[1,*], and Sheng Huang[2]

[1] School of Software, Beijing Institute of Technology
[2] School of Computer Science, Beijing Institute of Technology
{yangqing2005,20812036,cxzhang,zniu}@bit.edu.cn

Abstract. Link contexts have been applied to enrich document representation for a variety of information retrieval tasks. However, the valuable site-specific hierarchical information has not yet been exploited to enrich link contexts. In this paper, we propose to enhance link contexts by mining the underlying information organization architecture of a Web site, which is termed as logical sitemap to differ from sites supplied sitemap pages. We reconstruct a logical sitemap for a Web site by mining existing navigation elements such as *menus*, *breadcrumbs*, *sitemap* etc. It then enriches contexts of a link by aggregating contexts according to the hierarchical relationship in the mined logical sitemap. The experimental results show that our proposed approach can reliably construct a logical sitemap for a general site and the enriched link contexts derived from the logical sitemap can improve site-specific known item search performance noticeably.

Keywords: logical sitemap, site-specific known item search, link context.

1 Introduction

Well-designed Web sites usually organize hundreds or thousands of resources by human-made taxonomy schemes for navigation. The taxonomy schemes can be presented as *menus*, *breadcrumbs*, *sitemaps*, etc. For example, a typical site has a navigator bar at the top of page as Figure 1(a). It is easy for a user to recognize such organization information cues, and identify relationships between pages by navigator elements. Figure 1(b) is a typical online shopping site. Because there are a large number of categorized products to be presented, authors usually resort to multilevel menus to facilitate customers.

Although anchor text is the most direct descriptive information for the target page, it just predicts the target page and has no information for the relationship of pages. It also ignores the assigned categories of pages which are organized in the site-specific underlying organization structure.

In this paper, we mine the navigator information that are organized in a hierarchical structure to construct a logical sitemap. The logical sitemap contains the relationships between pages in the web site, therefore it is an effective source to enrich the link

---

* Corresponding author.

M. Wang (Ed.): KSEM 2013, LNAI 8041, pp. 36–47, 2013.
© Springer-Verlag Berlin Heidelberg 2013

contexts of pages. Different from our previous work [1], we propose an approach of Logical sitemap Construction from Menu Mining (LCMM) in a global view instead of previously limited tag patterns. The main contributions of a logical sitemap are two folds. First, it can enrich context of links by integrating site-specific categories. Second, contexts of parent pages can be propagated to their child pages to leverage anchor text sparsity problem.

To construct and integrate a logical sitemap to improve the performance of information retrieval, we propose an unsupervised approach to automatically extract the logical sitemap of a given Web site, and propagate contexts according to the hierarchical relationship in the logical sitemap to enrich the contexts of links. Each node in a logical sitemap is corresponding to a Web page that can be enriched with contexts of its parent nodes iteratively.

The rest of this paper is organized as follows: Section 2 describes our approach to the task to enhance link contexts from a logical sitemap by mining semantic hierarchical information of menus. Evaluation of our solution is presented in Section 3. In Section 4, we make a brief discussion of related work. Finally, Section 5 concludes the paper and discusses future work.

Fig. 1. Cropped Real-world Web Pages

2   LCMM: Logical Sitemap Construction from Menu Mining

From observations of a large number of typical sites, the most common patterns to organize the sites are menus and sitemap pages. To differ from the supplied sitemap pages, we name it as a physical sitemap. Because a supplied sitemap can be considered as a special kind of menu, in this paper, we focus on mining hierarchical relationship from menus, especially multilevel menus. The child menu items of a multilevel menu typically have the similar appearance and the whole of menus are visually aligned in a continuous area. Such visually similar appearances can be easily recognized by humans, and can also be accurately located by HTML tags.

Our hierarchical information from menu mining method comprises of three steps: (1) page representation (2) menu identification, and (3) menu level labeling. The first step segments the DOM tree of Web page into a tag path sequence and blocks. The second step identifies menus from the HTML blocks. The last step extracts the site-specific categories for pages in the site. The method is fully automatic and does not involve manually labeling. We describe the three steps in details in the next sections. Then, we construct a logical sitemap from the mined hierarchical information from menus.

2.1   Tag Path Sequence of Page

Instead of viewing the Web page as a DOM tree, we consider it as a sequence of HTML tag paths which are generated from the preorder traversal of the DOM tree. A *tag path* is constructed for each node as an ordered sequence of tags rooted from the *body* tag. We take *pos* to record a tag path occurrence position in the sequence of tag paths.

With the assumption that a same tag path will be rendered in the similar appearance, we count the frequency of distinct tag paths to get candidate separators of potential menus. To extract meaningful information, we only consider those content nodes which contain text contents. Formally, a content nodes is denoted as a triple $t_i =< pos, path, level >$, where *path* is a tag path, *pos* is the occurrence position of *path* in the tag path sequence, and *level* is to label the hierarchical level. For brevity, we denote those triples which have at least k occurrences as $T_k = \{|t| \geq k | t_i.path = t_j.path\}$. Accordingly, we define a *block* as the nodes which satisfy T_k and reside under the minimal common parent node p_{min}. The p_{min} can easily be identified by comparing tag paths forwardly from the end of the tag paths.

2.2   Menu Identification

Formally, a menu can be defined as

$$M^{(j)} = (M_{0,}^{(j)} ..., M_i^{(j)}), 0 < i, 0 \leq j \leq 4 \tag{1}$$

Without any loss of generality, every parent menu items may have different number of child menu items and they should have at least one. From the sampled sites, the level of menu does not exceed 5. A whole menu always resides under a sub tree of the DOM tree. That's to say, tag paths of menu items belonging to a menu should have the common prefix of tag paths. It is reasonable to assume that there are at least five items of the top level menu and child menu items have the similar appearance or it may have nothing meaningful for us.

With the notions of Section 2.1, the set of nodes $T_k \geq 5$ are potential menu items. Generally, for the multilevel menus ($j \geq 1$), the positions of menu items of different levels in a menu will be interleaved each other in the sequence of tag paths. By examining a large number of typical Web sites, we identify a set of features, based on the path, presentation, and layout of the Web pages, that are useful for identifying menus. Specially, we propose the following features for a block of tag paths of a given page p in menu identification.

URL Path. It is a conventional practice to group related files in the same folder. We assume that if the URL paths of links in a block have hierarchical relationship, the block is more likely to be a menu.

Tag. The *nav* is a new element in HTML 5 specification which is used to group together links to facilitate navigation. Therefore, a tree rooted from a *nav* element can be likely a candidate menu.

Uniformity of Child Nodes. Child nodes of menu items should appear similarly. We measure it with the average number of count of different tags between tag paths of child nodes. Accordingly, we define uniformity of the whole blocks as the total of uniformity of child nodes denoted as **uniformity of block**.

Naming Convention. Nodes of menus may have "id", "class" and "role" attributes. We sampled about 1000 sites to find that about 60% named their menus as "naviga-tion" and its variations such as shortened formats of "nav", and about 20% as "menu" or its variations such as "*menu". Besides, some names of menu items have a good indication of top level menu, such as "home", "help", "about " etc.

Link Node Ratio. The ratio of linked text to all text is a common method for recog-nizing navigation elements or link lists. It is named as "link node". Accordingly, the average of child node link text ratio is used to measure the link node ratio of the whole menu.

Occurrence Position. Menu often resides at the top part of a Web page. We measure the position feature with the position of the first menu item in the sequence of tag paths. It is more likely to be a menu for small position values than that of big ones.

In this paper, we consider the menu identification as a classification task. Those blocks which contain nodes ($T_k \geq 5$) are potential menus. we described six types of features for classifying tag path blocks into menu blocks and non-menu blocks. Each segmented block can be represented by a feature vector $\mathbf{x} =< x_1, x_2, \ldots x_6 >$, where x_i is the value for the ith feature for this block. Let c_1 and c_2 denote the two categories: menu block and non-menu block. In Section 3.1, we have investigated three kinds of classifiers for this specific classification problem, which are decision tree, naïve Bayes, and logistic regression. The experiments show that the decision tree algorithm gains the best performance.

2.3   Menu Level Labeling

Usually, hierarchical categories exist in multilevel menus. A multilevel menu is part of a Web page that contains multiple level menu items. The same level child items can be consecutive or non-consecutive because they may have different child items respectively. Furthermore, parent menu items may reside on the same level of the DOM tree with their child menu items and distinguish themselves from their child items with styles or other tags.

To identify menu levels, we propose an iterative division algorithm *LevelLabeler* from top to down to label the levels of menu items for above identified menus. For each menu, we use algorithm 1 to label the levels of menu items. The level of all menu items is set as 0 initially. After labeled by algorithm 1, the level of each child menu item is set as the position of its parent menu item. Algorithm 1 labels levels in an identified menu which is represented as a sequence of tag paths *block*(t) by the three steps:

1. Use the tag path of the first item in *block*(*t*) as the separator to split *block*(*t*) into several continuous segment *s*;
2. Filter out the split segments which have menu items less than *K* and store into *g*;
3. Split *block*(*t*) into small segments when *g* is empty;
4. For the segments in *g*, iterate to execute algorithm 1.

As described in Section 2.1, a page is represented into a sequence of tag path. We get a set of menu blocks identified by the menu identifying process, i.e. *block*(*t*), a

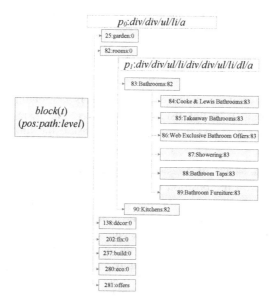

Fig. 2. Illustration of algorithm 1 from example of Figure 1(b)

segment from the position between 25 and 281. First we use the tag path $p_0(div/div/ul/li/a)$, of the first menu item $t_0 = (25: \text{garden}: 0)$, as the pivot point to split $block(t)$ into a several child menu segments, such as a segment with the position between 83 and 138 of the sequence of the page. Similarly, we repeat this process for each of the segments until they cannot be split any more. They are assumed to be parallel items and at the same level in the hierarchy. We describe our algorithm 1 in pseudo code as following.

Algorithm 1. *LevelLabeler*

Input: *block(t)*

Output: *labeled blocks block(t)*

Const: threshold, K

1 $g \leftarrow$ new List; $s \leftarrow$ new List;

2 $t_o \leftarrow block(0)$; $css \leftarrow false$;

3 split**:**

4 **for each** $t_i \in block(t)$ **do**

5   **if** $compare(t_i, t_o, css)$ **then**

6   last $\leftarrow t_i.\text{pos}$;

7   **if** !s.empty **then**

8   **if** s.count >k **then**

9   g.add(s);

10   s.clear();

              end

11   **end**

12   **else**

13         $t_i.level \leftarrow$ last;

14   s.add(t_i);

15   **end**

16 **end**

17 **if** $|g| = 0$ **then**

18      $css \leftarrow$ true;

        goto split;

19 **end**

20 **for each** $s \in g$ **do**

21   *LevelLabeler* (s);

22 **end**

In our experiments, the threshold parameter K is set to 2. To distinguish sub segments belong to which parent item, algorithm 1 use the position of the parent item to identify its child items. If a block has not been split into at least two segments, algorithm 1 try to split the block in fine-grained segments by considering CSS styles.

2.4   Two Stage Logical Sitemap Construction

With the labeled levels of menus, we can construct a logical sitemap to represent the underlying content organization structure of the site. The method above focuses on a single page. Because the scope of a menu item may consist of a group of pages, it requires additional information to assign them with suitable positions in the logical sitemap. Keller et al. [2] proposed a *navigation structure graph* to combine local and global navigation systems.

With the aim of enriching contexts of links in the setting of crawled part of pages of a specific site, we propose a two stage method to address this problem. First, it constructs a logical sitemap by mining menus. Then it detects the specific track of the pages in the site-specific categories schemes such "*breadcrumbs*".

First Stage. As described in Section 2.3, each content node has the position of its parent menu item as its level. If the level equals zero, it is assumed to be the top level. With the labeled level, we construct the initial logical sitemap from the top level with ascending order and identify every page with its URL. In order to build a tree-like logical sitemap, we add a virtual node as their root of top level menu items.

Second Stage. For a site with a large number of categories, it is a conventional practice to supply a location track to tell users where they are (as Figure 3.). Observed from sampled pages, it confirms this assumption. About 95% sites with a large number of categories supply "*breadcrumb*". Such location track can be exploited to identify a group of pages which menu items they belong to with reference to mined menus. In our experiments, we propose the following features to identify breadcrumb:1) a node which its class_ or _id contains "*breadcrumb*" or "*crumb*"; 2) links within a continuous area whose anchor texts and target links follow the order of mined menu levels.

Fig. 3. Cropped Example of breadcrumb[1]

Follow the hierarchy of the logical sitemap, it is trivial to aggregate contexts of links by hierarchical labels and contexts following the path from the root to the current node. For example, the contexts of *http://www.mdfarmbureau.com/Insurance.asp* are "Member Service, benefits".

---

[1] http://www.diy.com/nav/rooms/bathrooms/bathroom-taps

3   Experiments

This paper aims at enriching the context of links with the site-specific categories. Currently, there is no available dataset in the community for the evaluation of menu extraction task. Therefore, we manually label a test set as the ground truth. We collect 50 typical sites with multilevel menus from the Web to extract the menus. We use F_1 as evaluation metrics to measure the quality of menu mining.

3.1   Results and Discussions

Menu Blocks Classifiers. When evaluating the machine-learning methods, our objective is to find a classifier that is capable of classifying HTML blocks in some unseen Web sites. Therefore, we use the 10-fold cross-validation scheme to conduct the evaluation as follows: we train a classifier using 10% of the sites, and then use the trained classifier to extract the menu blocks for the remaining 90% sites, which guarantees the data used for training and testing are independent. Finally, we validate the results with the gold truth.

Table 1. Performance of menu blocks classifiers

Measure	Decision Tree	Naïve Bayes	Logic regression
F1	95.4%	90%	89.8%

For each split dataset, we trained a decision tree, a naïve Bayes, and a logistic regression classifier respectively. The final results are summarized in Table 1. As can be seen, the decision-tree classifier produces the best results, while the performance of naïve Bayes and logistics regression classifiers are close to each other and significantly lower than that of the decision tree classifier.

Hierarchy Detection. For each site a maximum of 200 pages were crawled. Results were classified into three kinds, i.e. accurate, partly correct and wrong. *Accurate* denotes the hierarchy detected is completely correct. *Partly correct* denotes there are at least one link's hierarchy is wrong. Results are classified as *Wrong* when the ratio of wrong links exceeds 30%. To be compared with our previous work [1], we refer it as TREC-BU which extract hierarchical information by using limited patters in a bottom-up way. We compare the two methods in the same test data. Results are shown in Table 2.

Table 2. Results of hierarchy detection

Result Type	LCMM	TREC_BU
Accurate	68.6%	54.4%
partly Correct	26.4%	35.2%
Wrong	5%	10.4%

Seen from Table 2, the performance of LCMM outperforms that of TREC_BU significantly. It is because TREC_BU just considers a limited hierarchical tag patterns and some hierarch information is missing. Another reason is that TREC_BU does not construct the logical sitemap globally. However, for LCMM, the rate of minor errors

26.4% is partly due to the mistakes of the combination process of main menus and local menus. It requires develop a more robust method to improve the combination process globally.

To demonstrate the performance of information retrieval of our proposed methods, we evaluate our method in site-specific know-item search.

3.2   Site-Specific Known Item Search

Site-specific known item search is the task where users expect a single, particular answer in response to a query within an indicated Web site [3]. We believe that contexts closer to the Web page ought to be given a higher weight in the hierarchical model. Their document frequency increases as the contexts move higher in the hierarchy because they are used to categorize more documents. Therefore, no explicit weighting is required.

Data Sets. To evaluate the reliability of the method and its potential for enriching contexts of links, we use 30 entity queries in related entity finding task in TREC 2010 Entity Track and disregard those queries which cannot be answered from the site where the supplied homepage belong to. We judge the relevance of results manually.

Primary Tests. The first benchmark is to use BM25 on the content of the Web pages only. In order to compare the expressive power of propagated site-specific hierarchical information to widely used anchor text document representation, a second benchmark test was run using only anchor text field. For all experiments, the BM25 parameters are $k_1 = 1.8$ and $b = 0.75$.

Table 2. MRR results for different fields

Fields	Content	Anchor	TREC_BU	LCMM
MRR	0.342	0.575	0.606	0.632

Table 3 shows that MRR results of BM25 on individual fields. It is not surprised to see that content-only performs poorly on know item queries. Anchor text representation received good results as expected. It is noted that enriched anchor text model

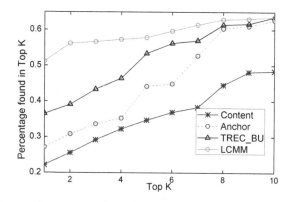

Fig. 4. Percentage of positive results found in top K results

(TREC_BU and LCMM) achieves noticeable improvement over the two baselines. LCMM is relatively better than our previous implemented TREC_BU.

To evaluate the ranking performance, we plot the percentage of positive results found in the top K results as K varies from 1 to 10. Figure 4 shows how the performance of the different fields vary as K increases. Though the three methods anchor, TREC_BU and LCMM achieve the similar results for the top 10, LCMM gets a much better early result. Also, all of them are much better than the content method because query words may not exist in contents.

4   Related Work

Previous work has widely studied and used anchor text as an important source of relevant information for many years. Craswell et al. [4] is among the earliest to demonstrate the effectiveness of anchor text on answering the information needs targeting at finding specific web sites. The following work on using anchor text to improve search gradually falls into three categories. One of them is to connect query intent with anchor text distribution on the web [5, 6]. The second category focuses on intelligent ways of anchor text importance estimation [7-9]. The third category focuses on solving anchor text sparsity problem [10, 11], i.e., only a few web pages has considerable amount of anchor text associated because page in-coming links follow power law distribution. The effort in this category is to incorporate appropriate complementary anchor text to enrich existing anchor text representation. Our method provides another data source to leverage anchor text sparsity.

Besides of using anchor texts extensively, the text surrounding a link also seems to describe the destination page appropriately. Pant [12] proposed a framework to evaluate various link-context derivation methods. They find that good link-context can be associated with the tag tree and demonstrate link contexts can be used in classifier-guided topical crawlers [13]. But they do not tell a method how to get the optimal aggregation node. We differ from them by propagating link contexts with the mined logical sitemap. In future, we will investigate our methods with the classification methods Xiang et.al [14].

Another line of related work is web structure mining. It can be roughly classified in two research directions according to their different objectives. The first direction aims at generating new structures such as rankings or topic hierarchies based on web documents and their structure (e.g. PageRank or HITS and their variations). The second aims at mining existing structures. Several approaches in this direction work solely on the web graph model, whose vertices represent pages and whose are arcs represent hyperlinks. Extracting a hierarchical structure with standard graph algorithms bases on the web graph is described in [15]. In [16] the edge weights for the web graph are computed based on text similarity and co-citation of hyperlinks. Using the hierarchical structure of URLs is a very common approach ([15, 17-19]). However, it is well known that the hierarchical structure of URLs does not reflect the site organization accurately [17]. Figure 1 is a typical example. Keller and Nussbaumer propose a MenuMiner algorithm to retrieve the original content organization of large-scale web sites by analyzing maximal cliques [20]. Our method belongs to the second direction to mine existing structure. Different from them, it mines the web structure of a given site from navigational elements, specially menus.

5   Conclusions and Future Work

We propose to build a logical sitemap to enhance link contexts by mining the underlying content organization architecture of Web sites from navigational elements. We demonstrate that the propagated link contexts from logical sitemap can be incorporated in the document representation to improve search performance noticeably in the site-specific known item search task. In future, we plan to apply this method to construct normalized domain-specific category hierarchies for a domain from different homogeneous sites such as online shops.

Acknowledgments. This work was supported by the National Natural Science Foundation of China (NO.61272361, NO.61250010).

References

1. Yang, Q., Jiang, P., Zhang, C., Niu, Z.: Reconstruct Logical Hierarchical Sitemap for Related Entity Finding. In: TREC 2010 (2011)
2. Keller, M., Nussbaumer, M.: Beyond the Web Graph: Mining the Information Architecture of the WWW with Navigation Structure Graphs. In: Proceedings of the 2011 International Conference on Emerging Intelligent Data and Web Technologies, pp. 99–106. IEEE Computer Society (2011)
3. Weninger, T., Zhai, C., Han, J.: Building enriched web page representations using link paths. In: Proceedings of the 23rd ACM Conference on Hypertext and Social Media, Milwaukee, Wisconsin, USA, pp. 53–62. ACM (2012)
4. Craswell, N., Hawking, D., Robertson, S.: Effective site finding using link anchor information. In: Proceedings of the 24th Annual International ACM SIGIR Conference on Research and Development in Information Retrieval, New Orleans, Louisiana, United States, pp. 250–257. ACM (2001)
5. Bron, M., et al.: The University of Amsterdam at TREC 2010 Session, Entity, and Relevance Feedback. In: TREC 2010 (2011)
6. Fujii, A.: Modeling anchor text and classifying queries to enhance web document retrieval. In: Proceeding of the 17th International Conference on World Wide Web, Beijing, China, pp. 337–346. ACM (2008)
7. Dou, Z., et al.: Using anchor texts with their hyperlink structure for web search. In: Proceedings of the 32nd International ACM SIGIR Conference on Research and Development in Information Retrieval, Boston, MA, USA, pp. 227–234. ACM (2009)
8. Lei, C., Jiafeng, G., Xueqi, C.: Bipartite Graph Based Entity Ranking for Related Entity Finding. In: 2011 IEEE/WIC/ACM International Conference on Web Intelligence and Intelligent Agent Technology, WI-IAT (2011)
9. Lu, W.-H., Chien, L.-F., Lee, H.-J.: Anchor text mining for translation of Web queries: A transitive translation approach. ACM Trans. Inf. Syst. 22(2), 242–269 (2004)

10. Metzler, D., et al.: Building enriched document representations using aggregated anchor text. In: Proceedings of the 32nd International ACM SIGIR Conference on Research and Development in Information Retrieval, Boston, MA, USA, pp. 219–226. ACM (2009)
11. Dai, N., Davison, B.D.: Mining Anchor Text Trends for Retrieval. In: Gurrin, C., He, Y., Kazai, G., Kruschwitz, U., Little, S., Roelleke, T., Rüger, S., van Rijsbergen, K. (eds.) ECIR 2010. LNCS, vol. 5993, pp. 127–139. Springer, Heidelberg (2010)
12. Talukdar, P.P., et al.: Weakly-supervised acquisition of labeled class instances using graph random walks. In: Proceedings of the Conference on Empirical Methods in Natural Language Processing, Honolulu, Hawaii, pp. 582–590. Association for Computational Linguistics (2008)
13. Venetis, P., et al.: Recovering semantics of tables on the web. Proc. VLDB Endow. 4(9), 528–538 (2011)
14. Xiang, S., Nie, F., Zhang, C.: Learning a Mahalanobis distance metric for data clustering and classification. Pattern Recogn. 41(12), 3600–3612 (2008)
15. Yang, C.C., Liu, N.: Web site topic-hierarchy generation based on link structure. J. Am. Soc. Inf. Sci. Technol. 60(3), 495–508 (2009)
16. Agarwal, A., Chakrabarti, S., Aggarwal, S.: Learning to rank networked entities. In: Proceedings of the 12th ACM SIGKDD International Conference on Knowledge Discovery and Data Mining, Philadelphia, PA, USA, pp. 14–23. ACM (2006)
17. Kumar, R., Punera, K., Tomkins, A.: Hierarchical topic segmentation of websites. In: Proceedings of the 12th ACM SIGKDD International Conference on Knowledge Discovery and Data Mining, Philadelphia, PA, USA, pp. 257–266. ACM (2006)
18. Kurland, O., Lee, L.: Respect my authority!: HITS without hyperlinks, utilizing cluster-based language models. In: Proceedings of the 29th Annual International ACM SIGIR Conference on Research and Development in Information Retrieval, Seattle, Washington, USA, pp. 83–90. ACM (2006)
19. Koschützki, D., Lehmann, K.A., Tenfelde-Podehl, D., Zlotowski, O.: Advanced Centrality Concepts. In: Brandes, U., Erlebach, T. (eds.) Network Analysis. LNCS, vol. 3418, pp. 83–111. Springer, Heidelberg (2005)
20. Talukdar, P.P., Crammer, K.: New regularized algorithms for transductive learning. In: Buntine, W., Grobelnik, M., Mladenić, D., Shawe-Taylor, J. (eds.) ECML PKDD 2009, Part II. LNCS, vol. 5782, pp. 442–457. Springer, Heidelberg (2009)

A Nonlinear Dimension Reduction Method with Both Distance and Neighborhood Preservation

Chao Tan, Chao Chen, and Jihong Guan

Department of Computer Science and Technology
Tongji University, Shanghai 201804, China
`tanchao222@gmail.com, aelodge@163.com, jhguan@tongji.edu.cn`

Abstract. Dimension reduction is an important task in the field of machine learning. Local Linear Embedding (LLE) and Isometric Map (ISOMAP) are two representative manifold learning methods for dimension reduction. Both the two methods have some shortcomings. The most significant one is that they preserve only one specific feature of the underlying datasets after dimension reduction, while ignoring other meaningful features. In this paper, we propose a new method to deal with this problem, it is called *Global and Local feature Preserving Embedding*, GLPE in short. GLPE can preserve both the neighborhood relationships and the global pairwise distances of high-dimensional datasets. Experiments on both artificial and real-life datasets validate the effectiveness of the proposed method.

Keywords: Manifold Learning, Dimension Reduction, Global Distance Preservation, Local Structures Preservation.

1 Introduction

Finding meaningful low dimensional representations of high dimensional data has been an important task in the field of machine learning. There are some classic methods that can handle well the problem of dimension reduction of linear data, such as principal component analysis (PCA) [1] and multiple dimension scale (MDS) [2]. A common disadvantage of these methods is that they can deal with only linear datasets well and have poor performance on nonlinear datasets. Since 2000, a number of manifold learning algorithms have been proposed for solving the problem of dimension reduction on nonlinear datasets, such as locally linear embedding (LLE) [3,4] and isometric feature mapping (ISOMAP) [5]. Unlike linear methods, these methods do not have the presumption that the original datasets are linear. They try to characterize a certain specific feature of the underlying datasets and keep it unchanged in low dimensional representations. For example, LLE tries to keep the neighborhood relationships of the data points, while ISOMAP tries to keep global pairwise geodesic distances.

These nonlinear methods have simple and intuitive ideas, are easy to implement, and are globally optimal. However, each of them can preserve only one

M. Wang (Ed.): KSEM 2013, LNAI 8041, pp. 48–63, 2013.

kind of features of the underlying datasets. That is, the other features will be omitted, even though these features are of similar significance. An example is shown in Fig. 1. The 3D "swiss roll" dataset is shown in Fig. 1(a). The low dimensional output of LLE preserves the neighborhood relationships of the original high dimensional data points very well, but distances between all data points are obviously not well maintained, as shown in Fig.1(b). On the other hand, the low dimensional output of ISOMAP, as shown in Fig.1(c)) preserves the distances between all data points pretty well, while the neighborhood relationships of data points are quite different from the original dataset.

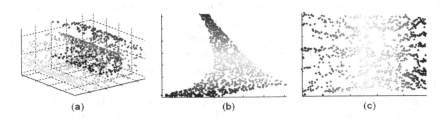

Fig. 1. An example of dimension reduction using LLE and ISOMAP. (a) 3-dimensional "Swiss Roll" dataset (1000 data points); (b) Output of LLE; (c) Output of ISOMAP.

One of the goals of dimension reduction is to preserve as many as possible the features hidden in the original datasets, so in this sense the algorithms above is not satisfactory. Furthermore, in many application scenarios the users usually have no idea of what features should be preserved, therefore they may be confused about which algorithms should be chosen for their applications. This situation urges us to explore new manifold learning approaches that can preserve multiple features of the underlying datasets after dimension reduction. This is the motivation of this paper.

In this paper, we try to develop a new manifold learning based dimension reduction approach that can simultaneously preserve neighborhood relationships and global pairwise distances in the low dimensional space. Since LLE and ISOMAP can fit to similar framework of kernel methods, the main difference between them lies in the "kernel" they choose. So if we can merge their kernels to a uniform kernel that can fully reflect our concerns on preserving the two different features, a method based on this idea will surely be able to address the problem mentioned above. Such a new method is exactly the contribution of this paper. Experiments on both artificial and real datasets show that the dimension reduction results of our method is better than LLE, ISOMAP and some other manifold learning methods.

The rest of the paper is organized as follows: Section 2 represents the related work in recent years. Section 3 introduces the two representative nonlinear dimension reduction methods LLE and ISOMAP. Section 4 presents LLE and ISOMAP in a kernel view. Section 5 gives a detailed description of our new method GLPE. In Section 6 we show experiments results on several different data sets. Conclusions and future works are given in Section 7.

2   Related Work

Manifold learning algorithms have developed for more than a decade since been proposed in 2000. In this period of time, manifold learning field has developed a number of different methods and research directions through continuous improvement and extension. We will introduce some related research status in this section.

First we discuss local linear embedding (LLE) [3] and isometric feature mapping (ISOMAP) [5], which can be said to be the most typical manifold learning algorithms. The core idea of LLE is to regard each local neighborhood of high dimensional data set in Euclidean sense, and try to reconstruct the low dimensional geometry structure information of each local neighborhood in the high dimensional data set. In this way, low dimensional coordinates are as close as possible with high dimensional data set in local sense, to achieve the purpose of dimension reduction. ISOMAP's core idea is to keep the distance between any two points of high dimensional data set remain unchanged in the low dimensional representation. Based on this thought the output of low dimensional coordinates are close to the high dimensional data sets in global sense. The two algorithms make dimension reduction of high dimensional data respectively from local and global perspectives. Many of following algorithms extended the two ideas and proposed more optimized manifold learning approaches.

Laplace eigenmaps (LEM) [6] algorithm is another kind of dimension reduction algorithm based on the local view. Different with LLE which is committed to preserving the geometry structure of each local neighborhood, LEM uses a more intuitive thought: the points nearby in high dimensional data set should still maintain nearby in their low dimensional representation. Thus the outputs of LEM give more expression to the keep of local distance.

Other manifold learning algorithms proposed recently include local tangent space alignment algorithm(LTSA) [7], maximum variance unfolding(MVU) [8], Hessian locally linear embedding eigenmaps(HLLE) [9] and Diffusion Maps algorithm(DFM) [10], etc. These algorithms not only follow the general framework of manifold learning, but also have distinctive features respectively, all have achieved good effect.

These new algorithms and their improvement have greatly enriched the field of manifold learning, enhanced its performance to satisfy different actual requirements and let the manifold learning gain more extensive application space. However, in the process of manifold learning's development, it's inevitable to encounter some defects. For example, in the process of dimension reduction, manifold learning algorithms usually reserve only a single characteristic of data set, cannot take different information of data sets which are multi-directional and multi-dimensional into consideration at the same time. This could lead to that the output in the process of dimension reduction cannot reflects the characteristics of high dimensional data set completely. If there are some improvements targeted to these deficiencies, the performance of these algorithms will enhance further, their application will be broadened and manifold learning field will develop continuously. This consideration is just the reason of the proposed article's research work.

3   Introduction to LLE and ISOMAP

In this section, we will give a short brief to two major manifold learning algorithms, whose advantages will be integrated to a new algorithm under kernel framework.

3.1   The LLE Algorithm

LLE was first proposed by Saul and Roweis in 2000 [3]. As an unsupervised machine learning algorithm that can compute low dimensional coordinates of high dimensional datasets, it linearly reconstructs the local neighbors of high dimensional datasets in low dimensional outputs. The outputs are obtained by solving a eigen-decomposition problem of a sparse matrix. The details of the LLE algorithm are summarized as follows:

1. Identify a neighborhood for each data point based on the Euclidean distance metric. This is done by the K-nearest neighbors method or the ε-radius method.
2. Evaluate the linear coefficients that reconstruct each point from its neighbors. This is done by minimizing the reconstruction error: $\varepsilon = |X_i - \sum_j w_{ij} X_j|^2$. Here, w_{ij} represents the coefficient of the i-th point reconstructed from the j-th point. Then the coefficient matrix W is constituted by all coefficients w_{ij} that minimize the reconstruction error.
3. Obtain the low dimensional output Y that mostly preserves the coefficient matrix W. The cost function is as follows:

$$\Phi(Y) = \sum_i |Y_i - \sum_j w_{ij} Y_j|^2. \tag{1}$$

Y is selected as the output of the algorithm to minimize the cost function. Equ. 1 can be further rewritten as: $\Phi(Y) = \sum_{ij} M_{ij}(Y_i \bullet Y_j)$. Here, matrix M is defined as $M = (I - W)^T (I - W)$, a symmetric, semi-positive definite matrix. To make the output unique, the translational, rotational and scaling degrees of freedom are removed from the output. This is done by adding two constrains:

$$\sum_i Y_i = 0, \tag{2}$$

$$\frac{1}{N} \sum_i Y_i \otimes Y_i = I. \tag{3}$$

Finally, the low dimensional output Y is computed by making eigen-decomposition of matrix M. Eigenvectors corresponding to the 2-nd to the $(d+1)$-th smallest eigenvalues of matrix M constitute the d-dimensional output Y, which is the final output of LLE algorithm. LLE is a dimension reduction method with the goal of retaining local neighborhood structure, attempts to join each neighborhood structure together to recover the whole data set. Accurate conservation of local structure information for high dimensional data set is LLE's biggest characteristic.

3.2   The ISOMAP Algorithm

ISOMAP was proposed by Tenenbaum and Silva. Unlike LLE that tries to preserve neighborhood properties, ISOMAP is targeted to preserve geodesic distances between all data points. The procedure of the algorithm can be outlined as follows:

1. Identify neighborhoods. This is similar to the first step of LLE, but ISOMAP needs to further compute the distances between each data point and its neighbors.
2. Compute pairwise geodesic distances of all points. With the neighborhood distances computed in Step (1), it is easy to compute the shortest distance between any two points. Then all these distances are used to construct a matrix D. As it has been proved that when the number of data points goes to infinity, the shortest distance between two points is approaching to their geodesic distance. So the shortest distances here are used as an approximation to real geodesic distances.
3. Compute the low dimensional output using the classic MDS method. This is done by minimizing the following cost function:

$$\varepsilon = \sqrt{\sum_{ij} (\tau(D) - Y_i \bullet Y_j)^2}, \tag{4}$$

where $\tau(D) = -HSH/2$, S is obtained by taking the squared elements in matrix D, $H = I - ee^T/n$, $e = [1, 1, ..., 1]^T$. And the constraints are:

$$Y^T Y = \Lambda \tag{5}$$

where Λ is a diagonal matrix of constants.

At last, the output Y is obtained by making eigen-decomposition of matrix $\tau(D)$. Put the eigenvalues of matrix $\tau(D)$ as $s_1, s_2, ...$(in descending order), and $v_1, v_2, ...$ are the corresponding eigenvectors. Then the matrix $Y = [\sqrt{s_1}v_1, \sqrt{s_2}v_2, ..., \sqrt{s_d}v_d]$ is the final d-dimensional output of the algorithm. The biggest characteristic of ISOMAP algorithm is keeping global distance between each point of high dimensional data set.

4   LLE and ISOMAP in a Kernel View

In this section we will reformulate the LLE and ISOMAP methods from the perspective of kernel, and explore more common properties of these two methods.

Definition 1. *For a binary function $k : X \times X \to R$, if there is an inner product space $(H, <, >)$ and mapping $\Phi : X \to H$, such that:$k(x, y) =< \Phi(x), \Phi(y) >$, then k is known as the kernel function. H is known as the feature space, Φ is known as the feature mapping.*

Definition 2. *Given a kernel function k and the input $x_1, ..., x_n \in X$, $n \times n$-order matrix $K := (k(x_i, x_j))_{ij}$ is known as the nuclear matrix of k in regard to $x_1, ..., x_n$ (or Gram matrix).*

Definition 3. *If the function $k : X \times X \to R$ satisfies the symmetrical characteristic, i.e. $k(x, y) = k(y, x)$, and for any $n \in N$ (positive integer), $x_1, x_2, ...x_n \in X$, $\alpha_1, \alpha_2, ...\alpha_n \in R$, such that $\sum_{i,j=1}^{n} \alpha_i \alpha_j k(x_i, x_j) \geq 0$, then the function k is known as positive definite, the corresponding matrix $K := (k(x_i, x_j))_{ij}$ is known as positive definite matrices.*

Kernel methods are a class of algorithms that map the data into a high dimensional feature space using kernel functions, and then process the data with common linear methods in that feature space. The key issue of kernel methods is the choice of kernel functions. Different kernel functions correspond to different features of the data, and the resulting kernel matrix is the characterization of a specific feature of the data. With the kernel matrix, some linear methods can be exploited, such as eigen-decomposition, to learn the original dataset. KPCA [11] and KISOMAP [13] are examples of kernel methods [12] [14].

From Section 3 we can see that, although LLE and ISOMAP have different objectives, their algorithms have many similarities. For example, they both need to identify neighbors for each data point; the specific information of the data they wish to preserve are both kept in a matrix; the final outputs are obtained by making eigen-decomposition of the previous matrix,etc. Not only these two methods, many other nonlinear dimensionality reduction methods share the same general steps. The most significant is that all these methods can be reduced to the problem of finding proper eigenvectors of a given matrix. These matrixes are semi-positive definite, and meet the conditions of kernel matrixes shown by Theorem 1 below. So these algorithms can be described from a kernel view.

Theorem 1. *The matrix K is a kernel matrix if and only if it is positive definite [15].*

This was once specifically discussed by Ham [16] et al. They believed that both LLE and ISOMAP have their kernel matrixes. For LLE, its kernel matrix can be written as : $K = (\lambda_{max} I - M)$, where M is defined as in Section 3.1, and λ_{max} is the biggest eigenvalue of M. So the original cost function 1 can be rewritten as:

$$\Phi(Y) = trace(Y^T KY) \tag{6}$$

Then the determining of LLE's output is equivalent to looking for eigenvectors corresponding to the biggest eigenvalues of matrix K. Since M is semi-positive definite and λ_{max} is the biggest eigenvalue of M, it is easy to see that K is semi-positive definite too. So K satisfies the requirement of nuclear matrix.

For ISOMAP, its kernel matrix is written as: $T = -HSH/2$, where H,S are defined as in Section 3.2. The objective of the algorithm is to maximize the following cost function:

$$\Phi(Y) = trace(Y^T TY) \tag{7}$$

What the author mentioned in the literature [16] is that if element of T is square of the point's Euclidean distance, T will be positive definite matrix. While element of T is actually the square of geodesic distance, so T is not guaranteed to be positive definite. However, in smooth manifold meeting the continuity condition, the geodesic distance between data points is proportional with their Euclidean distance in low dimensional space. So T is positive definite in the continuity conditions, also satisfying the condition of a kernel matrix.

So we can see that with a slight change in the original methods, LLE and ISOMAP can be viewed as kernel methods. With this kernel view, we can understand these methods in more depth. We can see that different kernel matrix represents different feature of the data, and is the most important factor that distinguish one algorithm from another. So a proper kernel matrix is the basis of the success of a algorithm.

5   GLPE: A Global and Local Feature Preserving Projection Approach

As described in the previous section, a kernel matrix represents the specific feature of data that a algorithm wishes to preserve. So, if we want to preserve multi features in one algorithm, a natural way is to design a new kernel that can simultaneously take different features of the data into consider. And our algorithm is just based on this idea.

From definitions in Section 4, K is the kernel matrix that preserves neighborhood relations of data and T is the kernel matrix that preserves the global distances. Before going any further, we first give the following theorem:

Theorem 2. If K_1, K_2 are nuclear matrixes, then: (1) $K_1 + K_2$ is a nuclear matrix; (2) $\alpha K_1 (\alpha \geq 0)$ is the nuclear matrix.

This theorem can be proved directly by Definition 1 and Definition 2.

According to Theorem 2, a simplest characteristic matrix which can maintain the two properties of the data set can be written as $K+T$. By Theorem 2, it is also a nuclear matrix. Since the two nuclear matrix is in two different reference frame, after normalization of K and T, a new kernel matrix that combines the two features can be proposed written as $K' + T'$, and the objective of the algorithm is to maximize the following cost function:

$$\Phi(Y) = trace(Y^T(K' + T')Y) \tag{8}$$

Then, the coordinate sets Y which maximizes the cost function is the low dimensional output that can preserve both global distances and local neighborhoods of the original data set.

To remove the translational, rotational and scaling degrees of freedom from the output, two constraints are still needed: $\sum_i Y_i = 0$ and $Y^T Y = \Lambda$. Then we make eigen-decomposition of the new kernel matrix to get the eigenvalues $s_1, s_2, \dots$ (in descending order) and corresponding eigenvectors $v_1, v_2, \dots$. The matrix $Y = [\sqrt{s_1}v_1, \sqrt{s_2}v_2, ..., \sqrt{s_d}v_d]$ is the output that maximize Formula (8) and thereby is the final d-dimensional output of the algorithm. Here we multiply each eigenvector by its corresponding eigenvalue's square root to meet the second constraint. And since both K and T are semi-positive definite, $K' + T'$ is also semi-positive definite. Then the first constraint is met too.

But what's easy to see is that if we simply use the matrix $K' + T'$ as the new kernel matrix, then the relative relations between the two kinds of features we want to preserve will be fixed. In real applications, there will always be some bias between preserving the neighborhoods construction or preserving the global distance. To handle with that, we need to introduce a balance factor that can flexibly adjust the relations between these two kinds of preservations. Define the balance factor as α, then the new kernel matrix can be written as $(1-\alpha)K' + \alpha T'$, and the cost function is :

$$\Phi(Y) = trace(Y^T((1-\alpha)K' + \alpha T')Y) \tag{9}$$

Then the output of the new algorithm is obtained by making eigen-decomposition of $(1-\alpha)K' + \alpha T'$. The bigger α is, the better the algorithm will do in preserving the global distances, otherwise the algorithm will show better performance in preserving the neighborhood relationships. When we try to decide the proper value for α, the difference of magnitude between the two kernels should be considered. So we need to carefully determine the value of α to ensure that the kernel matrix $(1-\alpha)K' + \alpha T'$ makes sense. How the value of α affects the algorithm output will be illustrated in the Experiments part.

The new algorithm with the constructed kernel $(1-\alpha)K' + \alpha T'$ is named "Global and Local Preservation Embedding", or GLPE for short. The detailed steps of the GLPE algorithm are listed as follows.

Algorithm GLPE:

Input: Input data set X, Neighborhood parameter k, Balance factor α.
Output: Low dimensional coordinates Y.

Step 1. **Identify the neighborhoods.** Identify the neighborhood of each data point using the $K-$nearest neighbors method.

Step 2. **Get kernel matrix K which preserves neighborhood relationships.** Determine the neighborhood matrix W which minimizes the cost function: $\varepsilon = |X_i - \sum_j w_{ij}X_j|^2$, then get the kernel K by $K = \lambda_{max}I - M$, where $M = (I-W)^T(I-W)$, λ_{max} is the largest eigenvalue of M.

Step 3. **Get kernel matrix T which preserves global distances.** Compute the shortest path distances between any two points and write it in matrix D, then get the kernel T by $T = -HSH/2$, where H is the centralized matrix and S is the matrix whose elements are the squared elements of matrix D.

Step 4. **Compute the low dimensional output Y.** Making eigen-decomposition of the kernel matrix $(1-\alpha)K' + \alpha T'$, obtaining eigenvalues $s_1, s_2, ...$ (in descending order), and the corresponding eigenvectors

$v_1, v_2, ...$, where K' and T' are kernel matrix K and T after normalization. Then the matrix $Y = [\sqrt{s_1}v_1, \sqrt{s_2}v_2, ..., \sqrt{s_d}v_d]$ is the final output of the algorithm.

Some of the properties of the GLPE algorithm can be summarized as follows:

1. The algorithm is guaranteed to have global optimum. As the final output is obtained by computing the eigenvectors of the kernel matrix and the kernel matrix is unique for a given data set, the algorithm can always find a global optimal low dimensional output, thus avoiding the problem of local minima.
2. The algorithm has definite time and space cost. The most time-consuming step of GLPE is the computing of the shortest path distances between all points, and the time complexity is $O(n^3)$, where n is the number of data points. So the time complexity of the algorithm is also $O(n^3)$. As for the space cost, GLPE only needs to do some basic matrix operations. Compared with those algorithms which need iterative calculations to converge to a final output, GLPE can guarantee to get the output within certain time and space constraints.
3. The algorithm provides flexibility in the process of dimensionality reduction. We can choose our preference in preserving either features of the data set to get the desired output, just by adjusting the value of the balance factor.

6   Performance Evaluation

In this section we will show GLPE's performance on tasks of dimensionality reduction and classification, and also compare it with some other related methods. The data sets we use include both artificial and real data sets.

6.1   Results of Dimensionality Reduction

First we use the very famous "Swiss Roll" data set as the input data set. This is also the one we use in Section 1 as a illustration. We reduce this data set to 2-dimensional space using LLE, ISOMAP and GLPE respectively, and the results are shown in Figure 2. Compared with LLE's result which merely preserves the local neighborhood relationships, GLPE can better recover the distribution of all data points; and compared with ISOMAP's result which completely focuses on the pairwise distances, GLPE gives consideration to local structures of data.

Further more, we make some extensions on the "Swiss Roll" data set. We let the "Swiss Roll" has more samples, as shown in Figure 3(a). For such a complex data set, if we try to characterize it from a single point of view, either the local neighborhoods preservation or the global distances preservation, the original data set cannot be well illustrated in a 2-dimensional plane. But as GLPE characterizes the data set from more perspectives, the 2-dimensional output is much more satisfactory, as shown in Figure 3(d).

As the performance of our algorithm has great relations with the balance factor α, our next experiment will show dimensionality reduction results with

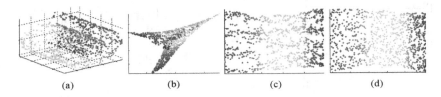

Fig. 2. Dimensionality reduction outputs of the "Swiss Roll" data set using 3 methods. (a) 3-dimensional "Swiss Roll" data set(1000 points); (b) Output of LLE; (c) Output of ISOMAP; (d) Output of GLPE, $\alpha = 10^{-6}$

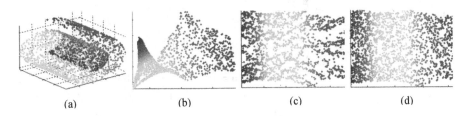

Fig. 3. Dimensionality reduction outputs of a more complicated "Swiss Roll" data set. (a) 3-dimensional extended "Swiss Roll" data set(2000 points); (b) Output of LLE; (c) Output of ISOMAP; (d) Output of GLPE, $\alpha = 10^{-8}$

different values of α. We can see that, as the value of α decreases, the output gradually changes from global distance preserving to local neighborhood preserving. The results are shown in Figure 4.

In another experiment we choose the "Spirals" data sets as input, also reducing them to 2-dimensional planes. The results shown in Figure 5 demonstrate that GLPE can get good performances on different data sets.

6.2 Results of Classifications on Low Dimensional Outputs

In many related tasks of machine learning, dimensionality reduction is often used as a preprocessing step of the following tasks, such as classification. So, a good dimensionality reduction output is the basis of the success of the following tasks. In this section we first do dimensionality reduction on two real data sets using four different algorithms, then divide the low dimensional outputs into two parts of the same size, one as the training set and the other as the testing set. Then we do classifications on the testing set using the classic $k - nn$ classifier. And finally we compare their classification accuracies.

First we apply the algorithms to the human face data set sampled from Olivetti Faces [17]. This data set contains face images of 8 individuals; each individual has 10 images of different facial expressions, each of size 64×64. And we select 5 images each into training and testing set. Some of the images are shown in Figure 6. Since we have already known the correct classifications of each image in the training set, then we'll do classifications on the testing set

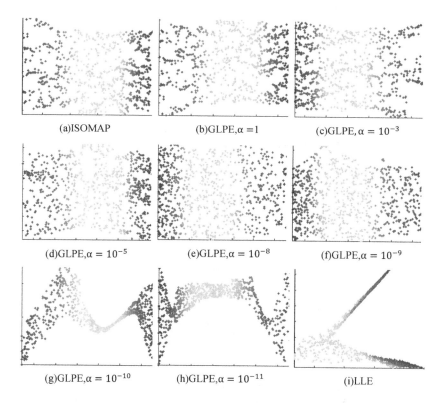

Fig. 4. Dimensionality reduction outputs of different algorithms and different choices of α. (a) 3-dimensional "Swiss Roll" data set(800 points); (b-i) Outputs of all kinds of settings

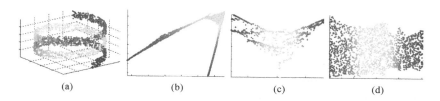

Fig. 5. Outputs of the "Spirals" data set. (a) 3-dimensional "Spirals" data set(1800 points); (b) Output of LLE; (c) Output of ISOMAP; (d) Output of GLPE, $\alpha = 10^{-6}$

and calculate the classification accuracy. We reduce the data set to 1,2,...,10 dimensions, then classify these different low dimensional points into 8 classes respectively. The classification accuracies are shown in Figure 7. It is easy to see that the average classification accuracy of GLPE is better than the other algorithms when the data set is reduced to different dimensions. Especially when reduced to higher dimensions, which are the usual target dimensions in real applications, the superiority of our method is more obvious.

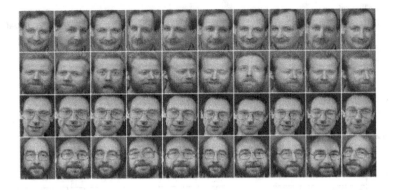

Fig. 6. Some of the images of the human face image data set

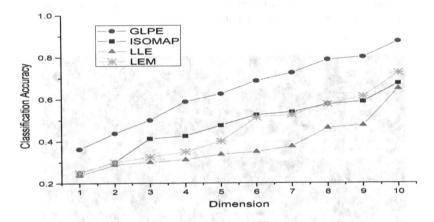

Fig. 7. Classification accuracies of Olivetti Faces when the data set is reduced to different low dimensions

Then we reduce the original data set to 3 dimensions, then classify the data points into 8 classes. The classification accuracies of all outputs with different neighborhood parameters K are shown in Figure 8, where K is used as the K-nearest neighbors method in step (1) of GLPE to identify the neighborhood of each data point. From Figure 8 we can see that GLPE has better average performance of classification effect than other algorithms with the change of different K.

Next we do experiments on the hand written digits data set sampled from MNIST Handwritten Digits [17]. The used data set contains 10 groups of hand written digits divided into 5 training sets and 5 testing sets, and each of the numerics: 0, 1, 3, 4 and 6 has 900 images. Each image is of size 28 × 28. Some of the images are shown in Figure 9. We first reduce the data set to 1,2,...,10 dimensions, then classify these different low dimensional points into 5 classes respectively. Figure 10 shows the classification accuracies of different algorithms.

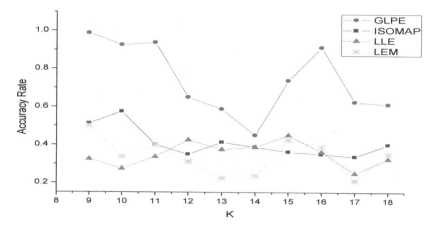

Fig. 8. Classification accuracies of Olivetti Faces with the change of neighborhood parameter K (all reduced to 3-dimensional space)

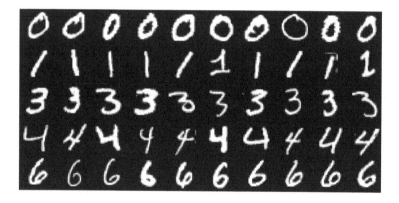

Fig. 9. Some of the images of the hand written digits data set

When the data set is reduced to different dimensions, the average classification accuracy of our method is outstanding obviously.

Then we reduce the data set to 3 dimensions, then classify the data points into 5 classes. The classification accuracies of all outputs with different neighborhood parameters K are shown in Figure 11. We can see that for the selection of different K, GLPE has better average performance of classification effect compared with other algorithms. But note that, when the neighborhood parameter K has increased to 10 or so, the classification accuracy becomes lower. It shows the neighborhood number of ten or so comes to a inflection point for the data set, some points which do not belong to the neighborhood will be included. Thus the algorithm's dimensionality reduction effect will be affected.

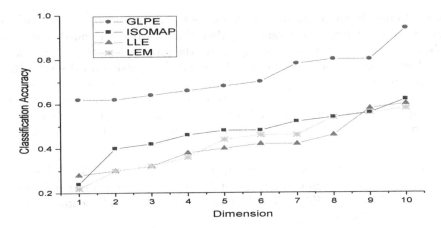

Fig. 10. Classification accuracies of hand written digits when the data set is reduced to different low dimensions

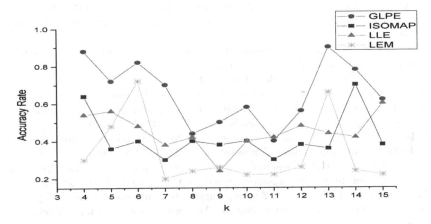

Fig. 11. Classification accuracies of hand written digits with the change of neighborhood parameter K (all reduced to 3-dimensional space)

7   Conclusion

Based on theories of kernel methods, we merge the kernels of two existing nonlinear dimensionality reduction methods - LLE and ISOMAP in a proper way, and propose a new method GLPE that can effectively characterize the overall properties of data. Experimental results show that our method can successfully preserve the local neighborhood structures and the global pairwise distances of the data set. The work of this paper is mainly to merge the kernels of two existing algorithms to form a new kernel, where the new kernel works better than both old ones. It reveals to us that, if we could preserve multi features of the high-dimensional data sets in one algorithm, then the resulted low-dimensional embedding results will more faithfully reflect the properties of the original data sets. The framework of the newly proposed kernel methods provides solid theoretical support for the idea, we just need to design the new kernels. How to design new kernels to better preserve properties of high-dimensional data sets from different perspectives, and to keep the low-dimensional representations as close to the original data sets as possible, is the focus of our future works.

Acknowledgement . This work was supported by the National Natural Science Foundation of China (NSFC) under grant No. 61173118, the Shuguang Program of Shanghai Municipal Education Commission, and the Fundamental Research Funds for the Central Universities.

References

1. Jolliffe, I.T.: Principal Component Analysis. Springer, New York (2002)
2. Cox, T.F., Cox, M.A.A.: Multidimensional scaling. Chapman and Hall, London (1994)
3. Roweis, S.T., Saul, L.K.: Nonlinear dimensionality reduction by locally linear embedding. Science 290, 2323–2326 (2000)
4. Saul, L.K., Roweis, S.T.: Think globally, fit locally: Unsupervised learning of low dimensional manifolds. The Journal of Machine Learning Research 4, 119–155 (2003)
5. Tenenbaum, J.B., De Silva, V., Langford, J.C.: A global geometric framework for nonlinear dimensionality reduction. Science 290, 2319–2323 (2000)
6. Belkin, M., Niyogi, P.: Laplacian Eigenmaps for Dimensionality Reduction and Data Representation. Neural Computation 15, 1373–1396 (2003)
7. Zhang, Z.Y., Zha, H.Y.: Principal Manifolds and Nonlinear Dimension Reduction via Local Tangent Space Alignment. Scientific Computing 26, 313–338 (2002)
8. Weinberger, K.Q., Saul, L.K.: An introduction to nonlinear dimensionality reduction by maximum variance unfolding. In: American Association for Artificial Intelligence (AAAI) (2006)
9. Donoho, D.L., Grimes, C.E.: Hessian eigenmaps: Locally linear embedding techniques for high-dimensional data. Proceedings of the National Academy of Sciences 100, 5591–5596 (2003)
10. Coifman, R.R., Lafon, S.: Diffusion Maps. Computational and Harmonic Analysis (2006)
11. Schölkopf, B., Smola, A., Müller, K.-R.: Kernel principal component analysis. In: Gerstner, W., Hasler, M., Germond, A., Nicoud, J.-D. (eds.) ICANN 1997. LNCS, vol. 1327, pp. 583–588. Springer, Heidelberg (1997)
12. Schölkopf, B., Smola, A., Müller, K.R.: Kernel principal component analysis. In: Advances in Kernel Methods-Support Vector Learning, pp. 387–352 (1999)
13. Choi, H., Choi, S.: Kernel isomap. Electronics Letters 40, 1612–1613 (2004)
14. Choi, H., Choi, S.: Robust kernel isomap. Pattern Recognition 40(3), 853–862 (2007)

15. Wang, G.S.: Properties and construction methods of kernel in support vector machine. Computer Science 33(6), 172–174 (2006)
16. Ham, J., Lee, D., Mika, S., Schölkopf, B.: A Kernel View of the Dimensionality Reduction of Manifolds. In: Proceedings of the Twenty-First International Conference on Machine Learning, pp. 47–54. IEEE, New York (2004)
17. http://www.cs.nyu.edu/~roweis/data.html

Consistent Ontologies Evolution Using Graph Grammars

Mariem Mahfoudh, Germain Forestier, Laurent Thiry, and Michel Hassenforder

MIPS EA 2332, Université de Haute Alsace
12 rue des Frères Lumière F-68093 Mulhouse Cedex
{mariem.mahfoudh,germain.forestier,laurent.thiry,
michel.hassenforder}@uha.fr

Abstract. Ontologies are often used for the meta-modelling of dynamic domains, therefore it is essential to represent and manage their changes and to adapt them to new requirements. Due to changes, an ontology may become invalid and non-interpretable. This paper proposes the use of the graph grammars to formalize and manage ontologies evolution. The objective is to present an a priori approach of inconsistencies resolutions to adapt the ontologies and preserve their consistency. A framework composed of different graph rewriting rules is proposed and presented using the AGG (Algebraic Graph Grammar) tool. As an application, the article considers the EventCCAlps ontology developed within the CCAlps European project.

Keywords: ontologies, graph grammars, evolution, rewriting, ontology changes, category theory, AGG.

1 Introduction

Designed as a response for vocabulary heterogeneity problems and semantic ambiguities of data sources, ontologies play the role of a semantic structure that represents and formalizes human knowledge of a specific domain. As they are often used for meta-modelling of dynamic domains, they constantly require to adapt to knowledge evolution. However, this evolution presents several challenges, especially in the changes definition and consistency preservation of the modified ontology. In fact, a misapplication of a change can alter the consistency of an ontology by affecting its structure and/or its semantic. This promotes the need to formalize the process of evolution.

This work proposes the use of graph grammars, based on algebraic approaches to express and manage ontologies evolution. Graph grammars are a rigorous formal method, allowing the verification of the feasibility of ontology changes. Thanks to their application conditions, they avoid the execution of changes that do not satisfy a set of constraints. They also offer many tools such as AGG (Algebraic Graph Grammar) which provides a simple environment for defining rewriting rules, helping the user to easily express his needs. Thus, the main objective of this work is to present a formal method for managing ontology changes and ensuring the consistency of the modified ontology.

M. Wang (Ed.): KSEM 2013, LNAI 8041, pp. 64–75, 2013.
© Springer-Verlag Berlin Heidelberg 2013

This paper is organized as follows: section 2 presents the ontologies changes and the graph grammars. Section 3 proposes the formalization of ontology changes with graph grammars. Section 4 presents an application using the ontology ECCAlps which has been developed in the context of the European project CCAlps. Section 5 shows some related works. Finally, a conclusion summarizes the presented work and gives some perspectives.

2   Ontology changes and Graph Grammars

2.1   Ontology Changes

Ontologies are commonly defined as a "specification of a conceptualization" [1]. They are composed by a set of classes, properties, individuals and axioms and they often need to evolve to integrate and reuse knowledge. Different classifications of ontology changes have been proposed [2,3]. One of the most widely used [4] distinguishes two types:

1. *Elementary/basic changes*: represent primitive operations which affect a single ontology entity, e.g. addition, deletion and modification;
2. *Composite/complex changes*: are composed of multiple basic operations that together constitute a logical entity, e.g. merge or split of classes.

Whatever its nature (basic or complex), an ontology change should be formalized in order to properly identify its inputs, its outputs and the inconsistencies that it is likely to generate. In this work, the ontology is considered as a graph G = (V, E) where V is a set of vertices which represent classes, individuals, etc. E is a set of edges which represent axioms. Thus, an ontology change can be expressed and formalized as a graph rewriting rule $r : G \rightarrow G'$.

2.2   Graph Grammars

Definition 1 (Graph Grammars). A graph grammar (GG) is a pair composed of an initial graph (G) called host graph and a set of production rules (P) also called graph rewriting rules or graph transformation.
A production rule P = (LHS, RHS) is defined by a pair of graphs:

- LHS (Left Hand Side) presents the precondition of the rule and describes the structure that has to be found in G.
- RHS (Right Hand Side) presents the postcondition of the rule and should replace LHS in G.

Graph grammars can be typed (TGG) and is defined as: $TGG = (G_T, GG)$ where $G_T = (V_T, E_T)$ is a type graph which represents the type information (type of nodes and edges). The typing of a graph G over G_T is given by a total graph morphism $t : G \rightarrow G_T$ where $t : E \rightarrow E_T$ and $t : V \rightarrow V_T$.

The graph transformation defines how a graph G can be transformed to a new graph G'. More precisely, there must exist a morphism (m) that replaces LHS by RHS to obtain G'.

There are different graph transformation approaches to apply this replacement, as described in [5]. The algebraic approach [6] is based on category theory with the *pushout* concept.

Definition 2 (Category Theory). A category is a structure consisting of:

1. a collection of objects O;
2. a set of morphisms M and a function $s : M \to O \times O$, $s(f) = (A, B)$ is noted $f : A \to B$;
3. a binary operation, called composition of morphisms $(\circ) : M \times M \to M$;
4. an identity morphism for each object $id : O \to O$.

The composition of $f : A \to B$ and $g : B \to C$ is associative and is written $g \circ f : A \to C$.

Definition 3 (Pushout). Given three objects A, B and C and two morphisms $f : A \to B$ and $g : A \to C$. The pushout of B and C consists of an object D and two morphisms $m_1 : B \to D$ and $m_2 : C \to D$ where $m_1 \circ f = m_2 \circ g$.

The algebraic approach is divided into two sub-approaches: the *Single pushout SPO* [7] and the *Double poushout DPO* [8]. In this work, only the SPO approach was considered as it is more general (e.g. without the gluing condition) and sufficient to represent the different ontology changes. Therefore, applying a rewriting rule (r) to an initial graph with the SPO method, consists in (Figure 1):

1. Find LHS in G using a morphism m : $LHS \to G$.
2. Delete $LHS - (LHS \cap RHS)$ from G.
3. Add $RHS - (LHS \cap RHS)$ to G. This operation is done by the construction of a pushout and gives a new version G' of G.

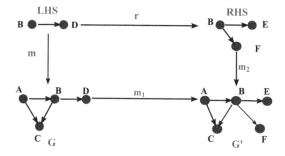

Fig. 1. Application of a rewriting rule graphs with the SPO approach

AGG Tool. Several tools have been proposed to support graph rewriting as AGG [9], Fujaba [10] or Viatra [11]. The AGG tool is considered as one of the most important tools. It supports the algebraic approach and typed attributed graphs. The AGG environment provides graphical editors for graphs and rules. It also allows to add the NACs (Negative Application Condition) which specifies a subgraph that may not occur when matching a rule. It is important to note that the internal graph transformation engine can also be used by a Java API and thus, be integrated into other custom applications.

3   Formalisation of Ontology Changes

This section introduces the definition and formalisation of ontology changes using typed graph grammars. The first step consists in creating the type graph which presents the meta-model of the ontology. The next step defines the ontology changes under the form of graph rewriting rules $(r_1, r_2, ...r_n)$.

3.1   Type Graph

In this article, OWL was chosen to describe ontologies since it is the standard proposed by the W3C, and the language usually adopted to represent ontologies. However, other languages can be considered by using converters[1].

Figure 2 shows the representation of OWL meta-model with AGG (G_T). The OWL meta-model [12] defines the basic conceptual primitives of OWL which are classes, properties (ObjectProperty and DataProperty), individuals, axioms (disjointWith, equivalentTo, etc.). The classes model the set of individuals and it can be primitive or complex (UnionClass, ComplementClass, IntersectionClass). The ObjectProperty models the relationship between classes (Domain and Range) whereas the DataProperty link a class (Domain) to a Datatype. All these primitives are represented as nodes and each of them have two attributes inherited from the class Entity. The attribute name specifies the name of the local entity, while the attribute iri (Internationalized Resource Identifier) allows to identify and to reference them. The G_T also defines the restrictions which are a particular type of class description. There are two types: restriction values (AllValuesFrom, SomeValuesFrom, HasValue) and cardinality (CardinalityRestriction). Axioms are represented as edges expressing the relationships between classes, properties and individuals. For example, the edge disjointWith represents the disjunction between two classes or two properties.

3.2   Ontology Changes with Graph Grammars

Adapting an ontology to new requirements consists in modifying its structure. However, these changes can cause inconsistencies which require the application of derived changes to correct them. This section describes how consistently express some ontology changes using graph grammars. In this paper, only elementary changes were considered (Figure 3).

---

[1] owl.cs.manchester.ac.uk/converter

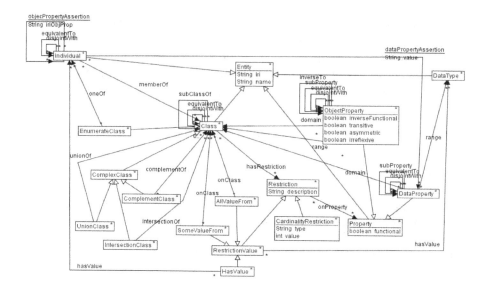

Fig. 2. Type Graph used for graph rewriting

Definition 4 (Ontology changes). An ontology change is formalized by 5-tuplet CH = (Name, NAC, LHS, RHS, CHD) where:

1. Name specifies the type of change;
2. NAC defines the condition which must not be true to apply the rewriting rule;
3. LHS presents the precondition of the rewriting rule;
4. RHS defines the postcondition of the rewriting rule;
5. CHD presents the derived changes. They are additional operations that could be attached to CH to correct its inconsistencies.

Inconsistencies addressed in this work are:

- Data redundancy can be generated following an add or rename operation. This type of inconsistency is corrected by the NACs.
- Isolated node, a node N_x called isolated if $\forall N_i \in N$, $\nexists V_i \in V | V_i = (N_x, N_i)$. This incoherence requires to link the isolated node to the rest of the graph. Depending of the type of node, derived changes are proposed.
- Orphan individual is an inconsistency which is generated as a result of removal of classes containing individuals.
- Axioms contradiction, the addition of a new axiom should not be accept if it contradicts an axiom already defined in the ontology. Such verification is necessary to maintain the semantics of the evolved ontology.

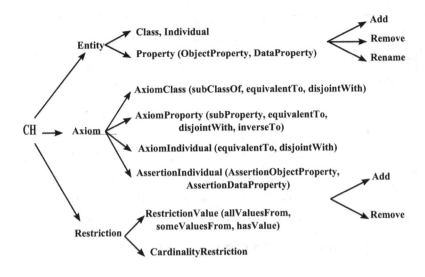

Fig. 3. Elementary changes

Thus, the *RenameObjectProperty(OBIRI, OBIRINew)* change consists in the re-naming of a node `ObjectProperty` (OB). Then, three graphs should be created: 1) the LHS consisting of a node OB where its attribute `iri` is equal to OBIRI; 2) the RHS consisting of a node OB where its attribute `iri` is equal to OBIRINew; 3) the NAC is equal to RHS to prevent the redundancy (Figure 4).

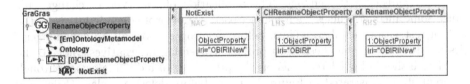

Fig. 4. Rewriting rules of the RenameObjectProperty change

The *AddClass(Cnew)* change allows the add of a new node of type `Class` in the host graph G ($Cnew \in G \wedge t : Cnew \rightarrow V_{Class}$). The rewriting rule consists of three graphs: 1) $LHS = \emptyset$; 2) $RHS = Cnew$; 3) $NAC = RHS = Cnew$; the NAC should be equal to RHS to prevent data redundancy. Besides, a node should not be isolated. To attach a node of type `Class` to the graph, two types of correction can be applied: `AddObjectProperty` or `AddAxiom`. The first one consists in adding a new property where the node *Cnew* is one of its member. The second inserts a new axiom to link *Cnew* to an existing property (`addDomain`, `addRange`) or connect it to another node of type `Class` applying the changes `AddEquivalentClass`, `AddDisjointClass`, `AddSubClass`, etc.

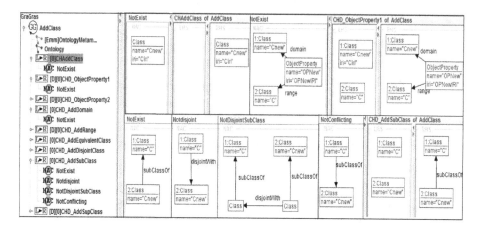

Fig. 5. Rewitting rules of the AddClass change

Figure 5 shows the rewriting rules of the `AddClass` change followed by some derived changes. They are classified by layers to define the sequence of their application: the user can select by a simple activation the derived changes which he wishes to apply.

The *AddDisjointClass (C1, C2)* change adds a disjunction axiom between two nodes of type `Class` (see Figure 6). Thus, three NACs are defined to verify the absence of edges of the type: 1) `disjointWith` to avoid redundancy; 2) `equivalentTo`, two classes can not be disjoint and equivalent at the same time; 3) `subClassOf`, two classes what share a subsumption relation can not be disjoints.

The *RemoveClass(C)* change. The application of this type of change may cause some inconsistencies such as the existence of orphans individuals or the lack of restriction members. Thus, before deleting a node, all its dependencies (its axioms) should be checked to propose correction alternatives. Indeed, the restriction should be deleted whereas the processing of individuals goes through different steps illustrated in Figure 7. Then, before deleting a class C defining individuals (I $memberOf$ C), it should check: 1) If C $subClassOf$ $C_p \wedge \forall C_i$ $subClassOf$ $C_p \wedge !disjointWith$ C. Then, I $memberOf$ C_p; 2) Else If $\exists C_i \in G$

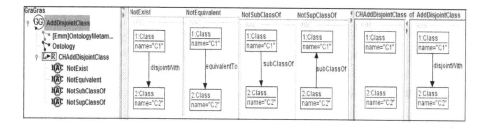

Fig. 6. Rewriting rule of AddDisjointClass change

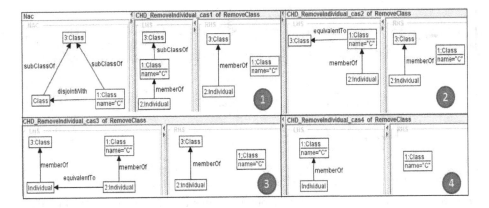

Fig. 7. Processing of RemoveIndividual of the RemoveClass change

where C_i *equivalentTo* C. Then, I *memberOf* C_i; 3) Else if $\exists I_i \in G$ where I_i *memberOf* $C_i \wedge I_i$ *equivalentTo* I. Then, I *memberOf* C_i; 4) If none of these cases is satisfied, the orphans individuals will be deleted from G.

4 Application

This work was developed in the frame of the CCAlps European project[2] which aims at providing an infrastructure to facilitate the collaboration between the creative industries and regions.

In this context, four OWL ontologies have been proposed: EventCCAlps, HubCCAlps, CompanyCCAlps and RegionCCAlps. The EventCCAlps ontology defines the concepts of the events. It presents the characteristics of an event (description, location, time, etc.) and its different relationships with other concepts (Company, Region, Hub, etc.). EventCCalps is based on the Event Ontology [13] and the Linking Open Descriptions of Events [14].

As an example of transformation, this section presents the deletion of the Employee class. In order to start the process of transformation and apply the rewriting rules, the ontology should be converted into an AGG graph. Indeed, two programs (*OWLToGraph* and *GraphToOWL*) have been developed to automate the transformation of OWL to AGG and vice versa. They are based on the AGG API and Jena library[3], an open source API to read and manipulate ontologies described in OWL, RDF and RDFS.

Figure 8 shows a result of the transformation of EventCCalps. Note that for reasons of readability the IRI have been removed from the figure.

The Employee class has different individuals: mariem, laurent, germain and michel. It is a subClassOf the Person class and there is no class in the ontology which inherits from Person class and it is, in the same time, disjoint with

---

[2] www.ccalps.eu, the project reference number is 15-3-1-IT
[3] jena.sourceforge.net

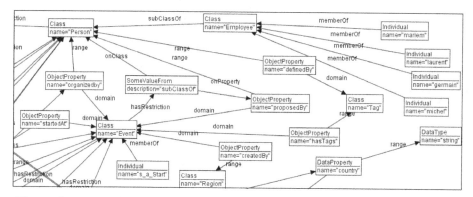

Fig. 8. An extract from EventCCAlps ontology after transformation to AGG graph

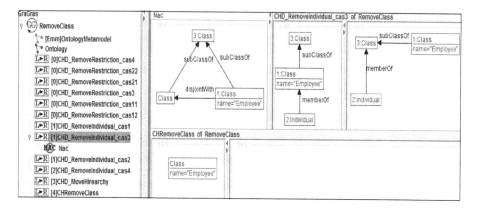

Fig. 9. Rewriting rules of deleting "Employee" class from EventCCAlps ontology

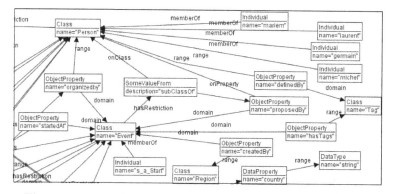

Fig. 10. EventCCAlps ontology after delete the "Employee" class

the `Employee` class. Then, the RemoveClass change invokes the derived change RemoveIndividual and attaches the individuals to the `Person` class. In this way, the individuals and the knowledge can be saved without affecting the consistency of the ontology. Figure 9 shows the definition of this change with AGG and the Figure 10 presents the result of the transformation. This simple example illustrates how the presented work could be used to manage ontology evolution.

5 Related Work

Ontologies evolution is often confused with the filed of database schemas evolution. In fact, many issues in ontology evolution are exactly the same as the issues in schema evolution. However, there are several differences between them. Instead of comparing directly the process of evolution, ontologies and database schemas evolutions are generally compared through an analyses of the differences between the ontology and the database schemas.

Noy and al. [15] have summarized this difference by the following points: 1) Ontologies themselves are data to the extent to which database schemas have never been. So, ontologies evolution must consider both the structure and instances of the ontologies; 2) Ontologies themselves incorporate semantics while database schemas do not provide explicit semantics for their data. Then, the restrictions must be considered in the ontology evolution process; 3) Ontologies are decentralized by nature so their content and usage are often more complex than a database schemas.

Ontologies evolution has been influenced by the research on schema evolution database [16] but it is a different area of research having its own characteristics.

The first proposed methods in the literature [17,4,18] have presented ontologies evolution process in general but they are considered as the basis of most current works. Thus, Hartung and al. [19] have studied the evolution and the difference between two versions of the same ontology. This work provided a *COnto-Diff* tool which can detect different basic changes, however, it has not presented any processing for inconsistencies. Khattak and al. [20] and Luong and al. [21] have proposed posteriori approaches to solve inconsistencies. This type of approach, unlike the a priori process that we propose, requires the implementation of changes to verify the alteration of the ontology and then cancel them if something went wrong. This causes a waste of time and resources. Dragoni and al. [22] have also addressed the impact of the ontologies evolution. They consider the ontology as a hierarchy of concepts and they ignore the conceptual and semantic relation which it models. Then, the proposed correction for monitoring changes have addressed only the subsumption relation. An interesting work has been presented in [23] which is based on pi-calculus. It manages the ontology changes with a formal method and it proposed some rules for preserving ontologies consistency.

The graph grammars allow the definition, formalization and application of ontology changes. Their ability to avoid the inconsistencies is the most important characteristics. It allows, due to application conditions, to verify the validity of each type of change and its effects on the graph.

6 Conclusion and Future Work

In this paper, we presented the use of the graph grammars to formalize and implement the ontology changes. We proposed an a priori approach of inconsistencies resolutions to adapt ontologies and preserve their consistency. The use of AGG allowed a simple definition of rewriting rules and it presented many advantages. Two programs were developed *OWLToGraph* and *GraphToOWL* to automate the back and forth process of transformation of the ontologies to graphs. They allow the user to work and avail the benefits of graph grammars even if his ontologies are defined by another representation language.

Many perspectives can be identified. Firstly, it is important to extend the study for the complex ontology changes. It would also be interesting to exploit ontology changes to define a formal approach of ontologies composition knowing that the composition is a combination of some basic changes (`AddClass`, `removeClass`, `AddAxiom`, `RemoveAxiom`, etc.). Integration of a query language (e.g. SPARQL) is envisaged in order to optimize the selection of ontologies entities.

References

1. Gruber, T.R., et al.: A translation approach to portable ontology specifications. Knowledge Acquisition 5(2), 199–220 (1993)
2. Stojanovic, L.: Methods and Tools for Ontology Evolution. PhD thesis, University of Karlsruhe, Germany (2004)
3. Qin, L., Atluri, V.: Semdiff: An approach to detecting semantic changes to ontologies. International Journal on Semantic Web and Information Systems (IJSWIS) 2(4), 1–32 (2006)
4. Klein, M.: Change Management for Distributed Ontologies. PhD thesis, Vrije Universiteit Amsterdam, Amsterdam, The Netherlands (2004)
5. Rozenberg, G.: Handbook of graph grammars and computing by graph transformation, vol. 1. World Scientific (1999)
6. Ehrig, H., Pfender, M., Schneider, H.J.: Graph-grammars: An algebraic approach. In: IEEE Conference Record of 14th Annual Symposium on Switching and Automata Theory, SWAT 2008, pp. 167–180. IEEE (1973)
7. Löwe, M.: Algebraic approach to single-pushout graph transformation. Theoretical Computer Science 109(1), 181–224 (1993)
8. Ehrig, H.: Introduction to the algebraic theory of graph grammars (a survey). In: Claus, V., Ehrig, H., Rozenberg, G. (eds.) Graph Grammars 1978. LNCS, vol. 73, pp. 1–69. Springer, Heidelberg (1979)
9. Ermel., C., Rudolf., M., Taentzer, G.: The agg approach: Language and environment. In: Handbook of Graph Grammars and Computing by Graph Transformation, pp. 551–603. World Scientific Publishing Co., Inc. (1999)
10. Nickel, U., Niere, J., Zündorf, A.: The fujaba environment. In: Proceedings of the 22nd International Conference on Software Engineering, pp. 742–745. ACM (2000)
11. Varró, D., Pataricza, A.: Generic and meta-transformations for model transformation engineering. In: Baar, T., Strohmeier, A., Moreira, A., Mellor, S.J. (eds.) UML 2004. LNCS, vol. 3273, pp. 290–304. Springer, Heidelberg (2004)

12. Object Management Group: Ontology definition metamodel (omg) version 1.0. Technical report, Object Management Group (2009)
13. Raimond, Y., Abdallah, S.: The event ontology. Technical report, Technical report, 2007 (2007), http://motools.sourceforge.net/event
14. Shaw, R., Troncy, R., Hardman, L.: Lode: Linking open descriptions of events. The Semantic Web, 153–167 (2009)
15. Noy, N.F., Klein, M.: Ontology evolution: Not the same as schema evolution. Knowledge and information systems 6(4), 428–440 (2004)
16. Rahm, E., Bernstein, P.A.: An online bibliography on schema evolution. ACM SIGMOD Record 35(4), 30–31 (2006)
17. Stojanovic, N., Stojanovic, L., Handschuh, S.: Evolution in the ontology-based knowledge management system. In: Proceedings of the European Conference on Information Systems-ECIS (2002)
18. Rogozan, D., Paquette, G.: Managing ontology changes on the semantic web. In: Proceedings of the 2005 IEEE/WIC/ACM International Conference on Web Intelligence, pp. 430–433. IEEE (2005)
19. Hartung, M., Groß, A., Rahm, E.: Conto-diff: Generation of complex evolution mappings for life science ontologies. J. Biomed. Inform. (1), 15–32 (2013)
20. Khattak, A.M., Latif, K., Lee, S.: Change management in evolving web ontologies. Knowledge-Based Systems 37(0), 1–18 (2013)
21. Luong, P.H., Dieng-Kuntz, R.: A rule-based approach for semantic annotation evolution. Computational Intelligence 23(3), 320–338 (2007)
22. Dragoni, M., Ghidini, C.: Evaluating the impact of ontology evolution patterns on the effectiveness of resources retrieval. In: 2nd Joint Workshop on Knowledge Evolution and Ontology Dynamics EvoDyn 2012 (2012)
23. Wang, M., Jin, L., Liu, L.: A description method of ontology change management using pi-calculus. In: Knowledge Science, Engineering and Management, pp. 477–489 (2006)

Social Welfare Semantics
for Value-Based Argumentation Framework[*]

Fuan Pu, Jian Luo, Yulai Zhang, and Guiming Luo

School of Software, Tsinghua University, Beijing, China
{pfa12,j-luo10,zhangyl08}@mails.tsinghua.edu.cn,
gluo@mail.tsinghua.edu.cn

Abstract. Logically intuitive properties of argument acceptance criteria have been paid great attention in the argumentation frameworks, however, which do not take into account the agents. Recently research has begun on evaluating argument acceptance criteria considering preferences of audiences, a kind of representative agents. In this paper we take a step towards applying social choice theory into value-based argumentation framework (VAF), and propose social welfare semantics, an extension of Dung's semantics incorporating social preference. Then the social welfare semantics and its relationship with the semantics of VAF are analyzed, and some non-trivial properties are proved based on Pareto efficiency.

Keywords: abstract argumentation, value-based argumentation, preferences, social welfare function, Pareto efficiency.

1 Introduction

Argumentation, as a growing interdisciplinary field of research, was conducted mainly in logic, philosophy, and communication studies in the beginning. It has now branched and become truly interdisciplinary as it has affected more and more fields, like cognitive science, where models of rational thinking are an essential part of the research program. At some point, argumentation methods and findings began to be imported into computing, especially in the area called artificial.

Starting from the abstract argumentation framework proposed by Dung in [4], to the important point here is that arguments are viewed as abstract entities, with a binary defeat relation among them. A series of semantics are given for an argumentation theory, and acceptable sets are defined based on these semantics. It shows this is effective for logic programming and non-monotonic reasoning.

Another early development of Dung's proposal is value-based argumentation frameworks (VAFs) [2], in which every argument is explicitly associated with a value promoted by its acceptance, and audiences are characterised by the preference they give to these values. The strength of an argument depends on the value it advances, and that whether the attack of one argument on another succeeds depends on the comparative

---

[*] This work is supported by the Funds 863 Project 2012AA040906, NSFC61171121, NSFC60973049, and the Science Foundation of Chinese Ministry of Education-China Mobile 2012.

M. Wang (Ed.): KSEM 2013, LNAI 8041, pp. 76–88, 2013.

strength of the values advanced by the arguments. The nature and source of knowledge about audience varies according to the context. Therefore, audiences may have different preferences or attitudes on values. Each audience has its own evaluation criteria on arguments. On this basis, an argument is objectively acceptable if and only if it is in every preferred extension for all audiences, while subjectively acceptable if and only if it is in some preferred extension for some audiences.

Social choice theory or social choice is a theoretical framework for analysis of combining individual preferences, interests, or welfares to reach a collective decision or social welfare in some sense [5]. A non-theoretical example of a collective decision is enacting a law or set of laws under a constitution. A very important concept in social choice theory is social welfare function, which maps a set of individual preferences for every agent in the society to a social preference.

We have known that there are a set of preferences in value-based argumentation framework given by a group of audiences. If we apply social welfare function on these preferences, then another line of thinking for evaluation of arguments is arised.

In this paper, we introduce social choice theory into value-based argument system for obtaining a social preference. Then a social value-based argumentation framework is formally defined. On this basis, a new perspective on analysing and designing argument acceptability criteria, that is, social welfare semantics is proposed. Lastly, we investigate social welfare semantics as well as its relationship with the semantics of VAF, and prove some important non-trivial properties based on Pareto efficiency.

This paper is organized as follows. Section 2 provides the definitions of basic concepts about abstract argument system and value-based argumentation framework. Section 3 introduces social choice theory into value-based argumentation framework and proposes social welfare semantics. Section 4 discusses and proves some properties concerning social welfare semantics as well as its relationship with the semantics of VAF, and provides an example of this semantics. In Section 5 we conclude and compares the proposed approach with related works.

2   Argumentation Preliminaries

In this section, we briefly outline key elements of abstract argumentation frameworks and value-based argumentation framework. We begin Dung's abstract characterization of an argumentation system:

Definition 1 (Argumentation Framework). *An abstract argumentation framework is a pair $AF = \langle \mathcal{X}, \mathcal{R} \rangle$ where $\mathcal{X}$ is a finite set of arguments and $\mathcal{R} \subseteq \mathcal{X} \times \mathcal{X}$ is a binary relation on $\mathcal{X}$. The pair $(a, b) \in \mathcal{R}$ means that the argument a attacks the argument b. Let $a\mathcal{R}b$ denote that $(a, b) \in \mathcal{R}$, and $a\bar{\mathcal{R}}b$ denote that $(a, b) \notin \mathcal{R}$.*

An argumentation framework can be represented as a directed graph in which vertices are arguments and directed arcs characterise attack (or defeat) relation among arguments.

Example 2. A simple example argument graph is shown in Fig. 1, in which the argumentation framework is $AF = \langle \mathcal{X}, \mathcal{R} \rangle$, where $\mathcal{X} = \{a, b, c, d, e\}$ and $\mathcal{R} = \{(a, b),$

$(c, b), (c, d), (d, c), (d, e), (e, e)\}$ or $\mathcal{R} = \{a\mathcal{R}b, c\mathcal{R}b, c\mathcal{R}d, d\mathcal{R}c, d\mathcal{R}e, e\mathcal{R}e\}$. Argument b has two attackers (*i.e.* counter-arguments) a and c. Since argument a and c has no attack relation, so we can write this as $a\overline{\mathcal{R}}c$.

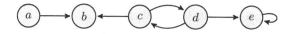

Fig. 1. A simple argument graph

Notation 3. *Let $\langle \mathcal{X}, \mathcal{R} \rangle$ be an argumentation framework, $b \in \mathcal{X}$ and $A, B \subseteq \mathcal{X}$. Define the following notations for union attack:*

- $A\mathcal{R}b$, *i.e. A attacks b, iff $\exists a(a \in A \wedge a\mathcal{R}b)$.*
- $b\mathcal{R}A$, *i.e. b attacks A, iff $\exists a(a \in A \wedge b\mathcal{R}a)$.*
- $A\mathcal{R}B$, *i.e. A attacks B, iff $\exists b(b \in B \wedge A\mathcal{R}b)$.*
- $A\overline{\mathcal{R}}b$, *i.e. A doesn't attack b, iff $\forall a(a \in A \wedge a\overline{\mathcal{R}}b)$.*
- $b\overline{\mathcal{R}}A$, *i.e. b doesn't attack A, iff $\forall a(a \in A \wedge b\overline{\mathcal{R}}a)$.*
- $A\overline{\mathcal{R}}B$, *i.e. A doesn't attack B, iff $\forall a(a \in A \wedge a\overline{\mathcal{R}}B)$.*

Notation 4. *Let $\langle \mathcal{X}, \mathcal{R} \rangle$ be an argumentation framework, $a \in \mathcal{X}$ and $S \subseteq \mathcal{X}$. Define the following notations:*

- $\gamma^-(a) \triangleq \{x \in \mathcal{X} : x\mathcal{R}a\}$ *denotes the set of (direct) attackers of a.*
- $\gamma^+(a) \triangleq \{x \in \mathcal{X} : a\mathcal{R}x\}$ *denotes the set of arguments (directly) attacked by a.*
- $\gamma^-(S) \triangleq \{x \in \mathcal{X} : x\mathcal{R}S\}$ *denotes the set of (direct) attackers of S.*
- $\gamma^+(S) \triangleq \{x \in \mathcal{X} : S\mathcal{R}x\}$ *denotes the set of arguments (directly) attacked by S.*

With these above notations, we firstly characterise the fundamental notions of conflict-free and defense.

Definition 5 (Conflict-free and Defense). *Let $\langle \mathcal{X}, \mathcal{R} \rangle$ be an argumentation framework.*

- *A set $S \subseteq \mathcal{X}$ is **conflict-free** iff $S\overline{\mathcal{R}}S$, i.e. there are no $a, b \in S$ such that $a\mathcal{R}b$.*
- *S **defends** an argument $a \in \mathcal{X}$ iff $\gamma^-(a) \subseteq \gamma^+(S)$, i.e. for each argument $b \in \mathcal{X}$: if b attacks a then b is attacked by S.*

Intuitively, a set of arguments is *conflict-free* if no argument in that set defeats another. A set of arguments *defends* a given argument if it defeats all its defeaters. In Fig. 1, for example, $\{a, c\}$ is conflict-free set and defends argument c.

Different semantics for collective acceptability of a set of arguments have been proposed by [4]. These are stated in the following definitions.

Definition 6 (Characteristic function). *Consider argumentation framework $AF = \langle \mathcal{X}, \mathcal{R} \rangle$ and let $S \subseteq \mathcal{X}$ be a conflict-free set of arguments. The characteristic function of AF is $\mathfrak{F}: 2^{\mathcal{X}} \mapsto 2^{\mathcal{X}}$ defined as $\mathfrak{F}(S) \triangleq \{a \in \mathcal{X} : S \text{ defends } a\}$.*

Definition 7 (Acceptability semantics). *Let S be a conflict-free set of arguments in framework $\langle \mathcal{X}, \mathcal{R} \rangle$.*

- *S is **admissible** iff it defends any element in S (i.e. $S \subseteq \mathfrak{F}(S)$).*
- *S is a **complete extension** iff $S = \mathfrak{F}(S)$.*
- *S is a **grounded extension** iff it is the (unique) least fixed-point of characteristic function $\mathfrak{F}$, i.e. $S = \mathfrak{F}(S)$ and there is no $S' \subsetneq S$ such that $S' = \mathfrak{F}(S')$.*
- *S is a **preferred extension** iff it is a maximal (w.r.t. set inclusion) admissible set.*
- *S is a **stable extension** iff it is a preferred extension that attacks all arguments in $\mathcal{X} \setminus S$.*

Acceptability semantics are solutions for argument systems, *i.e.*, concepts of what constitutes a set of mutually compatible arguments from $\mathcal{X}$ within a system $\langle \mathcal{X}, \mathcal{R} \rangle$. Typically, such subsets are defined via predicates $\sigma : 2^{\mathcal{X}} \mapsto \{\top, \bot\}$ so that $\sigma(S)$ holds of $S \subseteq \mathcal{X}$ in $\langle \mathcal{X}, \mathcal{R} \rangle$ the set S is viewed as acceptable with respect to the criteria defined by σ.

Given an acceptability semantics $\sigma \in \{admissible, complete, grounded, preferred, stable\}$ abbreviated as $\{adm, co, gr, pr, st\}$ we denote by $\psi_\sigma(\langle \mathcal{X}, \mathcal{R} \rangle)$ the set of subsets of $\mathcal{X}$:

$$\psi_\sigma(\langle \mathcal{X}, \mathcal{R} \rangle) \triangleq \{S \subseteq \mathcal{X} : \sigma(S)\}$$

Notice that, while all of these criteria are guaranteed to give some "answer", it is possible that the only answer they give is the empty set, *i.e.*, $\psi_\sigma(\langle \mathcal{X}, \mathcal{R} \rangle) \neq \emptyset$ but we may have $\psi_\sigma(\langle \mathcal{X}, \mathcal{R} \rangle) = \{\emptyset\}$. This is viewed as a key limitation of conventional systems.

Example 8. Consider once more the argument framework AF introduced in Example 2. The sets $\emptyset$ and $\{a\}$ are all admissible simply because they do not have any defeaters. The sets $\{c\}$, $\{d\}$, $\{a, c\}$ and $\{a, d\}$ are also admissible since they defend themselves. Thus, $\psi_{adm}(AF) = \{\emptyset, \{a\}, \{c\}, \{d\}, \{a, c\}, \{a, d\}\}$. Meanwhile, the set $\{a, c\}$, as well as $\{a\}$ and $\{a, d\}$, is also complete extensions for $\mathfrak{F}(\{a, c\}) = \{a, c\}$. Therefore, $\psi_{co}(AF) = \{\{a\}, \{a, c\}, \{a, d\}\}$. Accordingly, the preferred, grounded and stable extension of AF are $\psi_{pr}(AF) = \{\{a, c\}, \{a, d\}\}$, $\psi_{gr}(AF) = \{\{a\}\}$ and $\psi_{st}(AF) = \{\{a, d\}\}$ respectively.

Now that the set of acceptable arguments according to semantics σ is defined, we can define the status of any individual argument.

Definition 9 (Argument status). *Let $\langle \mathcal{X}, \mathcal{R} \rangle$ be an argumentation system, and $\psi_\sigma(\langle \mathcal{X}, \mathcal{R} \rangle)$ be its extensions under a given semantics σ. Let argument $x \in \mathcal{X}$,*

- *x is **credulously accepted** w.r.t. σ in $\langle \mathcal{X}, \mathcal{R} \rangle$ iff $\exists S \in \psi_\sigma(\langle \mathcal{X}, \mathcal{R} \rangle)$ such that $x \in S$.*
- *x is **sceptically accepted** w.r.t. σ in $\langle \mathcal{X}, \mathcal{R} \rangle$ iff $\forall S \in \psi_\sigma(\langle \mathcal{X}, \mathcal{R} \rangle)$ such that $x \in S$.*
- *x is **rejected** w.r.t. σ in $\langle \mathcal{X}, \mathcal{R} \rangle$ iff $\nexists S \in \psi_\sigma(\langle \mathcal{X}, \mathcal{R} \rangle)$ such that $x \in S$.*

The argument status show that the evaluation process in abstract argument systems concerns the justification state of arguments, not of their conclusions.

Now, we present the value-based framework as an extension of Dung's original argumentation framework. The basic idea underlying Value-based Argumentation Frameworks (VAF) is that it is possible to associate practical arguments with values, and that in order to determine which arguments are acceptable we need to consider the audience to which they are addressed, characterised in terms of an ordering on the values involved. The formal definition of such value-based argumentation frameworks is given below.

Definition 10 (Value-based argumentation framework). *A value-based argumentation framework is a 5-tuple VAF $= \langle \mathcal{X}, \mathcal{R}, \mathcal{V}, \omega, \succ \rangle$ where $\mathcal{X}$ and $\mathcal{R}$ are as for a standard argumentation framework, $\mathcal{V} = \{v_1, v_2, \ldots, v_k\}$ a set of k values, $\omega : \mathcal{X} \mapsto \mathcal{V}$ a mapping that associates a value $\omega(x) \in \mathcal{V}$ with each argument $x \in \mathcal{X}$, and $\succ = \{\succ_1, \succ_2, \cdots, \succ_n\}$ is the set of possible total orderings of n audiences, in which $\succ_i$ is a preference relation (transitive, irreflexive and asymmetric) on $\mathcal{V} \times \mathcal{V}$ reflecting the value preferences of audience i.*

Values are used in the sense of fundamental social or personal goods that are desirable in themselves, and should never be confused with any numeric measure of the strength, certainty or probability of an argument.

Audience in VAF can be seen as a representative agent (in multi-agent system), so we will not distinguish these concepts in remainder of this paper. When the VAF is considered by a particular audience, the ordering of values is fixed. We may therefore define an Audience specific VAF (AVAF) as follow.

Definition 11 (Audience specific VAF). *An audience specific value-based argumentation framework (AVAF) is a 5-tuple VAF$_i = \langle \mathcal{X}, \mathcal{R}, \mathcal{V}, \omega, \succ_i \rangle$ where $\mathcal{X}$, $\mathcal{R}$, $\mathcal{V}$ and ω are as for a VAF, $\succ_i$ is the value preferences of audience i.*

The AVAF relates to the VAF in that $\mathcal{X}$, $\mathcal{R}$, $\mathcal{V}$ and ω are identical, and $\succ_i$ is the set of preferences derivable from $\succ$ w.r.t. audience i. For any distinct $v, v' \in \mathcal{V}$ if v is ranked higher than v' in the total ordering defined by audience i, i.e. $(v, v') \in \succ_i$, we say that v is preferred to v' w.r.t. audience i, denoted $v \succ_i v'$.

Using VAFs, ideas analogous to those of admissible argument in standard argument system are given by relativising the concept of attack using that of successful attack with respect to an audience, defined in the following way. Note that all these notions are now relative to some audience.

Definition 12 (Successful attack). *Let $\langle \mathcal{X}, \mathcal{R}, \mathcal{V}, \omega, \succ_i \rangle$ be an audience i specific VAF. For any arguments $x, y \in \mathcal{X}$, x is a **successful attack** on y (or x **successfully defeats** y) w.r.t. the audience i if $(x, y) \in \mathcal{R}$ and it is not the case that $\omega(y) \succ_i \omega(x)$.*

Whether the attack succeeds depends on the value order of the audience considering the VAF. For any audience specific VAF, the successful attack relations w.r.t. audience i can be constructed by removing from the set of attacks $\mathcal{R}$ those attacks which fail because faced with a superior value, i.e.,

$$\mathcal{R}_i \triangleq \mathcal{R} \backslash \{(x, y) : \omega(y) \succ_i \omega(x)\}$$

or

$$\mathcal{R}_i \triangleq \{(x, y) : (x, y) \in \mathcal{R} \text{ and } \omega(y) \not\succ_i \omega(x)\}$$

where $\omega(y) \not\succ_i \omega(x)$ means $(\omega(y), \omega(x)) \notin \succ_i$.

Now for a given choice of value preferences $\succ_i$ we are able to construct an AF equivalent to the AVAF. Thus for any AVAF, $VAF_i = \langle \mathcal{X}, \mathcal{R}, \mathcal{V}, \omega, \succ_i \rangle$ there is a corresponding AF derived from it, *i.e.*, $AF_i = \langle \mathcal{X}, \mathcal{R}_i \rangle$.

Then we define acceptability semantics and argument status for VAF under a given semantics σ as follows.

Definition 13 (Acceptability semantics for AVAF). *Let $VAF_i = \langle \mathcal{X}, \mathcal{R}, \mathcal{V}, \omega, \succ_i \rangle$ be an audience i specific AVAF and $AF_i = \langle \mathcal{X}, \mathcal{R}_i \rangle$ be its equivalent AF. Given a semantics $\sigma \in \{adm, co, gr, pr, st\}$, the acceptability semantics for VAF is defined as $\psi_\sigma(\langle \mathcal{X}, \mathcal{R}_i \rangle)$, denoted $\psi_\sigma(VAF_i)$ or $\psi_\sigma(AF_i)$ for concision.*

Definition 14 (Argument status for VAF). *Let $\langle \mathcal{X}, \mathcal{R}, \mathcal{V}, \omega, \succ \rangle$ be a VAF and $\psi_\sigma (VAF_i)$ be the audience i's extensions under a given semantics σ. Let argument $x \in \mathcal{X}$,*

- *x is **subjectively acceptable** w.r.t. σ in VAF iff x is credulously accepted w.r.t. σ by some VAF_i but not all.*
- *x is **objectively acceptable** w.r.t. σ in VAF iff x is sceptically accepted w.r.t. σ by all VAF_i.*
- *x is **indefensible** w.r.t σ in VAF iff x is rejected w.r.t. σ by all VAF_i.*

Example 15. An example of value-based argument framework is illustrated in Fig. 2(a), in which the framework is $VAF = \langle \mathcal{X}, \mathcal{R}, \mathcal{V}, \omega, \succ \rangle$, where $\mathcal{X} = \{a, b, c, d\}$, $\mathcal{R} = \{a\mathcal{R}b, b\mathcal{R}a, a\mathcal{R}c, c\mathcal{R}a, a\mathcal{R}d, d\mathcal{R}a\}$, $\mathcal{V} = \{v_1, v_2, v_3, v_4\}$, ω mapping each argument to abstract value set $\mathcal{V}$ are illustrated in argument circles. In this example, there are three audiences A_1, A_2, A_3, with preferences $\succ_1 = \{v_1 \succ v_4 \succ v_2 \succ v_3\}$, $\succ_2 = \{v_2 \succ v_4 \succ v_1 \succ v_3\}$, and $\succ_3 = \{v_3 \succ v_4 \succ v_1 \succ v_2\}$ respectively. Thus, $\succ = \{\succ_1, \succ_2, \succ_3\}$ in VAF. The successful attacks of these three audiences are shown in Fig. 2(b), Fig. 2(c) and Fig. 2(d) respectively. When the semantics σ is preferred, the extensions of these audience are equal, *i.e.*, $\psi_{pr}(VAF_i) = \{\{a, b, c\}\}, i = 1, 2, 3$.

3 Applying Social Choice Theory into VAF

3.1 Social Choice Theory

Social choice theory [12] studies rules for aggregating audiences' beliefs and preferences. Given a value-based argument system and a group of audiences, the individuals may have divergent attitudes (or preferences) towards these values. If the group needs to reach a common position on the ordering of values of VAF, the question arises here is that how the individual preferences can be mapped into a collective one. To this end, we propose to apply the tools of social choice theory to value-based argument system.

Let $\langle \mathcal{X}, \mathcal{R}, \mathcal{V}, \omega, \succ \rangle$ be a VAF, where $\mathcal{V} = \{v_1, v_2, \ldots, v_k\}$ a set of k values, and $\succ = \{\succ_1, \succ_2, \ldots, \succ_n\}$ is a set of total orderings on $\mathcal{V}$ corresponding to a set of audiences $N = \{1, 2, \ldots, n\}$. Specifically, we assume that all audiences' preferences are *strict* total orderings on $\mathcal{V}$ rather than non-strict total orders to simplify the exposition. Let $\mathcal{P}$ represent the set of all possible strict preferences on $\mathcal{V}$. Obviously, the audience i's preference ordering $\succ_i$ satisfies $\succ_i \in \mathcal{P}$ and the orderings of n audiences (called preferences profiles) $\succ$ of VAF such that $\succ \subseteq \mathcal{P}^n$.

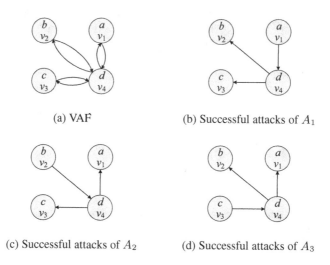

(a) VAF

(b) Successful attacks of A_1

(c) Successful attacks of A_2

(d) Successful attacks of A_3

Fig. 2. Example of VAF

Definition 16 (Social welfare function for VAF). *Let* $\langle \mathcal{X}, \mathcal{R}, \mathcal{V}, \omega, \succ \rangle$ *be a VAF, and* $\mathcal{P}$ *be the set of all possible preferences on* $\mathcal{V}$. *A social welfare function over VAF is a function* $\Gamma : \mathcal{P}^n \mapsto \mathcal{P}$.

Social welfare functions take preference profiles as input. Given preference profiles $\succ$, denote the preference ordering selected by the social welfare function Γ as $\succ_{\Gamma(\succ)}$. When the input ordering set $\succ$ is understood from context, we abbreviate our notation for the social preference as $\succ_{\Gamma}$. The element $(v, v') \in \succ_{\Gamma}$, also written as $v \succ_{\Gamma} v'$, can be interpreted as "v is socially preferred to v'", and $\succ_{\Gamma} \in \mathcal{P}$ is interpreted as social preference.

There exist many alternative ways to aggregate different preferences, most of them are based in some form of voting. In fact, the best known case of social welfare function is simple majority voting.

Definition 17 (Simple majority voting). *Let* $\succ$ *be preference profiles on* $\mathcal{V}$, $\succ_{\Gamma}$ *be social preference. For any* $v, v' \in \mathcal{V}$, $v \succ_{\Gamma} v'$ *iff* $|\{i : v \succ_i v'\}| > |\{i : v' \succ_i v\}|$.

For instance, if out of 100 individual relations, 60 are such that v is preferred v', while 40 verify that v' is preferred v, simple majority voting would yield that $v \succ_{\Gamma} v'$. That is, it only matters which alternative is verified by more individual relations than the other.

Now, let us review a crucial notion of fairness on the properties of social welfare function.

Definition 18 (Pareto efficiency (PE)). *Social welfare function* Γ *is Pareto efficiency if for any* $v, v' \in \mathcal{V}$, $\forall i(v \succ_i v')$ *implies that* $v \succ_{\Gamma} v'$.

In words, PE means that if all audiences suggest the same ordering among two values, then the social welfare function must select that ordering. Absolutely, The simple majority voting satisfies PE property. There are also some other fairness properties, such as, Independence of irrelevant alternatives (IIA), Nondictatorship, etc. which can found in [5].

In social choice theory, Arrows impossibility theorem, or Arrows paradox, says that if the decision-making body has at least two members and at least three options to decide among, then it is impossible to design a social welfare function that satisfies all these fairness properties at once. However, social choice theorists have investigated various possibilities ("ways out") to escape from the negative conclusion of Arrow's theorem, such as additional domain restrictions, limiting the number of alternatives, relaxing transitivity, etc..

Investigating those approaches to circumvent the impossibility is beyond the scope of our works, and related work can refer to the literature [11]. In this paper, we are merely concerned about the application of the social choice theory and the PE property. Therefore, we suppose that there always exists a social welfare function which satisfies PE property and can produce a socially desirable preference.

3.2   Social Welfare Semantics

Given a set of preferences and a social welfare function Γ, we can obtain a social preference $\succ_\Gamma$. Now we derive the concept of social VAF from social preference.

Definition 19 (Social VAF (SVAF)). *Let $\langle \mathcal{X}, \mathcal{R}, V, \omega, \succ \rangle$ be a VAF. The social VAF w.r.t. Γ is defined as $VAF_\Gamma = \langle \mathcal{X}, \mathcal{R}, V, \omega, \succ_\Gamma \rangle$, where $\mathcal{X}, \mathcal{R}, V$ and ω are also as for the VAF, and $\succ_\Gamma$ be the social preference. The SVAF's equivalent abstract argument system is denoted as $AF_\Gamma = \langle \mathcal{X}, \mathcal{R}_\Gamma \rangle$.*

In some sense, SVAF can be viewed an audience specific VAF, *i.e.* VAF_Γ, whose preference is the social preference $\succ_\Gamma$, and its equivalent abstract AF is AF_Γ constructed by Definition 12. Then we extend the semantics of abstract AF to SVAF to define a novel semantics of VAF, called social welfare semantics.

Definition 20 (Social acceptability semantics). *Given a semantics σ, the social acceptability semantics of a VAF w.r.t. σ is the acceptability semantics of VAF_Γ or AF_Γ, denoted as $\psi_\sigma(VAF_\Gamma)$ or $\psi_\sigma(AF_\Gamma)$.*

Definition 21 (Social argument status). *Let $\psi_\sigma(VAF_\Gamma)$ be social acceptability extensions of VAF under a given semantics σ. Let argument $x \in \mathcal{X}$,*

- *x is **socially credulously accepted** w.r.t. σ iff $\exists S \in \psi_\sigma(VAF_\Gamma)$ such that $x \in S$.*
- *x is **socially sceptically accepted** iff $\forall S \in \psi_\sigma(VAF_\Gamma)$ such that $x \in S$.*
- *x is **socially rejected** iff $\nexists S \in \psi_\sigma(VAF_\Gamma)$ such that $x \in S$.*

Example 22. Considering Example 15, there are three audiences with different preferences in the value-based argumentation framework. Applying social welfare function and simple majority voting, we know that $v_4 \succ_\Gamma v_1$ since most audiences think so. For the same reason, $v_4 \succ_\Gamma v_2$, $v_4 \succ_\Gamma v_3$, $v_1 \succ_\Gamma v_2$, $v_1 \succ_\Gamma v_3$ and $v_2 \succ_\Gamma v_3$. Therefore, $\succ_\Gamma = \{v_4 \succ v_1 \succ v_2 \succ v_3\}$, and its social VAF is shown in Fig. 3 and its preferred extension $\psi_{pr}(VAF_\Gamma) = \{\{d\}\}$.

Fig. 3. The successful attack graph of Example 15

4 Properties

In this section, we turn our attention to the properties of SVAFs and their semantics based on Pareto efficiency, and formally investigate the relationship of our semantics to the original semantics of VAF.

Lemma 23. *Let $\langle \mathcal{X}, \mathcal{R}, \mathcal{V}, \omega, \succ \rangle$ be a VAF and $\langle \mathcal{X}, \mathcal{R}, \mathcal{V}, \omega, \succ_\Gamma \rangle$ be its SVAF. For any $v, v' \in \mathcal{V}$, if $v \not\succ_i v'$ hold for all audience i, then $v \not\succ_\Gamma v'$ hold.*

Proof. Since $\succ_i$ is a strict ordering, so $v \not\succ_i v'$ means that $v' \succ_i v$. Using Pareto efficiency, This lemma is proved.

This lemma states that if everyone does not prefer v to v', then the social welfare function must not select that ordering.

Corollary 24. *Let $\langle \mathcal{X}, \mathcal{R}, \mathcal{V}, \omega, \succ \rangle$ be a VAF and $\langle \mathcal{X}, \mathcal{R}, \mathcal{V}, \omega, \succ_\Gamma \rangle$ be its SVAF. For any argument $x, y \in \mathcal{X}$, if each audience i satisfying $\omega(y) \not\succ_i \omega(x)$ then $\omega(y) \not\succ_\Gamma \omega(x)$.*

Proposition 25. *Let $\langle \mathcal{X}, \mathcal{R}, \mathcal{V}, \omega, \succ \rangle$ be a VAF and its SVAF's equivalent AF be $\langle \mathcal{X}, \mathcal{R}_\Gamma \rangle$. For any $x, y \in \mathcal{X}$, if $\forall i (x \mathcal{R}_i y)$, then $x \mathcal{R}_\Gamma y$.*

Proof. From the definition, we know that $x \mathcal{R}_i y \Leftrightarrow (x, y) \in \mathcal{R} \wedge \omega(y) \not\succ_i \omega(x)$. Since every VAF_i contain the same attack relations $\mathcal{R}$, merely consider $\omega(y) \not\succ_i \omega(x)$. By Corollary 24, it follows that $\omega(y) \not\succ_\Gamma \omega(x)$. Then $\mathcal{R}_\Gamma$ can be constructed by Definition 12, *i.e.* $\mathcal{R}_\Gamma = \{(x, y) : x \mathcal{R} y \text{ and } \omega(y) \not\succ_\Gamma \omega(x)\}$. Hence, $x \mathcal{R}_\Gamma y$.

Proposition 26. *Let $\langle \mathcal{X}, \mathcal{R}, \mathcal{V}, \omega, \succ \rangle$ be a VAF and its SVAF's equivalent AF be $\langle \mathcal{X}, \mathcal{R}_\Gamma \rangle$. For any $x, y \in \mathcal{X}$, if $\forall i (x \mathcal{\tilde{R}}_i y)$, then $x \mathcal{\tilde{R}}_\Gamma y$.*

Proof. If $x \mathcal{\tilde{R}}_i y$, two cases need to be considered:
Case 1: The first case is $x \mathcal{R} y$ in VAF. Obviously, $x \mathcal{\tilde{R}}_\Gamma y$ always holds since all VAF_i as well as VAF_Γ have same attack relations $\mathcal{R}$.
Case 2: The second case is $x \mathcal{R} y \wedge \omega(y) \succ_i \omega(x)$. All individual have same $\mathcal{R}$, it is just to consider the preference part $\omega(y) \succ_i \omega(x)$. When all audiences satisfy $\omega(y) \succ \omega(x)$ by Pareto efficiency, we have $\omega(y) \succ_\Gamma \omega(x)$. Thus, $x \mathcal{R} y \wedge \omega(y) \succ_\Gamma \omega(x)$ holds, that is, $x \mathcal{\tilde{R}}_\Gamma y$.

Proposition 27. *Let* $\langle \mathcal{X}, \mathcal{R}, V, \omega, \succ \rangle$ *be a VAF and its SVAF's equivalent AF be* $\langle \mathcal{X}, \mathcal{R}_\Gamma \rangle$. *For* $A, B \subseteq \mathcal{X}$ *and* $b \in \mathcal{X}$, *the following holds:*

(i). *if* $\forall i (A \mathcal{R}_i b)$, *then* $A \mathcal{R}_\Gamma b$.
(ii). *if* $\forall i (b \mathcal{R}_i A)$, *then* $b \mathcal{R}_\Gamma A$.
(iii). *if* $\forall i (A \mathcal{R}_i B)$, *then* $A \mathcal{R}_\Gamma B$.

Proof. (i). From Notation 3, $A \mathcal{R}_i b$ denote that for each argument $a \in A$ such that $a \mathcal{R}_i b$. If this holds for all audiences, by Proposition 26 we can easily prove that for each argument $a \in A$ such that $a \mathcal{R}_\Gamma b$, that is, $A \mathcal{R}_\Gamma b$.

Similar to the proof of (ii) and (iii). ∎

Corollary 28. *Let* $\langle \mathcal{X}, \mathcal{R}, V, \omega, \succ \rangle$ *be a VAF and its SVAF's equivalent AF be* $\langle \mathcal{X}, \mathcal{R}_\Gamma \rangle$. *For argument* $a \in \mathcal{X}$, *if* a *doesn't receive any attacks in* VAF_i *for all* i, *then the same situation in* VAF_Γ.

Proof. Prove its equivalence proposition, *i.e.* $\forall i (\mathcal{X} \mathcal{R}_i a) \rightarrow \mathcal{X} \mathcal{R}_\Gamma a$, by Proposition 27 (i). ∎

Proposition 29. *Let* $\langle \mathcal{X}, \mathcal{R}, V, \omega, \succ \rangle$ *be a VAF and its SVAF's equivalent AF be* $\langle \mathcal{X}, \mathcal{R}_\Gamma \rangle$. *For a set of arguments* $S \subseteq \mathcal{X}$, *if* S *is conflict-free in* VAF_i *(or* AF_i*) for each audience* i, *then* S *is conflict-free in* VAF_Γ *(or* AF_Γ*).*

Proof. S is conflict-free in $VAF_i \Leftrightarrow S \mathcal{R}_i S$. Hence, the primal problem can be represented as $\forall i (S \mathcal{R}_i S) \rightarrow S \mathcal{R}_\Gamma S$. Applying Proposition 27 (iii), this proposition is proved. ∎

From these above properties, we know that the social outcome must not go against any individual preference, *i.e.* if all individual like (*resp.*, dislike) the preference, then so does the social welfare function. More precisely, if each audience has the same relationship between two arguments, successful attack or unsuccessful attack, then the social argumentation framework also owns this relation.

Notation 30. *Let* $\langle \mathcal{X}, \mathcal{R}, V, \omega, \succ \rangle$ *be a VAF, and* a *be an argument belonged in* $\mathcal{X}$. *Let sign "*$*$*" be either "*$+$*" or "*$-$*". We define the following operators on* a:

$$\mathcal{M}^*(a) \triangleq \bigcup_{i=1}^{n} \gamma_i^*(a)$$

$$\mathcal{N}^*(a) \triangleq \bigcap_{i=1}^{n} \gamma_i^*(a)$$

where $\gamma_i^*(a)$ *is defined by Notation 4 corresponding with audience* i. *Accordingly,* $\mathcal{M}^-(a)$ *(resp.,* $\mathcal{M}^+(a)$*) denotes the set of all possible arguments that attack (resp., are attacked by)* a *for all audiences, and* $\mathcal{N}^-(a)$ *(resp.,* $\mathcal{N}^+(a)$*) denotes the set of arguments that commonly attack (resp., are attacked by)* a *for each audience.*

Proposition 31. *Let* $\langle \mathcal{X}, \mathcal{R}, V, \omega, \succ \rangle$ *be a VAF, and* VAF_Γ *be its SVAF. Let sign "*$*$*" be either "*$+$*" or "*$-$*". For any* $a \in \mathcal{X}$, *then* $\mathcal{N}^*(a) \subseteq \gamma_\Gamma^*(a) \subseteq \mathcal{M}^*(a)$ *holds.*

Proof. Firstly for each argument $x \in \mathcal{N}^*(a)$ denote x successfully attacks a *w.r.t* "−" or a successfully attacks x *w.r.t* "+" for all audiences. Therefore, x must be in $\gamma_\Gamma^*(a)$ respectively. Thus, $\mathcal{N}^*(a) \subseteq \gamma_\Gamma^*(a)$ holds.

Secondly, we can prove $\gamma_\Gamma^*(a) \subseteq \mathcal{M}^*(a)$ by contradiction. Assume $\gamma_\Gamma^*(a) \nsubseteq \mathcal{M}^*(a)$, then there must exist some argument x such that $x \in \gamma_\Gamma^*(a)$ and $x \notin \mathcal{M}^*(a)$. Now, $x \in \gamma_\Gamma^*(a)$, that is, $x\mathcal{R}_\Gamma a$ *w.r.t.* sign "−" or $a\mathcal{R}_\Gamma x$ *w.r.t.* sign "+" respectively. Hence by the Proposition 26, there exists an audience i such that $x\mathcal{R}_i a$ *w.r.t.* sign "−" or $a\mathcal{R}_i x$ *w.r.t.* sign "+" respectively, that is, $x \in \gamma_i^*(a)$. However, from the definition of $\mathcal{M}^*(a)$, we have $x \in \mathcal{M}^*(a)$. Contradiction.

This Proposition states that social welfare function will not produce new attack relations. In other words, the attack relations selected by social welfare function (in some sense) always depend on individual's preference. By the Proposition 31, we can easily prove the following corollary:

Corollary 32. *Let $\langle \mathcal{X}, \mathcal{R}, \mathcal{V}, \omega, \succ \rangle$ be a VAF, and VAF$_\Gamma$ be its SVAF. Let sign "$*$" be either "+" or "−", and $a \in \mathcal{X}$. If for each audience i such that $\gamma_i^*(a) = A$ implies that $\gamma_\Gamma^*(a) = A$, where $A \subseteq \mathcal{X}$.*

Now, let us investigate the relation of argument status between VAF and SVAF. Firstly, we define the following notation:

Notation 33. *Let $\langle \mathcal{X}, \mathcal{R}, \mathcal{V}, \omega, \succ \rangle$ be a VAF, and its SVAF be VAF$_\Gamma$. Given a semantics σ,*

- $\text{OA}_\sigma \triangleq \{x \in \mathcal{X} : x \text{ is objectively acceptable w.r.t. } \sigma\}.$
- $\text{SA}_\sigma \triangleq \{x \in \mathcal{X} : x \text{ is subjectively acceptable w.r.t. } \sigma\}.$
- $\text{I}_\sigma \triangleq \{x \in \mathcal{X} : x \text{ is indefensible w.r.t. } \sigma\}.$
- $\text{SSA}_\sigma \triangleq \{x \in \mathcal{X} : x \text{ is socially sceptically accepted w.r.t. } \sigma\}.$
- $\text{SCA}_\sigma \triangleq \{x \in \mathcal{X} : x \text{ is socially credulously accepted w.r.t. } \sigma\}.$
- $\text{SJ}_\sigma \triangleq \{x \in \mathcal{X} : x \text{ is socially rejected w.r.t. } \sigma\}$

Example 34. Consider an example from [1] in Fig. 4, in which there are eight arguments associated with a abstract value, red or blue. Suppose there are 100 audiences, whose preferences on values are as follows:

$$99 : red \succ blue$$
$$1 : blue \succ red$$

If $red \succ blue$, the preferred extension will be $\{e, g, a, b\}$, and if $blue \succ red$, $\{e, g, d, b\}$. Hence, $\text{OA}_{pr} = \{e, g, b\}$, $\text{SA}_{pr} = \{a, d\}$ and $\text{I}_{pr} = \{c, f, h\}$. If we employ simple majority voting, $red \succ blue$ will be social preference. Then $\text{SSA}_{pr} = \{e, g, a, b\}$, $\text{SCA}_{pr} = \emptyset$ and $\text{SJ}_{pr} = \{c, d, f, h\}$. From this example, we know the simple majority aggregation function will select the preference that most people like. In some sense, the social preference represent the willingness of most people.

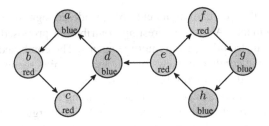

Fig. 4. VAF with values red and blue

5 Conclusions

In our work, we introduced a new persuasive on analysing and designing argument acceptability criteria in value-based argument system by using social welfare functions which provide a way of aggregating agents preferences in a systematic way in which to generate a socially desirable preference. Acceptability criteria can now be evaluated not only based on their logically intuitive properties, but also on their social welfare properties. Furthermore, we investigated the properties between these evaluation approaches base on Pareto efficiency, and we have the following significant findings:

- If all audience own some property, such as attack relation, conflict-free, etc., then the SVAF also have, that is, the social outcome must not go against all individual expectation.
- Social welfare function will not produce new attack relation, that is, social welfare function can not change logic connections among arguments.
- Simple majority aggregation function will select the preference that most people like.

Abstract argumentation frameworks have typically been analyzed without taking into account the audiences involved. This is because the focus has mostly been on studying the logically intuitive properties of argument acceptance criteria. Recently research has commenced on evaluating argument acceptance criteria taking into account agents strategic behaviour. [10,3,9] analyzed different argument evaluation rules taking into account the preferences of individual agent, and presented some operators for the aggregation of individual labelling of a given argumentation framework into a single one. They focused on a "compatibility" property: that the social outcome must not go against any individual judgment. However, all above of the listed works are based on the labeling approach [8], which have some limitations that only three labels are available, and the argument labelings must satisfy the condition that an argument is IN if and only if all of its defeaters are OUT, and an argument is OUT if and only if at least one of its defeaters is IN. Our work, based on value-based framework, is more flexible, and we can map arguments into different abstract values, and each agent can have their preference on these values. [7] took a step towards using argumentation in social networks incorporates social voting and propose a class of semantics for their new social abstract argumentation frameworks. Their main work is on the attacking relation and use voting method to determine the weight of argument. [6] considered logic-based argumentation with uncertain arguments. A more interesting contribution proposed in this paper is merging probability distributions from multiple agents. The combined probability distribution can reflect the conflicting views of two agents, which seems to related to our work. The difference is that our work bases on preferences.

This work presented in this paper is just the beginning of what we envisage to be a growing area at the intersection of group decision and formal argumentation theory. We envisage that our work, with its new approach to designing argumentation rules, will help bridge the gap between theory and application.

References

1. Bench-Capon, T., Atkinson, K.: Abstract argumentation and values. In: Simari, G., Rahwan, I. (eds.) Argumentation in Artificial Intelligence, pp. 45–64. Springer, US (2009)
2. Bench-Capon, T., Dunne, P.: Value based argumentation frameworks. In: Proc. 9th International Workshop on Non-Monotonic Reasoning, Toulouse, France, pp. 443–454 (2002)
3. Caminada, M., Pigozzi, G.: On judgment aggregation in abstract argumentation. Autonomous Agents and Multi-Agent Systems 1(22), 64–102 (2011)
4. Dung, P.M.: On the acceptability of arguments and its fundamental role in nonmonotonic reasoning, logic programming and n-person games. Journal of Artificial Intelligence 77(2), 321–357 (1995)
5. Gärtner, W.: A Primer on Social Choice Theory. Oxford University Press (2006)
6. Hunter, A.: A probabilistic approach to modelling uncertain logical arguments. International Journal of Approximate Reasoning 54(1), 47–81 (2013)
7. Leite, J.: a., Martins, J.a.: Social abstract argumentation. In: Proceedings of the Twenty-Second International Joint Conference on Artificial Intelligence (IJCAI), Barcelona, pp. 2287–2292 (2011)
8. Modgil, S., Caminada, M.: Proof theories and algorithms for abstract argumentation frameworks. In: Simari, G., Rahwan, I. (eds.) Argumentation in Artificial Intelligence, pp. 105–129. Springer, US (2009)
9. Rahwan, I., Larson, K.: Mechanism design for abstract argumentation. In: AAMAS 2008, International Foundation for Autonomous Agents and Multiagent Systems, Richland, SC, pp. 1031–1038 (2008)
10. Rahwan, I., Larson, K.: Pareto optimality in abstract argumentation. In: Proceedings of the Twenty-Third AAAI Conference on Artificial Intelligence, pp. 150–155 (2008)
11. Rahwan, I., Tohmé, F.: Collective argument evaluation as judgement aggregation. In: AAMAS 2010, International Foundation for Autonomous Agents and Multiagent Systems, Richland, SC, pp. 417–424 (2010)
12. Shoham, Y., Leyton-Brown, K.: Multiagent Systems: Algorithmic, Game-Theoretic, and Logical Foundations. Cambridge University Press (2009)

Intelligent Graphical User Interface for Managing Resource Knowledge in Cyber Physical Systems

Kaiyu Wan[1], Vasu Alagar[2], and Bai Wei[1]

[1] Xi'an Jiaotong-Liverpool University, Suzhou, PRC
[2] Concordia University, Montreal, Canada
kaiy.wan@xtjlu.edu.cn, alagar@cse.concordia.ca

Abstract. A Cyber Physical System (CPS) can be viewed as a large networked system, distributed geographically with each CPS site having hybrid computing power. A CPS node can be as simple as a sensor or as structured as a research lab with groups of participants who want to share resources in fulfilling a strategic task, or as complex as an autonomous system which procures and manages its resources with little human intervention. Because of the collective might of CPS nodes, the capabilities of CPS could have far reaching effects in finding solutions to grand challenges in fields that combine science, engineering, and economics. Thus, CPS applications will require resources from different domains. In the absence of central authority to moderate and manage resource requests for CPS services, there is a need to facilitate the coordination of services and cooperation to share resource knowledge among resource providing nodes in CPS. For a sustained development of economic value it is important that knowledge about resources are protected, shared, reused, and redistributed. Towards this purpose we propose a graphical user interface (GUI) at each CPS site. It is to facilitate the management of resource knowledge locally, and allow humans and systems at different CPS sites to interact, discover, query, and share resources globally. We discuss a generic model of resource in which resource data, resource information, and resource knowledge are structurally recorded. A three layered architecture is suggested for managing services hosted by resources. We describe the features of GUI that will assist the management activities at these three layers. The GUI is user-centric, designed on semantic principles, and will faithfully transform resource information into languages that can be shared and communicated securely across the CPS sites.

1 Introduction and Motivation

Cyber Physical Systems can be viewed as large networked systems of hybrid nature. The capabilities of such system are far reaching, as broadcast by the NSF program description [6]. It states that CPS initiative [2] is "to transform our world with systems that respond more quickly, are more precise, work in dangerous and inaccessible environments, and provide *large scale distributed services*." Environmental monitoring, high confidence medical systems for remote areas, transportation systems, energy distribution, and defense are some of the typical applications and instances of CPS [6,5]. The primary goal in these applications is the provision of *large scale distributed services*. CPS services will require resources from different domains. In the absence of central

M. Wang (Ed.): KSEM 2013, LNAI 8041, pp. 89–103, 2013.
© Springer-Verlag Berlin Heidelberg 2013

authority to moderate and manage resource requests for CPS services, there is a need to facilitate the coordination of services and cooperation among resource providing nodes in CPS to share resource knowledge. For a sustained development of economic value it is more important that knowledge about resources are protected, shared, reused, and redistributed. The main contribution of this paper fits this context. We propose a graphical user interface (GUI) at each CPS site, which facilitates the management of resource knowledge locally, and allow humans and systems at different CPS sites to interact, discover, query, and share resources globally.

The GUI that we describe in this paper is just the first phase in the development of a Graphical Resource Editor (GRE), which will have a wide-ranging set of capabilities for managing CPS services. In our research [11,12] we are following service-oriented design principles [1], and are investigating a three-tiered resource-centric service architecture for CPS. The GUI is intended to be a tool that will assist the management activities in these three tiers. That is, the GUI will be the front-end for different kinds of clients (humans and system components) to browse, discover, request, and allocate resources bound to service requests. The GUI is user-centric, designed on semantic principles, and will faithfully transform resource information into languages that can be shared and communicated securely across the CPS sites. Thus, the contributions are the following:

– GUI requirements, motivated by resource knowledge management issues
– GUI design principles for the three tiers corresponding to the three-tiered architecture [11,12]
– the functionalities that can be invoked in the three tiers of GUI and policy driven views of resource information, and
– the GUI generated XML versions of resource information for inter-operability across CPS sites.

2   GUI Requirements for Resource Management

The GUI is intended as a front-end for system designers, business leaders, and domain experts at each CPS site to structure resource knowledge and manage it. They are collectively responsible for protecting the site's resource knowledge and also publishing resource knowledge in selective ways so that other CPS sites may benefit from it. We view a CPS resource as a fundamental *economic or productive factor* required to accomplish an *activity*. Resources are means for a business enterprize to achieve its business goals. Resources are acquired, protected, sold, and shared according to the business policies of the resource owner and resource requester. These activities are also governed by legal rules set by the governments where the resource is found, sent, and used. The most significant aspect in resource management is the extent of protection imposed on securing sensitive resource information, as opposed to the extent of resource information that should be visible or shared in order to maximize resource utility. So we include resource data, resource information, resource knowledge, and rules governing their management in modeling resources at each CPS site.

Resource management is a multi-step activity. Roughly, we may consider *Resource Modeling* (RM), *Resource Publication* (RP), and *Resource Discovery and Allocation*

(RDA) as the distinct stages where IT can be used to advantage in resource management. This view leads to the important requirement that the GUI has three tiers, where each tier of the GUI enables the activities defined in that architecture tier through well-defined interfaces. We regard RM to be the lowest tier, RP the middle tier, and RDA the top tier. System designers, business leaders, and domain experts at each CPS site are the primary clients involved in RM tier for that site. Since resource model will include resource knowledge, some of which might be sensitive and of critical importance for the profitability of the enterprize, it is essential to safeguard them. At the same time, a large part of resource information and resource data should be made visible and shared with other CPS sites in order to maximize resource utility. Given these orthogonal aspects, and the fact the resources are distributed geographically across CPS sites as well as hierarchically within a CPS site, resource modeling stage is a complex task. The aim of GUI is to ease this complexity and serve the needs of different clients according to business and legal rules. We set access controls in the RM tier which are then propagated through the other two tiers. The *View Manager* of the GUI will project the publications of modeled resources in *Local Resource Publication* (LRP) and *Global Resource Publication* (GRP) parts of RP layer. The LRP is further split into three views, called *General LRP* (GLRP), *Preferred LRP* (PLRP) and *Secure LRP* (SLRP). Resources published in GLRP can be viewed by the agents who create the resource model and other agents who are authorized by the *Resource Security Administrator* (RSA). Resources published in PLRP can be viewed by *Resource Management Teams* (RMT). Resource published in SLRP have maximum security and can be viewed only by *Business Policy Managers* (BPM). The layer RA is split into *Local Resource Allocation* (LRA) and *Global Resource Allocation* (GRA). Resources required for services locally within the resource publication site are allocated through LRA. Resources required by other CPS sites are mediated and serviced through GRA. For the sake of simplicity, we assume that the GUI at every CPS site will have the same set of requirements.

3   GUI Design Principles

Since resource knowledge is not easy to disseminate the GUI must build resource model (in Tier RM) through an intensive interaction with the clients at that layer. Resource owner (producer) might start providing raw resource data, unstructured facts and figures. The GUI will try to get the data contextualized, by seeking answers from the resource owner and business leaders to questions that begin with *why?*, *who?*, *what?*, *when?*, and *where?*. In fact these are the five *dimensions* that Wan [10] identified for *specifying contexts*. The GUI will associate the resource data with context to produce resource information. This interaction between the GUI and the humans is facilitated by the *help* facility in GUI. Since information gathered in this manner may have to undergo many iterations and revisions, it is necessary that the GUI keeps track of many versions of a resource model. Resource knowledge is gathered from domain experts, such as scientists and engineers who have *experience* in using the resource. It is a mixture of experience reports, scientific predictions, and opinions on resource dependencies gathered from domain experts. From the content collected through user interaction the GUI builds a resource model in a *Resource Description Template (RDT)*. We discuss the syntax of RDT in Section 4.1, and the semantics of RDT contents in Section 4.2. The GUI builds RDTs for different resource types in the RM tier. The GUI will set resource protection features according to the access controls defined in Section 4.3, and offer interfaces that enforce a regulated view of resource information in a RDT. A detailed description of GUI features for RM tier is given in Section 5. In Section 6 we comment on the GUI design principles for the next two tiers, and our ongoing work in implementing them. The other general design principles are the following.

- The GUI components are loosely coupled, and will allow graphical plug-ins to view resource data in a manner that is most appropriate for the resource domain. This includes histograms, or graphs, or algebraic equations.
- Interfaces at the three tiers will allow a comprehensive and complete visibility of resource availability, resource requests, and resource allocation.
- Security settings of key aspects of resource knowledge may be distributed, if necessary, across the three tiers.
- Local and global pool of resources can be assessed for a project, regardless of physical location.
- Modifications to resource bookings can be monitored and dealt with by real-time reallocation of resources to other projects.

4   Resource Knowledge

In order to effectively deal with a multitude of CPS resources, we classify resource types into *human resources, biological resources, natural resources, man made resources*, and *virtual resources*. Our classification is certainly not exhaustive, however for the sake of a concise discussion we offer this classification. All resources in one category belong to one specific *type*. As an example, the category *human resources* consists of all human experts associated with their skills and other attributes. Resources required by a living being for survival and growth are biological resources. Examples include water and food. Natural resources are derived from the environment. Examples include trees, minerals, metals, gas, oil, and some fertilizers. Biological resource type is a subtype of natural resource type. Man made resources include physical entities such as bricks or mortar, books and journals for learning, and machineries. Any virtual component of limited availability in a computer is a virtual resource. Examples of virtual resource are *virtual memory, CPU time*, and the whole collection of Java resources [7]. Table 1 shows the structure of *Resource Description Template* (RDT), which is flexible and rich enough to describe resources from any of these types. Our goal is to explain the structure and semantics of RDT. GUI creates valid RDTs based on the discussion in this section.

4.1   Structure of RDT

In general, let $\mathcal{RT}$ be a finite set of resource types. A resource type $T \in \mathcal{RT}$ is a finite collection of resources of that type. As an example, *Metal* is a resource type, and $\{gold, platinum, iron, copper, zinc\}$ are resources of type *Metal*. The description of one resource r_T of type T is a RDT whose structure is shown in Table 1. The RDT table may be extended by adding more description elements. The tabular RDT format shown in Table 1 is meant for human agents. The GUI automatically generates the XML version of the RDT, once all resource information is gathered and assembled as an RDT. The XML version is used for resource propagation across CPS processing sites.

The human experts at each CPS site determine the essential features and properties to be specified in a resource model. The knowledge associated with these features are to be collected and recorded within the five sections of a RDT. In *Description* section, a general description of resource data and its attributes are stated. In *Functional Properties* section we list physical properties, chemical properties, temporal properties, and other properties that are inherent to the resource being modeled. This section captures resource information and parts of resource knowledge. In section *Nonfunctional and Trust* we list the nonfunctional properties, such as utility and cost, and the trustworthy properties such as availability, etc; that highlights the quality aspects of the resource. This section records resource knowledge, the opinion of the resource producer, and the team of experts at a site. In the *Rules* section we list the business rules and context rule! s that will bind resource data, information, and knowledge to a contract. The *Exceptions* section will list any adverse side effects in using the resource and the conditions for contract cancellation.

The main attributes of resources, especially when it comes to their *adaptation* for providing services, are *utility, availability, cost, sustainability, renewability, reuse*. The utility factor for a resource defines its relevance, and often expressed either as a numerical value u, $0 < u < 1$, or as an enumerated set of values {*critical, essential, recommended* }. In the former case, a value closer to 1 is regarded as critical. In the later case the values are listed in decreasing order of relevance. The representation $\{\langle a_1, u_1 \rangle, \langle a_2, u_2 \rangle, \ldots, \langle a_k, u_k \rangle\}$ shows the utility factor u_i for the resource in application area a_i for each resource produced by it. The utility factors published in a RDT are to be regarded as knowledge recommendations, based on some scientific study and engineering analysis of the resources conducted by the experts at the RP sites. Cost might depend upon duration of supply (as in power supply) or extent of use (as in gas supply), or in required measure (as in the supply of minerals). Dependency between resources can often be expressed as situations, in which predicate names are resources. Thus, RDT encodes resource data, resource information (including contractual information), and resource knowledge.

4.2 RDT Semantics

Below we explain the semantics related to the RDT shown in Table 1. The semantics for each resource type comes from the resource domain. Only domain experts can provide this semantic information.

- *Type:* The type of a resource is the *resource category*, as classified earlier or given in industries. We can include more resource types, such as Health (Medical) Resources.
- *Attribute:* The attribute section is used to provide the identity and contact information of resource producer. A general yet concise description of the resource may also be included. Some examples are the following.

Table 1. Resource Description Template

Description,	Resource(Name: r_T,Id: rid_T: ⟨*generic description of resource name*⟩ Type: ⟨ *resource type*: T⟩ Attribute: ⟨*producer, production facility profile, quality attributes, . . .*⟩
Functional Properties	Properties: {*physical properties, chemical properties, . . .*}
Nonfunctional, Trustworthy Properties	Utility: {⟨a_1, u_1⟩, ⟨a_2, u_2⟩, . . . , ⟨a_k, u_k⟩} Cost: *cost per unit* Availability: *available for shipment to all parts of the world or state constraints* Sustainability: *ratio of demand to supply for the next x years* Renewability: *YES/NO* Reuse: *list of applications for reuse of this resource* Recycling: *YES/NO* Other Resources in the Context of Use: *a list suggesting resource dependencies*
Legal Rules	Legal Rules for Supply: *URI to a web site* Context Information: *resource provider context* Context Rule: *for resource delivery*
Exceptions	Side Effects: *health and environmental protections* Contract Violation

(1) *human resources:* If the RDT is to describe the resource physician, an attribute might be 'GP' or 'SURGEON', or 'PSYCHIATRIST'.

(2) *biological resources:* If the RDT is to describe a vegetation, an attribute might be a statement on the specific vegetation type and the geographical region.

(3) *natural resources:* If the RDT is to describe a metal, an attribute might be 'PRECIOUS' or 'RARE EARTH'.

- *Properties:* The properties of a resource has many dimensions. Some of them are *physical* properties, *chemical* properties, *temporal* properties (persistent or change with time), and *trustworthiness* properties. Below are a few examples.

  (1) *human resources:* Academic and professional skills, as will be presented in a Curriculum Vita, constitute the properties of the human resource of interest. If the RDT describes administrator in a company, then the status, academic qualifications, professional expertise, and history of experience are properties.

  (2) *biological resources:* If the RDT describes vegetation, then a description of irrigated pasture lands, non-native annual grasslands including the vegetated sand mound, seasonal wetlands, sloughs, and drainage ditches may be included. Additional information on how these are used by a variety of wildlife and its potential for occurrence of special-status species may be included.

  (3) *natural resources:* For a metal typical properties include *physical properties* (such as physical state: solid, or liquid, luster, hardness, density, electro positivity, and ductility), *chemical properties* (such as reaction during heating, reaction with

water, reaction with acid), and *temporal properties:* (such as distribution characteristics in space (air) and time (seasonal)).

Trustworthiness properties are of three kinds. These are (1) *trustworthiness of the vendor*, expressed in terms of business policies, contractual obligations, and recommendations of peers; (2) *trustworthiness of the utility*, expressed in terms of safety guaranteed for workers, and environmental safety; and (3) *trustworthiness of the product*, expressed as a combination of quality guarantees, and rules for secure delivery of the product.

- *Utility:* The utility factor for a resource defines its relevance, and often expressed either as a numerical value u, $0 < u < 1$, or as an enumerated set of values $\{critical, essential, recommended \}$. In the former case, a value closer to 1 is regarded as critical. In the later case the values are listed in decreasing order of relevance. A RP may choose the representation $\{\langle a_1, u_1 \rangle, \langle a_2, u_2 \rangle, \ldots, \langle a_k, u_k \rangle\}$ showing the utility factor u_i for the resource in application area a_i for each resource produced by it. The utility factors published by a RP are to be regarded as recommendations based on some scientific study and engineering analysis of the resources conducted by the experts at the RP sites. Cost might depend upon duration of supply (as in power supply) or extent of use (as in gas supply), or in required measure (as in the supply of minerals). Dependency between resources can often be expressed as situations, in which predicate names are resources.

- *Cost:* The semantics of cost is the price per unit, where the unit definition might vary with resource type. For example, for natural gas the unit may be 'cubic feet', for petrol the unit may be 'barrel or liter'.

- *Availability:* This information may be provided under the three categories (1) Measured (provable), (2) Indicated (probable), and (3) Inferred (not certain).

- *Sustainability:* The semantics of sustainability is related to *Reserves, Contingent,* and *Prospective. Reserves* expresses a comparison between the measured amount of resource with the current demand. Possible ways to express this are: (1) sufficient: $currentdemand \leq measured$; (2) low: $currentdemand = measured$; (3) high: $currentdemand << measured.$ *Contingent* is an estimate (both amount and time period) of getting the reserves (this is a certainty). Is this a 'high' or 'low' estimate? Can the reserve meet the demand during this time period? How soon the 'Indicated' amount of resources will go into production? Possible ways to represent this are: (1) x amount within y days, and $demand$ after y days will satisfy $demand << measured + x.$ (2) similar inequalities to express sufficiency, low capacity. *Prospective* specifies the resource quantity determined, and an approximate time scale for its availability.

- *Renewability:* The semantics of renewability is related to the 'perpetual' or 'migratory' nature of the resource. For example, 'solar power' resource can be labeled 'perpetual'; however 'ground water' resource may not be available for ever.

- *Reuse:* The terms reuse and recycling are well understood both in technology and in environmental applications. A resource r_1 may be used in an application for a certain period of time. During this period the resource may decay, as in the case of iron used in producing trucks or aluminum used to manufacture aeroplanes. The recycling process re-produces the resource with its original specification. The reuse of a resource refers to the use of the resource in multiple applications. A robot might be used to act as a fire fighter, or just to flush away contaminated water. When a resource r_1 is used to produce a resource r_2, and in turn the resource r_2 is used to produce another resource r_3 we also have reuse of r_1.

- *Dependency:* The meaning of *Other Resources in the Context of Use* is to express 'resource dependency'. Examples of dependencies may be expressed using *before*, *during*, and *following* temporal operators. As an example, to extract coal from a mine, some natural resources such as power and other man-made resources are required *before* production commences. As another example, if a robot is used to pick the debris from a hazardous environment, an autonomous vehicle is required immediately *following* it in order to place the debris and move it.

- *Rules:* The semantics of legal rules include the business rules of the RP, the government regulations governing the distribution of resources, and international rules regarding quality of resources. The context information includes the resource provider context. Context rule is a predicate encoding the rules constraining resource delivery. We use the formal notation developed by Wan [10] to specify context information, and context rule. As an example, the context representation $[LOC : Shanghai, ADDRESS : 1235 NanjingStreet]$ gives the location information of service producer. The context rule $distance(Shanghai, LOC) \leq 100$ expresses the constraint that resource can be delivered to a city if it is within a distance of 100Kms from Shanghai. The justification for choosing this formalism is two-fold. First, the relational semantics given to context blends well with logical evaluation of legal rules and trustworthiness properties in contexts. Second, the set of context operators introduced by Wan [10] provides an abstract data type view of contexts, and can be imported as first class citizen in programming languages and in system design. That is, contexts can be communicated across CPS networks.

- *Exceptions:* The intent of this section is to list the impact and interference effects with environment. When the side effects contravene local environmental rules (as stated in the Rules section the contract can be canceled.

4.3   Protection and Projection

A RDT can be protected at atomic level as well as its entirety. At the atomic level each description element in RDT table can be protected. However, it is prudent only to protect the most valuable information, namely the one that is sensitive to economic leverage and national security. So, the domain experts in consultation with business managers and other government bodies should decide the aspects to be protected and also determine the level of protection necessary for each modeled element. As an example, not all functional properties may be allowed to be viewed by all users, and availability information of rare resources may not be published at all. To protect the information on a modeling element either it can be *encrypted* or *hidden* from viewing. Encryption method will require key distribution for those clients who are allowed to view the information. Key management in a large distributed system has several problems in itself. So, we pursue the hiding principle, based on access control methods. For a general discussion, let us assume that there are n modeling elements in a RDT. We can classify the users of resource information, based on the tiers where interact and their interaction roles.

In RM tier resource information is acquired, and filtered by domain experts (DE), and the information approved by domain experts are put into RDT format by system engineers (SE). In addition, the business leaders (BL) in the enterprise are also involved in formulating rules. Thus, there are three groups of agents in creating the n elements of RDT. Both DE and BL have a full set of rights on RDT, and this includes *read, write, modify, hide*. The group SE can only *create* the RDT by reading and writing in a *scratchpad*, a medium from which RDT can be written, but not modified. Our GUI is designed in such a way that RDT creation is a one-time project. It can be only *read* after its creation. In order to write on it or modify it, one has to go through the *scratchpad*. So, we provide a log-in authorization protocol for SEs to access the *scratchpad*. To access it a SE must be granted permission by a DE and by a majority of BLs. By locking the *scratchpad*, giving authentication by a majority consensus of BLs at the request of a DE, we ensure *information integrity* in RDT. No resource information that is not recommended by a DE and approved by a majority of BLs can be published.

In tier RP the RDTs constructed in the RM tier are published. The published RDTs can be viewed either locally at the site where it was created or globally at other CPS sites. Local view can be classified as *general view* (GV), *preferred view*, and *secure view*. Global view is constrained by *security contexts* that are constructed by monitoring resource requests from other CPS sites. A local client might be given access to have any combination of these three views. To accommodate these possibilities we use a binary vector v of length 3. If $v = (0, 0, 0)$ is attached to a RDT then it cannot be viewed by local clients. If $v = (0, 1, 1)$ is attached then both *preferred view* and *secure view* are possible for the RDT. Based on their profiles and roles, the local clients are formed into a number of groups, say $g_1, g_2, \ldots, g_k$. If a RDT X with vector v_X is assigned to group g_i, then every member of group g_i will have the views designated by vector v_X for the RDT X. Let us denote the set of RDTs that can be viewed by members of group g_i by σ_{g_i}.

In order to protect a RDT at atomic level, we consider all the clients in RP level. This is precisely the users in the set $G = \bigcup_{i=1,k} g_i$. Every client $p \in G$ belongs to only one group. Assume that there m clients in the set G. For each client, based on her profile and rank, a subset of elements in a RDT are hidden. That is, for a client $p \in G$ there exists a subset $\alpha_{p,X}$ of the elements in the description of RDT X that are hidden. Since p belongs to a group, say group g_i, client p can view every RDT X, $X \in \sigma_{g_i}$ and this view is constrained by the vector v_X. In summary, for every client $p \in g_i$, and for every RDT $X \in \sigma_{g_i}$ that the client can view the vector v_X sets the full view protection, and $\alpha_{p,X}$ sets the atomic view restriction.

5 Detailed Design of GUI

The GUI design is driven by the following hypotheses.

- the resource producer or its authorized agent, who may be the *domain expert*, provides the most authentic information about the resources,
- the graphical user interface is the only medium to create the RDTs,
- the agents who create a RDT have the valid security clearance from DEs and BLs,

Table 2. RDT for Robot

Description	Resource: ⟨*description of robot used in a rescue mission*⟩ Type: ⟨ *man made resource* ⟩ Attribute: *XYZ Manufactures* *link to production facility profile* *6 degrees of freedom, mobility in mountain terrain, forests, and in hazardous places*
Functional Properties	Properties: Weight: 30 Kg Height: 1.2 meters Arms: 2 Vision: *fitted with camera for night vision* Payload: *can lift and move 50 Kgs load at a time*
Nonfunctional Trustworthy Properties	Utility: ⟨*navigating hazardous environment, critical*⟩ ⟨*forest fire containment, critical*⟩ ⟨*emergency road service, recommended*⟩ ⟨*auto assembly plant, essential*⟩ Cost: $5,000 *per hour of rental - not for sale* Availability: *available for shipment to all parts of China* Sustainability: 100% Renewability: *YES* Reuse: *surveillance in forests* *clean atomic waste* Recycling: *YES* Other Resources in the Context of Use: *needs a separate vehicle for transporting to the work site* *needs a truck for storing the debris* *needs an ambulance to place survivors*
Rules	Legal Rules for Supply: *URI to a web site* Context Information: Provider Context: $[LOC : Shanghai]$ Context Rule: $distance(Shanghai, LOC) \leq 100$
Exceptions	Side Effects: *NIL*

- a RDT may be revised, and only latest revision will be published,
- a RDT is to be published only after a satisfactory analysis of its contents by a Trusted Authority who certifies that it meets business standards,
- protection based on role-based access control schemes, as discussed in Section 4.3, are enforced, and
- an XML version of every RDT is to be automatically generated, but not published.

Figure 1(a) is the High Level Class Diagram for creating a RDT. More elements may be included in *rdt.elements* section, whenever RDT structure is enriched. The Java implementation of the GUI has the following features. Every user must login and be authenticated. Based upon the access control policies, only those who are authenticated to create

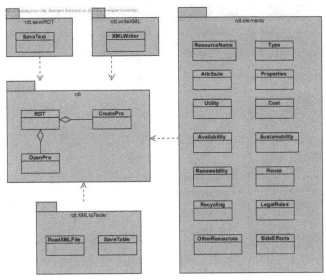

(a) High Level Class Diagram for RDT

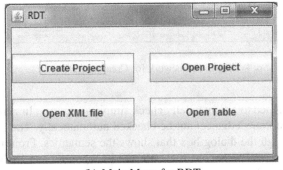

(b) Main Menu for RDT

Fig. 1. RDT Design

a RDT will be permitted to access the menu in Figure 1(b). Invoking the *Open Project* button will require more authentication to verify that the current user has the rights to view and/or edit an existing project. A successful validation at this stage will open the RDT creation framework shown in Figure 2. The user can select an item from those listed on the left side of this table. The GUI will guide the user to input the information corresponding to the selected item in easy steps through a dialog box(DB). Each DB is a *scratchpad* and is associated with a semantic box (SB) which is displayed to the user. The SB suggests the type of information expected to be input in the DB associated to it. This is semantic intelligence, extracted from domain experts. The user enters into the 'edit' mode of a DB after selecting a 'button' that reflects the name of a RDT element. The DBs can be saved, reclaimed for further editing, and saved. In Figure 2 the upper half of the right side of the window is the DB for inputting information that is semantically consistent with the selected item. The lower half of the right side table is the SB.

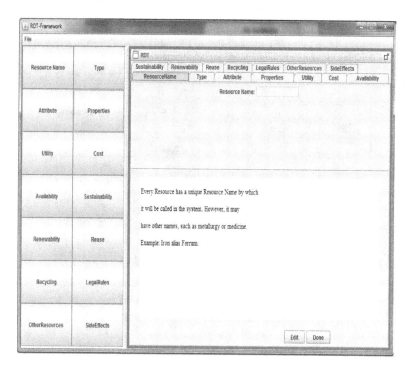

Fig. 2. Framework for Creating RDT

The system is designed to automatically check simple semantics. In general, properties related to resource knowledge cannot be checked automatically. The system interacts with the user through the dialog box that shows the semantics. From the information input by the user the RDT is constructed in the background and its XML version is generated automatically.

Figure 3(a) shows the RDT table created for a sample robot resource. The GUI automatically generates the XML file corresponding to it, and it is shown in Figure 3(b).

The design ensures that a RDT table, once created (and saved), can not be modified. Every time a RDT is to be modified, the *scratchpad* must be invoked through the GUI. A RDT can be viewed subject to access control policies that are enforced. The XML file cannot be viewed by clients, other than system components and those system engineers with supervisory roles. The only way to change a XML file is to change its corresponding RDT table, which in turn requires using the graphical user interface. The development chain 'GUI → RDT Table → XML' ensures the *integrity* of resource information, as it flows from its source (resource owner) to different service producers and service generating processes in the CPS.

6 Conclusion

We briefly compare our modeling methodology and then comment on our ongoing work. One of the earliest works on resource modeling is [13]. It is restricted to

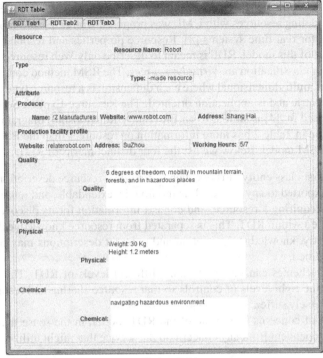

(a) Robot Table

```xml
<?xml version="1.0" encoding="UTF-8"?>
<RDT>
    <!--One RDT XML file-->
    <Resource>
        <ResourceName>Robot</ResourceName>
    </Resource>
    <Type>Man-made resource</Type>
    <Attribute>
        <Producer>
            <ProducerName>XYZ Manufactures</ProducerName>
            <ProducerWebsite>www.robot.com</ProducerWebsite>
            <ProducerAddress>Shang Hai</ProducerAddress>
        </Producer>
        <ProductionFacilityProfile>
            <WebsiteLinksToManufacturing>www.relaterobot.com</WebsiteLinksToManufacturing>
            <Address>SuZhou</Address>
            <WorkingHours>5/7</WorkingHours>
        </ProductionFacilityProfile>
        <Quality>6 degrees of freedom, mobility in mountain terrain, forests, and in hazardous
            places</Quality>
    </Attribute>
    <Properties>
        <Physical>Weight: 30 Kg Height: 1.2 meters Arms: 2 Vision: fitted with camera for
            night vision Payload: can lift and move 50 Kgs load at a time</Physical>
        <Chemical/>
    </Properties>
    <Utility>
        <ApplicationType>navigating hazardous environment</ApplicationType>
        <FactorType>critical</FactorType>
    </Utility>
    <Cost>
        <Amount>5000</Amount>
        <Currency>USD</Currency>
        <Unit>hour</Unit>
    </Cost>
    <Availability>
        <ExclusionList>Null</ExclusionList>
        <ConstraintList>available for shipment to all parts of China</ConstraintList>
    </Availability>
    <Sustainability>100%</Sustainability>
    <Renewability>YES</Renewability>
    <Reuse>survilence in forests clean atomic waste</Reuse>
    <Recycling>YES</Recycling>
    <LegalRules>www.robot.rules.html</LegalRules>
    <OtherResources>needs a separate vehicle for transporting to the work site needs a truck
        for storing the debris needs an ambulance to place survivors</OtherResources>
    <SideEffects>NIL</SideEffects>
</RDT>
```

(b) XML File for Robot Table

Fig. 3. Robot Example

workflow applications. For large distributed systems such as CPS this method is not applicable. Modeling resources with UML was proposed with respect to modeling run-time resources for real-time systems [8]. Resource properties and resource dependencies are not part of this model. RDF is meant to describe only Web resources [9], which according to our classification are *Virtual resources*. The RSM method considers the resource space as multi-dimensional where each dimension is a resource type [3]. RSM is not resource-centric and is application oriented. The Resource-Explicit Service Model (RESM) proposed in [4] is similar to an ER (Entity Relationship) diagram. The robot RDT description in Table 2 has more information on resources than the ER diagram of robot in the RESM model. Below we list the four distinct merits of RDT.

– RDT is a first class entity. It is encodable in a typed resource description language. It can be exported to any service. It is modifiable, extendable, and reusable.
– Rules for acquiring a resource, and context information for its discovery and use are published within RDT. This is separated from resource knowledge description. Consequently, knowledge description and contract descriptions may be independently modified.
– Protection schemes can be set at many different levels of RDT. The RP should determine the right levels of controls so that resource sharing and resource utility do not adversely suffer.
– Separation of concerns is a virtue of the RDT model, in the sense that a RDT is created independent from any concern on the service that might utilize it.

We are currently designing and implementing the next two tiers of the GUI. The precise syntax and semantics given in [12] will be followed in designing the tool at these tiers. Once we complete the tool, the GUI can be used to construct Resource Class Specifications (RCS), and Cyber Configured Services (CCS). We are also investigating methods, and tools based on them, for RDT validation, context-dependent resource allocation, and resource-centric service model for CPS.

Acknowledgement. This research is supported by Research Grants from National Natural Science Foundation of China (Project Number 61103029), Natural Science Foundation of Jiangsu Province, China (Project Number BK2011351), Research Development Funding of Xi'an Jiaotong-Liverpool University (Project Number 2010/13), and Natural Sciences and Engineering Research Council, Canada.

References

1. Georgakopolous, D., Papazoglou, M.P.: Service-oriented Computing. The MIT Press (2008)
2. C.S. Group. Cyber-physical systems: Executive summary. Report (2008),
   http://varma.ece.cmu.edu/summit/CPS-Executive-Summary.pdf
3. Zhuge, Y.H., Shi, P.: Resource space model, owl and database: Mapping and integration. ACM Transactions on Internet Technology 8(4) (2008)
4. Yen, I.-L., Huang, J., Bastani, F., Jeng, J.-J.: Toward a smart cyber-physical space: A context-sensitive resource-explicit service model. In: 33rd Annual IEEE International Computer Software and Applications Conference. IEEE Press (2009)

5. Networking, I. T. Research, and D. Program. High-confidence medical devices: Cyber-physical systems for 21st century health care. Technical report, NITRD (2009)
6. NSF. Usa nsf program solicitation, nsf-08-611. Report, NSF (2008)
7. Oracle. Java resources. Web report, Oracle (2003)
8. Selic, B.: A generic framework for modeling resources with uml. IEEE Computer 33(6), 64–69 (2000)
9. W3C. W3c recommendation. Technical report
10. Wan, K.: Lucx: Lucid Enriched with Context. Phd thesis, Concordia University, Montreal, Canada (2006)
11. Wan, K., Alagar, V.: Resource Modeling for Cyber Physical Systems. In: IEEE Proceedings of the 2012 International Conference on Systems and Informatics, Yantai, China (May 2012)
12. Wan, K., Alagar, V.: A Resource-centric Architecture for Service-oriented Cyber Physical System. In: Park, J.J.(J.H.), Arabnia, H.R., Kim, C., Shi, W., Gil, J.-M. (eds.) GPC 2013. LNCS, vol. 7861, pp. 686–693. Springer, Heidelberg (2013)
13. zur Muehlen, M.: Resource modeling in workflow applications. In: Proceedings of Workflow Management Conference

A Semantic Technology Supported Precision Agriculture System: A Case Study for Citrus Fertilizing

Ye Yuan[1], Wei Zeng[1], and Zili Zhang[1,2,*]

[1] Key Laboratory of Intelligent Software and Software Engineering, Southwest University, Chongqing 400715, China
[2] School of Information Technology, Deakin University, VIC 3217, Australia
{yuan7ye,zengiew,zhangzl}@swu.edu.cn

Abstract. Recently Semantic Technology (ST) has been widely applied in many fields such as health care, energy, environmental science, life sciences, national intelligence, and so on. ST can represent the meaning of information explicitly, which helps machines to process and integrate existing information automatically. In this paper, we discuss how to apply ST in precision agriculture. Citrus fertilizing is chosen as an example to show how to build the related semantic knowledge base as well as the citrus disease query system. With the help of ST, the precision agriculture system can play a greater role in increasing yield of citrus and avoiding plant diseases. A general framework of decision-making support system for citrus production is also provided.

Keywords: Decision-Making, The Internet of Things, Semantic Technology, Citrus Production.

1 Introduction

In recent years, researchers focus on the decision-making system for agricultural production, which has become a hot topic [1]. For example, in citrus production, the citrus growers often encounter a series of problems during their planting, such as picking, decision-making, application management, monitoring and controlling plant diseases and insect pests. It is hardly to meet the needs of our times if citrus plantation only depends on farmers' experience.

Recently, in order to solve the above problems, precision agriculture technology based on IoT(The Internet of Things) has been applied widely. Precision agriculture is supported by the information technology according to spatial variation, positioning, timing and quantization. The main idea of precision agriculture is to help farmers spend less money to obtain higher profit. Moreover, precision agriculture can also utilize agricultural resources efficiently and bring environmental benefit. It relies on new technologies like satellite imagery, information

---

* Corresponding author.

M. Wang (Ed.): KSEM 2013, LNAI 8041, pp. 104–111, 2013.

technology, and geophysical tools. Many research organizations are currently dedicating to improve the efficiency of agricultural production. The related researches mainly focus on: information collection on agriculture [2], simulation for crop growth [3], intelligent agricultural decision support system [4], precision agriculture equipment technology and products [5], etc.

The intelligent decision-making support system is the core technology in the precision agriculture. Traditional decision-making system uses a relational database, that has the simple functions of storage and query. Therefore, it is complex and difficult to infer new knowledge based on existing data. Meanwhile, it is not flexible on operations of adding, deleting and modifying. In order to overcome above shortcomings, Semantic Technology (ST) is adopted in the decision-making system. Two major components of ST are semantic database and semantic inference. Compared with relational database, it is flexible for semantic database to execute existing operations, e.g., adding, deleting and modifying. Inference is the other feature of ST, which can be used to interpret hidden knowledge from the displayed definitions and declarations by a processing mechanism [6].

In this paper, we develop a new framework based on ST which can help framers to improve the ability of decision making. What we have done is to build a semantic knowledge base that is applied to citrus knowledge storage and knowledge inference. The knowledge base is settled in cloud platform that is much more flexible and scalable. Farmers can browse our website or use our smart phone app to get recommendation for planting. What's more, suggestions for farmers are quite practical and specific.

This paper is organized as follows: Section 2 gives the architecture of our works and discusses ST. Section 3 shows how to build a semantic database and a citrus disease query system. Section 4 gives some conclusions and future work.

2   IoT Framework and Semantic Technology

2.1   Framework of IoT

The framework of our system is demonstrated in Fig. 1. The aim of the system is to support a decision-making for citrus planting. There are four components in our system: terminals (information collection), information transmission, cloud computing platform (knowledge storage and knowledge inference) and information distribution system. What's more, ST is contained in order to make this system more smarter.

Terminal aims to collect data, such as meteorological data, soil data. We have cooperated with China Mobile and made use of IoT system to perform basic needs. Now, the IoT system has realized some basic needs such as temperature data collection, wind speed collection and real time monitoring. We have tried to process these data and mined some important information to make the decision.

Information transmission contains a private channel and some basic networks. The private channel has been established by China Mobile and used for

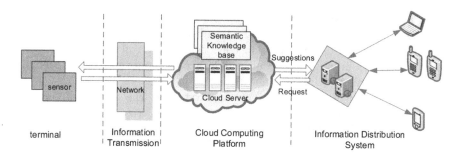

Fig. 1. The architecture of our works

transforming the data collected by terminals. Other basic networks perform some basic tasks such as information communication, and data transformation.

Cloud computing platform is the most important part of this system. A decision-making system is developed on this platform. And ST plays a core role in decision making. Because of cloud computing's stability and extensibility, ST is applied and all data are represented in triples. Based on these knowledge, the system can give farmers the appropriate instructions such as the yield of the past year, fertilization, and irrigation in order to help them make decisions.

Information distribution system is used to release information to users (such as farmers, experts). With the development of Internet, there are so many methods to transfer information, such as web page, cell phone. Thus, users can get information easily, for example, "When to fertilize", "Which type of the disease does citrus get?".

2.2   Semantic Technology

Researchers focus on the development of the next generation of the Web, namely Semantic Web, after Tim Berners put forward the concept and architecture of Semantic Web in 1998 [7]. Meanwhile, ST is the core of our intelligent decision-making system.

Decision-Making System. Comparing with the other applications of IoT, it is novel in agriculture field to build a decision-making system with ST. The key characteristics of the ST-based decision-making system are knowledge driven, individualization, active guidance and information integration. Schematic depiction of the system is shown in Fig. 2.

Some information collected by terminals of IoT are integrated by Sematic Technology, such as soil moisture, temperature, PH value, air temperature, humidity, light, artificial periodic acquisition the citrus growth factor information, and knowledge with existing citrus cultivation, nutrients and fertilizers. By using this knowledge to build subject-verb-object triples, the citrus ontology and semantic database are done. According to the RDF(Resource Description Framework) database and large-scale automatic semantic reasoning (e.g., domain and

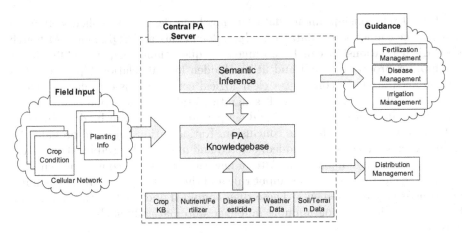

Fig. 2. Schematic depiction of the system

range restriction, subclass and subproperty restriction, transitivity and cardinality restriction), the decision support system for citrus plantation is built. Finally, our system can provide specific and personalized information service for citrus plantation.

Citrus Ontology and Semantic Database. An ontology explicitly describes the relationship between terminologies and concepts. It's the specification of the conceptual model of the objective world. The key issues for citrus production are precision fertilizer, irrigation, drainage drought forecasting, drought-resistant, and orchard maturity forecast of citrus. Therefore, the main work of this paper is to build citrus ontology and generate corresponding RDF semantic database. The RDF semantic database is not data collection simply, but integrates information from many fields. It has some specific features that can facilitate data merging even if the underlying schemas are different. Meanwhile, it also can specifically support the evolution of schemas over time without requiring all the data consumers to be changed. RDF extends the linking structure of the Web to use URIs to denote the relationship between things as well as the two ends of the link (this is usually referred to as a "triple") [8]. As mentioned before, the structure of RDF triples is different from traditional relational database tabular form, it can be added, deleted and modified flexibly. It also has good scalability and provides a strong support for large-scale automatic semantic reasoning, machine learning and new knowledge mining.

Semantic Reasoning. The core technology of semantic intelligent decision-making system for citrus plantation is large scale automated semantic ratiocination. On the basis of citrus ontology and semantic database [9], the semantic reasoning can express ratiocination rules of ontology language and reason the results from a group of relevant semantic data. In traditional relational database, the logical relationship among data has not been described explicitly. It often hides in the connection of database tables, documents and SQL codes. Although the logical relationship can be obtained by query language, it is difficult for a computer program to find and utilize hidden logical relationships and information accurately. In contrast, a clear logical connection is established among knowledgeable individuals in RDF semantic database. The semantic ratiocination on the database is a natural and simple process that can be done by computer to obtain valid inference conclusion. Ratiocination rules are provided by citrus experts , which can be added, deleted and modified by graphical operating interface and software platform. After a new rule is input, a consistency check is required to ensure that the new input rule and the existing rules satisfy the unity and consistency requirements. These rules compiled and used by the system can be converted into a programming language automatically such as Prolog and Jess [10].

3   Semantic Application

The first work for us is to build citrus ontology, especially the expert knowledge extraction for citrus fertilization and the establishment of semantic knowledge database. Similarly, collecting disease caused by less or above nutrition is also the focus for our study.

3.1   Establishment of Semantic Knowledge Base About Citrus Fertilization

First of all, recommendations on fertilization in the citrus production are raised by the experts in citrus research. Because RDF triples can be added, deleted and modified flexibly, these recommendations are made as any forms of chart and text. In order to attain this goal, we use TopBraid Composer [11] to spilt SVO triples based on the knowledge of these experts. TopBraid Composer is an enterprise-class modeling environment for developing Semantic Web ontologies and building semantic applications. Fully compliant with W3C standards, composer offers comprehensive support for developing, managing and testing configurations of knowledge models and their instance knowledge bases.

Figure 3 is infancy semantic mechanism topology diagram, which pick citrus growth infancy as example to establish ontology. According to the word of knowledge, "Bud fat: 10% -15% of the total annual quantity fertilizer from late February to early March", A class named as LifeCycle is established. After that, an instance named as infancy is created under this class. Final, a new property named as yearlyFertilization is established. Specially, infancy is subject, yearlyFertilization is predicate, 10% -15% is the object. Now a complete subject-verb-object has been done. In Fig. 4, fertilization ontology "Nitrogen" is the center of the structure of RDF triples topographies.

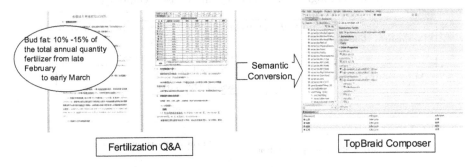

Fertilization Q&A   TopBraid Composer

Fig. 3. Infancy semantic mechanism topology diagram

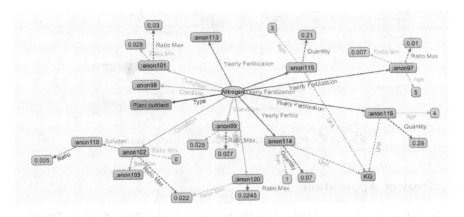

Fig. 4. Nitrogen RDF triples topographies

3.2 Disease Symptom Collecting

It is a very important task for citrus semantic intelligent decision-making system to manage disease ID. Therefore, counting symptoms of nutrient deficiency or excess is our top priority. Nearly 200 kinds of diseases are offered by citrus experts. After classifying and merging similar ones, there are more than 80 kinds of diseases. For example (Table 1), the features of magnesium deficiency for a citrus are yellow patches on both sides of the main vein, small fruits, less production. On the other hand, magnesium excess results in yellow leaves and the edges of the leaves like burning.

Because the symptoms are not classified in the previous semantic base, it is not convenient for farmers to operate the system. After classification, by choosing the symptoms in the client, farmers can obtain the diagnosis whether the citrus is lack of or rich of some certain elements from the decision-making system of the server. Meanwhile, the advices of corresponding solutions are coming. The result is shown in Fig. 5.

Table 1. Magnesium deficiency and excess symptoms

Magnesium deficiency	63 yellow patches on both sides of the main vein
	6 Small fruits
	32 Production decreases
Magnesium excess	70 Yellow leaves
	71 The edges of the leaves like burning

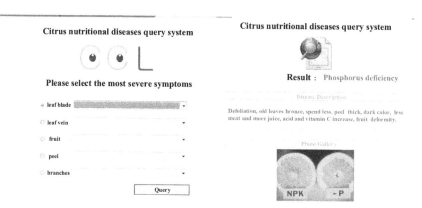

Fig. 5. Citrus nutritional diseases query system

3.3   Project Application

On one hand, the ultimate goal of this project is to start a corresponding business instance process when farmers request service. During the execution process, the system should start semantic intelligence ratiocination to obtain decision support. On the other hand, the system can perform regular and automatic ratiocination according to information obtained by system setting and real-time access. Making use of the existing information, the system can obtain useful opportunities for farmers by ratiocination and predict potential problems in citrus production. Then system would send timely service automatically to farmers when the farmers have not realized the problems such as when to fertilize on citrus in the current climate, soil and environmental conditions, recommending suitable citrus planting varieties and planting methods for farmers in terms of farming conditions [12].

4   Conclusion

This paper analyzes deeply citrus fertilizer application under the background of development of Internet of Things. It focuses on researching decision-making in order to explore ST on Internet of Things. It describes the decision-making system's structure, function and key technologies, and displays the case at the same time. Both aspects of data storage and reasoning new knowledge show that decision-making system based on ST is obviously superior to traditional decision-making system.

In the next step, we will perform a rule-based system to justify the ability of reasoning based on our conceptual framework.

Acknowledgment. This project was supported by the national science and technology support program (2012BAD35B08). The authors would like to thank Dr. Chao Gao for his great help in this paper.

References

1. Yang, Y., Wang, F., Zhao, J.: Intelligent fertilization decision support system based on knowledge model and webgis: Decision for fertilization. In: 2nd IEEE International Conference on Computer Science and Information Technology, ICCSIT 2009, pp. 232–235. IEEE (2009)
2. Hummel, J.W., Sudduth, K.A., Hollinger, S.E.: Soil moisture and organic matter prediction of surface and subsurface soils using an nir soil sensor. Computers and Electronics in Agriculture 32(2), 149–165 (2001)
3. Patow, G.: User-friendly graph editing for procedural modeling of buildings. IEEE Computer Graphics and Applications 32(2), 66–75 (2012)
4. Wang, Y., Jiang, P., Luo, Y.: Study on intelligent decision-making platform in the agricultural production. In: 2011 International Conference on Intelligent Computation Technology and Automation (ICICTA), vol. 2, pp. 224–227. IEEE (2011)
5. Zhou, L., Song, L., Xie, C., Zhang, J.: Applications of internet of things in the facility agriculture. In: Li, D., Chen, Y. (eds.) Computer and Computing Technologies in Agriculture VI, Part I. IFIP AICT, vol. 392, pp. 297–303. Springer, Heidelberg (2013)
6. Blanco, R., Halpin, H., Herzig, D.M., Mika, P., Pound, J., Thompson, H.S., Tran Duc, T.: Repeatable and reliable search system evaluation using crowdsourcing. In: Proceedings of the 34th International ACM SIGIR Conference on Research and Development in Information Retrieval, pp. 923–932. ACM (2011)
7. Meenachi, N.: Web ontology language editors for semantic web- a survey. International Journal of Computer Applications 53(12), 12–16 (2012)
8. Goasdoué, F., Manolescu, I., Roatiş, A.: Getting more rdf support from relational databases. In: Proceedings of the 21st International Conference Companion on World Wide Web, pp. 515–516. ACM (2012)
9. Huang, Z., van Harmelen, F.: Using semantic distances for reasoning with inconsistent ontologies. In: Sheth, A.P., Staab, S., Dean, M., Paolucci, M., Maynard, D., Finin, T., Thirunarayan, K. (eds.) ISWC 2008. LNCS, vol. 5318, pp. 178–194. Springer, Heidelberg (2008)
10. Pan, J.Z., Thomas, E.: Approximating owl-dl ontologies. In: Proceedings of the National Conference on Artificial Intelligence, vol. 22, pp. 1434–1439. AAAI Press (2007)
11. Weiten, M.: Ontostudio® as a ontology engineering environment. In: Semantic Knowledge Management, pp. 51–60. Springer (2009)
12. Riehl, S.: Archaeobotanical evidence for the interrelationship of agricultural decision-making and climate change in the ancient near east. Quaternary International 197(1), 93–114 (2009)

Cost-Sensitive Classification with k-Nearest Neighbors

Zhenxing Qin[1], Alan Tao Wang[1], Chengqi Zhang[1], and Shichao Zhang[1,2,*]

[1] The Centre for QCIS, Faculty of Engineering and Information Technology
University of Technology Sydney, Australia
[2] College of CS&IT, Guangxi Normal University, Guilin, 541004, China
alant.wang@gmail.com
{Zhenxing.Qin,Chengqi.Zhang,Shichao.Zhang}@uts.edu.au

Abstract. Cost-sensitive learning algorithms are typically motivated by imbalance data in clinical diagnosis that contains skewed class distribution. While other popular classification methods have been improved against imbalance data, it is only unsolved to extend k-Nearest Neighbors (kNN) classification, one of top-10 datamining algorithms, to make it cost-sensitive to imbalance data. To fill in this gap, in this paper we study two simple yet effective cost-sensitive kNN classification approaches, called Direct-CS-kNN and Distance-CS-kNN. In addition, we utilize several strategies (i.e., smoothing, minimum-cost k value selection, feature selection and ensemble selection) to improve the performance of Direct-CS-kNN and Distance-CS-kNN. We conduct several groups of experiments to evaluate the efficiency with UCI datasets, and demonstrate that the proposed cost-sensitive kNN classification algorithms can significantly reduce misclassification cost, often by a large margin, as well as consistently outperform CS-4.5 with/without additional enhancements.

1 Introduction

Classification aims at generating a classifier which minimizes classification errors, which is one of the main research topics in machine learning and data mining. Therefore, there are great many classification algorithms developed, typical examples include Decision Tree, Naïve Bayes, Instance based learning (e.g. kNN) and SVM. Twenty years ago, motivated by imbalance data in clinical diagnosis that contains skewed class distribution, these popular classification algorithms had been extended to deal with the classification tasks with non-uniform cost, called cost-sensitive classification, and attracted vast interest of data mining researchers [1, 2, 8, 16, 26]. These approaches incorporate misclassification cost and other costs in the classification technique, thus provide practical classification result in a multiple-cost environment. However, it is only unsolved to extend k-Nearest Neighbors (kNN) classification, one of top-10 datamining algorithms, to make it cost-sensitive to imbalance data.

---

* Corresponding author.

M. Wang (Ed.): KSEM 2013, LNAI 8041, pp. 112–131, 2013.
© Springer-Verlag Berlin Heidelberg 2013

KNN classification is one of the most popular and widely used instance-based learning algorithms. Different from model-based classification algorithms (i.e. training models from a given dataset and then predicting test examples with the models), it needs to store the training data in memory in order to find the "nearest neighbours" to answer a given query. Despite of the popularity of KNN, very little work's been reported on KNN in the area cost-sensitive learning. Hence some challenges are must out there to stop us. However, regarding to the potential of KNN, it is still worth a good try to clarify the challenges and give some solutions.

To make KNN cost-sensitive, in this paper, two simple but effective approaches, Direct-CS-KNN and Distance-CS-KNN, are proposed to minimise the misclassification cost in cost sensitive learning. In order to clarify challenges and difficulties encountered in our new algorithms, completed performance studies are conducted and compared with the benchmark method cost-sensitive C4.5, by bunches of experiments with various cost settings and multiple typical datasets.

Furthermore, we also propose several additional enhancement methods - smoothing, minimum-cost K value selection, feature selection and ensemble selection, to further improve the performance of our new cost-sensitive KNN algorithms. The smoothing improvement is also applied to cost-sensitive C4.5 for performance study.

Our experiment results show that the proposed new cost-sensitive KNN algorithms can effectively reduce misclassification cost, often by a large margin. And they consistently outperform CS-4.5 (cost-sensitive C4.5) on the selected UCI data sets in case of with or without enhancements.

The rest of paper is organized as follows: Section 2 provides a brief review of the cost-sensitive learning, KNN classification Section 3 describes our two new cost-sensitive KNN algorithms and several improvements which can effectively reduce misclassification cost in KNN classification. Section 3 describes some additional enhancement methods: feature selection, direct cost-sensitive learning/probability calibration and ensemble selection methods. Experimental results for six UCI data sets are presented in Section 5. Finally, in Section 6 we conclude the work with a discussion of future improvements.

2   Related Work

2.1   Cost-Sensitive Learning

Cost-sensitive learning is an extension of traditional non-cost-sensitive data mining. It is an important research area with many real world applications. For example, in medical diagnosis domain, diseases are not only very expensive but also rare; for a bank, an error of approving a home loan to a bad customer is more costly than an error of rejecting a home loan to a good customer. Traditional data mining methods aimed at minimizing error rate will perform poorly in these areas, as they assume equal misclassification cost and relatively balanced class distributions. Given a naturally much skewed class distribution and costly faulty predictions for the rare class, an error based classifier may very likely ends up building a useless model.

Cost-sensitive learning is an advanced form of data mining that satisfies these special needs. Research in cost-sensitive learning is still in an early stage and there are different methods to it. Most cost-sensitive learning methods are developed based on the existing non-cost-sensitive data mining methods. To make an error-based classifier cost-sensitive, a common method is to introduce biases into an error based classification system in three ways: 1) by changing the class distribution of the training data, 2) by modifying the learning algorithms, 3) and by taking the boosting approach (Li et al. 2005). An alternative method is called direct cost-sensitive learning which uses the conditional probability estimates provided by error based classifiers to directly compute the optimal class label for each test example using cost function [2].

Most of the cost-sensitive learning methods assume that for an M-class problem, an M by M cost matrix C is available at learning time, and does not change during the learning or decision making process. So it is static. The value of C(i, j) is the cost involved when a test case is predicted to be class i but actually it belongs to class j.

A static cost matrix always has the following structure when there are only two classes:

Table 1. Two-Class Cost Matrix

	Actual negative	**Actual positive**
Predict negative	$C(0, 0) = C_{00}$	$C(0, 1) = C_{01}$
Predict positive	$C(1, 0) = C_{10}$	$C(1, 1) = C_{11}$

In above cost matrix, the cost of a false positive is C_{10} while the cost of a false negative is C_{01}. Conceptually, the cost of labeling an example incorrectly should always be greater than the cost of labeling it correctly. Mathematically, it should always be the case that $C_{10} > C_{00}$ and $C_{01} > C_{11}$ [2].

As per Elkan, if a cost matrix C is known in advance, let the (i, j) entry in C be the cost of predicting class i when the true class is j. If $i = j$ then the prediction is correct, while if $i \neq j$ the prediction is incorrect. The optimal prediction for an example x is the class i that minimizes:

$$L(x,i) = \sum_{j=1}^{n} p(j \mid x) C(i, j) \qquad (1)$$

The most popular base algorithms used in cost-sensitive learning include Decision Tree, Naïve Bayes and SVM.

2.2   KNN Classification

KNN classification is an instance based learning algorithm which stores the whole training data in memory to compute the most relevant data to answer a given query. The answer to the query is the class represented by a majority of the K nearest neighbours. There are three key elements of this approach: a set of labelled examples, a distance function for computing the distance between examples, and the value of K - the number of nearest neighbours. To classify an unlabelled example, the distance of this example to the labelled examples is calculated, its K-nearest neighbours are identified, and the class labels of these nearest neighbours are then used to classify the unlabelled example [22].

The choice of the distance function is an important consideration in KNN. Although there are other options, most instance based learners use Euclidean distance which is defined as below:

$$Dist(X,Y) = \sqrt{\sum_{i=1}^{D} (Xi - Yi)^2}$$

Where X and Y are the two examples in data set, and Xi and Yi (i = 1 .. D) are their attributes.

Once the nearest-neighbour list is selected, the test example can be classified based on the following two voting methods:

1. Majority voting: $y' = \arg\max_{v} \sum_{(xi,\,yi) \in Dz} \delta(v, yi)$

2. Distance-Weighted Voting: $y' = \arg\max_{v} \sum_{(xi,\,yi) \in Dz} w_i \delta(v, yi)$,

    Where $w_i = 1/d(x', xi)^2$

There are several key issues that affect the performance of KNN. One is the choice of K. If K is too small, then the result could be sensitive to noisy data. If K is too large, then the selected neighbours might include too many examples from other classes. Another issue is how to determine the class labels, the simplest method is to take a majority vote (method 1), but this could be an issue if the nearest neighbours vary widely in their distance and the closer neighbours more reliably indicate the class of the test example. The other issue of method 1 is that it is hard to deal with cost-sensitive learning and imbalanced data sets. A more sophisticated approach, which is less sensitive to the choice of K, weights each example's vote by its distance (method 2), where the weight factor is often taken to be the reciprocal of the squared distance ($w_i = 1/d(x', xi)^2$).

KNN classifiers are lazy learners, unlike eager learners (e.g. Decision Tree and SVM), KNN models are not built explicitly. Thus, building the model is cheap, but classifying unknown data is relatively expensive since it requires the computation of the K-nearest neighbours of the examples to be labelled. This, in general, requires computing the distance of the test examples to all the examples in the labelled set, which can be expensive particularly for large training sets.

2.3   Direct Cost-Sensitive Classification

Any learned classifier that can provide conditional probability estimates for training examples can also provide conditional probability estimates for test examples. Using these probability estimates we can directly compute the optimal class label for each test example using the cost matrix. This cost-sensitive learning method is called direct cost-sensitive learning [26].

All direct cost-sensitive learning algorithms have one thing in common: They do not manipulate the internal behavior of the classifier nor do they manipulate the training data in any way. They are based on the optimal cost-sensitive decision criterion that directly use the output produced by the classifiers in order to make an optimal cost-sensitive prediction.

3   Making KNN Cost-Sensitive - The Proposed Approach

In this paper, we focus on binary classification problems. We propose two approaches to make KNN classifier sensitive to misclassification cost, and several additional methods to further improve the cost-sensitive KNN classifier performance.

3.1   Direct Cost-Sensitive KNN

Direct Cost-sensitive approach is quite simple and intuitive which has been studied with decision tree C4.5. In this paper, we use KNN algorithm to train a traditional non-cost-sensitive classifier as the baseline. After the K nearest neighbors is selected, we calculate class probability estimate using below formula:

$$\Pr(i \mid x) = \frac{Ki}{K}$$

Where Ki is the number of selected neighbors whose class label is i. Using the above probability estimate and Eq. 1, we can directly compute the optimal class label for each test example. We call this approach DirectCS-KNN.

In traditional cost-blind KNN classification, the K value is either fixed or selected using cross validation. When the K value is fixed, most of times it is quite small, such as 3, 5, 7 etc. the KNN classifier performance is not impacted a lot by the variation of the K value. However, the aim of our DirectCS-KNN is minimizing misclassification cost, the probability estimate (not the error rate) generated by the original KNN classifier is more important. In this case, the selection of an appropriate K value plays a critical role in terms of building a statistically stable cost-sensitive KNN classifier which can produce better probability estimate and reduce misclassification cost.

In this paper, we will test the following three ways of selecting the K value:

- Fixed value
- Cross validation
- Choose the *K* value which minimizes the misclassification cost in training set

Although our DirectCS-KNN approach is straightforward and easy to implement, it has the following shortcomings:

- If the K value selected is too small, the probability estimation provided by the KNN (Ki/K) is statistically unstable, it will cause data over-fitting and increase the misclassification cost to the test examples
- In many real world data sets, noise is often expected in the training examples, and KNN algorithm is particularly sensitive to the noisy data, therefore generate very poor probability estimate

As we described in Section 2, several methods have been proposed for obtaining better probability estimate from traditional cost-blind classifiers in direct cost-sensitive learning. Zadrozny and Elkan [2] proposed to use an un-pruned decision tree and transform the scores of the leaves by smoothing them. They call this method m-estimation. The similar method can be applied to our DirectCS-KNN approach. In order to make DirectCS-KNN more stable and further reduce misclassification cost, in this paper, we propose two changes to the original m-estimation: First we use cross valuation to determine the m value for different data sets. It is more proper than a fixed value. Second, we use the smoothed probability estimate (together with cost matrix) at the step of determining the K value.

Now let's look at the m-estimation formula we specified in section 2.2 again. In order to explain the impact of m-estimation to the performance of DirectCS-KNN approach, we represent the formula in a slightly different way:

$$\Pr(i \mid x) = \left(\frac{K}{K+m}\right) \times \left(\frac{Ki}{K}\right) + \left(\frac{m}{K+m}\right) \times b$$

As we can see from this formula, the value m controls the balance between relative frequency and prior probability. It has the following impacts:

- To the noisy data, with m-estimation, m can be set higher so that the noisy value for Ki/K plays less important role in the final estimation and the impact of noisy data is reduced.
- When the K value is small, without smoothing, the probability estimation provided by the selected neighbors (Ki/K) is statistically unstable. However, with m-estimation, $K/(K+m)$ closes to 0 and $m/(K+m)$ closes to 1, so that the probability estimation is shifted towards the base rate (b). It works particularly well on data sets with skew class distribution.

In the experiment section, we apply this smoothing method to the Direct-CS-KNN and Distance-CS-KNN (specified in section 3.2), and compare this approach to the other proposed variations of Cost-sensitive KNN algorithms.

3.2   KNN with Cost-Sensitive Distance Function

The second approach involves modifying the distance function of the KNN algorithm. Let's review the distance-weighted voting function in section 2.2:

$$y' = \arg\max_{v} \sum_{(xi, yi) \in Dz} w_i \delta(v, yi)$$

Where $w_i = 1/d(x', xi)^2$

Based on the second formula, in a binary decision case, we assume that the distance-weight of a test example to a positive training example is Wp, and the distance-weight of the same test example to a negative training example is Wn.

When misclassification cost is not considered, the training examples with the highest W values will be selected as the nearest neighbors regardless of their class labels. However, in a cost-sensitive situation, the cost of false positive (FP) and the cost of false negative (FN) might be very different. Selecting a nearest neighbor with different class labels incurs different potential cost. To simplify the case, we assume that the cost of true positive (TP) and true negative (TN) are both 0. In cost-sensitive learning, the purpose is to minimize the misclassification cost instead of errors. Now we calculate the potential cost (C_p) of selecting a positive nearest neighbor is $FP *$ Wn. And the potential cost (C_n) of selecting a negative nearest neighbor is $FN * Wp$. The training examples with the lowest potential cost should be selected as the nearest neighbors.

We replace the distance-weighted voting function in the original KNN algorithm with this new cost-sensitive approach. The last step is to use the modified classifier to predict class labels for all test examples. Be aware that for each test example, after all its nearest neighbors are selected, the above cost-sensitive distance-weighted voting approach and the cost matrix will still be used to calculate the class label which minimizes the misclassification cost. The detail of the algorithm is described below:

For each test example, the total distance-weight of all positive neighbors is Wpa, and the total distance-weight of all negative neighbors is Wna, where:

$Wpa = W1 + W2 + W3 + \ldots\ldots + Wk$ (k is the number of positive neighbors)
$Wna = W1 + W2 + W3 + \ldots\ldots + Wj$ (j is the number of negative neighbors)

In this case we can define the probability of labeling an unlabeled example to P is $P_p = Wpa/(Wpa+Wna)$.And the probability of labeling an unlabeled example to N is $P_n = Wna/(Wpa+Wna)$.

Where $P_p + P_n = 1$. Now we calculate the potential cost (C_p) of labeling this test example to P is $FP * P_n$. And the potential cost (C_n) of labeling this test example to N is $FN * P_p$.

If $C_p > C_n$, the unlabeled example is classified as N, and the probability of this prediction is $C_p/(C_p + C_n)$. Otherwise, the unlabeled example is classified as P, and the probability of this prediction is $C_n/(C_p + C_n)$. We call this approach **Distance-CS-KNN**.

3.3   Potential and Challenges

We conduct performance for **DirectCS-KNN** and the **CS-C4.5** on different datasets. In most of datasets, the best result of **DirectCS-KNN** outperforms the **CS-C4.5** on both of the minimal cost and AUC measurements. A typical result is shown in Table 4 of section 5.2 (marked with gray background in the table). As we predicted before, KNN does has the great potential to cost-sensitive learning.

However, we do find some challenges when apply KNN into cost-sensitive learning environment. Firstly, the choice of K, which is also an open question in KNN study. Secondly, the large working loads to tuning the K while we change the cost ratio and datasets.

Good news is there are lot of work reported to tackle the challenges in general KNN. On the other hand, we introduce some additional enhancements on DirectCS-KNN which make us easier to search a good enough K and reduce the tuning work-load. The enhancements are described in next section.

4   Additional Enhancements

4.1   Calibration Methods for Improving Probability Estimates in Cost-Sensitive Learning

Any learned classifier that can provide conditional probability estimates for training examples can also provide conditional probability estimates for test examples. If the learned model does not explicitly compute these values, most classifiers can be modified to output some value that reflects the internal class probability estimate [9]. Using these probability estimates we can directly compute the optimal class label for each test example using the cost matrix.

However, it is well known that the probability estimates provided by classifiers from many error based learners are neither unbiased, nor well calibrated. Decision tree (C4.5) and instance based learners (KNN) are well known examples. Two smoothing methods, Laplace Correction and M-Smoothing, have been proposed for obtaining better probability estimates from decision tree classifiers [2, 13].

- **Laplace Correction.** Using C4.5, [13] evaluated different pruning methods and recommended not pruning the tree, instead, using Laplace correction to calculate class probabilities at leaves. The Laplace correction method basically corrects the probabilities by shifting them towards 0.5, in a two-class problem.

- **M-Smoothing.** [26] proposed to use an un-pruned decision tree and transform the scores of the leaves by smoothing them. They point out that the Laplace correction method doesn't work well for datasets with a skewed class distribution. They suggest using a smoothing method called m-estimation. According to that method, the class probabilities are calculated as follow.

$$\Pr(i \mid x) = \frac{N_i + b*m}{N+m}$$

Where b is the base rate of the class distribution and m is a parameter that controls the impact of this correction. The base rate is the relative frequency of the minority class. They recommended choosing the constant m so that b×m = 10. The experiments conducted by Zadrozny and Elkan [24] show that using C4.5 decision tree as the base learner, the direct cost-sensitive learning approach with m-estimation achieved less misclassification cost than that of MetaCost [1] on the KDD-98 data set. They call this method m-smoothing. For example, if a leaf contains three training examples, one is positive and the other two are negative, the raw C4.5 decision tree score of any test example assigned to this leaf is 0.33. The smoothed score with m = 200 and b = 0.05 (the base rate of KDD-98 data set) is:

$$P' = (1 + 0.05 \times 200) / (3 + 200) = 11 / 203 = 0.0542$$

Therefore, the smoothed score is effectively shifted towards the base rate of KDD-98 data set.

Furthermore, Niculescu-Mizil and Caruana (2005) experimented two other ways of correcting the poor probability estimates predicted by decision tree, SVM and other error based classifiers: Platt Scaling [12] and Isotonic Regression. These methods can also be used in directly cost-sensitive learning algorithms.

- **Platt Scaling.** Platt (1999) proposed a calibration method for transforming SVM predictions to posterior probabilities by passing them through a sigmoid. A sigmoid transformation is also used for boosted decision trees and other classifiers. Let the output of a learning method be $f(x)$. To get calibrated probabilities, pass the output through a sigmoid:

$$P (y = 1 \mid f) = 1 / (1 + exp (Af + B))$$

Where the parameters A and B are fitted using maximum likelihood estimation from a fitting training set $(f_i ; y_i)$. Gradient descent is used to find A and B. Platt Scaling is most effective when the distortion in the predicted probabilities is sigmoid-shaped [10].

- **Isotonic Regression.** Compared to Platt Calibration, Isotonic Regression is a more powerful calibration method which can correct any monotonic distortion [10 24, 25] successfully use this method to calibrate probability estimates from SVM, Naive Bayes and decision tree classifiers. Isotonic Regression is more general than other calibration methods we discussed above. The only restriction is that the mapping function must be isotonic (monotonically increasing). This means that given the predictions f_i from a model and the true targets y_i, the basic assumption in Isotonic Regression is that:

$$y_i = m (f_i) + \epsilon_i$$

Where m is an isotonic function. Then given a train set (f_i, y_i), the Isotonic Regression problem is to find the isotonic function m' that:

$$m' = \operatorname{argmin} z \sum (yi - z(fi))^2$$

4.2   Feature Selection

Feature selection (also called attribute selection) is an important research area of data mining and machine learning. It is a kind of data pre-processing strategy. Many classification algorithms such as nearest neighbor and decision tree can be benefited from an effective feature selection process. The reason is in practice, the real world data sets often contain noisy data and irrelevant/distracting/correlated attributes which often "confuse" classification algorithms, and results in data over-fitting and poor predication accuracy on unseen data.

Many feature selection methods were developed over the years in practical data mining and machine learning research, such in Statistics and Pattern Recognition. The two commonly used approaches are the filter approach and the wrapper approach. The filter approach selects features using a pre-processing step which is independent of the induction algorithm. The main disadvantage of this approach is that it totally ignores the effects of the selected feature subset on the performance of the induction algorithm. On the contrary, in the wrapper approach, the feature subset selection algorithm conducts a search for a good subset using the induction algorithm itself as a part of the evaluation function. The accuracy of the induced classifiers is estimated using accuracy estimation techniques [4].

Most previous feature selection research focuses on improving predication accuracy. To the best of our knowledge, the impact of using feature selection to improve cost-sensitive classifier performance is not well studied. We believe, due to the fact that if properly designed, the feature selection approach can effectively remove the noisy data and irrelevant/distracting/correlated attributes from training set so that our cost-sensitive KNN algorithms can find "better" neighbors with the most relevant attributes which minimizes misclassification cost.

4.3   Ensemble Method

Ensemble data mining method, also known as Committee Method or Model Combiner, is a data mining method that leverages the power of multiple models to achieve better prediction accuracy than any of the individual models could on their own. The algorithm works as below:

Firstly, base models are built using many different data mining algorithms.

Then a construction strategy such as forward stepwise selection, guided by some scoring function, extracts a well performing subset of all models. The simple forward model selection works as follows:

1. Start with an empty ensemble;
2. Add to the ensemble the model in the library that maximizes the ensemble's performance to the error (or cost) metric on a hill-climb set;
3. Repeat Step 2 until all models have been examined;
4. Return that subset of models that yields maximum performance on the hill-climb set.

Ensemble learning methods generate multiple models. Given a new example, the ensemble passes it to each of its multiple base models, obtains their predictions, and then combines them in some appropriate manner (e.g., averaging or voting). Usually, compared with individual classifiers, ensemble methods are more accurate and stable. Some of the most popular ensemble learning algorithms are Bagging, Boosting and Stacking.

Evaluating the prediction of an ensemble typically requires more computation than evaluating the prediction of a single model, so ensembles may be thought of as a way to compensate for poor learning algorithms by performing a lot of extra computation.

4.4   KNN with Cost-Sensitive Feature Selection

Real world data sets usually contain noisy data and irrelevant/disturbing features. One of the shortcomings of instance based learning algorithm such as KNN is that they are quite sensitive to noisy data and irrelevant/disturbing features, especially when the training data set is small. This issue can cause poor classification performance on unseen data. Feature selection strategy can certainly help in this situation.

As I mentioned in Section 4.2, there are many different feature selection methods developed. They fall into two categories: the filter approach and the wrapper approach. As per the study of [4], the wrapper approach generally perform better than the filter approach, and some significant improvement in accuracy was achieved on some data sets for Decision Tree algorithm and Naïve Bayes algorithm using the wrapper approach.

The wrapper approach proposed by Kohavi [4] conducts a search in the space of possible parameters. Their search requires a state space, an initial state, a termination condition and a search engine. The goal of the search is to find the state with the highest evaluation, using a heuristic function to guide it. They use prediction accuracy estimation as both the heuristic function and the evaluation function. They've compared two search engines, hill-climbing and best-first, and found that the best-first search engine is more robust, and generally performs better, both in accuracy and in comprehensibility as measured by the number of features selected.

In this paper, we apply this wrapper approach to both of our Direct-CS-KNN and Distance-CS-KNN algorithm to improve classifier performance. In cost-sensitive learning, the ultimate goal is to minimize the misclassification cost. We cannot simply apply Kohavi and John's wrapper approach directly to our cost-sensitive KNN algorithms. Therefore, we propose a variation of Kohavi and John's feature selection wrapper. The main difference is using misclassification cost instead of error rate as both the heuristic function and the evaluation function.

The other settings in our experiment are similar: The search space we chose is that each state represents a feature subset. So there are n bits in each state for a data set with n features. Each bit indicates whether a feature is selected (1) or not (0). We always start with an empty set of features. The main reason for this setup is computational. It is much faster to find nearest neighbours using only a few features in a data set. We chose the best-first search algorithm as our search engine. The following summary shows the setup of our cost-sensitive feature selection problem for a simple data set with three features:

State:	A Boolean vector, one bit per feature
Initial state:	A empty set of features (0,0,0)
Search space:	(0,0,0) (0,1,0) (1,0,0) (0,0,1)
	(1,1,0) (0,1,1) (1,0,1) (1,1,1)
Search engine:	Best first
Evaluation function:	Misclassification Cost

4.5   KNN with Cost-Sensitive Stacking

KNN classifier is very popular in many real world applications. The main reason is that the idea is straightforward and easy to implement. It is simple to define a dissimilarity measure on the set of observations. However, handling the parameter K could be tricky and difficult, especially in cost-sensitive learning. In this paper, we propose an ensemble based method, more specific, **Cost-sensitive Stacking**, to handle the parameter K.

As we mentioned in section 2.5, ensemble selection is a well-developed and very popular meta learning algorithm. It tends to produce a better result when there is a significant diversity among the classification models and parameters. Stacking is an ensemble technique whose purpose is to achieve a generalization accuracy (as opposed to learning accuracy) , and make it as high as possible. The central idea is that one can do better than simply list all predictions as to the parent functions which are consistent with a learning set. One can also use in-sample/out-of-sample techniques to find a best guesser of parent functions. There are many different ways to implement stacking. Its primary implementation is as a technique for combining generaliser, but it can also be used when one has only a single generaliser, as a technique to improve that single generaliser [21].

Our proposed Cost-sensitive Stacking algorithm works as below: Adding multiple cost-sensitive KNN classifiers with different K values to the stack, and learning a classification model on each; estimating class probability for each example by the fraction of votes that it receives from the ensemble; using Equation 1 to re-label each training example with the estimated optimal class; and reapplying the classifier to the relabeled training set. The idea is similar to bagging approach used in MetaCost [1].

5   Experimental Evaluation

5.1   Experiment Setup

The main purpose of this experiment is to evaluate the performance of the proposed cost-sensitive KNN classification algorithms with feature selection and stacking, by comparing their misclassification cost and other key performance measurements such as AUC across different cost ratios (FN/FP) against another popular classification algorithm, C4.5 with Minimum Expected Cost and its enhanced version with smoothing. All these algorithms are implemented in Weka (Witten and Frank 2000), and they are listed in Table 2 below.

Table 2. List of Cost-sensitive Algorithms and Abbreviations

#	Method	Abbreviation	Base Classifier
1	Direct Cost-sensitive KNN	DirectCS-KNN	KNN
2	Direct Cost-sensitive KNN with Smoothing	DirectCS-KNN-SM	KNN
3	Direct Cost-sensitive KNN with K value selection	DirectCS-KNN-CSK	KNN
4	Distance Cost-sensitive KNN	DistanceCS-KNN	KNN
5	Distance Cost-sensitive KNN with Feature Selection	DistanceCS-KNN-FS	KNN
6	Distance Cost-sensitive KNN with Stacking	DistanceCS-KNN-STK	KNN
7	C4.5 with Minimum Expected Cost	CS-C4.5	C4.5
8	C4.5 with Minimum Expected Cost and Smoothing	CS-C4.5-SM	C4.5

Please note that we will be conducting three experiments using the above algorithms and six data sets chosen from UCI repository. The details of these data sets are listed in Table 3.

Table 3. Summary of the Data set Characteristics

Dataset	No. of attributes	No. of examples	Class distribution (P/N)
Statlog (heart)	14	270	120/150
Credit-g	21	1000	300/700
Diabetes	9	768	268/500
Page-blocks	11	5473	560/4913
Spambase	58	4601	1813/2788
Waveform-	41	3347	1655/1692

These data sets are chosen based on the following criteria:

- Data sets should be two-class because the cost-sensitive KNN classification algorithms we are evaluating currently can only handle two-class data sets. This condition is hard to satisfy and we resorted to converting several multi-class data sets into two-class data sets by choosing the least prevalent class as the positive class and union all other classes as the negative class. For two-class data sets, we always assign the minority class as the positive class and the majority class as the negative class.

- This experiment does not focus on the data sets with many missing values, so all the data sets we selected do not have many missing values. If any examples have missing values, we either remove them from the data sets or replace them using Weka's "ReplacingMissingValues" filter.

- These data sets include both balanced and unbalanced class distributions. The imbalance level (the ratio of major class size to minor class size) in these data sets varies from 1.02 (Waveform-5000) to 8.8 (Page-blocks).

We conduct three experiments on above datasets with different cost matrixes:

In the first experiment, we use the UCI Statlog(heart) data set. We evaluate the classifier performance by calculating the misclassification cost and AUC generated by the different variations of the Direct Cost-sensitive KNN algorithm and CS-C4.5. Below is a brief description of the Statlog(heart) data set:

- The Statlog(heart) data set is one of a few data sets in UCI library with recommended cost matrix. The cost matrix is normalized and the cost ratio (FN/FP) is set to 5. The cost of TP and TN are both set to 0. This data set has been used extensively in cost-sensitive learning research previously.

In the second experiment, we still use the Statlog(heart) data set to conduct the test. We evaluate the performance of our two new cost-sensitive algorithms, DirectCS-KNN and DistanceCS-KNN by calculating their misclassification cost and AUC.

In the third experiment, five UCI data sets are used to perform the test. The misclassification cost FP is always set to 1, and FN is set to an integer varying from 2 to 20 (2, 5, 10, 20 respectively). We assume that the misclassification of the minority class always incurs a higher cost. This is to simulate real-world scenarios in which the less frequent class is the more important class. The cost of TP and TN are both set to 0. We evaluate our DistanceCS-KNN classifier (and its variations) performance against CS-C4.5 by comparing their average misclassification cost.

All of the three experiments are repeated for 10 times and ten-folder cross validation method is used in all tests to prevent over-fitting data.

5.2   Experiment Results and Discussion

In this section, we present an experimental comparison of the cost-sensitive KNN and other competing algorithms presented in the previous section. For easy reading and discussing, the results without enhancements are marked with gray background on each table.

The first experiment aims to show the performance of enhancements. Results are listed in Table 4. It lists the key performance measurements such as average misclassification cost and AUC for the Statlog(heart) data set. The experiment shows that on this popular UCI data set and with the recommended misclassification cost, we can achieve better performance on DirectCS-KNN algorithm through smoothing (DirectCS-KNN-SM) and K-value selection with minimum-cost (DirectCS-KNN-CSK) approach. In this experiment, we chose a fixed K value (K=5) for both DirectCS-KNN and DirectCS-KNN-SM, and automatically select K value with minimum-cost (on training data) for DirectCS-KNN-CSK.

Table 4. Key performance measurements on Statlog(heart)

Table 4.1. Average Misclassification Cost

Data Set	DirectCS-KNN	DirectCS-KNN-SM	DirectCS-KNN-CSK	CS-C4.5	CS-C4.5-SM
Statlog(heart)	0.3815	0.3605	0.3556	0.6704	0.4938

Table 4.2. Area Under ROC (AUC)

Data Set	DirectCS-KNN	DirectCS-KNN-SM	DirectCS-KNN-CSK	CS-C4.5	CS-C4.5-SM
Statlog(heart)	0.744	0.763	0.768	0.759	0.776

From the first experiment, we can draw some conclusions. First, in term of reducing misclassification cost, our three new methods have achieved lower cost than CS-C4.5 (with or without enhancements), all by a large margin. Compared to CS-C4.5, DirectCS-KNN reduced the misclassification cost by 43%, and both DirectCS-KNN-SM and DirectCS-KNN-CSK reduced the misclassification cost by more than 46%. Second, by using AUC, a well-recognized measurement in cost-sensitive learning, we can see our two new methods, DirectCS-KNN-SM and DirectCS-KNN-CSK achieved higher AUC than CS-C4.5, but the AUC of our DirectCS-KNN is slightly lower than CS-C4.5. Third, among the four algorithms we tested, DirectCS-KNN-SM and DirectCS-KNN-CSK always perform better in terms of achieving lower misclassification cost and higher AUC. Overall, DirectCS-KNN-CSK is the best of the four algorithms.

In the second experiment, we still use Statlog(heart) data set with the recommend cost matrix. This time we focus on our two new algorithms, DirectCS-KNN and DistanceCS-KNN. We evaluate their performance by comparing their misclassification cost and AUC. Since the first experiment showed that smoothing and K-value selection with minimum-cost methods can reduce the misclassification cost of the DirectCS-KNN classifier, we apply both methods to the new cost-sensitive

KNN algorithms, DirectCS-KNN and DistanceCS-KNN to achieve better performance. The test results are shown in Table 5 below.

Table 5. Key performance measurements on Statlog(heart)

Table 5.1. Average Misclassification Cost

Data Set	DirectCS-KNN	DistanceCS-KNN
Statlog(heart)	0.3512	0.344

Table 5.2. Area Under ROC (AUC)

Data Set	DirectCS-KNN	DistanceCS-KNN
Statlog(heart)	0.7642	0.7688

The second experiment is simple and straightforward, it shows that our modified cost-sensitive KNN algorithm, DistanceCS-KNN performs better than the more naïve, straightforward DirectCS-KNN algorithm. It reduces misclassification cost and increases AUC. This experiment sets up a good foundation for the next experiment in this pager. In our last experiment, we will mainly focus on the DistanceCS-KNN algorithm and its variations.

The test results of our last experiment are shown in Table 6 and 7. Table 6 lists the average misclassification cost on selected five UCI data sets. Table 7 lists the corresponding results on the t-test. Each *w/t/l* in the table means our new algorithm, DistanceCS-KNN and its variations, at each row wins in *w* data sets, ties in *t* data sets and loses in *l* data sets, against CS-C4.5. Similar to the second experiment, we applied both smoothing and *K*-value selection with minimum-cost methods on our DistanceCS-KNN algorithm and its variations.

Table 6. Average misclassification cost on selected UCI data sets

Table 6.1. Cost Ratio (FP=1, FN=2)

Data Set	DistanceCS-KNN	DistanceCS-KNN-FS	DistanceCS-KNN-STK	CS-C4.5	CS-C4.5-SM
Diabetes	0.3758	0.3596	0.3633	0.3828	0.3722
Credit-g	0.428	0.417	0.402	0.435	0.408
Page-blocks	0.0607	0.0592	0.0585	0.0422	0.0397
Spambase	0.1091	0.1044	0.0993	0.1052	0.0973
Waveform-5000	0.1341	0.1315	0.1298	0.1951	0.1933

Table 6.2. Cost Ratio (FP=1, FN=5)

Data Set	DistanceCS-KNN	DistanceCS-KNN-FS	DistanceCS-KNN-STK	CS-C4.5	CS-C4.5-SM
Diabetes	0.5573	0.536	0.5352	0.6003	0.5789
Credit-g	0.598	0.5815	0.582	0.77	0.681
Page-blocks	0.0965	0.0896	0.0838	0.0846	0.0767
Spambase	0.2006	0.1937	0.1864	0.2121	0.1905
Waveform-5000	0.1637	0.1596	0.1562	0.3756	0.3254

Table 6.3. Cost Ratio (FP=1, FN=10)

Data Set	DistanceCS-KNN	DistanceCS-KNN-FS	DistanceCS-KNN-STK	CS-C4.5	CS-C4.5-SM
Diabetes	0.5898	0.5832	0.5869	0.8268	0.7642
Credit-g	0.717	0.6756	0.62	1.043	0.832
Page-blocks	0.1214	0.1163	0.1031	0.1297	0.1108
Spambase	0.3019	0.2836	0.2712	0.3512	0.3122
Waveform-5000	0.1933	0.1896	0.1815	0.611	0.5752

Table 6.4. Cost Ratio (FP=1, FN=20)

Data Set	DistanceCS-KNN	DistanceCS-KNN-FS	DistanceCS-KNN-STK	CS-C4.5	CS-C4.5-SM
Diabetes	0.7591	0.725	0.717	0.9635	0.8281
Credit-g	0.939	0.822	0.8161	1.258	1.035
Page-blocks	0.1782	0.1665	0.1546	0.1838	0.1633
Spambase	0.4027	0.3817	0.3552	0.5781	0.4842
Waveform-5000	0.1963	0.1915	0.1848	1.0678	0.9912

Table 7. Summary of t-test

Cost Ratio (FP:FN)		CS-4.5-CS
1:2	DistanceCS-KNN	3/0/2
	DistanceCS-KNN-FS	4/0/1
	DistanceCS-KNN-STK	4/0/1
1:5	DistanceCS-KNN	4/0/1
	DistanceCS-KNN-FS	4/0/1
	DistanceCS-KNN-STK	5/0/0
1:10	DistanceCS-KNN	5/0/0
	DistanceCS-KNN-FS	5/0/0
	DistanceCS-KNN-STK	5/0/0
1:20	DistanceCS-KNN	5/0/0
	DistanceCS-KNN-FS	5/0/0
	DistanceCS-KNN-STK	5/0/0

From the last experiment, we can also draw several conclusions. First, for all the data sets we have tested, our cost-sensitive KNN algorithms generally perform better than CS-C4.5, the higher the cost ratio, the better our new algorithms perform. This is because CS-C4.5 ignores misclassification cost when building decision tree, it only considers cost at classification stage, while our cost-sensitive KNN algorithms consider misclassification cost at both classification stage and the stage of calculating distance weight. Second, our two new improvements, DistanceCS-KNN-FS and DistanceCS-KNN-STK outperform the original DistanceCS-KNN algorithm on most of the selected UCI data sets across different cost ratios. Third, DistanceCS-KNN-STK is the best among all the four algorithms we tested, it is very stable and performs better than other competing algorithms across different cost ratios.

6   Conclusion and Future Work

In this paper, we studied the KNN classification algorithm in the context of cost-sensitive learning. We proposed two approaches, DirectCS-KNN and DistanceCS-KNN, to make KNN classifier sensitive to misclassification cost. We also proposed several methods (smoothing, minimum-cost K value selection, cost-sensitive feature selection and cost-sensitive stacking) to further improve the performance of our cost-sensitive KNN classifiers. We designed three experiments to demonstrate the effectiveness and performance of our new approaches step by step. The experimental results show that compared to CS-C4.5, our new cost-sensitive KNN algorithms can effectively reduce the misclassification cost on the selected UCI data across different cost ratios.

In the future, we plan to test the new approaches on more real world data sets which are relative to cost-sensitive learning, and compare them to more cost-sensitive learning algorithms. We also would like to extend our cost-senstive KNN algorithms to handle multi-class data sets and evaluate the effectiveness of the new algorithms on real world multi-class data sets.

Other possible improvements include using calibration methods such as Platt Scaling or Isotonic Regression to get better class membership probability estimation, trying other search algorithms such as Genetic algorithm with cost-sensitive fitness function in our feature selection wrapper.

Acknowledgement. This work is supported in part by the Australian Research Council (ARC) under large grant DP0985456; the China "1000-Plan" National Distinguished Professorship; the China 863 Program under grant 2012AA011005; the Natural Science Foundation of China under grants 61170131 and 61263035; the China 973 Program under grant 2013CB329404; the Guangxi Natural Science Foundation under grant 2012GXNSFGA060004; the Guangxi "Bagui" Teams for Innovation and Research; the Guangxi Provincial Key Laboratory for Multi-sourced Data Mining and Safety; and the Jiangsu Provincial Key Laboratory of E-business at the Nanjing University of Finance and Economics.

References

1. Domingos, P.: MetaCost: a general method for making classifiers cost-sensitive. In: Proceedings of the Fifth ACM SIGKDD International Conference on Knowledge Discovery and Data Mining, pp. 155–164 (1999)
2. Elkan, C.: The foundations of cost-sensitive learning. In: Nebel, B. (ed.) Proceeding of the Seventeenth International Joint Conference of Artificial Intelligence, Seattle, August 4-10, pp. 973–978. Morgan Kaufmann (2001)
3. Greiner, R., Grove, A.J., Roth, D.: Learning cost-sensitive active classifiers. Artificial Intelligence 139(2), 137–174 (2002)
4. Kohavi, R., John, G.H.: Wrappers for feature subset selection. Artificial intelligence 97(1-2), 273–324 (1997)
5. Kotsiantis, S., Pintelas, P.: A cost sensitive technique for ordinal classification problems. In: Vouros, G.A., Panayiotopoulos, T. (eds.) SETN 2004. LNCS (LNAI), vol. 3025, pp. 220–229. Springer, Heidelberg (2004)
6. Kotsiantis, S., Kanellopoulos, D., Pintelas, P.: Handling imbalanced datasets: A review. GESTS International Transactions on Computer Science and Engineering 30(1), 25–36 (2006)

7. Li, J., Li, X., Yao, X.: Cost-Sensitive Classification with Genetic Programming. In: The 2005 IEEE Congress on Evolutionary Computation, vol. 3 (2005)

8. Ling, C.X., Yang, Q., Wang, J., Zhang, S.: Decision trees with minimal costs. In: Brodley, C.E. (ed.) Proceeding of the Twenty First International Conference on Machine Learning, Banff, Alberta, July 4-8, vol. 69, pp. 69–76. ACM Press (2004)

9. Margineantu, D.D.: Methods for Cost-sensitive Learning. Oregon State University (2001)

10. Niculescu-Mizil, A., Caruana, R.: Predicting good probabilities with supervised learning. Association for Computing Machinery, Inc., New York (2005)

11. Oza, N.C.: Ensemble Data Mining Methods, NASA Ame Research Center (2000)

12. Platt, J.C.: Probabilities for SV machines. In: Advances in Neural Information Processing Systems, pp. 61–74 (1999)

13. Provost, F., Domingos, P.: Tree Induction for Probability-Based Ranking. Machine Learning 52, 199–215 (2003)

14. Quinlan, J.R.: C4.5: Programs for machine learning. Morgan Kaufmann, San Mateo (1993)

15. Sun, Q., Pfahringer, B.: Bagging Ensemble Selection. In: Wang, D., Reynolds, M. (eds.) AI 2011. LNCS, vol. 7106, pp. 251–260. Springer, Heidelberg (2011)

16. Turney, P.: Types of cost in inductive concept learning. In: Workshop on Cost-Sensitive Learning at the Seventeenth International Conference on Machine Learning, p. 1511 (2000)

17. Wang, T., Qin, Z., Jin, Z., Zhang, S.: Handling over-fitting in test cost-sensitive decision tree learning by feature selection, smoothing and pruning. Journal of Systems and Software (JSS) 83(7), 1137–1147 (2010)

18. Wang, T., Qin, Z., Zhang, S.: Cost-sensitive Learning - A Survey. Accepted by International Journal of Data Warehousing and Mining (2010)

19. Wettschereck, D., Aha, D.W., Mohri, T.: A review and empirical evaluation of feature weighting methods for a class of lazy learning algorithms. Artificial Intelligence Review 11(1), 273–314 (1997)

20. Witten, I.H., Frank, E.: Data Mining: Practical Machine Learning Techniques with Java Implementations, 2nd edn. Morgan Kaufmann Publishers (2000)

21. Wolpert, D.H.: Stacked generalization. Neural Networks 5, 241–259 (1992)

22. Wu, X., Kumar, V., Ross Quinlan, J., Ghosh, J., Yang, Q., Motoda, H., McLachlan, G.J., Ng, A., Liu, B., Yu, P.S.: Top 10 algorithms in data mining. Knowledge and Information Systems 14(1), 1–37 (2008)

23. Zadrozny, B., Elkan, C.: Obtaining calibrated probability estimates from decision trees and naive bayesian classifiers. In: Proceedings of the 18th International Conference on Machine Learning, pp. 609–616 (2001)

24. Zadrozny, B., Elkan, C.: Learning and making decisions when costs and probabilities are both unknown. In: Proceedings of the Seventh ACM SIGKDD International Conference on Knowledge Discovery and Data Mining, pp. 204–213. ACM Press, San Francisco (2001)

25. Zadrozny, B., Elkan, C.: Transforming classifier scores into accurate multiclass probability estimates, pp. 694–699. ACM, New York (2002)

26. Zadrozny, B.: One-Benefit learning: cost-sensitive learning with restricted cost information. In: Proceedings of the 1st International Workshop on Utility-Based Data Mining, pp. 53–58. ACM Press, Chicago (2005)

27. Zhang, J., Mani, I.: kNN approach to unbalanced data distributions: a case study involving information extraction (2009)

28. Zhang, S.: KNN-CF Approach: Incorporating Certainty Factor to kNN Classification. IEEE Intelligent Informatics Bulletin 11(1) (2003)

A Thesaurus and Online Encyclopedia Merging Method for Large Scale Domain-Ontology Automatic Construction

Ting Wang, Jicheng Song, Ruihua Di, and Yi Liang

College of Computer Science and Technology, Beijing University of Technology,
Beijing, China
S200807007@emails.bjut.edu.cn

Abstract. While building the large-scale domain ontology, the traditional manually-based construction method is low efficient and not feasible. In order to construct the large scale domain-ontology automatically; therefore, proposed a two stage method for large scale Chinese domain-ontology automatic construction by merging the thesaurus and online encyclopedia together. On the first stage, fuzzy mapping the thesaurus to OWL and construct the domain rough-ontology; on the second stage, including the following steps: (i) Concept Mapping and Pruning, (ii) deciding attribute from Chinese encyclopedia and (iii) rough-ontology self-Adaptive and expansion. Practice has proved that the system has the feasibility and robustness on achieving the large scale domain ontology automatic and rapid construction.

Keywords: domain ontology, thesaurus, two stage method, ontology learning, linked open data.

1 Introduction

The vision of the Semantic Web is to create a "web of data", so that the machine can understand the semantic information on the internet [1]. In order to achieve the goal of Semantic Web, many areas of semantic data sets are published on the web, and achieve semantic linked data.

Large-scale domain ontology and knowledge base construction is the basis and prerequisite of the linked data. Linking open data (LOD) [2] based on domain knowledge base has been widely used in the field of e-government, such as U.S. government has carried on Open Government Data Project [3] (http://www.data.gov/) and the British Government Data Project [4] (http://data.gov.uk/) has published a linking government semantic data sets and its applications platform on the web.

At present, the extent of the various government departments' Open Data in Chinese government information resources is limited; sharing degree and the comprehensive utilization rate is not high. The government data format is not unified and the lack of large-scale e-government ontology has brought a great deal of inconvenience on linking open government data between different knowledge sources.

M. Wang (Ed.): KSEM 2013, LNAI 8041, pp. 132–146, 2013.

While building large-scale domain ontology, traditional manual methodology need domain experts to participate, and it is time-consuming and difficult to adapt. Therefore, in order to solve large-scale domain ontology automation construction, this paper proposes a two stage method for large scale Chinese domain-ontology automatic construction by merging the thesaurus and online encyclopedia together.

2   Related Work

Many scholars and institutions have made efforts to build domain ontology with clear semantic structure, and rich relationship between entities. EU has developed and implemented "E-Europe 2002 action plan", and they established the e-government ontology (OntoGov) [5]; Chinese Taiwan scholars built e-government knowledge base for sharing ontology and refining synonyms [6].

Some scholars research on how to convert domain-specific thesaurus to ontology. In the conversion of Agrovoc, Food and Agriculture Organization(FAO) of United States mapped the thesaurus hierarchical relationships directly to the inheritance relationship of ontology [7], and in Chinese "Agricultural Thesaurus", which may correspond to the inheritance, property, and equivalent relationship [8]. Unified Medical Language System(UMLS) [9] contains 100 biomedical vocabularies and classification table, 750,000 concepts and their 10 million links, but its semantic quality is limited.

Wordnet [10] was built by the domain experts manually with the help of Princeton University's psychologists, linguists and computer engineers according to the principles of cognitive psychology, it is a dictionary database, and supports multi-language. HowNet [11] is the concept of a representative of the Chinese and English words to describe the object, as the basic common sense knowledge base, it is to reveal the relationships between concepts and the concept's attributes.

Based on thesaurus to build domain ontology, the conversion is mainly according to fixed semantic relations containing in the OWL specification, but this level and the level of semantic relationships reveal simple and weak for semantic retrieval and relationship reasoning, which cannot meet the reasoning requirement in application.

With the development of Web 2.0, the collaborative creation of a specific domain or open domain appears more and more in the online encyclopedia such as Wikipedia [12], Baidu Baike [13] and Hudong Baike [14]. YAGO [15] and DBpedia [16] are based on Wikipedia to build large-scale knowledge base. They extract knowledge from structured information of Wikipedia pages, implying that the "is-a" relationship between concepts of hierarchical relationships in the taxonomy system. The InfoBox information on the entry's page contains some knowledge <S,P,O> triples. Wu and S.Weld. developed Kylin [17,18] system which is not only use of structured content, but also try to extract knowledge triples from unstructured text in Wikipedia's articles.

Constructing Chinese e-government ontology is the basis of linked open government data. E-government thesaurus is the set of domain vocabulary, which covering 21 sub-areas of the e-government, containing a large number of fixed simple semantic relationships and having concepts in classification and hierarchy. Baidu Baike as the largest Chinese encyclopedia involves and covers 13 areas. The online open encyclopedia and the e-government thesaurus have the characteristics of interdisciplinary coverage, but each of them has the different special emphases, which provide the possibility for automatically building the large-scale domain ontology.

Therefore, this paper proposes a two stage method for large scale Chinese domain-ontology automatic construction by merging the thesaurus and online encyclopedia together for the Chinese linked open government data building and domain knowledge sharing.

3   Preliminary

This section first gives some related definitions, and then briefly introduces the General E-government Thesaurus and the Baidu encyclopedia.

3.1   Related Definitions

Below, based on the idea of set theory, giving the definition of the domain thesaurus, ontology and the relationship between two of them.

Definition 1 knowledge triples: a factual statement that can be expressed as a Semantic Web triples: <subject, predicate, object>, and semantic data can be formalized as a triple set. The knowledge of the set of triples K, $K \subseteq S\ P\ O$, where S is the set of subjects, P is the set of predicates, O is a set of objects.

Definition 2 Ontology: ontology is a formal description of the shared important concepts in a particular domain. An ontology model can be described by a four-triple: $O=\{C, P, H^C, H^P\}$. Where C and P represent the set of ontology concepts and properties, H^C and H^P expressed hierarchical semantic relationships between the elements in the set of concepts and attributes.

Definition 3 Knowledge Base: denote instances' set corresponding to domain ontology O is I, the domain knowledge base KB is constituted by the ontology O and instances' set I, that is, $KB=\{O, I\}$.

Definition 4 Attributes: the OWL specification describes ontology O, datatype property is used to express the semantic relationship between the instance and the literal value (value attributes); attributes used to express the semantic relationships between different instances are called object property. The set of concepts described by Property p is called the domain of property p, denoted as D(p); the set of concepts that represent the legitimate property values of the object attributes are called attribute's range, denoted as R(p).

Definition 5 THESAURUS: Domain thesaurus = (T, H^T), which refers to a series of terms with similar relations in a particular area, and all the term $t_i \in T$ construct the vocabulary set T. T manifested a tree; H^T includes hierarchical relationship between terms.

Definition 6 Encyclopedia concept system: In the open taxonomy system B, all the concept $C_i \in C$ in the encyclopedia composed of a set of encyclopedia concept C.

Form of taxonomy tree to organize themselves, denoted as $B=(C, H^C)$, where H^C contained the hierarchical relationships between encyclopedia concepts.

Definition 7 Union of encyclopedia predicates: All predicates appear in entries' Infobox constructs the union, denoted as *Info_Attr*. That is, if *ARTICLE$_i$*'s Infobox appears in its web page, the set of predicates appear in an Infobox given P, $P=\{p_1, p_2, p_3 .... p_n\}$. Where p_i is a predicate that appears in the entry's Infobox. Hence, *Info_Attr* $= P_1 \cup P_2 \cup ... \cup P_n$, n is the number of online encyclopedia articles' pages that contain Infobox.

Definition 8 Intersection of concepts: all the vocabularies both appear in thesaurus set T and encyclopedia concept set C constructs the set T and set C's concept intersection, denoted as *Concept_M = T∩C*.

Definition 9 Intersection of attributes: all the vocabularies both appear in thesaurus set T and encyclopedia attributes set *Info_Attr* constructs the set T and set *Info_Attr*'s attributes intersection, denoted as *Attr_M = T∩Info_Attr*.

3.2   Chinese General E-Government Thesaurus

The thesaurus is a kind of semantic dictionary, which constructing by the various relationships between terminologies and can reflect the semantical concepts in a certain subject domain. ANSI's (American National Standards Institute) thesaurus standard (Z39.19-1980) provides 13 kinds of vocabulary, which generally appear in the form of "U, D, S, F, and C" structure. The main purpose of the thesaurus is to make the user's query more accurate. It is to improve the recall, precision, and is an important way to achieve multilingual retrieval and intelligent concept retrieval [19].

Chinese General E-Government Thesaurus (Trial) "(Alphabetical table) contains a total of 20,252 items, containing 17,421 descriptors and 2831 non-descriptors. The category list is a three-level hierarchical structure, including 21 first level categories, 132 second level items, and 37 class headings. Chinese General e-government thesaurus (trial) covers various fields of Chinese e-government and its related knowledge areas, including the *activities of political parties, organizations, theoretical studies, political and ideological work,* and some other related aspects [20,21].

In order to make effective use of existing human knowledge systems, we can convert thesaurus into domain ontology directly and automatically. The converting can not only speed up the construction process of the domain ontology, but also can increase the scientific of the ontology; thus, we can make it become a complete system and overcome the thesaurus's drawback in specific area's application.

3.3   Baidu Encyclopedia Knowledge Base

Based on Baidu encyclopedia [13], some organizers have constructed the Chinese encyclopedia knowledge base. Until March 2012, Baidu encyclopedia has more than 4.4 million Chinese terms. The open taxonomy system contains the 13 top categories and more than 1300 sub-concepts.

CASIA has built an ontology by crawling down the Baidu Baike wiki pages. According to the taxonomy of the Chinese Encyclopedia, the ontology has a four-level structure, and each term belongs to a concept as its individual.

Infobox appearing on term's web page can provide various semantic knowledge; that is, Infobox imply a lot of triples by taking the term name as the subject, attribute name as the predicate, and attribute value as the object. Therefore, CASIA also extracts the well-structured knowledge from the Infobox automatically. For example, the fact that "Beijing is the capital of China" is represented in the Infobox of the term: "China" as "Capital: Beijing". Hence, a knowledge triple as "<China, Capital, Beijing>" can be extracted directly. After crawling all the wiki pages with Infoboxes in Baidu Baike encyclopedia, CASIA-KB has resolved out 1.698 million knowledge triples represented in each Infobox. All of them are described in Chinese literals.

In this paper, "the knowledge triples from Baidu encyclopedia are provided by CASIA Knowledge Base (http://imip.ia.ac.cn/casia-kb/)"[22].

4   Architecture

The architecture of domain ontology automatic building system as shown in Figure 1, the two stage method for large scale Chinese domain-ontology automatic construction, specifically include the following two steps:

(1) **The first stage:** fuzzy mapping the thesaurus to ontology. As the first stage of the two-stage method, proposed the fuzzy Mapping Algorithm for crude converting the domain thesaurus to OWL, the algorithm mapped the thesaurus' F, S, C, D and Y relationships roughly to the inheritance relations and equivalent relations of concept. Hence, construct the domain rough ontology.

(2) **The second stage:** merging rough ontology with encyclopedia knowledge base, mainly including the following steps: (i) concept mapping and pruning, (ii) deciding attribute from Chinese encyclopedia and (iii) rough-ontology self-adaptive and expansion. Section 6.1 to 6.3 will give the detail's algorithm description of these steps.

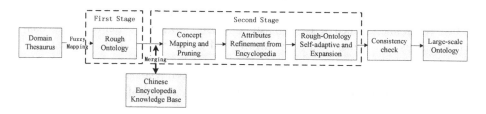

Fig. 1. Architecture of the two stage method

5   Fuzzy Mapping from Thesaurus to OWL

Generally speaking, the F, S, C, D and Y relationships in domain thesaurus can implicitly correspond to N types of different semantic relationships in OWL standard. This 1: N mapping situation bring difficulties and challenges to the convert from thesaurus to large-scale ontology automatically. In the thesaurus' semantic relations, the hierar-chical-relationship and equivalent-relationship have clearer semantic relation than the others; thus, these two relations are easier to set up the mapping rules in the conversion procedure. Extracting the ontology concepts from thesaurus' terms may reduce the dependence on domain experts in ontology building process.

As the first stage of the two-stage method, using the thesaurus to OWL mapping algorithm to convert the F, S, C, D and Y relations to the inheritance relations and equivalent relations roughly.

Table 1. Fuzzy mapping rules from thesaurus to OWL

Relationship between Terms	Reference symbol in Chinese(English)	Reference meaning	Possible semantic relationship	Fuzzy mapping relationship
Hierarchical relationship	S(BT)	Broader terms	\<be part of\> \<be kind of\> \<be attribute of\> \<be instance of\>	\<subClassOf\>
	F(NT)	Narrower terms		
Equivalent relationship	Y(USE)	Descriptors	\<be similar to\> \<equivalentClass\>	\<equivalentClass\>
	D(UF)	Non-descriptors		
Related relationship	C(RT)	Related terms	\<be part of\> \<be attribute of\> \<equivalentClass\> \<be similar to\>	\<equivalentClass\>

Through the simple and rough mapping on the first stage, we can fill the thesaurus' 20252 vocabularies into the ontology efficiently and quickly; thus, containing a certain scale and some simple semantic relations' rough ontology model will be established. Rough ontology can provide the foundation for the ontology framework's enrich and improve on the second stage. The fuzzy mapping algorithm is implemented with Jena API, and the specific mapping rules are shown in Table 1.

After the rough mapping from thesaurus to OWL, we can get the rough e-government ontology. Concrete, the *rdfs:subClassOf* modeling primitives characterize the parent-child semantic relationship between concepts; and the *owl:equivalentClass* primitives express class equivalence relationship between concepts. The description snippet of Chinese e-government rough ontology as shown in Figure 2.

6   Merging Chinese Thesaurus with Online Encyclopedia

"Chinese General E-government thesaurus" contains 20252 terms, while the number of concepts in Baidu Baike' taxonomy system (ontology) is only about 1380.

```
<owl:Class rdf:about="http://www.owl-ontologies.com/Onto/e_gov.owl#PartySchool">
    <dc:title>PartySchool</dc:title>
        <rdfs:subClassOf rdf:resource="http://www.owl-ontologies.com/Onto/e_gov.owl#School"/>
            <dc:title>School</dc:title> </owl:Class>
<owl:Class rdf:about="http://www.owl-ontologies.com/Onto/e_gov.owl# PartyGroupings ">
    <dc:title>PartyGroupings</dc:title>
        <rdfs:subClassOf rdf:resource="http://www.owl-ontologies.com/Onto/e_gov.owl# Group"/>
            <dc:title>Group</dc:title> </owl:Class>
        <owl:Class rdf:about="http://www.owl-ontologies.com/Onto/e_gov.owl#StyleoftheParty">
<dc:title>StyleoftheParty</dc:title>
<owl:equivalentClass rdf:resource="http://www.owl-ontologies.com/Onto/e_gov.owl#PartyStyle"/>
    <dc:title>PartyStyle</dc:title>
        <rdfs:subClassOf rdf:resource="http://www.owl-
ontologies.com/Onto/e_gov.owl#StyleofWork"/>
            <dc:title>StyleofWork</dc:title> </owl:Class>
```

Fig. 2. Description snippet of Chinese e-government rough ontology

In the first stage, we propose the fuzzy mapping rules of thesaurus model, the generated rough ontology framework after this stage of processing seems too simple and not accurate, and semantic relationship between some concepts is uncertain.

Therefore, in the second stage, we put forward to merging the rough ontology model and online Encyclopedia knowledge base together. By utilizing the Baike knowledge base's architecture, rough ontology model will be enriched and improved.

6.1   Concept Mapping and Pruning

Firstly we retrieve the HIT's "Tongyici Cilin"(extended version) [23] to find the synonym set of the encyclopedia concepts and predicates in Infobox. Then finding the *Concept_M* and *Attr_M* set by mapping the thesaurus set T with encyclopedia concept set C and encyclopedia attributes set *Info_Attr* on the concept-level. The definition of *Concept_M* and *Attr_M* has given in section 3.1.

According to the statistical analysis, there are 446 concepts in thesaurus and encyclopedia's intersection: *Concept_M*, which has a percentage of about 32% in encyclopedia concept set C. Similarly, the number of predicates in the encyclopedia Infobox appears in e-government thesaurus is 1071, that is, the number of predicates in *Attr_M* is 1071, which has a percentage of about 4.5% in Baike Infobox attributes' set: *Info_Attr*. The distribution of *Concept_M*'s concepts in the encyclopedia ontology as shown in the table 2:

Table 2. Distribution of the co-occurrence concepts in Baidu encyclopedia taxonomy

Top-level Concept	Person	Technology	Art	Geography	Sport	Science	Culture	History	Life	Social	Nature	Economics	Education	Total
sub-concepts	120	77	84	133	165	157	132	118	116	102	104	36	36	1380
co-occurrence	23	29	35	30	59	54	46	7	34	69	24	21	15	446
percentage	19%	38%	42%	23%	36%	34%	35%	6%	29%	68%	23%	58%	42%	32%

According to concept's hierarchy and inheritance in ontology, if terms have the hypernym or hyponym relation between them, then these terms can be defined as Concept and these concepts' semantic relations can be defined as *rdfs:subclassof*.

But in the rough ontology, some of them are more suitable for appearing as the Property or Individual style but not the Concept in ontology. For this reason, we propose a rough ontology pruning algorithm to delete these unsuitable terms (Term2Concept Pruning Algorithm). The description of the algorithm is as follows:

Step 1: Taking each of the 446 co-occurrence concepts in the *Concept_M* set as the "entry point". The "entry point" is for pruning operation on the rough ontology.

Step 2: Getting the simple path from each "entry point" to thesaurus' highest level term (Level 1: Subject Headings), and defining all terms on this simple path as the concept of OWL.

Step 3: Traversal all sub-nodes of the "entry point" term, and get the "entry point" term's set of sub-tree: $Sub\_T = \{TR_1, TR_2, TR_3, ..., TR_n\}$. Denote TR_i as one of the sub-tree of the "entry point".

Step 4: if a term in TR_i also appears in *Attr_M* set, then all nodes (concepts) on the TR_i will be pruned from the rough ontology. That is, if attribute $p_i \in TR_i$, and $p_i \in$ *Attr_M*, then delete TR_i.

After pruning the rough ontology by the algorithm, the semantic relationships between concepts are more clear and explicit, which provide the foundation for the attribute and instance's automatic filling and expansion to ontology.

6.2   Automatic Refinement of Attributes from Encyclopedia's Infobox

After the concepts' mapping, then we use the large scale knowledge Chinese encyclopedia to enrich semantic relations for e-government ontology; that is, attributes' refinement and filling.

After a rough mapping on the first stage, we converted the thesaurus' words into ontology in the form of concept; however, semantic relationships between concepts are too simple. The structured information of Infobox representing on a term's web page provides a mount of semantic relationship knowledge between entities.

Each Baike's concept contains a number of entities (terms), about only 10 percent of them contain the Infobox information on their web page but others not [22]. We carry on the statistics of the predicate in Baidu encyclopedia's Infobox template and specifically count the appearing times for each property in Infobox while this Infobox represent on a term's web page and contains this property. We ignore all the web pages that do not contain the Infobox. Finally, we compute each property's appearing frequency of each concept.

We have done the statistics of all the 13 top-level classes and their 1380 subclasses' property appearing frequency in Baike taxonomy.

For different concept in Baike open taxonomy, the appearing frequency of predicate in Infobox can reflect the characteristics of different subclasses. Meanwhile, the semantic relations of e-government ontology are too simple; therefore, we consider some strategies for enriching and filling the Baike's structured information (Infobox) semantic relations to e-government ontology.

We compute and census all the predicates' appearing frequency corresponding to the 1380 subclasses of Infobox template, and have carried them on sorting. The results show that the number of predicates in Baidu Baike's Infobox is a total of 22442; part of the Infobox subclass predicates' appearing frequency can reflect the semantic features of the class.

However, because the encyclopedia is geared to the collaborative users and the taxonomy system is open and free. Therefore, from the statistical results, many kinds of predicate frequency sorting is unable to reflect the feature of concept, (take the concept: "trade" as an example) as shown in table 3. That is, some predicates with obvious semantic features appear in the low frequency zone or middle frequency zone mostly, while a lot of non-feature predicates appear in the high-frequency zone.

In order to solve this problem, based on the main idea of the traditional TF-IDF algorithm [24], we propose a new algorithm: PF-ICF (predicate frequency - inverse class frequency), which is used to calculate the importance of each predicate for its corresponding subclass. Below we describe the PF-ICF algorithm in detail.

Table 3. Attributes of concept "trade" sorted by PF value

Sorting	Attribute Name	Appear Times	PF Value
1	Chinese name	689	0.431974922
2	nationality	455	0.285266458
3	birthday	443	0.277742947
4	occupation	433	0.271473354
5	birthplace	407	0.255172414
6	price	388	0.243260188
7	graduated university	347	0.217554859
8	company name	345	0.21630094
9	Date of Establishment	341	0.213793103
10	headquarter address	328	0.205642633
11	main contribution	311	0.194984326
12	taxonomy	285	0.178683386
13	enterprise nature	272	0.170532915
14	scope of business	271	0.169905956
15	foreign name	229	0.143573668
16	nation	189	0.118495298
17	representative articles	184	0.115360502
18	the number of employees	92	0.057680251
19	area	87	0.054545455
20	population	56	0.035109718

The main idea of PF-ICF is: assuming that a predicate has a high appearing frequency in its corresponding class instances' Infobox of the web page, then the PF value is higher. And if this predicate appears in its sibling class instances' Infobox rarely, it shows that the predicate has good ability of concept distinguishing; that is, the value of ICF is high.

So the higher PF$\times$ICF value can reflect the semantic features of a concept. We denote the concept C_j's instances containing Infobox as the set: I_j. In particular:

$$pf_{i, j} = \frac{n_{i, j}}{\sum_k n_{k, j}} \tag{1}$$

PF (predicate frequency) indicate the appearing frequency of predicate p_i in the set I_j (does not contain C_j's sibling class). For the Infobox predicate p_i appearing in a class instance set I_j, its importance definition is shown in formula (1).

In the formula (1), $n_{i,j}$ is the number of the predicate p_i's appearing times in the Infobox of instances in set I_j, the denominator is the number of instances in set I_j.

(inverse class frequency, ICF) ICF defines the metric of universal importance on predicate p_i. That is, if fewer instances of C_j's sibling concepts contain the predicate p_i, the ICF value is higher; and the predicate p_i has good ability of distinguishing categories. For a specific predicate p_i's ICF; firstly, $|I|$ divided by $|\{j: p_i \in I_j \}|$ gives the quotient, then use the Log function on the quotient, and define the result as icf_i:

$$icf_i = \log \frac{|I|}{|\{j : P_i \in I_j\}|} \tag{2}$$

Among them, $|I|$ represent the number of sibling class' instances which contain Infobox, $|\{j: p_i \in I_j \}|$ is the number of sibling class' instances that contains predicate p_i appearing in their Infobox. If the value of $|\{j: p_i \in I_j \}|$ is zero, then take the denominator as $1+|\{j: p_i \in I_j \}|$. Then, according to the formula (1) and the formula (2), we can get each Infobox predicate p_i's PF-ICF value, which is given by the formula (3):

$$pf\text{-}icf = pf_{i, j} \times icf_i \tag{3}$$

Therefore, a high frequency of a predicate in the particular concept, as well as the low or middle frequency of this predicate in the set of sibling concepts can produce high weight of PF-ICF. Therefore, PF-ICF tends to filter out the common and high frequency predicates, and then can select out the predicates which can reflect a certain concept's semantic features.

In order to not lose the generality; firstly, taking on the appearing frequency statistics for the predicates in Infobox of each concept, secondly, according to the PF-ICF algorithm, calculate each predicate's value of *pf-icf* and rank from high to low, the results as shown in Table 4 (take the concept of "trade" as an example).

We randomly select and sample about 100 concepts' *pf-icf* value in *Concept_M*, all the TOP-20 results list of attributes are submitted to the e-government expert for the correctness verification, *pf-icf* value results show that in each concept property list, more than 70% of attributes can reflect the semantic features and relationship; but the PF value ranking results are generally lower than 50%.

Through the result authentication and analysis with Beijing e-government experts, conclusion can be given: after reordering the ranking results by using the PF-ICF algorithm, the correct rate is obviously better than just based on the results of attribute's appearing frequency. That is to say, more attributes that can reflect the semantic features of concept are ranked in high frequency zone, and precision of the attribute automatic refinement has been improved by the PF-ICF algorithm.

For each attribute list of a concept, we fill properties of TOP-20 into e-government ontology.

Table 4. Attributes of concept "trade" sorted by PF-ICF value

Sorting	Attribute Name	Appear Times	PF-ICF Value	PF Value
1	primary service	8	0.718391586	0.005015674
2	origin	4	0.324985384	0.002507837
3	competent organization	5	0.312973835	0.003134796
4	publication cycle	6	0.302905734	0.003761755
5	editing unit	6	0.302905734	0.003761755
6	host unit	9	0.280996658	0.005642633
7	language	13	0.260794979	0.008150 47
8	incumbent schoolmaster	14	0.256634805	0.008777429
9	foreign name	229	0.243172266	0.143573668
10	Chinese name	87	0.235363595	0.054545455
11	area	87	0.235363595	0.054545455
12	abbreviation	27	0.221356501	0.0169279
13	Group address	1	0.200172543	6.27E-04
14	main business	1	0.200172543	6.27E-04
15	service provider	1	0.200172543	6.27E-04
16	game features	1	0.200172543	6.27E-04
17	representative country	1	0.200172543	6.27E-04
18	earliest nation	1	0.200172543	6.27E-04
19	start time	1	0.200172543	6.27E-04
20	International Trade taxonomy	2	0.181341145	0.001253918

6.3   Rough-Ontology Self-adaptive and Expansion for Attribute

After the above operations, we have filled the properties in Infobox that can reflect concept's semantic feature into e-government ontology. But in e-government thesaurus, there are still many domain vocabularies, which can represent much more rich semantic relations between concepts. For example: the term "TV presenter" can be used as the *OWL:DataProperty* to express concept of "TV programs" and "individual of person" between semantic relations of instances, which can be described as the knowledge triples: < TV programs, TV presenter, individual of person >.

The merging of thesaurus and encyclopedia knowledge gives the self-adaptive ability to thesaurus. While the co-occurrence terms in the thesaurus and encyclopedia

Infobox's intersection *Attr_M* provide us the possibility to discover and mapping semantic relationships automatically in the domain thesaurus. Through this process, we can extend the semantic relationships for each concept except the TOP20 attributes in section 6.2.

Specifically, we take the following algorithm to map and extend the semantic relationships automatically:

Step 1: firstly, for the thesaurus and Infobox predicates intersection set *Attr_M*. Each term in intersection *Attr_M* is the potential one that might to be adjusted.

Step 2: for each property p_i in the intersection *Attr_M*, if it already exists in the ontology, then continue to the next attribute p_{i+1}. If not, then add the current attribute p_i into ontology. The domain of p_i is the concept C_i corresponding to encyclopedia, and C_i belongs to the intersection *Concept_M*.

Step 3: if the attribute p_i adjusted into the ontology is not a top-level vocabulary in thesaurus, and the concept of C_i in thesaurus is also not the top-level vocabulary, that is: Both p_i and C_i have hypernyms in thesaurus, then fill p_i's hypernym: $S(p_i)$ as the attribute into the ontology. The domain of $S(p_i)$ is C_i's hypernym $S(C_i)$, denoted as: $D(S(p_i)) = \{S(C_i)\}$.

Repeat step 3, until there is no hypernym of p_i or C_i.

Step 4: Because of the inheritance of concept's attribute, the concept C_i's attribute p_i filled into the ontology is also the attribute of C_i's subclasses. Similarly, for the equivalent classes of concept C_i: C_i', the domain of property p_i is also the C_i'. Therefore, the semantic relationships   between the subclasses have been enriched too.

By merging encyclopedia knowledge and e-government thesaurus together, we achieve the goal of extraction and enrich on the semantic relations of concept in thesaurus. On one hand, we fill the semantic relations in encyclopedia Infobox into the e-government ontology automatically based on PF-ICF algorithm. On the other hand, because the concepts of encyclopedia and the thesaurus have the intersection, we discover and exploit the self-adaptive ability of thesaurus; thus, some thesaurus' terms can be automatically adjusted to a concept's attribute.

6.4   Definition of Property in E-gov Ontology

The extracted attributes used to describe the semantic relationship between individual and attribute value, the definition of properties are given below.

(1) Domain

We call all the properties merging from encyclopedia Infobox as Infobox-Property. For each property p_i of the Infobox-Property, we firstly find all the encyclopedia's pages (individuals) that it appears on, and get the page set $W_p = \{w_1, w_2,... w_n\}$, then according to the title of each entry in page set W_p, we can get the concept C_i corresponding to this entry from Baike's taxonomy concept set C, if C_i appears in intersection *Concept_M*, then add C_i to the property p_i's domain. These concepts are denoted as set $D_p = \{C_1, C_2, ... , C_m\}$, and $D_p \subseteq Concept\_M$.

We define the set D_p as the Infobox-Property p_i's domain, denoted as $D_p = D(p_i)$.

(2) Range

The attributes in government ontology are unified defined as *Datatype Property* type. Therefore, attribute range is denoted as "*xsd:string*".

6.5   Filling the Individuals into E-gov Ontology

As mentioned above, we obtain the intersection *Concept_M* of encyclopedia open taxonomy and e-government thesaurus. At the same time, we have crawled down the instances of encyclopedia concept from the encyclopedia's web pages; each encyclopedia concept contains a large number of instances. Therefore, we fill the instances belonging to encyclopedia concepts in intersection *Concept_M* into the e-government ontology directly.

Each instance has the two general attributes, which is the title and URL source link. The title is the entry's name on the encyclopedia's web page, using Dublin Core to describe it: *dc:title*. The URL source link is the encyclopedia's hyperlink of this entry, the primitive: *rdfs:seealso* is used to describe it. The two kinds of general-attributes can be obtained directly while crawling down the instance's web page.

For the Infobox-Property set P generated from Infobox template in section 6.2, if the instance adding to ontology belongs to the concept C_i in intersection *Concept_M* and C_i's encyclopedia web page contains Infobox, then mapping all the properties in this Infobox to the attributes of C_i in e-government ontology, and filling the knowledge triple $<S, P, O>$ corresponding to the mapped attribute into the e-government ontology directly.

Through this step, based on the intersection of e-government ontology concepts and encyclopedia concepts, we merge and fill a total of 3820821 instances belonging to concepts in *Concept_M* into the e-government ontology automatically.

7   Results

After the above steps, based on the two stage method, we merge and adjust the rough ontology containing 20252 concepts with encyclopedia knowledge. Finally, we construct a well-formed large-scale e-government ontology containing a total of 21 top-level concepts, 29367 sub-concepts, 10089 attributes and 4281278 instances.

Detailed statistical information of e-government ontology as shown in Table 5.

8   Conclusions and Future Work

Compared to the traditional hand-built method, the thesaurus and online encyclopedia knowledge base merging method not only can fetch up the lack of description on the semantic relationship between terms in thesaurus, but also can make use of the amount of knowledge in the online encyclopedia. Hence, the domain-ontology can be filled quickly, and use the advantages of the two different knowledge sources, the system as a whole can be improved. Finally, in order to achieve the vision of semantic web, building the large scale domain-ontology base can provide the foundation and support for the linking open data.

In future work, based on the machine learning theory, we will fill the value for attributes in ontology automatically and efficiently. At the same time, because of the openness on encyclopedia taxonomy; therefore, many instances and Infobox's triples have been classified improperly; in the future we will find the appropriate clustering algorithm to classify these triples accurately, so as to optimize the ontology model.

Table 5. Chinese E-government Ontology information

TOP Concept	# Concepts	#Related Properties	#Individuals	#RDF Triples
Integrated E-government	3231	467	129274	133906
Economic management	1397	312	88173	90506
Territorial resources, Energy	396	319	149437	150790
Industry, Transportation	2318	575	233610	237653
Information industry	739	397	115835	117765
Urban-rural development, Environment Protection	440	252	65468	66664
Agriculture, Water conservancy	476	128	38000	38860
Finance	1174	267	30593	32568
Commerce, Trade	1812	269	90202	92821
Tourism, Service	94	122	73903	74363
Meteorological, Hydrological, Mapping, Earthquake	347	341	85194	86564
Foreign Affairs	173	0	2791	2964
Polity, Supervision	1233	268	55564	57601
Science, Education	1279	1357	600571	605921
Culture, Sanitation, Sport	1485	2523	1029040	1038094
Military affairs, National defense	1513	491	105528	108514
Laboring, Personnel system	447	83	76103	76799
Civil Affairs, community	231	26	76809	77118
Administration	506	0	20010	20516
Comprehensive Caucus	692	354	66596	68350
Phraseology	9384	1538	1088943	1102941
Total	29367	10089	4221644	4281278

Acknowledgments. This work was supported in part by a grant from National Natural Science Foundation of China (No. 61202075) and IBM Shared University Research (SUR) Project (E-government Data Semantic Integration and Knowledge Discovery on the Cloud). The knowledge triples from Baidu encyclopedia are from CASIA Knowledge Base (http://imip.ia.ac.cn/casia-kb/).

References

1. Berners-Lee, T., Hendler, J., et al.: The Semantic Web. Scientific American (2001)
2. Bizer, C., Heath, T., et al.: Linked data on the web. In: Proceeding of the 17th International Conference on World Wide Web, pp. 1265–1266. ACM, New York (2008)
3. Ding, L., Franzo, D.D., et al.: Data-gov Wiki: Towards Linking Government Data. In: AAAI Spring Symposium (2010)
4. Nigel, S., Kieron, O., Berners-Lee, T., et al.: Linked open government data: Lessons from data.gov.uk. IEEE Intelligent Systems, 16–24 (2012)
5. Tambouris, E., Gorilas, S., Kavadias, G., Apostolou, D., Abecker, A., Stojanovic, L., Mentzas, G.: Ontology-Enabled E-Gov Service Configuration: An Overview of the OntoGov Project. In: Wimmer, M.A. (ed.) KMGov 2004. LNCS (LNAI), vol. 3035, pp. 122–127. Springer, Heidelberg (2004)
6. Chen, C.-C., Yeh, J.-H., Sie, S.-H.: Government Ontology and Thesaurus Construction: A Taiwanese Experience. In: Fox, E.A., Neuhold, E.J., Premsmit, P., Wuwongse, V. (eds.) ICADL 2005. LNCS, vol. 3815, pp. 263–272. Springer, Heidelberg (2005)
7. Liang, A.C., Boris, L., Margherita, S., et al.: From AGROVOC to the Agricultural Ontology Service. In: Proceedings of the International Conference on Dublin Core and Metadata Applications (2006)
8. Xian, J.G.: The Study and Implementation of the Conversion System from Chinese Agricultural Thesaurus to Agricultural Ontology. Chinese Academy of Agricultural Sciences (2008)
9. Woods, J.W., Sneiderman, C.A., et al.: Using UMLS metathesaurus concepts to describe medical images: Dermatology vocabulary. Comp. Biol. Med. 36, 89–100 (2006)
10. Stark, M.M., Riesenfeld, R.F.: Wordnet: An electronic lexical database. In: Proceedings of 11th Eurographics Workshop on Rendering. MIT Press, Cambridge (1998)
11. Dong, Z.D., Dong, Q.: Hownet, http://www.keenage.com
12. http://www.wikipedia.org
13. http://baike.baidu.com
14. http://www.hudong.com
15. Suchanek, F.M., Kasneci, G., et al.: YAGO: A Large Ontology from Wikipedia and WordNet. J. Web Sem. 6(3), 203–217 (2008)
16. Bizer, C., Lehmann, J., Kobilarov, G., Auer, S., Becker, C., et al.: DBpedia - A Crystallization Point for the Web of Data. Journal of Web Semantics 7(3), 154–165 (2009)
17. Wu, F., Weld, D.S.: Automatically Refining the Wikipedia Infobox Ontology. In: Proceeding of 17th International Conference on World Wide Web, pp. 635–644. ACM (2008)
18. Wu, F., Weld, D.S.: Autonomously semantifying Wikipedia. In: Proceedings of the 16th ACM International Conference on Information and Knowledge Management, pp. 41–50. ACM Press, New York (2007)
19. Huang, L.X.: Concept of thesaurus and its application in Web Information Retrieval. Modern Information 8, 171–172 (2005)
20. Zhao, X.L.: Chinese General E-Government Thesaurus (Category table). Scientific and Technical Documentation Press (2005)
21. Zhao, X.L.: Chinese General E-Government Thesaurus (Alphabetical table). Scientific and Technical Documentation Press (2005)
22. Zeng, Y., Wang, H., Hao, H.W., Xu, B.: Statistical and Structural Analysis of Web-based Collaborative Knowledge Bases Generated from Wiki Encyclopedia. In: Proceedings of the IEEE/WIC/ACM International Conference on Web Intelligence (2012)
23. Mei, J.J., et al.: Tongyici Cilin. Shanghai Lexicographical Publishing House (1983)
24. Salton, G., Buckley, C.: Term-weighting approaches in automatic text retrieval. Information Processing & Management 24(5), 513–523 (1988)

Towards a Generic Hybrid Simulation Algorithm Based on a Semantic Mapping and Rule Evaluation Approach

Christoph Prackwieser, Robert Buchmann,
Wilfried Grossmann, and Dimitris Karagiannis

University of Vienna, Research Group Knowledge Engineering
Währingerstrasse 29, 1090 Vienna, Austria
{prackw,rbuchmann,dk}@dke.univie.ac.at,
wilfried.grossmann@univie.ac.at

Abstract. In this paper we present a semantic lifting methodology for heterogeneous process models, depicted with various control flow-oriented notations, aimed to enable their simulation on a generic level. This allows for an integrated simulation of hybrid process models, such as end-to-end models or multi-layer models, in which different parts or subprocesses are modeled with different notations. Process simulation outcome is not limited to determining quantitative process measures as lead time, costs, or resource capacity, it can also contribute greatly to a better understanding of the process structure, it helps with identifying interface problems and process execution requirements, and can support a multitude of areas that benefit from step by step process simulation - process-oriented requirement analysis, user interface design, generation of business-related test cases, compilation of handbooks and training material derived from processes.

Keywords: business process modelling, simulation, hybrid process models, semantic lifting.

1 Introduction

Process simulation is often seen as a method to determine such quantitative process measures as lead time, costs, or resource capacity. However, simulating a process and graphically animating or highlighting its relevant paths, also contributes to a better understanding of the process, helps with identifying interface problems, and supports communication of altered process implementations (possibly deriving from the introduction of new software, underlying technology or from organizational changes). Despite the fact that process simulation is a well-known analysis method, it is surprisingly not very widely used in practice, on one hand due to quantity and quality of input data, on the other hand due to the models themselves. This may be caused by the diversity of business process modeling languages generating heterogeneity of the process repository even within the same company, and thus requiring extensive alignment in order to setup the input for simulation. Each language tries to find the right level of abstraction, expressing concepts that are fit to different contexts, while

M. Wang (Ed.): KSEM 2013, LNAI 8041, pp. 147–160, 2013.

ignoring others. Some modelers may see shortcomings where others see strengths as a recent argumentation of Prof. Reisig [1] proves.

With our approach, we aim to overcome the legacy of notation heterogeneity, under the assumption that different individuals or units contribute with fragments using different control flow-oriented process modeling notations to a hybrid model encompassing a cross-organizational, cross-functional or multilayered process. Best practices from supply chain management, such as SCOR [2] recommend a tight integration and visibility of processes, from the customer's customer to the supplier's supplier.

The presented approach allows for simulating such heterogeneous models, while preserving their graphical representation, thus reducing additional re-modeling efforts and increasing the original model creator's identification with the simulation result.

With the rise of quite universally usable modeling notations like Business Process Modeling Notation (BPMN) or Unified Modeling Language (UML) this problem may be defused over time (under the assumption of universal acceptance) but there are still a lot of legacy models in companies and not every modeling requirement is satisfied with such standardized modeling notations [3]. Models that were created for special project goals in another modeling notation could be linked and simulated together to form, for example, a cross-functional end-to-end process, including business-focused and IT-related process models. This also reduces the initial effort for model creation.

The structure of the paper is as follows: After this introduction we discuss related literature and similar approaches in Section 2, where the problem is also formulated in the context of hybrid process modeling. In Section 3, we describe our approach, both conceptually and with a proof of concept, along with some implementation notes regarding the model transformation required for running a simulation, and the simulation itself. A conclusive SWOT analysis closes this paper.

2   Problem Statement and Background

2.1   Problem Positioning

In literature the term "hybrid" is used for a variety of model characteristics or simulation functionalities [4]. In this paper, a process model is called a "hybrid process model" when at least one of its process fragments or sub-processes is modeled in a different modeling notation. The analysis of end-to-end processes concerns a variety of domains and, therefore, involves different stakeholders. Although every domain has to contribute to the overall enterprise goals, each one has its own circumstances and objectives that lead to the application of specialized approaches and instruments to deal with each domain's requirements [5]. This general observation is also true for modeling languages. Although there are standardized modeling methods like BPMN or UML, special process-oriented application scenarios such as compliance management, risk management, or platform-specific process automation are not sufficiently supported. Hybrid process models can appear in a horizontal-oriented form as an end-to-end process, a vertical-oriented multi-layer process, or a combination of both.

Horizontal-oriented hybrid process models can come up when domain specific process models of two or more departments/units are linked together subsequently to form a cross-functional end-to-end process. Vertical hybrid models can be a result of a classical functional decomposition method wherein different modeling notations are used to depict various levels of granularity or stakeholder perspectives on the model. A strict vertical orientation means that the whole model is built up by sub-process models that are linked to form a hierarchy and are not aligned subsequently. The link is established from a process step to a corresponding sub-process, which describes the step in greater detail. Another application example described by Muehlen and Recker as "emerging tendency in industry to separate business-focused process modeling from implementation-oriented workflow implementation" [6] where workflow models typically use implementation platform-specific modeling languages or special dialects of standardized languages, such as BPMN. Also for classical, not workflow oriented software implementation a business focused process model could have links to a user interface flow model to show the software support of specific business-oriented tasks. In this context, we state the problem through its following components:

Assumption: A manufacturing company coordinates a virtual enterprise with multiple possible configurations of its supply chain. **Open issues:** The coordinating company needs to evaluate end-to-end processes for specific virtual enterprise instances. The participating business entities are willing to provide process visibility, but they use different tools and languages to model their processes;

Assumption: A manufacturing company has its process map designed according to a multi-layered framework such as SCOR. **Open issues:** The company aggregates processes from its multiple SCOR layers to simulate their cost but uses different notations for high level supply chain and low level workflows, as required by auditing and documentation purposes.

Proposed solution: A methodology (enabled at meta-modeling level) of annotating process elements represented with different notations but having overlapping semantics, in order to enable the detection of work order along simulated process paths (while preserving the original notations).

2.2   Literature and Related Work

A starting point for our research interest was the "Hybrid Method Engineering" approach introduced by Karagiannis and Visic [3] as part of the "Next Generation Modeling Framework" (NGMF). On a conceptual level the approach deals with the merging of two or more modeling methods into a single hybrid metamodel, which is described in more detailed [7] using an encapsulation mechanism. They also present a corresponding technical realization platform and an organizational structure. Their approach is based on Karagiannis and Kühn's [8] proposed structure of modeling method components. Therein, different modeling languages of a hybrid method form together with modeling procedures a modeling technique. The modeling technique and dedicated mechanism and algorithms are the main parts of a modeling method in turn. These mechanism and algorithms are automated functionalities, which are performed on the models in order to use or evaluate them. To define a hybrid metamodel they suggest the usage of a metamodeling language which is defined by a meta-metamodel. In this approach, a generic algorithm such as universal simulation, which should work on all involved modeling languages, must be implemented on a meta^2-model whereof all language specific sub metamodels are instantiated.

Whereas there are several publications describing transformation from one modeling language to another often motivated by Model Driven Architecture (MDA) [9] approaches, quite few sources are concerned with the conceptualization of generic (simulation) algorithms for hybrid models. A good overview of modeling language transformations involving block-oriented languages, especially BPEL, is provided in [10]. Other papers describe, for example, transformations from BPMN to BPEL [11], from UML to BPEL [12], and from BPEL to EPC [13]. Because of the numerous analysis methods and simulation algorithms applicable on Petri Nets, there are also some publications concerning transformations from BPEL to Petri Nets [14]. To deal with simulation tools that are merely capable of processing models in one specific language, different modeling method transformation approaches are presented by various authors. Rozinat, Wynn et al. [15] use a simulation tool that is based on Colored Petri Nets (CPN) so they describe how to convert YAWL models by preserving simulation relevant data into Petri Nets.

3   The Proposed Approach

3.1   The Simulation Core Concepts

As stated earlier, we strive to enable a generic simulation algorithm which has **the key requirement** of being applicable on control flow oriented process models while leaving the simulated model notations unchanged in order to improve the user's identification with (and understanding of) simulations results. To simulate different process modeling languages using one generic algorithm there is a need for a semantic mapping of specific modeling language object types to the metamodel on which the simulation algorithm relies.

There are alternative realization options to fulfill the key requirement. One could hardcode all possible object types of different predefined modeling languages into the simulation algorithm. However, this alternative lacks flexibility, as with every newly added domain-specific language program code of the simulation algorithm has to be changed. Another possibility would be to transform the graphical modeled process each time immediately before a simulation run takes place into a "super notation" known by the simulation algorithm. The problem hereby is that the user has to depend on this automatic transformation step and cannot check if the meaning of the process model is translated correctly in the intermediate format. A token animation, for example, to depict waiting queues is difficult to implement because the simulation and animation algorithm are not executed on the basic models directly.

In our approach, the algorithm works directly on the graphical model, which is enhanced by a layer of generic formulated information about the objects' semantics, flow control, and resource assignment. This information is generated during a separate transformation step and stored in specific attributes of each modeling object (however this transformation is much less complex than a complete redesign of models in another notation). Thus, an experienced user is able to alter the output of the transformation step before the simulation takes place; the user therefore, has the ability to formulate the input for the simulation specific to individual needs.

To identify necessary semantic core concepts of a generic simulation algorithm which have influence on the control flow of a process, we analyzed various graphical process modeling notations. To be sure to support both horizontal and vertical oriented hybrid models, we selected modeling languages that are used for at least one of the following domains: a) Strategic core processes; b) Business processes; and c) IT-specification, process automation.

We compiled the list of the major business process modeling methods with graphical notations out of a structured literature study [16-19]. Included notations are Process Maps and Value Added Chain Diagrams (VACD), Business Process Modeling Notation (BPMN), Event-Driven Process Chains (EPCs), UML Activity Models, ADONIS Business Process Management Systems (BPMS) [20], Process Schematics of the Integrated Definition Method 3 (IDEF3), Flow Chart Notation, Task Flow User Interface Modeling and Business Process Execution Language for Webservices (BPEL4WS).

Table 1. Semantic concepts of flow objects relevant for simulation

Notation	Semantic Concepts of Flow Objects Relevant for Simulation								
	Activity	Sub Process	Start	End	XOR	AND	Merge	Event	Neutral
Value Chain Models	x	x	x	x					x
UML 2.0 Activity Diagrams	x		x	x	x	x	x		
EPC	x	x	x	x	x	x	x	x	x
ADONIS BPMS	x	x	x	x	x	x	x	x	x
BPMN 2.0	x	x	x	x	x	x	x	x	
IDEF3	x	x			x	x	x		
Flow Charts	x	x	x	x	x	x	x		
Mask Flow Diagrams	x	x	x	x	x			x	
BPEL	x	x	x	x	x	x	x	x	

Table 2. Simulation-related semantic concepts

Semantic Concept	Keyword	Description
Activity	ACT	Describes an executable task in a process. Each execution consumes execution time of a resource such as a role or IT system and causes costs.
Sub Process	SUB	Is a placeholder for a linked process fragment. It is used for structuring of process models or reuse of process fragments.
Start	STA	Depicts the arrival of new entities into the process model. The simulation generates corresponding process token which are transfered through the model
End	END	Marks the end of a process path. Each arriving token will be consumed.
XOR	XOR	Depicts an exclusive or decision of different outgoing control flow branches
AND	AND	Depicts the possible parallel execution of the same process instance. Therefore the token will be cloned at the AND object. Outgoing control flow branches can be annotated with condition which leads to an OR functionality
Merge	UNI	All at the latest AND object cloned tokens are merged to one token. This synchronisation can lead to waiting time.
Event	EVE	Depicts an intermediate event which controls the flow of tokens regarding predefined condtions. A flow interruption can lead to waiting times
Neutral	NTR	It is within the controlflow but has no effekt on token flow or simulation result itself. It can be used for example for cross-references between processes.

In this simulation approach, we consider the control flow only. Information flows like message handling in BPEL4WS are not supported so far. Furthermore, the simulation algorithm is currently applicable for event-based graphical modeling notations only where transitions and connecting control flows are modeled explicitly.

To extract the semantic core pattern of simulation perspective, we analyzed each modeling notation and extracted all of the simulation relevant patterns in regard to flow control and influence on simulation results. As a result of this study we compiled a list of nine core patterns, as we have illustrated in Table 1. In Table 2, all identified simulation-related sematic concepts are explained. The listed keywords are used for rules formulation. The mapping of object types to semantic simulation concepts is not sufficient. To gain more flexibility and cover as much semantics as possible, we add a rule mechanism to determine the next object in the control flow.

One of the preconditions of our approach is that each modeled object is mapped to exactly one simulation core concept. The introduction of another semantic concept would solve this problem, but we want to keep the number of semantic simulation concepts as low as possible. Therefore we developed a rule language that enhances the semantic meaning of an object with annotation indicating the next object in the control flow, to be taken as input by the simulation algorithm. The basic syntax is the following:

_KEYWORD__RuleForDeterminationOfSubsequentObject_

3.2   The Rules Syntax

We use the rules to identify the subsequent object in a process, as an input for the simulation algorithm. However, the rules mechanism is not limited to the representation of the control flow only, so we can also utilize it to express conditions for the assignment of executing resources to an activity or to formulate a distribution of entry arrival in start objects.

To represent a rule in our concept we use the OMG Production Rule Representation as a basic syntax where a production rule is defined as a "**statement of programming logic** that specifies the execution of one or more actions in the case that its conditions are satisfied" [21]. A production rule consist of one or more conditions in the form of logical formulas and one or more produced actions in the form of action expressions [22]. A typical conditional rule is represented as:

```
IF condition THEN producedAction ELSE alternativeAction
```

This syntax is capable of expressing rules of the following types:

Flow rules. In control flow-oriented process models the chronological subsequence of tasks is typically depicted by connecting arrows (directed edges) from a predecessor to its immediate successor object. A flow rule that is annotated at the preceding object reproduces the information about the successor given by the outgoing arrow. The modeled arrows remain in the process model; although they are not necessary for the simulation anymore, they are very important for a quick understanding of the process structure and the simulation results. We are aware, though, that this enhancement is redundant information, but by introducing an additional layer, we gain the flexibility to describe the control flow independent of the modeling languages, to be used by the generic simulation algorithm. Therefore, the evaluation of the flow rules has to lead to the designated successor regarding the control flow of a process model. For the identification of the subsequent object a unique identifier, we use the unique internal ID of an object provided by the meta-modeling platform.

One of the most frequently used control flow structures in process management is a sequential flow. To express this structure in a simple form, we introduce the following short form syntax here in conjunction with a keyword (see Table 2):

_KEYWORD_*IDOfSubsequentObject*

This is the same as:

_KEYWORD_IF *TRUE* THEN *IDOfSubsequentObject*

In **Fig. 1**, there are two examples for sequential flows and the respective mapping results shown. On the left side, a BPMN task with ID 101 has the semantic meaning of an activity and a subsequent object ID 102. The right side shows a Value Chain that has no connecting arrows at all, but the subsequence is given by the horizontal position of the objects. Also, the process objects have a semantic meaning of activities. The flow rule subsequence is derived from the object's graphical position.

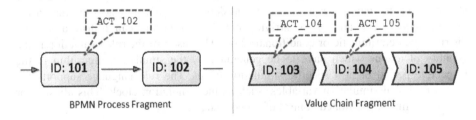

Fig. 1. Flow rule examples for a sequential control flow

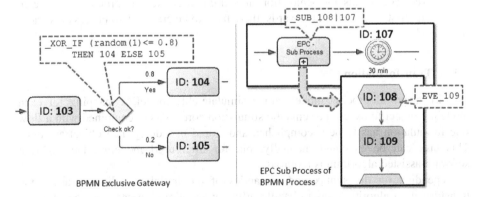

Fig. 2. Flow rule examples for gateways and subprocesses

A typical example for the application of a flow rule is the occurrence of a XOR in a process model. As depicted in **Fig. 2** on the left side the Exclusive Gateway in BPMN notation corresponds to an XOR and is expressed in Flow Rule notation as a programmatic condition. The statement "random(1)" generates a randomly distributed number which is used to decide on the subsequent object, according to the given branch probability for each token at simulation runtime. On the right side, there is a sub-process object in a BPMN process, which calls for a process in EPC notation. This is an example of a hybrid process model. BPMN sub-process objects corresponds to the semantic concept Sub Process "SUB". The syntax '*IDOfSubProcessObject | IDOfSubsequentObject*' commands the simulation algorithm to simulate the sub-process beginning with the object *IDOfSubProcessObject* first. After running through the sub-process and returning to the calling process, the process will then proceed with the object *IDOfSubsequentObject.* These IDs can also be the results of (nested) rules.

Resource rules. Some process modeling notations allow for the declaration of activity executing resources (for example, EPCs and; ADONIS BPMS). Resources are often concepts like roles, persons, organizations, tools, or IT-systems, such as software applications or hardware. A simulation analysis that yields to quantitative results like lead time, capacity consumption, or process costs should include these resources. Therefore, every activity has to be evaluated with an average execution time. Every time this activity is executed, the assigned resource is occupied for the whole execution time. Rules can be used to assign resource IDs to activities.

Start rules. Objects that map to the semantic simulation concept "Start" are the starting point for new process instances and depict the arrival of new entities in the process model. The amount of arriving entities per period can follow a random distribution, deterministic or conditional [23]. The result of the rule evaluation is the quantity of tokens which have to be generated by the simulation algorithm in the respective period or at a specific simulation event. The rule language supports the access to global simulation variables, such as the simulation clock. This allows, for example, a time-conditioned control of process starts.

```
IF (Period <= 20) THEN 1 ELSE 2
```

This rule commands the simulation algorithm to create one process instance in each of the first 20 periods; afterwards, there have to be created two process instances per period.

3.3   Transformation

To turn a hybrid process model into a simulate-able model, a mapping for each modeling object to exact one semantic simulation core concept combined with a flow rule formulation has to be accomplished and stored in a designated object attribute. This task can be executed manually, but for often-used modeling languages, a software-assisted algorithm is required.

Depending on the syntax and semantics of the initial modeling notation, the transformation algorithm has to handle different situations. In case the semantics of the modeling notations matches the simulations core semantic, like in some flowchart notations, only a 1:1 mapping of modeling object types to simulation concepts is necessary. However, with increasing complexity and expressiveness of initial modeling notations, the transformation algorithm has to not just recognize the object type, but it must also interpret the context of an object. As an easily understandable example is that an AND from the EPC notation could be of the semantic type AND if it has more than one outgoing connectors, or the semantic concept of a merge if there is just one outgoing connector.

In some cases, further user input is required. For example, in a value chain, a process object might have more than one outgoing connectors; this could be interpreted as either the start of a parallel or an alternative path. Besides the transformation of objects and control flows, there might also be a need for transforming a resource assignment. In the EPC notation, resources are modeled as separate objects connected by a relation to the specific function. In ADONIS BPMS, there are object references linked from specified attributes of activities to resource objects like roles. In BPMN, there is no standardized resource assignment, but a user can decide to adopt the organizational object assigned to the respective lane.

As there are no defined modeling standards, the formulation of start-rules has to be made manually.

3.4   Simulation

Static modeling objects, like activities or events, are traversed by a temporary entity's act on them. The simulation algorithm's main task is to control the time advance, to manage the interaction of static and temporary objects, and to collect results. During a simulation run, temporary objects handled by the algorithm are called tokens. The algorithm provides data for the visualization component to visualize the token path throughout the model, and updates the simulation log with collected data. The result component relies on the log file to create reports on various aggregation levels.

Thus, validity of results is dependent on a) the correctness of the original models; b) the availability and correctness of the data to be collected from the model (which should be execution data stored in attributes of the modeling objects, like time, cost); c) the successful transformation of the hybrid process model, as described in the previous section. There has to be at least one object with a semantic meaning of "Start" including a valid start rule. To execute the algorithm, the start model has to be selected, and the number of intervals or process instances that should be simulated has to be provided.

In this paper, we concentrate on the method of enabling a generic simulation via the semantic mapping. Therefore, we will not describe the functional principles of the algorithm in deep detail. As state changes in business processes happen at precise time points, we use a Discrete Event Simulation Algorithm [24] as a basis for our approach. To provide a suggestive token animation, we chose the time slicing method, with fixed intervals to move the model forward in time. Roughly, for each time interval, the following steps are executed by the algorithm:

— At the beginning of each time interval, the algorithm identifies all objects that are mapped to the semantic simulation concept "Start." The algorithm evaluates in cooperation with the rule engine each start-rule of these objects and creates a corresponding number of new tokens, which represent process instances;

— Afterwards, the algorithm traverses each active token in the model. The status of a token can be "idle" or "ready." A status of "idle" indicates that the corresponding work order is still being executed by an activity or that the token has to wait due an event, synchronization in hand, or occupied resource. A token with the "ready" status will be moved by the algorithm to the subsequent object, according to the flow rule of the token's current object, until the token status changes to "idle" or the token is terminated by an end object;

— For report generation and further applications, every step is logged during the simulation run;

— All token movements are depicted by a visualization component.

3.5   Proof of Concept

Simulation results are not only useful for time, capacity, or cost based analysis. A study of the dynamic behavior of a process model helps to understand the wanted and unwanted interactions between process instances better and creates an additional time-based and animated visualization on the otherwise static models.

One advantage of simulation in contrast to analytic methods is the possibility to identify and investigate specific process paths throughout the end-to-end process model. Each distinct chronological sequence of modeling objects that are traversed by at least one process instance (case) is a process path.

As a first evaluation step of this approach we integrated different modeling notations such as Value Chain, BPMN and EPC in a meta-modeling platform (the list from Table 2 should be supported with minimal additional effort once the transformation is in place). Further on, we implemented the transformation, simulation and visualization algorithms by using the scripting language of the platform.

Fig. 3 shows screenshots of this implementation. The different models are connected, building a vertical hybrid model from subprocesses using different notations. The application of the simulation algorithm results in a number of possible paths, whereas three are depicted in the result list. Path 2 is highlighted throughout the process models, and all traversed objects and the resulting execution, waiting and cycle times are shown in the table.

In combination with these attributes, the path information could be used to generate auxiliary value:

— process-oriented handbooks or training material;

— business-related test cases;

— process-oriented requirement analysis;

— analysis of IT systems used throughout specific process paths for business continuity management;

— user interface design, especially to present applications behavior regarding specific processes.

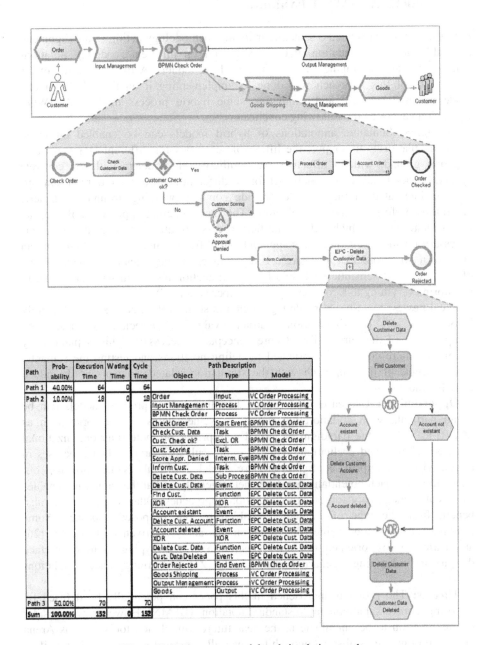

Fig. 3. Hybrid process model and simulation results

4   Conclusive SWOT Evaluation

In this paper, we presented a semantic mapping methodology acting as an alignment, to enable generic simulation algorithms to run on different modeling notations without altering the graphical representation of the models. A first evaluation of the practicality of this approach was done by implementing it on the ADOxx meta-modeling platform [25] and simulating some hybrid process models. A SWOT evaluation generated the following conclusions:

Strengths: Semantic annotations of hybrid models can be enabled on meta-modeling platforms to support the lifting of semantics for different languages to a common ground, without affecting the familiarity of modelers with the concrete syntax. Legacy models, or models that follow different notations due to requirements, may be integrated in this way to provide common working ground for generic algorithms performing model evaluation. The metamodeling approach builds up an intermediate layer which allows unified access to all core semantic concepts necessary for the simulation. This approach simplifies the mapping function from an order of n^2 to an order or n (where n is the number of languages to be mapped) and offers a common semantic backbone for the simulation algorithm without having to transform everything to a common syntax ("super notation").

Weaknesses: State-based modeling languages such as Petri Nets are not currently involved in our experimentation. Further evaluation, especially regarding the completeness of the current list of core concepts, is necessary. This is planned by integrating more control flow-oriented modeling notations and testing the semantic core concepts of this approach with the simulation related concepts of the added modeling notations.

Opportunities: Use cases that require handling hybrid process models can be found particularly in collaborative supply or value chains. Frameworks such as SCOR, as well as the paradigm of virtual enterprises, promote inter-organizational process visibility and evaluation. While it is difficult to guarantee that processes in ad-hoc virtual enterprise configurations are modeled with the same language, our annotation approach potentially preserves heterogeneity and at the same time enables sufficient homogeneity for certain simulation approaches. Other domains may also benefit from the type of simulation enabled by our proposed approach. Some examples are software engineering (requirements analysis driven by processes [26] automatic business-oriented test case generation or prototyping in user interface design) and e-learning (scenario-based training [27] or learning space simulations [28]).

Threats: The problem approached in this paper can potentially be defused by imposing a universal alignment to standard notations (BPMN). However, we consider this assumption to be unrealistic in the near future. Simulation tools such as Arena also support notations abstract enough to cover all control flow concepts in a unified way, but our approach aims to preserve the original models and to provide the semantic lifting in a more transparent manner (enabled on the metamodeling level, by the annotating rule attributes).

References

1. Reisig, W.: Remarks on EgonBörger: "Approaches to model business processes: a critical analysis of BPMN, workflow patterns and YAWL, SOSYM 11: 305-318". Software and System Modeling 12(1), 5–9 (2013)
2. Supply Chain Council, The Supply Chain Operations Reference Model, http://supply-chain.org/resources/scor
3. Karagiannis, D., Visic, N.: Next Generation of Modelling Platforms. In: Grabis, J., Kirikova, M. (eds.) BIR 2011. LNBIP, vol. 90, pp. 19–28. Springer, Heidelberg (2011)
4. van Beek, D.A., Rooda, J.E.: Languages and applications in hybrid modelling and simulation: Positioning of Chi. Control Engineering Practice 8, 81–91 (2000)
5. Clark, T., Sammut, P., Willans, J.: Applied metamodelling: a foundation for language driven development, Ceteva, Sheffield (2008)
6. zur Muehlen, M., Recker, J.: How Much Language Is Enough? Theoretical and Practical Use of the Business Process Modeling Notation. In: Bellahsène, Z., Léonard, M. (eds.) CAiSE 2008. LNCS, vol. 5074, pp. 465–479. Springer, Heidelberg (2008)
7. Zivkovic, S., Kühn, H., Karagiannis, D.: Facilitate Modelling Using Method Integration: An Approach Using Mappings and Integration Rules. In: Proceedings of the ECIS 2007, vol. 122 (2007)
8. Karagiannis, D., Kühn, H.: Metamodelling platforms. In: Bauknecht, K., Tjoa, A.M., Quirchmayr, G. (eds.) EC-Web 2002. LNCS, vol. 2455, p. 182. Springer, Heidelberg (2002)
9. OMG Model Driven Architecture, http://www.omg.org/mda/
10. Mendling, J., Lassen, K.B., Zdun, U.: Transformation Strategies between Block-Oriented and Graph-Oriented Process Modelling Languages. Vienna University of Economics and Business (2005)
11. Ouyang, C., Van Der Aalst, W.M.P., Dumas, M., Ter Hofstede, A.H.M.: From Business Process Models to Process-oriented Software Systems: The BPMN to BPEL Way (2006)
12. Gardner, T.: UML Modelling of Automated Business Processes with a Mapping to BPEL4WS. In: Proceedings of the First European Workshop on Object Orientation and Web Services at ECOOP (2003)
13. Mendling, J., Ziemann, J.: Transformation of BPEL processes to EPCs. In: 4th GI Workshop on Event-Driven Process Chains (EPK 2005), CEUR Workshop Proceedings, pp. 41–53 (2005)
14. Hinz, S., Schmidt, K., Stahl, C.: Transforming BPEL to Petri Nets. In: van der Aalst, W.M.P., Benatallah, B., Casati, F., Curbera, F. (eds.) BPM 2005. LNCS, vol. 3649, pp. 220–235. Springer, Heidelberg (2005)
15. Rozinat, A., Wynn, M.T., van der Aalst, W.M.P., ter Hofstede, A.H.M., Fidge, C.J.: Workflow simulation for operational decision support. Data & Knowledge Engineering 68, 834–850 (2009)
16. Wei, W., Hongwei, D., Jin, D., Changrui, R.: A Comparison of Business Process Modeling Methods. In: IEEE International Conference on Service Operations and Logistics, and Informatics, SOLI 2006, pp. 1136–1141 (2006)
17. Fu-Ren, L., Meng-Chyn, Y., Yu-Hua, P.: A generic structure for business process modeling. Business Process Management Journal 8, 19–41 (2002)
18. Wenhong, L., Alex, Y.T.: A framework for selecting business process modeling methods. Industrial Management & Data Systems 99, 312–319 (1999)

19. List, B., Korherr, B.: An evaluation of conceptual business process modelling languages. In: Proceedings of the 2006 ACM Symposium on Applied Computing, Dijon, France, pp. 1532–1539. ACM (2006)
20. BOC Information Technologies Consulting AG: Method - BPMS (Business Process Management System), `http://www.boc-group.com/products/adonis/method-bpms/`
21. Production Rule Representation (PRR). Version 1.0, vol. formal/2009-12-01 (2009)
22. Wagner, G.: Rule Modeling and Markup. In: Eisinger, N., Małuszyński, J. (eds.) Reasoning Web. LNCS, vol. 3564, pp. 251–274. Springer, Heidelberg (2005)
23. Tumay, K.: Business process simulation. In: Proceedings of the 1995 Winter Simulation Conference, pp. 55–60 (1995)
24. Banks, J.: Introduction to simulation. In: 1999 Winter Simulation Conference Proceedings, WSC 1999. 'Simulation - A Bridge to the Future' (Cat. No.99CH37038) (1999)
25. BOC Group, ADOxx product page, `http://www.adoxx.org/live/`
26. Li, G., Jin, Z., Xu, Y., Lu, Y.: An Engineerable Ontology Based Approach for Requirements Elicitation in Process Centered Problem Domain. In: Xiong, H., Lee, W.B. (eds.) KSEM 2011. LNCS, vol. 7091, pp. 208–220. Springer, Heidelberg (2011)
27. Liew, B.Y.T., Tsui, E., Fong, P.S.W., Lau, A.S.M.: Rapid Authoring Platform for Instructional Design of Scenarios (RAPIDS). In: Proceedings of IEEE 12th International Conference on Advanced Learning Technologies, Rome, Italy, pp. 473–475. IEEE (2012)
28. Gao, S., Zhang, Z., Hawryszkiewycz, I.: Supporting adaptive learning in hypertext environments: a high level timed Petri net-based approach. International Journal of Intelligent Systems Technologies and Applications 4(3/4), 341–354 (2008)

Micro-blog Hot Event Detection
Based on Dynamic Event Model

Hua Zhao and Qingtian Zeng[*]

College of Information Science and Engineering, Shandong University of Science
and Technology, Qingdao, China, 266590
doctorhuazhao@yahoo.com.cn

Abstract. Micro-blog is a rapid developing platform, where users can share information about the event happened in the real life, and sometimes many hot
events even originate in this platform. So it is very important to detect the hot
event in the micro-blog texts. Micro-blog text is a kind of the short, fractional
and grass-root text. Based on the analysis of the dynamic evolvement property
of the event, we propose a micro-blog event detection method based on dynamic event model, where the event model is created based on the two centroids.
This paper firstly analyzes the dynamic evolvement property, secondly introduces the creation and modification process of the two-centriods based event
model, and finally gives the event detection method based on the proposed
dynamic event model. Experimental results showed that our proposed model
improves the performances of the micro-blog event detection greatly.

Keywords: micro-blog, event detection, dynamic event model, dynamic
evolvement.

1 Introduction

With the development of Web 2.0, micro-blog has become more and more popular
and important in the people's daily life. It is an information publication, spread and
achievement platform, where the users can add his/her interested users as his/her
friends, and then can receive the information published from his/her friends. The
length of micro-blog is limited to 140, which is very short, which cause some difficulties for event detection because of the data sparseness. The limited length and the
huge user group make large contribution to the rapid development of micro-blog,
which makes micro-blog an important place for monitoring. So it is very important
for the decision-maker to detect their interested events from the micro-blog texts rapidly and accurately.

Event detection is a task of the Topic Detection and Tracking (TDT), and originally is defined to be the task of automatically detecting new events in the news stream
and associating incoming stories with topics created so far[1]. So many researchers
adopt event detection to detect the hot event in the micro-blog texts, but the traditional

---

[*] Corresponding author.

M. Wang (Ed.): KSEM 2013, LNAI 8041, pp. 161–172, 2013.

event detection can not be applied to the micro-blog text directly because the micro-blog texts have its own properties which will be given in section 3.

Dynamic evolvement property is an important property of the event, which means the focus of the event will change as time goes on, and the property will be analyzed in Section 3. In order to detect the hot events in the micro-blog text effectively, we propose a micro-blog topic detection method based on dynamic event model, where we create an event model based on two centroids. Experimental results showed that our proposed methods improve the performances of the micro-blog event detection greatly.

The structure of the paper is as follows. Section 2 gives a short overview of the current approaches in the micro-blog event detection research. Section 3 analyzes the properties of the microblog and the event. Section 4 laid an emphasis on the creation and the modification of the dynamic event model. Section 5 presents the event detection method based on the dynamic event model Section 6 discusses the experimental results and analysis. Section 7 gives the conclusions inferred from our work.

2   Related Work

The event detection research has aroused many researchers' interest. Most of the researches orient the Twitter, which is the most big microblog platform [2]. Takeshi proposed to monitor the event based on the analysis of Twitter texts which is based Bayesian, and they can detect more 80% earthquake [3]. Sasa and Miles proposed a new event detection method which can deal with 1.6 billion quickly [4].

The event detection research for Chinese microblog has just begun. Tong Wei used Vector Space Model to express the text model when detected hot topic from the Sina micro-blog [5]; Ma Bin proposed a thread tree-based microblog topic detection method, compared the performance of TFIDF and LDA, and they found that LDA model is better than TFIDF [6]. Due to the length of microblog text, the data sparseness problem is serious in microblog topic detection. The traditional method to solve this problem is to expand the text with external resources. For example, Ishikawa, S. [7] expanded the micro-blog texts based on Wikipedia, which has got better performance. More researchers expanded the micro-blog texts based on the semantic dictionary [8].

The micro-blog topic detection algorithm is still based on the clustering method, for example Single-Pass [9], K-Means, HMM, and so on. Many users explored to combine the structural property on the basis of the traditional topic detection algorithm. Based on the dialogic property of micro-blog, Ma Bin proposed a thread-based two stage clustering method [6]. Rui Long proposed a micro-blog topic detection method based on the word co-occurrence graph [10], which firstly creates the word co-occurrence graph, and then regard the disconnected cluster as a news topic. Lin Hongfei proposed an emotional language model, which realized the hot event detection based on the analysis of the difference between emotional language models in adjacent periods [11].

3   Property of Micro-blog and Topic Dynamic Evolvement

3.1   Property of Micro-blog

Before we carry out our works, we make a deep analysis to the properties of micro-blog texts, which as follows:

- It is a kind of dialogic text. So many micro-blog texts are the replies or the transmissions of other micro-blog texts, which cause the micro-blog text include many meaningless data.
- It is a kind of grass-roots text. Once a user registers in micro-blog, he/she then can publish his/her any interested information. Different users have the different word-styling, and they will usually adopt different words to express the same idea.
- It is a kind of short text. The length of the micro-blog text is limited to 140 words, which is short. The short property will cause data sparseness which will bring much difficulty to event detection.
- It is a kind of text including certain political ideas. Once an event happens, there will have many discusses about it in the micro-blog, which will include much sensitive information. So it is important for the decision-maker to monitor the micro-blog information to detect these discusses about this event.

There are also other properties, such as mass data, real-time updates, and so on. These properties make the micro-blog event detection more difficulties than the traditional topic detection.

3.2   Pre-process of Micro-blog Text

Based on the above analysis to the micro-blog texts, we can see that the text usually include much noise data, so before the topic detection, we must pre-process the downloaded micro-blog texts, which is very important to the success of event detection system. Now, our adopted pre-processes are as follows:

(1) Micro-blog texts cleaning. The noise data in the micro-blog text will make a bad influence on the accuracy of the similarity computation between two micro-blog texts during event detection, so the cleaning process is to delete the noise data:

- Delete URL. There are many URLs in the micro-blog, which is mainly because that the length of a micro-blog text is limited to be no more than 140 words, in order to give detailed introduction to a certain topic, users usually add a URL in its current micro-blog text, where will include the detail information. Some researchers make an effort to explore the usage of these URLs. Because the URLs are usually have the unified format, so we choose to delete them using regular expression.
- Remove user account. In the Sina Micro-blog, there are many user accounts, which are usually used in two cases, and we adopt different operations to these two cases. For the first case, when the current micro-blog A is the reply to the

micro-blog B, the user account is usually used as "//@user account" to cite the content of B. We then will delete the string "//@user account" completely for this case. The second case is when a user writes his/her micro-blog, he (she) sometimes mentions another micro-blog user. The user account will be used as "@user account" in this case. We will only delete the symbol "@" for this case because the user account is the actual content of the micro-blog.

- Delete emoticons. Emoticons are common in the Internet, also in the micro-blog text. Users usually use these emoticons to express their ideas about a certain topic. These emoticons are very useful for the sentiment analysis, but they are of lesser significance in keyword extraction. The emoticons are usually converted into strings with unified format, that is, the strings are enclosed by bracket, for example [sunlight], [doubt] and so on. According to the above foundlings, we delete the emoticons based on the predefined rules.

(2) Because we mainly detect the hot event from Chinese micro-blog, so we secondly carry out Chinese Word Segment (CWS). In our experiments, we adopt the ICTCLAS2011 to segment and tag, which is shared by Chinese Academy of Sciences.

(3) Remove the stop words.

3.3   Property of Topic Dynamic Evolvement

In this section, we will analyze the topic dynamic evolvement property, and the dynamic evolvement process is shown in Fig.1.

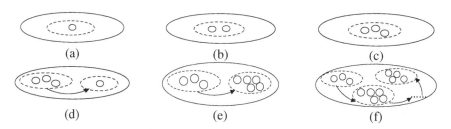

Fig. 1. Dynamic Evolvement Property of Event

In Fig.1, ⬯ represents a certain event, ⬯ represents a focus of the event, and ○ represents a micro-blog text, ➤ means there appears a dynamic evolvement between two focuses. (a) means that the event is just created, and only has one related micro-blog text. (b) and (c) mean that the second and the third related micro-blog texts are detected, and they have the same focus with the first text. But when the fourth text is detected (shown in (d)), it has the different focus with the former three texts, which means that appears the second focus. With the detection process going on, there detects more and more texts about the second focus, shown in (e). (f) represents all the focuses when the detection process is over. Now, we firstly give the following definitions.

- **Definition 1**: Evolvement Inflection Point (EIP): we call the arrow($\rightarrow$) between two focuses as the Evolvement Inflection Point (EIP).
- **Definition 2**: Centroid: each focus is called a centroid, and the newest centroid is called Current Centroid (CC), and all the other centroids are called Original Centroids (OC). Centroid is represented by Vector Space Model.

We must make clear that the process shown in Fig.1 means that the all the texts related to the first focus are detected, and then the texts related to the second focus, and then the third focus...But the dynamic evolvement process of the event in real life may not evolve strictly in accordance with the process, that is to say the texts related to different focus may comes crossways. But all the focus will roughly appear in a certain order. Take an air disaster as example, the first focus is usually about the introduction about the time, the place of the disaster, and the following focuses are usually about the rescue, the losses and the reasons and so on. So the process shown in Fig.1 is a much general process.

4   Dynamic Event Model Based on Two Centroids

4.1   Creation of Evolvement Inflection Point

4.1.1   When to Create Evolvement Inflection Point

The basic and critical process of the two-centroids based dynamic event model is to create the EIP correctly. Ideally, once there appears a dynamic evolvement in the focus of the event, a new EIP should be created. Based on the analysis of the event dynamic evolvement property, we find that during the evolvement, the words appearance accords with regular rule, so in order to describe the rule accurately, we propose a definition of distribution density and an EIP creation method based on the distribution density.

- **Definition 3**: Distribution density: the distribution density of the word w of the event E at time t is referred as $DD(t,w,E)$, and is computed by the following formula:

$$DD(t,w,E) = \frac{DF(t,w,E)}{TextNum(t,E)} \tag{1}$$

Where $DF(t,w,E)$ is the document frequency of the word w of the event E at time t, and $TextNum(t,E)$ is the number of the micro-blog texts which is detected to related to E at time t.

Our experimental results show that the distribution density of a word accord with the following three rules:

(1) Some words have the relatively higher distribution densities, shown in (a), which means these words are the common words that most of the centroids include.

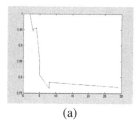 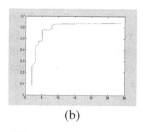 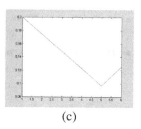

(a)   (b)   (c)

Fig. 2. Distribution Density

(2) Some words do not appear at the beginning of the event, and they appear during the evolvement of the event, shown in (b). The appearance of these words represents the appearance of the new focus, which is helpful to the creation of the EIP.

(3) Some words have the relatively lower distribution densities during all the evolvement of the event, shown in (c), and these words usually appear occasionally.

Based on the above analysis, we adopt these words shown in (2) as the EIP creation criterion.

4.1.2   The Method to Create Evolvement Inflection Point

Before we give the algorithm to create the EIP based on distribution density, let's give the following assumptions:

- **Assumption 1**: we assume that there has not evolvement within the first $\theta_{\neg evo}$ related micro-blog texts.

From the analysis in section 4.1.2, we can find that the assumption is reasonable; this is because, for a certain event, the first several texts are usually about the brief introduction to the basic information of the event.

Based on the assumption, we give the following EIP creation process:

(1) Define an EIP creation threshold, θ_{Num}

(2) When the system detects a related text, we do the following operations:

- If a word of the text appear in the event for the first time, then tag this word as candidate word, and record the appearance time, distribution density, set the appearance number ($appnum$) of this word to 1, that is $appnum=1$.

- If a candidate word cw has appeared for more than one day, disqualify it from candidate word; else compute the new distribution density of cw at this time, if the new distribution density is larger than the current recorded distribution density, then we modify   the distribution density of cw,   and modify appearance number: $appnum=appnum+1$.

- If the appearance number of a word within one day is up to θ_{Num}, then create a new EIP, and at the same time disqualify the word from candidate word.

4.2   Creation and Modification of Original Centroid

During the event detection process, the original centroid is created and modified along with the creation of EIP, as follows:

(1) When an event is just detected, there is no original centroid;

(2) When we create the second EIP is created, we create the original centroid of this event, and set the original centroid equal to the current centroid.

(3) When we create the follow-up EIP, we modify the original centorid, and set OC to the sum of OC and CC. Suppose v_{oj} and w_{oj} ($1 \leq j \leq n$) is the feature and the corresponding weight of OC , that is $OC = (v_{o1}, w_{o1}; v_{o2}, w_{o2}; ...; v_{on}, w_{on})$ and the current centriod $CC = (v_{c1}, w_{c1}; v_{c2}, w_{c2}; ...; v_{cm}, w_{cm})$, the sum of OC and CC is referred as $Sum(OC, CC) = (v_{s1}, w_{s1}; v_{s2}, w_{s2}; ...; v_{sl}, w_{sl})$, which meet the following conditions:

$$v_{sr} \in F(OC) \cup F(CC) \tag{2}$$

$$w_{sr} = \begin{cases} W(v_{sr}, OC) + W(v_{sr}, CC) & if \ v_{sr} \in F(OC) \cap F(CC) \\ W(v_{sr}, OC) & if \ v_{sr} \in F(OC) - F(OC) \cap F(CC) \\ W(v_{sr}, CC) & if \ v_{sr} \in F(CC) - F(OC) \cap F(CC) \end{cases} \tag{3}$$

Where $1 \leq r \leq l$, l is the number of features after the sum, $W(v_{sr}, OC)$ ($W(v_{sr}, CC)$)is the weight of the feature v_{sr} in the centroid OC (CC), $F(OC)$ ($F(CC)$) is set of the features of the centroid OC (CC).

4.3   Creation and Modification of Current Centroid

• **Definition 4:** we call the set of the related texts which are detected during the creation of the current EIP and the next EIP as the Current Text Set (CTS).

During the event detection, CC will be created when a new EIP is created, and it will be modified when a new related text is detected. The process is as follows:

(1) When the event is just created, we create the first current centorid, and when we create a new EIP, we will modify it.

(2) When the system detects a related text, then counts the document frequency of all the words in the CTS, and rank these words in descending order according to their document frequency.

(3) Taking the first ω words as the features of CC, and then compute the weight of the features by averaging, which as follows:

$$Weight(\delta, t, CTS) = \frac{\sum_{txt_i \in CTS} tf_{\delta} \times \log(\frac{N}{n_{\delta}} + 0.01)}{TextNum(t, CTS)} \tag{4}$$

Where $Weight(\delta, t, CTS)$ is the weight of the feature δ in CTS at time t, tf_δ is the term frequency of δ in the text txt_i $1 \le i \le TextNum(t, CTS)$, N is the number of all the texts, and n_δ is the number of the texts which include the feature δ.

5   Event Detection Method Combined Time Information

5.1   Similarity Computation Between Micro-blog Text and Event

In this paper, we adopt the Single-Linkage method to compute the similarity between the micro-blog text and the event, which as follows:

$$Score(text, E) = Max(Co\sin e(text, OC), Co\sin e(text, CC)) \tag{5}$$

Where $Score(text, E)$ is the similarity between the micro-blog text $text$ and the event E, OC and CC are the original centroid and the current centroid, respectively, the function $Max(x, y)$ returns the bigger value of x and y.

We must make clear that before we create the first EIP, there is no original centroid, so we set $Co\sin e(text, OC)$ before the creation of the first EIP.

5.2   Event Detection Based on Single-Pass Clustering

Our detection algorithm is based on the Single-pass clustering. The algorithm processes each micro-blog text sequentially, until all the texts are processed, which as follows:

(1) Apply pre-processing to the micro-blog text and build the text's vector space model;

(2) If the text is the first text of the stream, then create an event using the text as the seed, create the first current centroid, and decide if it need to create a EIP;

(3) Else compute the similarity between the text and all the events created so far, and record the highest similarity and the event which achieved the highest similarity. If the highest similarity is lower than the pre-defined threshold, then create a new event, and create the first current centroid of this event; else add the text to the event, and if it need to create a new EIP, then modify the original centroid, create a new current event, else only modify the current of the event;

(4) repeat the above steps until all the stories in the news stream were done.

6   Experiments and Results Analysis

6.1   Corpus and Evaluation Metrics

Now, the micro-blog topic detection research is just beginning, so there is no public micro-blog corpus. In order to evaluate the methods proposed in our paper, we collect

15,850 pieces of micro-blog manually from Sina, which include 11 topics. The topics in the corpus include the "limited purchasing of milk", "divorce for the house" and "two sessions", and so on. In order to simulate the real environment, we arrange these pieces of micro-blog in chronological order.

We adopt the evaluation metrics used in TDT to evaluate our systems, which include the miss rate, the false rate and the normalized cost. If the detection results of the ith topic are listed in Table 1, then the miss rate ($Miss(i)$) and the false rate ($Fallout(i)$) of the ith topic can be computed using (6) and (7), respectively.

Table 1. Meanings of the Evaluation Parameters

	Related Micro-blog	Not Related Micro-blog
Detected Micro-blog	a	b
Not Detected Micro-blog	c	d

$$Miss(i) = \frac{c}{a+c} \tag{6}$$

$$Fallout(i) = \frac{b}{b+d} \tag{7}$$

Once we obtain the performance (miss rate and fall rate) for every topic, we can achieve the system performance by averaging these metrics. The average miss rate (P_{Miss}), average false rate (P_{Fall}) and the normalized cost ($(C_{Det})_{Norm}$) can be computed by (8), (9) and (10), respectively.

$$P_{Miss} = \frac{1}{n}\sum_{i=1}^{n} Miss(i) \tag{8}$$

$$P_{Fall} = \frac{1}{n}\sum_{i=1}^{n} Fall(i) \tag{9}$$

$$(C_{Det})_{Norm} = \frac{C_{Miss} \times P_{Miss} \times P_{tar} + C_{FA} \times P_{Fall} \times P_{\neg tar}}{\min(C_{Miss} \times P_{tar}, C_{FA} \times P_{\neg tar})} \tag{10}$$

Where P_{tar} is the probability of seeing a new story in the stream; C_{Miss} is the cost of missing a new story; $P_{\neg tar}$ is the probability of seeing an old story, $P_{\neg tar} = 1 - P_{tar}$; CFA is the cost of a false alarm; The values of C_{Miss}, C_{FA}, and P_{tar} are usually predefined according to the application, in our experiments, C_{Miss}, C_{FA}, and P_{tar} are set to 1.0, 0.1, and 0.02, respectively. The smaller the normalized cost $(C_{Det})_{Norm}$, the better the performance of the systems.

6.2   Experiments and Results

We firstly carry out the experiment about our baseline system on the training corpus. Our baseline system is also based on Single-pass clustering, but the event is modeled as a single centroid. From the training results, we find that when the similarity threshold is equal to 0.4 and the number of features of the event centroid is 100, the performance of the baseline system is best, where the performance of this system is: P_{Miss} =0.3766, P_{Fall} =0.0520, and $(C_{Det})_{Norm}$ =0.6314.

In order to verify the effectiveness of the two centroids event model, we carry out three groups of experiments.

(1) SYSTEM1. This system tests the performance of the baseline on the test corpus, where the performance is: P_{Miss} =0.3023, P_{Fall} =0.0789, and $(C_{Det})_{Norm}$ =0.6889.

(2) SYSTEM2. We do several experiments when θ_{Num} and $\theta_{\neg evo}$ are set to different values, and the experimental results are shown in table 2 (limited by the space, we only list the value of $(C_{Det})_{Norm}$ in table 2):

$\theta_{\neg evo}$ \ θ_{Num}	2	3	4	5	6
5	0.7657	0.7639	0.7253	0.7342	0.7653
6	0.7865	0.7750	0.7081	0.6902	0.6897
7	0.7690	0.7589	0.7198	0.6865	0.7752
8	0.6987	0.6867	0.6205	0.6478	0.7123
9	0.6576	0.6590	**0.5727**	0.5986	0.6589
10	0.6754	0.6981	0.6124	0.6347	0.6983
11	0.7809	0.7674	0.7321	0.7210	0.6973
12	0.7823	0.7705	0.7289	0.7187	0.7256

From the above results, we can see that the micro-blog event detection method based on the proposed two-centroid event model is successful, which can decrease the value of the $(C_{Det})_{Norm}$ effectively. When we take deep analysis into the experimental results, we can find that the reason is the event model based on the two centroids can decrease the miss rate greatly.

In order to comparison the results in detail, we give the comparison between the experimental results of these two systems on the different event, where the event number is form 1 to 20, shown in Figure 3. From the figure, we can see that the two centroid based system perform better than the baseline system on the most of the events. But for some certain events, for example event 3 and event 16, the two centroids based system performs worse than the baseline system. One reason is that the system detects a not related text, and at just this time, the system decides to create an EIP according to our creation algorithm. So, the current centroid created based on the wrong text will lead the focus of the event to evolve to a wrong (which means not accordance with the reality) direction. But, taken together, the proposed two centorids event model improve the performance of the detection system successfully.

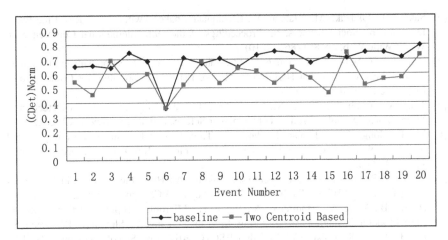

Fig. 3. Comparison between the Experimental Results

7   Conclusions

Microblog is an information publication platform, and has aroused many researchers'' interests. More and more users chose to share their information through the micro-blog, which make the microblog an important place for monitoring, so it is important to detect the hot event for the decision-makers.

Taken the dynamic evolvement property into account, we propose a dynamic event model based on two centroids, which include an original centroid and a current cent. These two centroids are created and modified along with the creation of the EIP. Then, we give a microblog event detection method based on the event model. Overall, from the experimental results, we can see that the proposed event model is successful, but we also find that EIP is critical to the performance of the detection system. So in the future work, we will continue to study the more effective creation method of EIP.

Acknowledgment. This work was supported by the China Postdoctoral Science Foundation (2011M501155); NSFC under Grant (No.61170079 and No.61202152); the Special Fund for Fast Sharing of Science Paper in Net Era by CSTD (2012107).

References

[1] Kaur, K., Gupta, V.: A survey of Topic Tracking Techniques. International Journal of Advanced Research in Computer Science and Software Engineering 5(2), 383–392 (2012)

[2] Kwak, H., Lee, C., Park, H., et al.: What is Twitter, a Social Network or a News Media? In: Proceedings of the 19th International Conference on World Wide Web (WWW 2010), Raleigh, North Carolina, USA, pp. 591–600. ACM (2010)

[3] Sakaki, T., Okazaki, M., Matauo, Y.: Earthquake Shakes Twitter User: Real-time Event Detection by Social Sensors. In: Proceedings of the 19th International Conference on World Wide Web, pp. 851–861 (2010)

[4] Petrovi, S., Osborne, M., Lavrenko, V.: Streaming First Story Detection with ap-plication to Twitter. In: Proceedings of HLT-NAACAL, pp. 181–189. Association for Computational Linguistics, Stroudsburg (2010)

[5] Tong, W., Chen, W., Meng, X.: EDM: An Efficient Algorithm for Event Detection in Micro-blogs. Journal of Frontiers of Computer Science and Technology, 1–12 (2012)

[6] Ma, B., Hong, Y., Lu, J., Yao, J., Zhu, Q.: A Thread-based Two-stage Clustering Method of Micro-blog Topic Detection. Journal of Chinese Information Processing 26(6), 121–128 (2012)

[7] Ishikawa, S., Arakawa, Y., Tagashira, S., Fukuda, A.: Hot topic detection in local areas using Twitter and Wikipedia. In: ARCS Workshops (ARCS), pp. 1–5 (2012)

[8] Huang, B., Yang, Y., Mahmood, A., Wang, H.: Microblog Topic Detection Based on LDA Model and Single-Pass Clustering. In: Yao, J., Yang, Y., Słowiński, R., Greco, S., Li, H., Mitra, S., Polkowski, L. (eds.) RSCTC 2012. LNCS, vol. 7413, pp. 166–171. Springer, Heidelberg (2012)

[9] Du, Y., He, Y., Tian, Y., Chen, Q., Lin, L.: Micro-blog bursty topic detection based on user relationship. In: 2011 6th IEEE Joint International Conference on Information Technolog and Artificial Intelligence, pp. 260–263 (2012)

[10] Long, R., Wang, H., Chen, Y., Jin, O., Yu, Y.: Towards Effective Event Detection, Tracking and Summarization on Microblog Data. In: Wang, H., Li, S., Oyama, S., Hu, X., Qian, T. (eds.) WAIM 2011. LNCS, vol. 6897, pp. 652–663. Springer, Heidelberg (2011)

[11] Yang, L., Lin, Y., Lin, H.: Micro-Blog Hot Event Detection Based on Emotion Distribution. Journal of Chinese Information Processing 26(1), 84–90 (2012)

Research on Fuzzy Intelligent Recommendation System Based on Consumer Online Reviews

Narisa Zhao, Quan-He Wang, and Jia-Feng Zhong

Institute of Systems Engineering, Dalian University of Technology, Dalian, 116024, China
nmgnrs@dlut.edu.cn

Abstract. Many consumers only have fuzzy requirement for products, because they are not the experts of the domain who have much experience for products. The system mines explicit attributes and implicit attributes of products from on-line reviews. Through using semantic analysis technology and building the fuzzy inference rules based on these products attributes, the system can understand the sentiment of the consumers' review which shows system's intelligence. The sentiment words of implicit product attributes are expressed by the fuzzy function, which is the foundation of the sentiment calculation. Finally the experiment proves that our recommendation method is effective and the system can satisfy consumers' requirement.

Keywords: online review, fuzzy recommendation, sentiment computing, fuzzy inference.

1 Introduction

The recommendation and mining system based on consumers' online reviews has been gradually concerned by scholars in recent years [1-4]. These systems utilize the techniques of text mining and natural language processing to extract word-of-mouth of products from online reviews and then quantify them to provide a useful basis for consumers making purchase decisions. Liu [1] quantifies the positive and negative information in user reviews and visually feeds the analysis results back to users. Dave [2] mines sentiment in user reviews and classifies text tendency. Both of them achieve good recommendation effects. Popescu [3] mines product information in online reviews using rules and templates. Pang [4] comprehensively expounds the techniques involved in opinion mining and text sentiment analysis.

At present, most of online shopping search engines do not support the user demand of fuzzy input. In fact, for the majority of consumers, the requirements for products are generally not clear and can't be precisely described. Lots of information about product evaluations exists in online reviews and is reflected by evaluation words in them, while other information is related to specific parameters provided by manufactures. Traditional fuzzy recommendation systems ignore these important fuzzy evaluation information existing in online reviews. Therefore, with the integration of online reviews, product specific parameters and fuzzy demands of users for products, we can achieve a

M. Wang (Ed.): KSEM 2013, LNAI 8041, pp. 173–183, 2013.

better personalization recommendation system for consumers. For these reasons, this paper studies how to achieve intelligent personalization recommendation based on consumer online reviews and fuzzy demands.

2   Architecture Design of Product Recommendation System

The fuzzy intelligent recommendation system based on consumer online reviews consists of three parts. One part is a database including a product database and an online review database, the second part is a fuzzy rule base, the third part is an interactive fuzzy recommendation module(see Figure 1). First, we download related parameters and online reviews using Web crawler from the network and separately import them into the product database and review database. Second, we input fuzzy inference rules into the recommendation system as priori knowledge in advance. Finally, users input fuzzy demands into the system and the recommendation system will calculate the approximate degree between user input and the existing products in the system and return several satisfactory products for users to choose.

3   Product Attribute Mining and Fuzzy Sentiment Computing

3.1   Product Attribute Mining

When searching for products, consumers always make demands from the perspective of product attributes. Therefore, it's crucial to search for product attributes matching user requirements. Online reviews contain a massive amount of fuzzy information of product attributes, so how to mine the information of product attributes rapidly based boundary, Chinese reviews first need to be segmented and POS(part of speech) tagged. This takes natural language processing system of LTP(Language Technology Platform) [5] to preprocess reviews. The system includes word segmentation, POS and dependency parsing. The results are saved as XML documents, which is better for further analyzing. This paper mainly adopts the following method to mine product attributes.

Step 1. Count frequent items of nouns after word segmentation, remove Chinese stop words as well as the noises and extract alternative attributes. In this step, the noises include one-word noun, personal, the nouns unrelated to products and so on. And we remove all one-word attributes(one-word nouns are usually not attributes in Chinese) and filter personal based on the pre-written personal dictionary. Some products' attributes may be obtained by the combination of several nouns, and it will take more time and slow down the system to extract attributes of these products. So in this paper, this case is not studied.

Step 2. Express the correlations between products and alternative attributes by mutual information formula [4]. A higher value of PMI indicates a stronger relation:

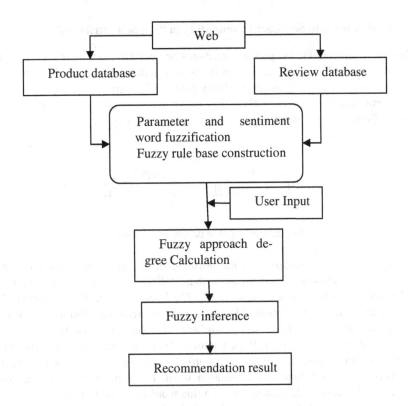

Fig. 1. The Architecture of product fuzzy recommendation system

$$PMI = \log \frac{Hits(C,F)+1}{\sqrt{(Hits(C)+1)*(Hits(F)+1)}} \tag{1}$$

In the formula above, Hits(C) stands for the number of products returned from search engine, Hits(F) represents the number of alternative attributes of products and Hits(C,F) means the number of the two kinds that appear together. Then sort product attributes by the value of PMI. But there is a problem between the accuracy and recall rate of attributes. If the threshold of PMI is taken a large value, then the accuracy of mining attributes rises, while recall rate of attributes declines, which causes that system can't give the matching attributes after user input. According to paper [6], the thresholds of PMI for different products are also different and no unified one can be given, so we choose alternative attributes in the top 30 percent of PMI as product attributes.

The extracted product attributes can be divided into two categories: those related to product parameters provided by manufacturers and those only related to review information. In this paper, the first kind is called "explicit attribute" of product and the other kind "implicit attribute". Both of them constitute set of product attributes.

3.2   Evaluation and Sentiment Computing in "implicit attributes"

In Chinese Online Reviews, product attributes are not far from their corresponding evaluation words. So two methods can be used to rapidly extract evaluation words: the sliding window method and dependency parsing [7]. This paper adopts the second one. For example："手机电池很耐用(*The battery of mobile phone is very durable*)" is shown in Figure 2 after word segmentation and dependency parsing, which is based on Apriori algorithm.

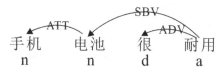

Fig. 2. A example of dependency parsing

In Figure 2, "电池(*battery*)" is the attribute of "手机(*mobile phone*)", "耐用 (*durable*)" is the evaluation word of this attribute, "很(*very*)" is the modifier of durable. "耐用(*durable*)" is the parent node of "电池(*The battery*)" in XML documents. ATT stands for the noun-noun relation, ADV the modifier-predicate relation and SBV the subject-predicate relation. Because the attribute "电池(*battery*)" has been mined, we need to find its parent node next. If the parent node is adjective and the dependency is SBV, then it can be judged as evaluation word. Then try to find adverbs and negative words between the attribute and evaluation word, if the parent node of adverb or negative word is this evaluation word, then this word is a modifier of this evaluation word. The process is expressed by pseudo code in Figure 3.

```
    for each Word in Sentence:
bi=index(w)
    if Word is Product Attributes:
      if WordParent is 'a' and WordLink is 'sbv':
        wpi=index(WordParent)
          for each ModifierWord in range(bi, wpi):
            if ModifierWord is 'd' and LinkType is 'adv'
              c=tuple(Word, ModifierWord, WordParent)
            else:
              c=tuple(Word, WordParent)
```

Fig. 3. Pseudo code extraction of evaluation word and modifier

In Figure 3, Word stands for the word in the sentence, bi represents the sequence number of this word, *WordParent* means the parent node of this word, *WordLink* means dependency between *Word* and *WordParent* and *WordParentIndex(wpi)* means the sequence number of *WordParent*. Try to find *ModifierWord* in the distance(*bi, wpi*), if it is found, then *Word, ModifierWord* and *WordParent* can be grouped together into a *tuple*, or only Word and *WordParent* can be grouped. Then write syntactic rules to make the system identify this sentence pattern.

Different sentiment words and evaluation words have fuzzy attributes and different degrees, so we need to predefine the levels. Select sentiment word base shown in Hownet (http://www.keenage.com) and take (-6, 6) as the interval. Then mark the words. The higher the score is, the greater positive emotion is, otherwise, the greater negative emotion is. Then input the scored sentiment words into Hash dictionary and query scores corresponding to the evaluation words as the fuzzy level in this dictionary according to practical knowledge (see Table 1).

Table 1. Fuzzy levels of words

Fuzzy Level	Very low	low	A little low	Moderate	A little high	high	Very high
Score	-6	-4	-2	0	2	4	6

Modifiers and negative words result in the migration of meanings. Thus, splitting this up into cases is required according to word combinations. Set modifier as M, negative word as N and evaluation word as W. See Table 2.

Table 2. Word combination and corresponding migration

Word combination	corresponding migration
M+W	Strengthen word meaning
N+W	Reverse word meaning
M+N+W	Reverse and strengthen word meaning
N+M+W	Reverse and weaken word meaning

Set the reviews where both attribute A_i and evaluation words appear as review set $S=\{s_1, s_2...s_i...s_n\}$. Average the scores of evaluation words that appear in this set, and the value is represented by V_i as the fuzzy attribute input of A_i.

The calculation above is aimed at "implicit attribute" because these attributes are only in online reviews and can't be acquired from the information provided by manufactures.

4   The Realization of Personalization Recommendation System

4.1   Computing Multi-attribute Product Matching Degree

Suppose that there is a product which has n explicit attributes and m implicit attributes, then make eigenvectors of the attributes as follows:

$$X_i = \left\{ x_{i1}, x_{i2}, x_{i3}, \cdots x_{it} \cdots x_{in} \right\}; \quad X_i^{'} = \left\{ x_{i1}, x_{i2}, x_{i3} \cdots x_{it} \cdots x_{im} \right\} \quad (2)$$

x_{it} means attribute item after fuzzy operation and can be shown by the Gaussian membership function.

$$\mu(x) = gaussmf(x) = \exp\left(\frac{-(x-b)^2}{\sigma^2}\right) \tag{3}$$

Each degree corresponds to a Gaussian function. This paper takes (-6, 6) as interval. The requirements of users include parts of demands for explicit attributes and implicit attributes of products which can be also expressed as eigenvectors as follows:

$$Y = \{y_1, y_2, y_3, \cdots y_t \cdots y_s\}; \quad Y' = \{y_1, y_2, y_3, \cdots y_t \cdots y_k\} \tag{4}$$

Y and Y' respectively represents requirement eigenvectors of explicit attributes and implicit attributes. y_t can be also expressed by the Gaussian membership function.

We need to calculate the matching degree between users' demands for explicit and implicit attributes of products and the corresponding attributes of some products.

For explicit attributes:

$$(Y, X_i) = \sum_t^s w_t(y_t, x_{it}), \quad \sum_{t=1}^s w_t = 1 \tag{5}$$

Among this, y_t and x_{it} represent explicit attributes of products corresponding to the input by users. The result is the matching degree of explicit attributes as follows:

$$Val_d = (Y, X_i) \tag{6}$$

With the same method, the matching degree of implicit attributes is expressed as Val_h.

$$(y_t, x_{it}) = (y_t \bullet x_{it}) \wedge \overline{(y_t \oplus x_{it})} \tag{7}$$

Take the formula above to calculate the matching degree. $(y_t \bullet x_{it})$ means inner product and $(y_t \oplus x_{it})$ means cross product. Then input Val_d and Val_h of each product into fuzzy inference system to do further reasoning.

4.2   Fuzzy Rule Construction and Product Recommendation

The reasoning process is shown in figure 4. The most common way is to express knowledge as the rules of natural language form:

IF premise (antecedent), THEN conclusion (consequent).

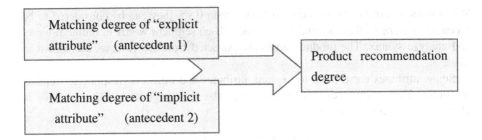

Fig. 4. Reasoning process

Both antecedents and consequent are represented by Gaussian fuzzy membership function, and we choose Mamdani implication as the inference method.

$$\mu_{B^k}(y) = \max_k \left[\min \left[\mu_{A_1^k}\left(input(t_1)\right)...\mu_{A_i^k}\left(input(t_i)\right)\right]\right], k = 1, 2..., r \qquad (8)$$

In this formula, k represents the kth rule, A represents antecedents, t represents the input value of the antecedent, and B represents the consequent of the fuzzy reasoning.

In this fuzzy inference system, antecedents are explicit matching degree and implicit matching degree and the consequent is the recommendation degree of a product. After computing the recommendation degree of each product, we use the centroid method to de-fuzzy the recommendation degree fuzzy function and rank the products according to the defuzzification values.

5   Calculation and Analysis of Numerical Cases

Choose processors in Jingdong Mall as research subject and download 64143 reviews of 38 processors. But there are a certain number of repeated reviews, which influences the effect of the recommendation system. Therefore we remove this kind of reviews using Jaro distance[9] to identify the similarity of two reviews:

$$d_j = \frac{1}{3}\left(\frac{m}{|s_1|} + \frac{m}{|s_2|} + \frac{m-t}{m} \right) \qquad (9)$$

If the distance between two characters is less than,

$$\left\lfloor \frac{\max\left(|s_1|,|s_2|\right)}{2} \right\rfloor - 1 \qquad (10)$$

then it means the two characters match each other. In Chinese, the smallest unit is the word. m means the number of matching characters, t means transposition character.

When the similarity of two reviews is higher than 0.95, they can be considered to be repeated and one of them should be removed. Then segment words in online reviews and analyze syntax. The product attributes extracted from the reviews are shown in Table 3.

Some attributes correspond to explicit attributes and others correspond to implicit attributes. The attribute that appears both in the web page and online reviews is classified as explicit attribute.

Table 3. Product attributes

The mined product attributes
dominant frequency, pin, performance, bus, interface, cache, fan, temperature, accessory, power consumption, thread, radiator, platform, heating value, server, voltage, manufacturer, kernel, specification, economy, game, multimedia, image, power, instruction, controller, channel, compatibility, technique, price, architecture, sound, quality, air-cooling, longevity, rotation speed, video, stability, noise

User input is always described as natural language and the forms of natural language are various, which makes it difficult to extract evaluations of attributes after parsing. So predefine description form as follows:

<search product>: [product attribute]+[product evaluation]

Separate products from product attributes with colons and different attribute evaluations with commas. Search interface is shown as Figure 5.

Fig. 5. Product fuzzy search engine

In the user interface, user input fuzzy requirements for products and click "search", Then the system automatically recommend products for users. The result is shown in figure 6 including brand, price, model and the rank.

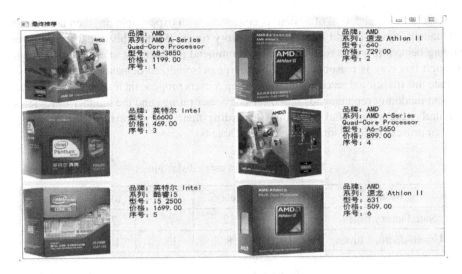

Fig. 6. The recommendation result

6 processors closest to user requirements are listed for selecting. The recommendation values of them are shown in Table 4. Among this, the recommendation value is the result of removing fuzziness with the centroid method.

Table 4. The models and recommendation values of products

Models	Intel I5 2500	AMD A8-3850	AMD 640	AMD FX-8120	Intel E6600	Intel i7 2600K
Recommendation values	3.86	3.28	2.66	2.10	2.07	1.99

5 persons familiar with processors are chosen to assess the recommendation system. Each person proposes requirements for processors at random and clicks 20 times, so 100 times totally. Here these 5 users can be regarded as experts in this field and the recommendation system uses fuzzy rule inference, so the process of inference can be viewed as making buying decisions guided by experts. It shows the intelligence of the system. If 5 users can find a satisfactory processor among these 6 recommended ones, then the system is successfully recommended, otherwise it fails. Statistics are shown in Table 5.

Table 5. Statistics of user satisfaction

users	user1	user2	user3	user4	user5
Satisfactory times	16	18	16	15	17
Unsatisfactory times	4	2	4	5	3

If the recommendation of system is consistent with expert opinion, then add one in satisfactory times, or add one in unsatisfactory times. Without historical data of user buying behavior, online malls can only recommend products by sales or user ratings. Though this way is reasonable, it still ignores personalized demands of users. To simulate this traditional recommendation, when 5 users propose their fuzzy demands, the system randomly recommends based on sales, user ratings and the number of reviews. Similarly, every one clicks 20 times according their fuzzy demands and another 6 chosen processors. Statistics are shown in Table 6.

Table 6. Statistics of user satisfaction

users	user1	user2	user3	user4	user5
Satisfactory times	10	12	7	7	11
Unsatisfactory times	10	8	13	13	9

In Table 5, the average satisfaction degree is 82%, however, only 47% in Table 6. Both of these are expected. The first recommended way mines personalized demands from online reviews. The second one makes recommendation based on interest and overall quality of products and seldom considers personalized demands. So the satisfaction degree of the first one is much higher than the second one.

6   Conclusion

Taking full advantage of consumers' evaluation information in online reviews, this fuzzy intelligent recommendation system extracts information of product attributes which consumers have concerns together on. Users make demands in the form of natural language and the system can calculate fuzzy approach degree between the demands and product information. Construct fuzzy inference rules to compute recommendation degree, defuzzy results and rank them. 6 processors closest to user requirements are recommended. The experimental result shows the system has a good recommendation effect.

This system utilizes the techniques of fuzzy and text mining and combines user demands with product information. Finally, it achieves personalization recommendation. But several aspects should be further improved for better recommendation including adding other mining techniques into the system, expanding the product database and improving fuzzy inference system.

Acknowledgment. This work is partially supported by a research grant from the Program of National Natural Science Foundation of China (No. 61072128), Natural Science Foundation of Liaoning Province (No.201102024) and Technology and Project Foundation of Dalian (No.2011A17GX078).

References

1. Liu, B., Hu, M., Cheng, J.: Opinion observer: Analyzing and comparing opinions on the web. In: The 14th International Conference on World Wild Web, Chiba, Japan, pp. 342–351 (2005)
2. Dave, K., Lawrence, S., Pennock, D.M.: Mining the peanut gallery: Opinion extraction and semantic classification of product reviews. In: 12th International Conference on World Wide Web, Budapest, Hungary, pp. 519–528 (2003)
3. Popescu, A.M., Etzioni, O.: Extracting product features and opinions from reviews. In: The Conference on Human Language Technology and Empirical Methods in Natural Language Processing, Vancouver, B.C., Canada, pp. 339–346 (2005)
4. Pang, B., Lee, L.: Opinion mining and sentiment analysis. Foundations and Trends in Information Retrieval 1, 1–135 (2008)
5. Che, W., Li, Z., Liu, T.: LTP: A Chinese Language Technology Platform. In: 23rd International Conference on Computational Linguistics, Demonstrations, Beijing, China, pp. 13–16 (2010)
6. Li, S., Ye, Q., Li, Y.: Mining features of products from Chinese customer online reviews. Journal of Management Sciences in China 2, 142–152 (2009)
7. Somprasersri, G., Lalitrojwong, P.: Mining Feature-Opinion in Online Customer Reviews for Opinion Summarization. Journal of Universal Computer Science 16, 938–955 (2010)
8. Jaro, M.A.: Probabilistic linkage of large public health data files. Statistics in Medicine 5, 491–498 (1995)

Motivation to Innovation – The Driving Force of Reward-Based Organizational Knowledge Management for New Product Development Performance

Shanyu Lei[*] and Weixiao Ma

School of Management, Dalian University of Technology,
No.2 Linggong Road, Ganjingzi District, Dalian, 116024, P.R. China
leishanyu@gmail.com

Abstract. The paper is to reveal the effects of the knowledge management (KM) process motivated by reward systems on new product development (NPD) performance. Based on key concepts in NPD, KM, and reward systems, the study first proposed a theoretical framework and then conducted a quantitative analysis by structural equation modeling (SEM), using data from a sample of 120 innovative firms in 3 Chinese cities. The study found most of the hypotheses supported. Knowledge innovation had a positive impact on NPD performance. In KM system, the acquisition of knowledge had a strong positive association with knowledge integration. Moreover, higher level of organizational reward system could bring higher level of knowledge acquisition and innovation.

Keywords: New product development, Knowledge management, Knowledge acquisition, Knowledge integration, Knowledge innovation, Reward system.

1   Introduction

With the increasing complexity of the environment of technology and market, innovation is a black hole that requires a firm to invest a huge amount of resources while financial return is increasingly uncertain. This makes a large number of firms unwilling to innovate and thus intensifies the "can-but-does-not" tension between their innovative capabilities and behaviors (Srivastava & Gnyawali, 2011). However, innovation is promising in introducing new products, and the rapid shift in technology and the market makes NPD a critical determinant of organizational sustainable competitive advantage and superior long-run organizational performance. Nonaka (1994) proposed that a key premise on new product innovation is that organizational NPD is a function of a firm's capability to manage, maintain, and create knowledge. Wallin and Krogh (2010) also described innovation as a process that covers the combination and use of knowledge for the development and introduction of something new and useful. A firm cannot create, sustain and renew competitive advantage without relentless pursuit of knowledge development for sustainable innovation.

---

[*] Corresponding author.

M. Wang (Ed.): KSEM 2013, LNAI 8041, pp. 184–194, 2013.

Reflecting the importance of organizational NPD, many works indicate the key role of KM, which is the management of a firm's internal and external knowledge by means of acquisition, integration and innovation. Knowledge acquisition is expected to enhance a firm's ability to exploit new opportunities towards production enrichment through individual learning and organizational absorptive capacity. With new knowledge imported into the organizational knowledge base by different units of a team, knowledge integration is the transferring of a knowledge unit to another and the interaction of new and prior knowledge. Thus, learning and sharing can be viewed as preceding elements of dynamic capabilities towards innovation. Previous research provides significant insight into NPD that surrounds its relationship with KM.

Though managers are supposed to be aware of importance of KM including knowledge acquisition, integration, and innovation , KM would not likely be a contribution to a firm with a "can-but-does-not" tension, since the KM process is mainly dependent on each employee's effort towards the whole organization. In such a firm, active governing systems would not be taken seriously, because employees tend to regard their unique knowledge as private advantage to secure their positions under internal competition for rewards, status, and promotions. Additionally, when inevitable costs occur in the process of KM, such as knowledge acquisition from outside, the "can-but-does-not" attitude becomes employees' safe way to avoid loss and protect themselves. Some research addresses findings that organizational reward influences behavior and performance of organization's members in a positive way and enhance organizational performance and competitive advantage .

Much research stresses the importance of mobilization to innovation, and organizational incentive systems are useful for employees' KM covering acquisition, integration and innovation. But three still remain unclear, (1) How reward systems relate to the process of KM; (2) The interrelationship of the three sub-processes of KM; and (3) How these factors combine to influence NPD performance. This study contributes to the strategic innovation management literature by focusing on the linking role of reward systems on KM and organizational NPD performance. The remainder of the paper is organized as follows: Section 2 constructs the research framework and develops theoretical hypotheses; Section 3 presents samples, research models, and methods; Section 4 demonstrates the statistical results; Section 5 discusses the results and draws several conclusions; Section 6 is about research limitations and future works.

2   Research Framework and Hypotheses

2.1   Knowledge Innovation and NPD Performance

NPD performance has been a topic focused on for several decades, Moorman (1995) defined it as the degree to which organizational goals involving new product profit, sales, and share have been reached, successful NPD is critical for the renewal, survival, and success of organizational and economic growth. New product introduction will help organizations to diversify, adapt, and reinvent themselves in changing environment, and an adequate frequency of new product introduction indicates the significance of a firm's competitiveness and profitability. New products are regarded essentially as the externalized form of new knowledge. To develop new product is considered an essential complex process of knowledge innovation, which refers to the creation of knowledge distinguishing the competitiveness of products. For any innovative firm, it has to rely on knowledge creation to ensure business and constant growth; the successful firms usually have the capability to enlarge the scope of innovation for NPD by including all main sources of knowledge and by fostering the overall process of knowledge innovation. Thus:

Hypothesis 1. A firm's NPD performance is positively associated with its knowledge innovation.

2.2   Influential Mechanism of Three Sub-processes of KM

Following Nonaka and Takeuchi (1995), organizational knowledge is defined as the validated understanding and belief in a firm about the relationship between the firm and its environment, and organizational knowledge is composed of two types: explicit knowledge, defined as codified and easily translated facts and information; and tacit knowledge, defined as personal know-how that may be hard to confirm and convey. These conceptions that help clarify organizational knowledge reflect collective viewpoints on how existing resources should be configured and exploited for an advantage, and a firm's KM effort.

Staples *et al.* (2001) demonstrated the benefit of KM: managing organizational knowledge effectively brings firms with better product popularity; they then suggested that firms should take KM seriously as an important strategy. In line with Lee and Yang's (2000) research that KM process is composed of knowledge acquisition, integration, and innovation. In such a complicated process, knowledge acquisition is the organizational absorptive capability to identify and obtain externally generated knowledge that was critical to its operations. It mainly characterizes with the activity of accepting a unit of knowledge from the external environment and transforming it into a representation that can be internalized within the organization. Knowledge integration has been defined as transferring, sharing and maintaining information and knowledge (Chang & Ahn, 2005), its main task being identifying how new and prior knowledge interact with each other while incorporating new information into a knowledge base by disseminating. According to Nahapiet & Ghoshal (1998), the integration of knowledge refers to the process of bringing together "elements previously unconnected or developing new ways of combining elements previously associated". Knowledge innovation refers to the process of converting tacit knowledge to explicit knowledge. It is a key role in knowledge creation and product innovation.

Requirement of knowledge is a precondition of knowledge sharing, exchanging and transferring, and only a deep understanding and a broad base of related knowledge can provide a better chance for the combination of previously unconnected aspects and recombination of previously associated aspects. This is because knowledge integration is better achieved through dynamic interaction and feedback. No matter how explicit or tacit knowledge is involved, firms can depend not only on their internal knowledge resources, but also on their employees' performance in acquiring information from outside and integrating existing knowledge and ideas to upgrade and stock organizational knowledge. Knowledge acquisition would thus enhance both the depth and breadth of external knowledge available to the focal firm, and this sufficient depth of knowledge will be more helpful for the focal firm to develop some new knowledge that can be transformed into new products . Acquired knowledge has also been proven to be significant in realizing products differentiation. Yang (2005) found that firms that practice effective knowledge acquisition are more flexible and therefore more able to seize strategic opportunities and implement innovation; In real practice, NPD teams usually seek knowledge from external sources and generate knowledge internally; members in this team must have an appreciation for the importance of their teammates and a willingness to share their knowledge base in their own disciplines for the productivity of the whole team. Following this, integration of all accumulated knowledge has great benefits on improving the creation of new knowledge and products themselves.

From the discussion above, three hypotheses is proposed as:

Hypothesis 2. A firm's knowledge integration is positively associated with knowledge acquisition in the KM process.

Hypothesis 3. A firm's knowledge innovation is positively associated with knowledge acquisition in the KM process.

Hypothesis 4. A firm's knowledge innovation is positively associated with knowledge integration in the development of new products.

2.3   Reward Systems and KM

To effectively manage knowledge, it is urgent for firms to realize their strategic NPD plan. Much research has been conducted on the role of reward systems in KM process; because acquiring, integrating and innovating knowledge are all tough work that employees are usually unwilling to undertake unless they are properly rewarded (Sarin & Mahajan, 2001). A reward system entails the deliberate use of the pay system to guide and direct the behaviors and efforts of various individuals and departments towards the achievement of organizational goals. This has been becoming a crucial contributor to strategy implementation, operational effectiveness, and competitive advantage. Two types of rewards, including intrinsic and extrinsic, are suggested to be useful to motivate organizational members: the extrinsic incentives related to monetary rewards are commonly used, while intrinsic rewards motivate individuals when they seek enjoyment, interest, satisfaction, or self-expression in the work itself. Tampoe (1993) demonstrated that individuals who are occupied with knowledge creation and innovation are very likely to be influenced by personal promotion, project achievement, and autonomy. The reward systems play a key role in improving the level of human resources by motivating knowledge employees to share and transfer their knowledge. Sarin and Mahajan (2001) proposed a positive relationship between outcome-based rewards and innovative team performance, but intrinsic rewards are more helpful for knowledge sharing, compared to just rewarding financial success, because they can promote an atmosphere that expedites both formal and informal communication which involves more learning behaviors.

Though some literature proves that reward systems substantially motivate employees and improve organizational performance, the success of such a system depends on its design. For organizational KM, the most difficult aspect is that it requires many departments or teams with different knowledge background to collaborate, accompanied by firms' strategies of either creating new knowledge or combining existing ones or both. Moderate reward systems can encourage more employees to participate in the implementation of KM strategy, because its orientation has been found to increase intelligence generation, dissemination and the responsiveness of an organization (Jaworski & Kohli, 1993). Therefore, the relationship between reward systems and organizational KM is supposed as:

Hypothesis 5. A firm's knowledge acquisition is positively associated with its reward systems.
Hypothesis 6. A firm's knowledge integration is positively associated with its reward systems.
Hypothesis 7. A firm's knowledge innovation is positively associated with its reward systems.

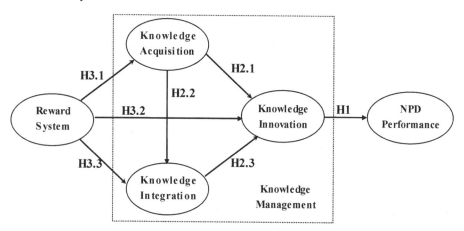

Fig. 1. Hypothesized model of the analysis

3   Methodology

3.1   Sample

To examine the above hypothetical relationship between organizational reward systems, KM and NPD performance, data of 120 innovative enterprises practicing NPD from China's three cities including Dalian, Shenyang, and Shenzhen, were collected through questionnaires. The first two cities were targeted because their GDP are the largest in China's northeastern region; Shenzhen was targeted because it is representative of rapid developing cities in Southern China, and it is also a top-tier city of China. Many innovative companies located in these three cities are remarkable for their significant growth in NPD, it meets our research purpose to investigate China's innovative enterprises' general situation.

The survey was conducted from October 2011 to March 2012, and 250 firms' senior-level managers were contacted via telephone to solicit participation or to recommend knowledgeable respondents, as top managers are adequate informant for the study, and so they usually have sufficient knowledge about the organizational incentive policies, KM strategies and NPD performance. As a result, 193 firms agreed to participate and 135 surveys were returned, and after following up the non-responding companies, 27 additional surveys were added. A total of 162 surveys were collected, and after removing 42 invalid samples, the remaining valid samples for research were 120, among which 64 (53.33%) were from Dalian, 15 (12.50%) from Shenyang and 41 (50.83%) from Shenzhen.

3.2 Questionnaire and Variables

The survey questionnaire was designed based on previous research and summaries of our in-depth interviews. After two professors familiar with this field checked the content validity of the items, a pretest with 9 managers was conducted to validate the measures in terms of their face validity, clarity, and appropriateness to China's context.

All scales of the items were measured on a seven-point Likert scale ranging from "Strongly Disagree =1" to "Strongly Agree =7" and the items were adapted and reworded to fit the present context as follows: (1) Knowledge acquisition: Three items were used to measure knowledge acquisition, the first two of which were adapted from Norman (2002), greatly emphasizing the acquisition of administration and technical skills, and the third one was developed to tap the tacit and explicit knowledge acquisition in accordance with research by Yang (2005). (2) Knowledge integration: Three items of knowledge integration were adapted from Yang's (2005) scales, which had much of their groundings from Norman (2002), focusing on the extent to which firms relate their knowledge base with external knowledge sources (suppliers, consumers, and cooperators, respectively). (3) Knowledge innovation: It was measured by three items drawn from several studies (Yang & Rui, 2009), reflecting the degree of proclivity and ambitiousness for strengthening a firm's ability for knowledge creation and renewal. (4) Reward systems: Two items were adapted to measure reward systems from research by Jaworski and Kohli (1993), reflecting the degree to which firms motivate individuals and teams by means of intrinsic and extrinsic rewards. (5) NPD performance: Four items of NPD performance were adopted from Moorman (1995), indicating the new product's financial contribution and market competitiveness.

3.3 Measures Validation

With SPSS 12.0 and Amos 18.0, the validity of the measures was initially assessed by examining the reliability of the constructs. Cronbach's alpha coefficients are presented in parenthesis besides each construct in Table 1. An inspection of the results reveals that the smallest among all alpha coefficients are 0.77 (knowledge integration). Next, the entire set of items was subjected to confirmatory factor analysis (CFA) by structural equation model (SEM) to verify unidimensionality. Items that load on multiple constructs and that have too low item-to-construct loadings should be deleted. As a result, factor loadings of items to corresponding constructs range from 0.67 to 0.91, all in sufficient significance ($p < 0.01$), which therefore supports the convergent validity. In addition, several measures of model fit are acceptable: $x^2 = 140.86$; $df = 80$; $x^2 / df = 1.76$; NFI = 0.88; RFI = 0.82; IFI = 0.94; TFI = 0.91; CFI = 0.94; RMSEA = 0.06. These all suggest that the model represents a good fit to the data. Table 1 summarizes these analytic results.

Table 1. Construct Measurement and Confirmatory Factor Analysis

Constructs and Items (Cronbach's Alpha)	Mean (S. D.)	Standardized loadings (t)
Reward systems (0.87)		
1-1 No matter which department they are in, individuals in this business unit get recognized for being sensitive to competitive moves.	4.08 (1.49)	0.77 [a]
1-2 Formal rewards (i.e. pay raise, promotion) are forthcoming to anyone who consistently provides information/knowledge.	4.79 (1.32)	0.91(5.08)
Knowledge acquisition (0.85)		
2-1 Your firm encourages all employees to pursue new knowledge pertinent to the project.	5.14 (1.39)	0.85 [a]
2-2 User-friendly tools are available to aid you to acquire new knowledge pertinent to your project.	4.93 (1.34)	0.82 (10.29)
2-3 As a result of the project, you have improved existing/developed new management/technical skills.	4.81 (1.17)	0.76 (9.37)
Knowledge integration (0.77)		
3-1 Cooperative agreements are in place to facilitate knowledge integration between your organization and your suppliers.	4.48 (1.31)	0.80 [a]
3-2 Interactions with market in forms of interviews and questionnaire survey are frequently conducted.	4.67 (1.50)	0.70 (7.42)
3-3 Close bonds have been established with universities or relative institutions.	4.14 (1.50)	0.67 (6.46)
Knowledge innovation (0.86)		
4-1 Your firm producing knowledge as a product.	5.06 (1.25)	0.89 [a]

Table 1. *(Continued)*

Constructs and Items (Cronbach's Alpha)	Mean (S. D.)	Standardized loadings (t)
4-2 Utilize knowledge innovation in order to maintain a competitive advantage in the market.	5.08 (1.27)	0.811 (11.04)
4-3 Your firm introduced new ideas about your process, product, and project.	4.64 (1.34)	0.753 (9.81)
New product performance (0.90)		
5-1 Market share relative to its stated objective.	4.43 (1.20)	0.88[a]
5-2 Sales relative to its stated objective.	4.45 (1.22)	0.89 (13.24)
5-3 Profit margin relative to its stated objective.	4.62 (1.18)	0.75 (9.98)
5-4 The ratio of new products sales on total sales to its stated objectives.	4.48 (1.24)	0.81 (11.23)

CFA goodness of fit: x^2 = 140.86; df = 80; x^2 / df = 1.76; NFI = 0.88; RFI = 0.82; IFI = 0.94; TFI = 0.91; CFI = 0.94; RMSEA = 0.06

[a] Fixed parameter.

4 Analysis and Results

The analytic results of path coefficients related to hypothetical models are summarized in Table 2.

Table 2. Results of Path Coefficients

Hypothesis No.	Paths	Standardized parameter(t-value)
1	Knowledge innovation → NPD performance	0.70 (7.77)
2	Knowledge acquisition → Knowledge innovation	0.29 (1.60)
3	Knowledge acquisition → Knowledge integration	0.72 (5.35)
4	Knowledge integration → Knowledge innovation	0.21 (1.12)
5	Reward systems → Knowledge acquisition	0.57 (4.40)
6	Reward systems → Knowledge integration	0.17 (1.45)
7	Reward systems → Knowledge innovation	0.36 (2.69)

Path analysis goodness of fit: x^2 = 155.56; df = 83; x^2 / df = 1.87; NFI = 0.87; RFI = 0.81; IFI = 0.93; TFI = 0.90; CFI = 0.93; RMSEA = 0.07

Table 2 indicates a good fit of the model. Path coefficient from knowledge innovation to NPD performance is proved to be positive and statistically significant (0.70, *t*=7.77, *p*<0.01); Path coefficients from knowledge acquisition to integration and innovation are 0.64 (*t*=5.35, *p*<0.01), and 0.29 (*t*=1.60, *p* ≈0.10), respectively. These results are believed to support hypotheses 2 and 3. Path coefficient from knowledge integration to innovation is proved to be positive but not statistically significant as 0.21 (*t*=1.12, *p*>0.1), it cannot support hypothesis 4, reminding us that integration is more likely but cannot be certain for its contribution to innovation. Results of this analysis also proves significant role of reward systems on KM, because the coefficients from reward systems to acquisition and innovation are positive and statically significant respectively (0.56, *t*=4.40, *p*<0.01 and 0.33, *t*=2.694, *p*<0.05), indicating that reward systems have a significantly positive effect on some (not all) sub-processes of KM, supporting hypotheses 5 and 7. Coefficient from reward systems to knowledge integration is also positive (0.17, *t*=1.45, *p*≈0.15), but its p-value is a little higher that could not support hypothesis 6 definitely. All of the above analytic results of path coefficients and their relationship are depicted in Fig.2.

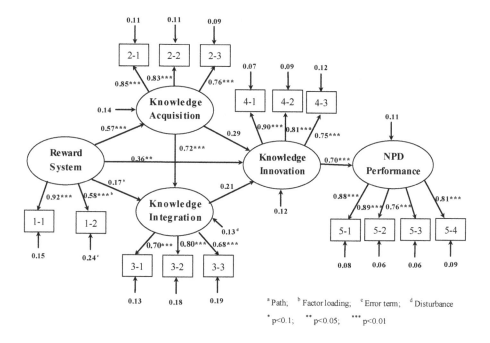

Fig. 2. Path analysis results

5   Conclusions and Implications

This research endeavored to demonstrate that successful knowledge innovation leads to excellent NPD performance, which was supported by Hypothesis 1. In this sense, for any firm with NPD practice, sustainable development tends to depend on its knowledge innovation, while innovation strategy for these firms should depend on the creation of new knowledge rather than merely on innovative resources acquisition. Many Chinese firms are eager to strengthen competitiveness by mergers or acquisitions of foreign firms for enlarging innovative resources. The fact is that there are very few successful cases of mergers or acquisitions in recent years, which imply to top managers that they should review the importance of building up a solid internal structure to manage knowledge-creation process for NPD performance.

This study also explored the influential mechanism among three sub-processes of KM covering knowledge acquisition, integration, and innovation. The results revealed that knowledge acquisition was proved to promote integration and innovation, which supported Hypotheses 2.1 and 2.2, while path coefficient from integration to innovation was also proved to be positive, but it was not statistically significant, thus failing to support Hypothesis 2.3.

In accordance with some recent research highlighting reward systems' role on motivation, this study also adopted intrinsic and extrinsic reward systems to investigate their motivation for three sub-processes of KM. Results 3.1 and 3.2 suggest that firms should design a convincing reward and punishment policy by regulating details and boundaries explicitly and clearly. Hypothesis 3b was not significantly supported, failing to show that higher level of rewards intrigues effective knowledge integration. A possible explanation for this result would be that the sampled firms generally have a short involvement in the cause and effect of knowledge integration.

References

1. Chang, S.G., Ahn, J.H.: Product and Process Knowledge in the Performance-Oriented Knowledge Management Approach. Journal of Knowledge Management 9(4), 114–132 (2005)
2. Gross, S.E., Friedman, H.M.: Creating an Effective Total Reward Strategy: Holistic Approach Better Support. Benefits Quarterly 20(3), 7–12 (2004)
3. Jaworski, B.J., Kohli, A.K.: Market Orientation: Antecedents and Consequences. Journal of Marketing 57(3), 53–70 (1993)
4. Lee, C.C., Yang, J.: Knowledge Value Chain. Journal of Management Development 19(9), 783–793 (2000)
5. Moorman, C.: Organizational Market Information Processes: Cultural Antecedents and New Product Outcomes. Journal of Marketing Research 32(3), 318–336 (1995)
6. Nahapiet, J., Ghoshal, S.: Social Capital, Intellectual Capital and the Organizational Advantage. Academy of Management Review 23(2), 242–266 (1998)
7. Nonaka, I.: A Dynamic Theory of Organizational Knowledge Creation. Organization Science 5(1), 14–38 (1994)
8. Nonaka, I., Takeuchi, H.: The Knowledge-Creating Company: How Japanese Companies Create the Dynamics of Innovation. Oxford University Press, New York (1995)
9. Norman, P.M.: Protecting Knowledge in Strategic Alliances: Resource and Relational Characteristics. Journal of High Technology Management Research 13(4), 177–202 (2002)
10. Sarin, S., Mahajan, V.: The Effect of Reward Structures on the Performance of Cross-Functional Product Development Teams. Journal of Marketing 65(2), 35–53 (2001)

11. Sirvastava, M.K., Gnyawali, D.R.: When Do Relational Resources Matter? Leveraging Portfolio Technological Resources for Breakthrough Innovation. Academy of Management Journal 54(4), 797–810 (2011)
12. Staples, D.S., Greenaway, K., Mckeen, J.D.: Opportunities for Research About Managing the Knowledge-Based Enterprise. International Journal of Management Reviews 3(1), 1–20 (2001)
13. Tampoe, M.: Motivating Knowledge Workers: The Challenge for the 1990s. Long Range Planning 26(3), 49–55 (1993)
14. Wallin, M.W., Krogh, G.V.: Organizing for Open Innovation: Focus on the Integration of Knowledge. Organizational Dynamics 39(2), 145–154 (2010)
15. Yang, J.: Knowledge Integration and Innovation: Securing New Product Advantage in High Technology Industry. Journal of High Technology Management Research 16(1), 121–135 (2005)
16. Yang, J., Rui, M.: Turning Knowledge Into New Product Creativity: An Empirical Study. Industrial Management & Data Systems 109(9), 1197–1210 (2009)

Size-Constrained Clustering
Using an Initial Points Selection Method

Kai Lei, Sibo Wang, Weiwei Song, and Qilin Li

Shenzhen Key Lab for Cloud Computing Technology & Applications (SPCCTA),
School of Electronics and Computer Engineering, Peking University, Shenzhen, P.R. China
leik@pkusz.edu.cn, {wangsibo,songweiwei}@sz.pku.edu.cn,
zirin.lee@gmail.com

Abstract. Size-Constrained clustering tries to solve the problem that how to classify dataset into groups based on each document's similarity with additional requirement which each group size is within a fixed range. By far, adding constraints to assignment step in K-Means clustering is a main approach. But the performance of the algorithm also depends highly on the initial cluster centers like standard K-Means. We propose an initial points selection method by recursively discovering the point with large density around it. Root Mean Square Error and convergence speed (iteration times) are the two most important evaluation standards for clustering using an iterative procedure. Our experiments are conducted on about ten thousand research proposals of National Natural Science Foundation of China and the results show that our method can reduce the iteration times by over 50% and get smaller Root Mean Square Error. The method is scalable and can be coupled with a scalable size-constrained clustering algorithm to address the large-scale clustering problem in data mining.

Keywords: size-constrained clustering, initial cluster centers, density around point.

1   Introduction

Clustering algorithm is often viewed as an unsupervised method for data analysis and it has been applied to many fields, such as data mining, statistical data analysis and knowledge discovery. K-Means clustering [1] has become a very famous method for clustering.

However, one drawback to the K-Means algorithm is that the algorithm often converges with one or more clusters which either are empty or summarize very few data points. In some cases, there are constraints about the cluster size requiring that the size of each cluster must be in a range. The solution to this problem was first introduced in the Bradley's paper [2]. In their work, they proposed adding constraints to the underlying clustering optimization problem requiring that each cluster has at least a minimum number of points based on K-Means clustering. They transformed the cluster assignment step into the Minimum Cost Flow (MCF) problem [4] and solved it by linear network optimization [3]. Jianmin Zhao [5] proposed two constrained

M. Wang (Ed.): KSEM 2013, LNAI 8041, pp. 195–205, 2013.
© Springer-Verlag Berlin Heidelberg 2013

K-Means algorithms: Linear Programming Algorithm (LPA) and Genetic Constrained K-Means Algorithm (GCKA) in his Ph.D. Thesis. Linear Programming Algorithm modified the K-Means algorithm into a linear programming problem with constraints requiring that each cluster has m or more subjects. The most significant difference between Bradley's constrained K-Means algorithm and his LPA is that he ran the algorithm with a large number of random sets of initial points and chose the one with minimal root mean squared error (RMSE) as their final solution. Shunzhi Zhu also proposed a heuristic algorithm to transform size constrained clustering problems into integer linear programming problems in their work [6].

It is known that the iterative algorithms such as K-Means are especially sensitive to initial starting condition. As size-constrained clustering is commonly based on K-Means clustering, it also needs initial centers. Thus, the selection of initial centers has significant impact on the final result and it is also an important factor to improve the clustering solutions.

From the literature and the following experiment, we find that if initial selected centers are close to the final cluster centers, it will reduce the iteration times and get better global minimum for size-constrained clustering. While bad initial centers which with frequent change may lead to bad solution.

In this paper, we propose a method of choosing k points from dataset as the initial centers for size-constrained clustering with k cluster. This method can get the initial points nearer to the optimal result centers in the starting stage of clustering. We can decrease the final RMSE and get less iteration times to reduce the clustering time.

The remaining portion of the paper is organized as follows. In the Section 2, we provide some related work about initialization methods for K-Means clustering and size-constrained clustering. In Section 3, we discuss size-constrained clustering algorithm and our proposed method for initial points selection. Section 4 presents experiment results and discussion of the proposed method in comparison with the other two methods for initial points selection which are widely used on real datasets. Finally, Section 5 concludes the paper.

2   Related Work

2.1   Initialization Methods for K-Means Clustering

In the past, several methods were proposed to solve the cluster initialization for K-Means algorithm. A recursive method for initializing the means by running k clustering problems is discussed by Duda and Hart [7]. A variation of this method takes the entire data into account and then randomly perturbs it k times. For the initial cluster center, Jain and Dubes [8] proposed a method that selects initial values randomly with several times and selected the average of these final cluster centers at the starting stage of K-Means clustering.

The refinement algorithm, proposed by Bradley and Fayyad, builds a set of small random sub-samples of the data and clusters data in each sub-sample by K-Means [9]. All centroids of all sub-samples are then clustered together by K-Means using the k-centroids of each sub-sample as initial centers. The centers of the final clusters giving minimum clustering error are to be used as the initial centers for clustering the original set of data using K-Means algorithm.

Deelers and Auwatanamongkol [10] proposed an algorithm to compute initial cluster centers for K-Means algorithm. They partitioned the data set in a cell using a cutting plane that divides cell in two smaller ones. The plane is perpendicular to the data axis with the highest variance and is designed to reduce the sum squared errors of the two cells as much as possible, keeping the two cells far apart as possible. Also they partitioned the cells once at a time until the number of cells equals to the predefined number of clusters k. In their method the centers of the k cells become the initial cluster centers for K-Means algorithm.

Khan and Ahmad [11] proposed Cluster Center Initialization Algorithm (CCIA) to solve cluster initialization problem. CCIA is based on two observations, with similar patterns to each other. It begins with calculating mean and standard deviation for data attributes, and then separates the data with normal curve into certain partitions. CCIA uses K-Means and density based on multi scale data condensation to observe the similarity of data patterns before finding out the final initial clusters. The experiment results of the CCIA performed the effectiveness and robustness to solve the several clustering problems.

D. Steinley, J. Michael and Brusco indicate several options for initializing the algorithm, compare the procedures, and make several recommendations [12]. J.A. Lozano, J.M. Pena, P. Larranaga compare empirically four initialization methods for the K-Means algorithm: random, Forgy, MacQueen and Kaufman [13]. The results of their experiments illustrate that the random and the Kaufman initialization methods outperform the rest of the compared methods, which make the K-Means more effective and more independent on initial clustering and on instance order.

2.2   The difference between Initialization for K-Means and Size-Constrained Clustering

To the best of our knowledge, there is limited work on initial selection method for size-constrained clustering. In Jianmin Zhao's Thesis [5] he adopted the method of selecting different random points and ran many times to find the best solution. But this will increase the total time to get the final result, and it hasn't gotten the best result yet.

At first thought we may apply the initialization method for K-Means to size-constrained clustering, but there are some differences between K-Means clustering and size-constrained clustering. In K-Means clustering we mainly consider assigning the similar points into the same group without noticing the size constraint of each group, which result in local optimal. While in size-constrained clustering, we need to consider the problem of cluster size in order to get a global optimal result. Sometimes, we need to take two similar points as different initial centers in size-constrained clustering, which will not be selected in K-Means clustering. Therefore we propose our method for initialization in size-constrained clustering.

3   Initial Points Selection Method

3.1   Size-Constrained Clustering

Problem description: Given a dataset $D = \{X^i\}_{i=1}^{m}$ of m points in R^n and cluster size constraint range [min, max], find cluster centers $C^1, C^2, \dots C^k$ in R^n and the assgin array $T_{i,j}$ ($T_{i,j}=1$ means that X^i is assigned to cluster C^j) to minimize the sum of squared distance between every X^i and its assigned center.
     specifically:

$$\underset{C,T}{minimize} \sum_{i=1}^{m} \sum_{j=1}^{k} T_{i,j}(\| X^i - C^j \|^2)$$

$$\sum_{i=1}^{m} T_{i,j} \in [min, max]; j = 1, 2, \dots k$$

subject to:

$$\sum_{j=1}^{k} T_{i,j} = 1; i = 1, 2, \dots m$$

$$T_{i,j} \geq 0, i = 1, 2, \dots m; j = 1, 2, \dots k$$

Like K-Means clustering, size-constrained clustering problem is also solved in two step recursively, cluster assignment and cluster update. It adds constraints in the step of cluster assignment which requires each cluster size in a given range. The cluster assignment problem can be solved by LP(linear programming) or Simplex Network [2].
     The problem can be solved iteratively and is described as follows:
     Suppose cluster centers $C^{1,t}$, $C^{2,t}$, ... $C^{k,t}$ at iteration t, compute $C^{1,t+1}$, $C^{2,t+1}$, ... $C^{k,t+1}$ at iteration t + 1 in the following 2 steps:
     Cluster Assignment: For each data record $X^i \in D$, assign X^i to cluster j such that specifically:

$$\underset{T}{minimize} \sum_{i=1}^{m} \sum_{j=1}^{k} T_{i,j}(\| X^i - C^{j,t} \|^2)$$

$$\sum_{i=1}^{m} T_{i,j} \in [min, max]; j = 1, 2, \dots k$$

subject to:

$$\sum_{j=1}^{k} T_{i,j} = 1; i = 1, 2, \dots m$$

$$T_{i,j} \geq 0, i = 1, 2, \dots m; j = 1, 2, \dots k$$

Cluster Update: Compute $C^{j,t+1}$ as the mean of all points assigned to cluster j.

$$C^{j,t+1} = \begin{cases} \dfrac{\sum_{i=1}^{m} T_{i,j}^{t} X^{i}}{\sum_{i=1}^{m} T_{i,j}^{t}} & if \ \sum_{i=1}^{m} T_{i,j}^{t} > 0. \\ C^{j,t} & otherwise. \end{cases}$$

Stop when $C^{j,t+1} = C^{j,t}$, j = 1, 2, ... , k, else increase t by 1 and go to step 1.

In this paper, we focus on the initialization method for the clustering. That is to say, our effort is made on finding the cluster centers at iteration 0 : $C^{1,0}$, $C^{2,0}$, ... $C^{k,0}$.

3.2   Density around the Point

We represent the k-th nearest point to X^{i} as $KNP(i,k)$, and represent all the k nearest neighbors to X^{i} as $KNN(i,k)$, where

$$KNN(i,k) = \{KNP(i,j)\}_{j=1}^{k}$$

We use k-nearest points radius (KNR) to represent the density around a point, which means the average distance of the k nearest points to it and is computed as follows:

$$KNR(i,k) = \frac{\sum_{X^{j} \in KNN(i)} distance(X^{i}, X^{j})}{k}$$

where

$$distance(X^{i}, X^{j}) = \sqrt{\sum_{t=0}^{n} (X_{t}^{i} - X_{t}^{j})^{2}}$$

3.3   Initial Points Selection Algorithm Description

Based on the assumption that the points with large density around them are most likely to be the final centers, we propose our method of selecting points from dataset as the initial centers.

Our algorithm is recursively finding the point which has the largest density around it. The algorithm is described as following steps.

As figure 1 described, RP represents the set of the remaining points which has not been selected or removed, and IC contains the result of selected initial centers prepared for the clustering. We assume that each cluster has $(min + max)/2$ points. We get the distance matrix of every two points in step 1, and the distance matrix is computed only once. We get the number of clusters k by dividing the number of points by assumed cluster size, that is $k = 2m/(min + max)$. In step 2, we compute the KNR of all points in RP, and find the minimum in step 3. Then we add the point with minimum KNR to IC, and remove the point and its KNN from RP. In step 4 we refresh RP, if RP is not empty, we go on to step 1, and repetitively find the next initial point.

```
Input
   Dataset D = {X^i}^m_{i=1}, Cluster Size Constraint
   [min, max]
Output
   Initial CentersIC = {X^i}^k_{i=1}, k represents the
   number of clusters
Begin
   step 1)
        Get cluster number k = 2m/(min + max), get
        distance matrix, and initialize RP with
        D.
   step 2)
        For each point X^i in RP, Get its
        KNR(i, (min + max)/2).
   step 3)
        Find the point which has the minimum KNR,
        add it to IC, remove it and its all
        (min + max)/2 nearest neighbors from RP.
   step 4)
        Go to step 2 until RP is empty
```

Fig. 1. The proposed algorithm

Note that we get one point at each iteration. At the last iteration, if there are not enough points to get $KNR(i, (min + max)/2)$, we can calculate the KNR by using the all remaining points, choose one point which has the minimum KNR added to IC and remove all remaining points to stop this algorithm.

3.4   The Time Complexity and Stability of the Algorithm

The complexity of getting the distance array is $O(n^2)$ and the complexity of getting all points' KNR is $O(n^2 log(n/k))$.

So the total complexity is

$$O(n^2 + kn^2 log(n/k)) = O(kn^2 log(n/k))$$

Our method is deterministic, and is independently on the instance order, for each step of the algorithm is dependent on all the remaining points instead of part of them.

4   Experiments and Discussion

4.1   Dataset

Our experiments consist of two parts. In the first part, we use two real datasets : the Johns Hopkins Ionosphere dataset and the Wisconsin Diagnostic Breast Cancer dataset(WDBC), which are commonly used in clustering and data mining. The Ionosphere dataset contains 351 data points in R^{33} and values along each dimension are normalized to have mean 0 and standard deviation 1. The WDBC data subset consists of 683 normalized data points in R^9. The second part is conducted on four real dataset in our real project: proposals on Electronics & Information System and Computer Science from 2008 to 2010 of NSFC(National Natural Science Foundation of China), and they are described in table 1. The size column is the number of proposals in each dataset.

Table 1. Four proposals dataset

| Dataset | Category | | | |
	Discipline	year	size	dimensions
1	Electronics and information system	2008	2077	438
2	Computer science	2008	2120	402
3	Computer science	2009	2777	489
4	Electronics and information system	2010	3161	603

We get the feature space of each dataset by removing stop words and noise words, and the number of features of each dataset is shown in dimensions column in table 1.

To evaluate our method, there are two other methods to compare. One is the simplest method by selecting the first k points as the initial centers. The other method is random selection which is widely used in K-Means clustering [13].

4.2   Evaluation Metrics

We use RMSE (Root Mean Square Error) [5] and iteration times as evaluation metrics.

RMSE represents the quality of the final clustering, where $RMSE = \sqrt{\sum_{i=1}^{m} \sum_{j=1}^{k} T_{i,j}(\| X_i - C_j \|^2)/m}$. The less RMSE is, the nearer the result is to the optimal result.

Iteration times describe the convergence speed. The less the iteration times is, the less time it needs to finish the clustering.

4.3   Result and Discussion

We use network simplex method in the step of cluster assignment [3].

In the following tables and figures, sequential represents the method of selecting the first k points as the initial points; random represents randomly selecting k points from the dataset D; select represents our proposed method.

We illustrate the iterating process of clustering on Ionosphere and WDBC datasets with different cluster sizes in figure 2. The variable K in figure 2 represents the number of cluster during the clustering. For a specific dataset, the smaller K is, the bigger cluster size is. The figure shows the change of RMSE during the iterating process of clustering using different initial points selection methods.

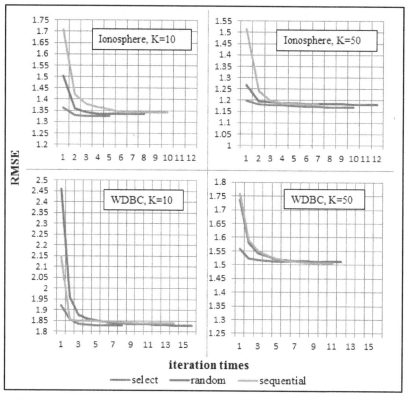

Fig. 2. The iterating process of clustering on Ionosphere and WDBC with different cluster size

As the figure 2 above shows, it's apparent that clustering on the same dataset with bigger cluster size, which uses our proposed method of initial points selection, needs less iteration times and smaller RMSE. That means the convergence is faster and the clustering is more effective. In contrast, the clustering with smaller cluster size needs almost the same iteration times with the other two initial points selection methods. These experiments prove our assumption that the points with large density are most likely to be the final centers. That's because, it's more probable to separate most points into k clusters using our proposed method on dataset with big cluster size. However, clustering with small cluster size may take more noise points into the initial points. Although it can reduce the RMSE at the beginning of the clustering, the centers of clusters may change frequently as size-constrained added while clustering. Therefore more iteration times are required to converge.

We arrange another experiment using four real datasets of NSFC in real project. We also illustrate the iterating process and RMSE changing of clustering on each of the four dataset using different points selection methods in figure 3. We can directly compare the differences in total iteration times and RMSE between the three methods through the figure. This is different from the normal K-Means clustering in initial points selection.

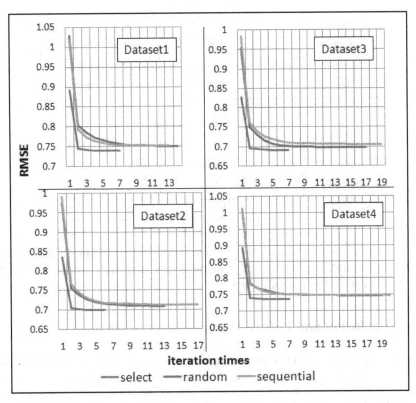

Fig. 3. The iterating process of clustering on each 4 real datasets in our real project

In figure 3, it's apparent that the effect of our proposed initial points selection method is better than the other ones. We can find that our proposed method reduces the iteration times of clustering to only 6 or 7 times, while other two random and sequential methods result in 13 to 21 iteration times. The extra time is mainly in iteration, finishing the clustering and clustering large data. We aim at high dimensions of data in the selection method. We also notice that random selection does not always get better result than sequential method. Sometimes it needs more iteration times to converge. From the table we can see that experiment using sequential method on dataset 1 only needs 13 iteration times to get the final result while random method needs 14 iteration times, as the uncertain initial centers selected by random selection method may be worse than sequential selection method.

Table 2 shows the final RMSE of size-constrained clustering using three different initial points selection methods on the four dataset. The results show that, on all four dataset, our method can get less final RMSE than the other two methods. The result also shows that the problem, which the clustering result depends highly on initial centers in K-Means, also appears in size-constrained clustering.

Table 2. Final RMSE

Dataset	1	2	3	4
sequential	0.752	0.7118	0.7042	0.747
random	0.7502	0.7091	0.6974	0.7463
select	0.7395	0.6991	0.691	0.7362

Table 3. RMSE at first iteration

Dataset	1	2	3	4
sequential	1.0187	0.9914	0.985	1.0129
random	1.0277	0.9756	0.9555	1.0134
select	0.8925	0.8354	0.8274	0.8945

Table 3 shows that the initial centers of our method are nearer to the final optimal centers, for the RMSE after the first iteration is less than other two methods, about 0.8~0.9, while other two values about 1.

5   Conclusion and Future Work

The clustering result depends highly on initial centers in K-Means clustering method and also in size-constrained clustering. Thus improving the convergence speed and reducing RMSE are two important considerations to optimize size-constrained clustering. In this paper, we propose a method which recursively finds the initial center, which is a point in the max density group of its closest neighbors, during the initial points selection stage. By comparing our method with the other two methods, the sequential one and the random one, we can conclude that that our method can reduce the iteration times by over 50% and get smaller RMSE.

There are two things to consider in our future work. First of all, we need to reduce the time complexity of our points selection algorithm by removing the redundant calculations. Secondly, we would like to conduct experiments on more complex data sets from real applications.

Acknowledgments. This work was supported by NFSC project (Grant No. 61103027), 973 project (No. 2011CB302305) and Shenzhen Gov Projects (JCYJ20120829170028558 and ZYA201106080025A).

References

1. Dubes, R.C., Jain, A.K.: Algorithms for Clustering Data. Prentice Hall (1988)
2. Bradley, P.S., Bennett, K.P., Demiriz, A.: Constrained K-Means Clustering. MSR-RT-2000-65, Microsoft Research (2000)
3. Bertsekas, D.P.: Linear Network Optimization. MIT Press, Cambridge (1991)
4. Kelly, D.J., O'Neill, G.M.: The Minimum Cost Flow Problem and The Network Simplex Solution Method (September 1991)
5. Zhao, J.: Optimal Clustering: Genetic Constrained *K-Means* and Linear Programming Algorithms (2006)
6. Zhu, S., Wang, D., Li, T.: Data clustering with size constraints. Knowledge-Based Systems 23(8), 883–889 (2010)
7. Duda, R.O., Hart, P.E.: Pattern Classification and Scene Analysis. John Wiley and Sons, NY (1973)
8. Jain, A.K., Dubes, R.C.: Algorithms for Clustering Data. Prentice Hall, Englewood Cliffs (1988)
9. Bradley, P.S., Fayyad, U.M.: Refing initial points for *K-Means* clustering. In: 15th Internat. Conf. on Machine Learning (1998)
10. Deelers, S., Auwatanamongkol, S.: Enhancing K-Means algorithm with initial cluster centers derived from data partitioning along the data axis with the highest variance. Internat. J. Comput. Sci. 2, 247–252 (2007)
11. Khan, S.S., Ahmad, A.: Cluster center initialization algorithm for *K-Means* clustering. Pattern Recognition Letters 25(11), 1293–1302 (2004)
12. Steinley, D., Brusco, J.M.: Initialization *K-Means* Batch Clustering: A Critical Evaluation of Several Techniques. Journal of Classification 24(1), 99–121 (2007)
13. Lozano, J.A., Pena, J.M., Larranaga, P.: An empirical comparison of four initialization methods for the *K-Means* algorithm. Pattern Recognition Lett. 20, 1027–1040 (1999)

Handling Incoming Beliefs

Philippe Besnard[1] and Éric Grégoire[2],[*]

[1] IRIT UMR CNRS 5505
118 route de Narbonne F-31062 Toulouse France
besnard@irit.fr

[2] CRIL
Université d'Artois - CNRS UMR 8188
rue Jean Souvraz SP18 F-62307 Lens France
gregoire@cril.univ-artois.fr

Abstract. Most logic-based approaches to knowledge and belief change in artificial intelligence assume that when a new piece of information comes up, it should be merely added to the current beliefs or knowledge when this does not lead to inconsistency. This paper addresses situations where this assumption does not hold. The focus is on the construction of Boolean standard-logic knowledge and belief bases in this context. We propose an approach to handle incoming beliefs that can require some formulas reconstruction or a form of preemption to be performed.

Keywords: Knowledge Engineering, Knowledge Representation, Logic-based Artificial Intelligence.

1 Introduction

These last two decades, a fertile domain of research in knowledge representation and reasoning has concerned the way knowledge and beliefs should change in light of new information. Especially, so-called (logic-based) belief revision, belief update [1,2,3,4,5] and knowledge fusion [6,7] have become Artificial Intelligence research fields in their own rights [1,8,9,10,11,12]. Their main focus is on situations where a new piece of information is logically contradicting the pre-existing knowledge or beliefs. When no logical contradiction arises, these approaches assume that this new piece of information should be adopted as such, without any specific treatment. However, this latter assumption does not hold in frequent situations. Accordingly, in this paper we adapt the construction of Boolean standard-logic belief and knowledge bases[1], allowing them to encompass situations of that kind in an adequate way. Let us give some motivating examples.

---

[*] This work has been supported in part by the *Région Nord/Pas-de-Calais* and the EC through a FEDER grant.

[1] From now on, we do not distinguish between the words belief and knowledge.

M. Wang (Ed.): KSEM 2013, LNAI 8041, pp. 206–217, 2013.

A first example is about the necessity to *merge pieces of knowledge them-selves*. Assume that a standard-logic knowledge base Δ is under development and already contains $R_1 = $ *If the switch is on and the lamp bulb is ok then the light is on*. Later on, we are given another rule or belief $R_2 = $ *If the switch is on and the switch is not broken then the light is on*. R_2 does not contradict R_1 from a logical point of view. If R_2 is merely inserted within Δ then we can deduce *Light is on* from Δ when the switch is broken, according to R_1 and provided that both *The switch is on* and *The lamp bulb is ok* are established. This is clearly counter-intuitive. Actually, to match our intuitions, both rules should be merged to form $R_3 = $ *If the switch is on and the lamp bulb is ok and the switch is not broken then the light is on*, which must replace both rules R_1 and R_2.

If we had represented the rules in a converse manner like *If the light is on then the switch is on and the lamp bulb is ok*, the above problem would have been avoided; however, we would have lost the intended capacity to infer that the light is on based on the observations that the switch is on and the lamp bulb is ok. Hence, the selected way to represent rules. Actually, the problem addressed here is not founded on a specific way to represent rules; the necessity to merge rules can occur whenever rules do not a priori capture all their possible exceptions. Although this basic example is straightforward, its treatment in the general case is not direct. First, not all rules should be merged. In the example, R_1 and R_2 must be merged because each of them captures a kind of *compulsory* condition for a shared conclusion to be derived. We should thus be able to distinguish between compulsory and non-compulsory conditions for a rule to apply. Moreover, rules are not necessarily explicit in Δ but can be mere implicit deductive consequences of Δ, only. Also, it must be investigated to which extent this issue might also concern formulas that do not encode rules but simply share some variables. Especially, when a syntax-independent logic-based representation of knowledge is adopted, there is no way to distinguish between a rule and e.g. its representation as a clause, i.e. as a disjunction of signed variables. Finally, what must be dropped from Δ after the merged formulas is inserted must also be defined carefully. We address these questions in this paper.

The issue of *preempting subsuming knowledge* is a slightly related problem. Assume now that Δ contains only the rule $R_0 = $ *If the switch is on then the light is on*. From a logical point of view, R_1 is a mere logical deductive consequence of R_0 and is thus subsumed by R_0. As such, from a logical point of view, R_1 does not bring any actual additional information to R_0. Indeed, in the presence of both rules R_0 and R_1, whenever the switch is on, the light is on (not depending on whether the lamp bulb is ok or not). In order to enforce R_1 over R_0 and thus require *The lamp bulb is ok* for the light to be on, we need the ability to make some subsumed information (namely, R_1) prevail over (or say, preempt) the subsuming knowledge (R_0). Remark that this issue of *preempting subsuming knowledge* also occurs as a sub-part of the *merging pieces of knowledge* one: when the merged formula has been inserted within Δ, things must be settled so that the latter formula is not subsumed by any of the initial ones, which by construction subsume it.

Let us stress that in the general case situations where subsumption links must be annihilated are neither specific to rules, nor due to the selected form of implication connective and its use, nor due to causality issues. On the contrary they can concern any logical formula. To illustrate that, assume that a new piece of information $office \lor home \lor bar$ comes up and must prevail over $office \lor home$ that is already present [13]. From a logical point of view the new piece of information is a deductive consequence of the former one and there is no inconsistency involved. However, we need the subsumed information to replace or prevail over the subsuming one.

In this context, our contribution in this paper builds on and extends some previous works. First, an approach to characterize and solve the subsumption issue has been introduced in [14]. Candidate rationality postulates for belief change operators that allow beliefs to be preempted by subsumed ones have been presented in [15]. The formal characterization of this specific handling of subsumption has been extended to a general non-monotonic setting in [16] and applied to the legal domain in [13]. Starting from this, the contribution of this paper is at least twofold. First, we present a new formal solution to the subsumption issue that avoids rules to be preempted when their preconditions are not satisfied and that also accommodates the problem of merging pieces of knowledge themselves (none of these issues was encompassed in any of the aforementioned pieces of work). Then, the focus is on knowledge engineering issues. More precisely, we study the insertion of knowledge from a pragmatic point of view, taking the above issues into account, when a distinction is made between explicit and implicit information in a knowledge base.

The logical setting in this paper is standard (clausal) Boolean logic. On the one hand, it is the simplest possible framework for presenting and addressing the above subsumption-related issues. On the other hand, recent dramatic progress in Boolean search and reasoning has now revived Boolean logic as a realistic and attractive framework for representing large knowledge bases and solving numerous complex reasoning tasks in artificial intelligence [17].

The paper is organized as follows. In the next section, basic notions about standard Boolean logic are recalled. The question of merging pieces of knowledge and the subsumption issue are presented and solved in section 3. In section 4, a concept of compulsory clauses is introduced. Section 5 focuses on an adapted prime implicate representation. A concept of restrictive clauses is presented in section 6. Main issues about the interaction of a new belief with a preexisting base are addressed in section 7, together with some computational issues. The paper ends with perspectives and promising paths for further research.

2   Logic-Based Framework

To concentrate on the aforementioned conceptual problems, we consider the simple framework of standard (clausal) Boolean logic. Let $\mathcal{L}$ be a language of formulas over a finite alphabet $\mathcal{P}$ of Boolean variables, also called *atoms*. Atoms are noted $a, b, c, \ldots$ The $\land, \lor, \neg, \Rightarrow$ and $\Leftrightarrow$ symbols represent the standard conjunctive, disjunctive, negation, material implication and equivalence connectives, respectively. A *literal* is an atom or a negated atom. Formulas are built in the usual way from atoms, connectives and parentheses; they are noted f, g, h, etc. A formula is in conjunctive normal form (CNF) when expressed as a conjunction of *clauses*, where a clause is a disjunction of literals. For convenience, clauses can be represented by their set of involved literals. The empty clause represents *false*. Also for convenience, the set of involved literals of a clause can be enriched by the value *false*, while still representing the clause. Also for convenience, the disjunction forming a clause can be safely enriched by a disjunct representing *false*.

Interpretations are functions assigning either *true* or *false* to every atom. A *model* of a set of formulas Δ is an interpretation that satisfies every formula of Δ. Δ is *consistent* (also said *satisfiable*) when its set of models is not empty. $\Delta \vdash f$ expresses that the formula f can be deduced from Δ, i.e., that it is *true* in all models of Δ.

A *knowledge base* Δ is a consistent finite set of (non-tautological) clauses and the incoming belief f is a consistent non-tautological clause. We distinguish between Δ, which represents the *explicit* clauses of the base, from the set of all the deductive conclusions of Δ, noted $Th(\Delta)$: $\Delta \vdash f$ iff $f \in Th(\Delta)$.

A word of caution can be needed for readers who are familiar with rule-based systems but not with logic. We exploit the sound and complete deductive capabilities of Boolean logic. Especially, we do not only simply allow for mere forward and backward chaining on $\Rightarrow$ as in traditional rule-based systems. For example, from the rule $a \Rightarrow b$ and $\neg b$, we derive $\neg a$ using contraposition. Also, keep in mind that a rule of the form $(a \wedge b \wedge \neg c) \Rightarrow (d \vee e)$ is logically equivalent to $\neg a \vee \neg b \vee c \vee d \vee e$ (which is also represented by $\{\neg a, \neg b, c, d, e\}$) and will be treated as such.

3   Preempting Subsuming Knowledge and Merging Clauses

In the following, two central concepts are *strict implicant* and *subsumption*.

Definition 1. *Let f and g be two formulas. f is a strict implicant of g iff $f \vdash g$ but $g \nvdash f$. Δ strictly subsumes (in short, subsumes) g iff $\Delta \vdash f$ for some strict implicant f of g.*

Interestingly, when f and g are clauses under their set-theoretic representation, f is a (strict) implicant of g when f is a (strict) non-empty subset of g. Moreover, when f is an implicant made of $n - 1$ literals from the n different variables involved in g (i.e., when f is one longest sub-clause of g), f is said to be a *prime implicant* of g. For example, $\Delta = \{office \vee home\}$ subsumes $office \vee home \vee bar$ and $office \vee home$ is a prime implicant of $office \vee home \vee bar$.

3.1. Preempting Subsuming Knowledge

All strict implicants g of f must be expelled after f has been introduced inside Δ when f must prevail. Interestingly, as formulas of Δ are under CNF format, it is sufficient to expel the prime implicants of f. For example, when we ensure that the prime implicant *office∨home* of *office∨home∨bar* is expelled, we are guaranteed that the smaller implicants *office* and *home* are expelled, too. Indeed, when any of these implicants remains, *office∨home* remains derivable, too.

Actually, the problem of making a formula prevail over all its strict implicants in Δ must sometimes be adapted by replacing Δ by one of its subsets, say Δ', in Definition 1. Typically, Δ' is selected as containing the permanent information involving generic rules or other permanent knowledge, whereas the rest of Δ contains facts that are temporary or related to a specific case or result from the instantiation of the generic rules to a specific situation. When using such Δ' only, we reason about generic rules independently of specific facts and the process of transforming and expelling formulas considers the generic rules, only. Note that facts subsume any rule that contains them as part of its conclusion. For example, *Light is on* subsumes *If the switch is on and the lamp bulb is ok then light is on.* By considering rules only, we do not consider facts that are related to a specific situation and avoid expelling these facts although they subsume rules. In the rest of the paper, for notational convenience, we assume that any formula from Δ can be expelled in the process of accommodating an incoming belief.

3.2. About Merging Pieces of Knowledge themselves

Consider now the first example from the introduction, which requires some rules to be merged. Assume Δ contains the rules R_1 and R_2 in clausal form; namely, $\neg switch\text{-}on \lor \neg bulb\text{-}ok \lor light\text{-}on$ and $\neg switch\text{-}on \lor \neg switch\text{-}ok \lor light\text{-}on$. R_1 and R_2 need be replaced by $R_3 = \neg switch\text{-}on \lor \neg bulb\text{-}ok \lor \neg switch\text{-}ok \lor light\text{-}on$. In this example, the operation is straightforward but more complex situations are also to handled.

First, assume R_1 is the clausal representation of *If A and condition-1 and condition-2 then B* whereas R_2 is intended to represent *If A and condition-3 then B.* Thus, $R_3 = $ *If A and condition-1 and condition-2 and condition-3 then B* is inserted in clausal form within Δ. Clearly, retracting R_2 and R_1 might not been enough to make this new rule prevail. Indeed, Δ might also entail for example *If A and condition-1 and condition-3 then B*, which would subsume R_3. Thus, after having inserted the newly formed clause R_3 in Δ, we must make it prevail over all its strict implicants using the aforementioned process of preempting subsuming knowledge.

Then, not every clause should be merged. Intuitively, we need to merge two clauses when they are about a same subject and when at least one of them translates a form of compulsory condition about this subject. In the last example, we merged R_1 and R_2 because they (both) express compulsory conditions for having B on the basis of A.

In the next sections, concepts of compulsory and restrictive clauses are proposed. They will help us distinguish, capture and solve the various situations about merging pieces of knowledge themselves and about making knowledge prevail over the subsuming one.

4   Compulsory Clauses about f'

In the following, Δ is a consistent finite set of non-tautological clauses that has accommodated an incoming non-contradictory and non-tautological clause f. Unless explicitly indicated, when a clause is referred to, it is not tautological.

Definition 2. *When f can be rewritten as $f = g \vee f'$ where g and f' are two clauses with an empty intersection and where g is not the empty clause. f is said to be about f'. g is called the anti-condition for f' in f.*

When the anti-condition g for f' in f is *false*, f entails f': hence the "anti" prefix. However, by convenience, we will write "condition" instead of "anti-conclusion". Please, note that the definition entails that f is not about itself: f is about *false* and about any of its strict subsets. In our framework, the incoming clause f can be asserted together with the additional information that f is *specifically intended* to be about one specific given f', or without any information about such a possible intent. In the first, case this does not prevent the above definition from concluding that f is also about other f' when f is not just a literal. As motivated earlier, $f = g \vee f'$ might have been asserted together with some clause-labeling information expressing that f need be compulsory about f' in the final Δ.

A first family of such situations occurs when f is not allowed to be subsumed, i.e. when no strict sub-clause of f is allowed to belong to $Th(\Delta)$. In the motivating example, *office∨home∨bar* was such a subsumption-free clause in Δ. To express this, our proposed convention is to select $f' = false$ (and thus $g = f$) and provide some labeling indication that f must be compulsory about f' in Δ.

In the example encoding rules, we might express that $f = g \vee f' = \neg switch-on\vee\neg switch-ok\vee light-on$, which encodes the rule *If the switch is on and the switch is ok then the light is on*, is about $f' = light-on$. We might additionally express that f translates compulsory conditions about f' and that the following requirements must be met.

Requirements (for $f = g \vee f'$ about f' to be compulsory for f' in Δ)

1. $f \in \Delta$. We require any compulsory clause to be explicit, i.e. to belong to Δ.
2. $f' \notin Th(\Delta)$. Otherwise, f could be interpreted as being a mere deductive consequence of f'. Remember that we have assumed that Δ is intended to contain the generic rules and clauses; more generally, we have assumed that any formula in Δ can be expelled to make f must prevail over the knowledge subsuming f. Moreover, when $f' = false$, $f' \in Th(\Delta)$ would contradict the prerequisite that Δ is consistent.
   From these two first requirements, we conclude that $\neg g \notin Th(\Delta)$.
3. When $f' \neq false$, we have $\neg f' \notin Th(\Delta)$. Otherwise, f' could never be derived because the prerequisite that Δ is consistent must be satisfied. If f' were always blocked from derivation then there could not exist at the same time any additional condition g that would authorize such a derivation.

4. When $f' = false$, $\nexists h = g' \vee f'$ about f' in $Th(\Delta)$ such that h is a strict subset of f, like $office \vee bar$ in the motivating example.
5. For any $h = g' \vee f'$ about f' where $h \in Th(\Delta)$, we have $g \subseteq g'$ unless $f' = false$ or $g' \in Th(\Delta)$. This requirement does not exclude from $Th(\Delta)$ all clauses containing a strict subpart of g. For example, if we are able to derive that $in\text{-}Dalian\text{-}KSEM$ is true, a standard-logic reasoner is justified to also conclude e.g. $h = in\text{-}Dalian\text{-}KSEM \vee light\text{-}on$, or $h = in\text{-}Dalian\text{-}KSEM \vee \neg switch\text{-}on \vee light\text{-}on$. These last inferences can be made independently of the actual truth value of $light\text{-}on$ and despite any possible compulsory clause about $light\text{-}on$.
6. A compulsory clause about f' is *also* implicitly expressing conditions about any f'' where $\emptyset \subset f'' \subset f'$. More precisely, when a clause $f = g \vee f'$ is compulsory about f', $h = g \vee f''$ is compulsory about f'' for any f'' such that $\emptyset \subset f'' \subset f'$, *provided that* h belongs to $Th(\Delta)$. For example, when $\neg vacation \vee go\text{-}to\text{-}beach \vee go\text{-}to\text{-}mountain$ is compulsory about $go\text{-}to\text{-}beach \vee go\text{-}to\text{-}montain$, we have that $h = \neg vacation \vee go\text{-}to\text{-}beach$ is compulsory for $go\text{-}to\text{-}beach$, provided that $h \in Th(\Delta)$.

Hence, the formal definition.

Definition 3. *When $f = g \vee f'$ is about f'. f is compulsory about f' in Δ iff*

1. $f \in \Delta$, and
2. $f' \notin Th(\Delta)$ and, when $f' \neq false$, $\neg f' \notin Th(\Delta)$, and
3. when $f' = false$, no strict implicant of f belongs to $Th(\Delta)$, and
4. $\forall (g' \vee f') \in Th(\Delta)$, $g \subseteq g'$ unless $g' \in Th(\Delta)$ or $f' = false$.

Example. Let $f = a \vee b \vee c1 \vee c2$ be compulsory about $c1 \vee c2$ in Δ. According to 3.1, $f \in \Delta$. According to 3.2, $c1 \vee c2 \notin Th(\Delta)$ and thus $c1 \notin Th(\Delta)$ and $c2 \notin Th(\Delta)$. Also, $\neg c1 \wedge \neg c2 \notin Th(\Delta)$. As a consequence of both 3.1 and 3.2, $\neg g = \neg a \wedge \neg b \notin Th(\Delta)$. According to 3.4, $f = a \vee d \vee c1 \vee c2 \in Th(\Delta)$ requires that $d \in Th(\Delta)$.

Property 1. When $f = g \vee f'$ is compulsory about f' in Δ, $g \vee f''$ is compulsory about f'' for any f'' s.t. $\emptyset \subset f'' \subset f'$.

Indeed: first note that $f'' \neq false$. Then, assume $\exists g' \vee f'' \in Th(\Delta)$: $g \not\subseteq g'$ unless $g' \in Th(\Delta)$. Since $g' \vee f'' \vdash g' \vee f'$, all this would contradict $\forall (g' \vee f') \in Th(\Delta)$, we have $(g \subseteq g')$ unless $g' \in Th(\Delta)$ or $f' = false$.

Clearly, this property can be of practical computational importance since none of the conditions in Definition 3 needs to be checked to decide whether $g \vee f''$ is compulsory about f'' when we know that $g \vee f''$ is compulsory about f'' and that $g \vee f'' \in Th(\Delta)$.

Example (Cont'd). Assume also that $a \vee b \vee c1 \in Th(\Delta)$. Then, $a \vee b \vee c1 \in \Delta$ and this clause is compulsory about $c1$.

Property 2. When a clause $f = g \vee f'$ is compulsory about f' in Δ, we have that

1. all the aforementioned *Requirements (for $f = g \vee f'$ about f' to be compulsory for f' in Δ)* are satisfied,
2. when $f' \neq false$, there is no other compulsory clause about f' in Δ,
3. $\neg g \notin Th(\Delta)$,
4. when f is compulsory about f' in Δ, f is also compulsory about f' for any $\Delta' \subseteq \Delta$ such that $f \in \Delta$.

The last three properties are also of a practical computational importance. For example, the last one allows us to retract information from Δ without altering the compulsory status of a clause f when f remains derivable.

A labeling is used to mark clauses in order to recognize compulsory clauses in Δ that were actually *intended to be so*. We do not label formulas in $Th(\Delta) \setminus \Delta$ since all compulsory clauses belong to Δ.

Definition 4. *Let $f = g \vee f' \in \Delta$. f is either explicitly labeled "required-compulsory" (in short, RC) or implicitly labeled "not-required-compulsory" (in short, NRC) about f' in Δ. By default, f is implicitly labeled NRC about f' in Δ. Only clauses that are compulsory about f' can be labeled RC about f'.*

For convenience, when the context does not make it ambiguous, we say that f is RC (resp. NRC), implicitly referring to Δ, f' and to the condition g for f' in f.

By default, a clause in Δ is thus (implicitly) marked NRC about any of its sub-clauses, including the empty one, as we expect RC clauses to be outnumbered by NRC ones. Let us stress again that all compulsory clauses about f' are not necessarily labeled RC about f': they are marked NRC when the agent/user did not require them to be compulsory.

Condition 4 in Definition 3 does not provide any hint about which g' should be checked in that condition. To circumvent the latter issue, we turn to a *prime implicate* representation of Δ.

5   Prime Implicate Representation

Definition 5. *A prime implicate of a finite set Δ of formulas is any clause h that satisfies both conditions below*

(1) $h \in Th(\Delta)$
(2) $h' \Leftrightarrow h \in Th(\Delta)$ for every clause h' s.t. $\Delta \vdash h'$ and $h' \vdash h$

Δ_{PI} denotes the set of all prime implicates in Δ.

Accordingly, h is a prime implicate of Δ iff h is a minimal (w.r.t. $\subseteq$) non-tautological clause amongst the set formed of the clauses l such that $\Delta \vdash l$.

Prime implicates have already been investigated in belief revision and change mainly because they provide a compact and syntax-independent yet complete representation of a belief base (see e.g. [18] and [19]) and because interesting computational tasks (like satisfiability checking and entailment) are tractable in this framework [20]. In the worst case, computing the set of prime implicates of Δ containing a clause l (a task that we will often refer to) is however not in polynomial total time unless P=NP (it is in polynomial total time when for example the clause is positive and Δ is Horn) [21]. Although the compactness and some of the computational features of a prime implicates representation happen to be welcome properties, the motivation for focusing on prime implicates is here different and stems from their intrinsic epistemological nature, as shown by the following properties.

The first two properties are straightforward and well-known. The third one shows how this prime implicate representation will help us in dealing with compulsory clauses.

Property 3. Let f be a consistent non-tautological clause.

1. $f \in Th(\Delta)$ iff $f \in Th(\Delta_{PI})$.
2. f is not subsumed in Δ iff $f \in \Delta_{PI}$.
3. $f = g \vee f'$ is the only clause about f' in Δ_{PI} iff f is compulsory in Δ.

It would be tempting to adopt Δ_{PI} as the actual representation for Δ. However, Δ can contain formulas that are not prime implicates of Δ and that thus do not belong to Δ_{PI} but in $Th(\Delta_{PI}) \setminus \Delta_{PI}$. Accordingly, we assume that Δ is a superset of Δ_{PI} and that $\Delta = \Delta_{PI} \cup \Delta_{nonPI}$, where Δ_{nonPI} contains the explicit clauses that are not prime implicates of Δ. As a consequence, $Th(\Delta) = Th(\Delta_{PI}) = Th(\Delta_{PI} \cup \Delta_{nonPI})$.

6   Restrictive Clauses about f'

Being compulsory about f' is sometimes a too strong requirement. It must sometimes be softened as follows: any strict implicant of $f = g \vee f'$ containing f' is not allowed to belong to $Th(\Delta)$ while, at the same time, other clauses about f' are allowed to exist in $Th(\Delta)$. For example, we might require that *If first-class-passenger and valid-boarding-pass then fast-lane* does not coexist with any shorter rule containing *fast-lane* in Δ whereas *If VIP then fast-lane* is allowed to exist in $Th(\Delta)$. The first clause is called restrictive about f' (where $f' = fast\text{-}lane$) in Δ.

Definition 6. *When $f = g \vee f'$ is about f', f is restrictive about f' in Δ iff*

1. $f \in \Delta$, and
2. $f' \notin Th(\Delta)$ and, when $f' \neq false$, $\neg f' \notin Th(\Delta)$, and
3. no strict implicant of f containing f' belongs to $Th(\Delta)$.

Useful properties of restrictive clauses are as follows.

Property 4.

1. Any compulsory clause about f' in Δ is restrictive about f' in Δ.
2. When Δ contains a restrictive clause about f' that is not compulsory about f' and when $f' \neq false$, there is no compulsory clause about f' in Δ.

3. Assume a clause $f = g \vee f'$ is restrictive about f' in Δ. Let f'' s.t. $\emptyset \subset f'' \subset f'$ and $g \vee f'' \in Th(\Delta)$, we have that $g \vee f''$ is restrictive about f'' in Δ.
4. $f = g \vee f'$ is restrictive about f' in Δ iff $f \in \Delta_{PI}$.

Since permissive clauses are expected to outnumber restrictive ones, and since not all restrictive clauses are required to be so by the agent or user, the following labeling convention is followed.

Definition 7. *Let f be a clause about f' in Δ. f is marked either "required-restrictive" (in short RR) about f' in Δ, or implicitly "not-required-restrictive" (in short NRR) about f' in Δ. By default, clauses labeled NRC about f' are marked NRR about f'. Clauses marked RC about f' are marked RR about f'.*

For convenience, NRR clauses about f' are also called permissive clauses. Note that restrictive clauses are thus permissive when they are not intended to be required restrictive by the user or the agent.

7 Handling an Incoming Belief

As in most belief change approaches, we adopt a form of preference for more recent information and apply a principe of minimal change. A total ordering of RC or RR clauses (based on the time-stamp expressing when the labeling occurred) is assumed available to direct the selection of clauses to be expelled or that must have their labeling changed, when such a choice among several candidates occurs. The treatment is also intended to be well-suited for iteration, when a succession of incoming beliefs occurs and when Δ is built incrementally, at least in the sense that $\Delta = \Delta_{PI} \cup \Delta_{nonPI}$ is assumed to comply with the $\{RC, NRC, RR, NRR\}$ labeling before and after an incoming belief shows up. Due to space limitation, we only sketch the main steps of the approach when the incoming belief is a clause $f = g \vee f'$ that is intended to be compulsory about one given f'. This is the most complex situation and it allows us to illustrate principles that also apply to other cases.

Two main approaches can be distinguished based on the understanding of the precise role of f in that respect: f can be intended to either *replace* any formula about f' in Δ' (actually, as we have seen, all prime implicates containing f' in Δ'_{PI}), or *weaken* those clauses by enforcing g within their condition about f'. Consider the second situation as a case study.

Several situations can occur, depending on the current $\{RC, NRC, RR, NRR\}$ labeling of clauses in Δ'. Interestingly, we take advantage of the aforementioned properties of the prime implicate representation and of compulsory/restrictive clauses. For example, we know that when a clause is compulsory about f', it is unique in that respect and belongs to Δ_{PI}. When a clause is retracted from Δ, the remaining compulsory/restrictive clauses do not change their status (and thus the corresponding labeling of clauses does not change). When a clause $g \vee f'$ is compulsory/restrictive about f' then a same status is derived for any $g \vee f''$ where $\emptyset \subset f'' \subset f'$ when $g \vee f'' \in Th(\Delta)$.

However, let us stress on the following points. Δ_{PI} can need some updating operations when an incoming clause must belong to it. When an incoming clause is to be RC or NRC, other RC or RR clauses might need to be expelled. When they can co-exist together, their status can however need to be downgraded into RR or NRR with respect to the concerned sub-clause. In the following, we do not mention how Δ_{nonPI} is handled, because this does not involve any technical difficulty or complexity.

The specific focus is on when $f = g \vee f'$ about f' (when $f' \neq false$) is intended to be compulsory about f' with respect to a pre-existing Δ', in order to deliver a final base, noted Δ. Consider the case where there exists a clause $h = k \vee f'$ that is compulsory about f' in Δ. There is thus no other restrictive clause about f' in Δ.

First, Δ is initialized to Δ'. h is the unique clause about f' in Δ_{PI}. $m = g \vee k \vee f'$ must become compulsory about f' in Δ. When $m = h$ this means that m is already compulsory about f' in Δ: the procedure ends. Otherwise, h is retracted from Δ_{PI}. Clauses about f'' where $\emptyset \subset f'' \subset f$ are retracted from Δ_{PI}, too. Δ_{PI} is updated with the constraint that $m \in \Delta_{PI}$. Let $N = \{n$ s.t. $n = g \vee k \vee f''$ where $\emptyset \subset f'' \subset f'$ and $n \in Th(\Delta_{PI})\}$. Mark all elements of N by RC about f''. As a clause has been introduced within $Th(\Delta_{PI})$, it might happen that RC and RR clauses about some sub-clauses are no longer compulsory (restrictive) but only restrictive (permissive) about those sub-clauses. Accordingly, all RC and RR clauses must be checked again, according to the aforementioned total order translating a preference for more recent information. When a clause cannot be compulsory, it is checked whether it can be downgraded to RR and, in the negative case, it becomes merely permissive about the concerned sub-clause. Also, clauses that remain in Δ but cannot be any longer RR become permissive about the concerned sub-clause.

8   Conclusion and Perspectives

Most research efforts about knowledge and belief change have taken the consistent case for granted. On the contrary, we claim that taking into account a new piece of information that does not contradict the preexisting knowledge is not always a straightforward issue; it might actually involve complex reasoning paradigms. This paper intends to be a contribution to the study of these paradigms by focusing on situations where the novel information can need to prevail over the existing knowledge. In the future, we plan to extend this work to the first-order case and to non-monotonic logics. In this last respect, a first result is that the approach in the paper directly applies to fragments of non-monotonic logics that include forms of negation as failure, provided that Δ does not entail any literal. In this case, negation as failure can be replaced by standard negation to analyze in an adequate manner the interactions between generic rules, independently of any specific concrete case or data.

References

1. Fermé, E., Hansson, S.: AGM 25 years. twenty-five years of research in belief change. J. of Philosophical Logic 40, 295–331 (2011)
2. Alchourrón, C., Gärdenfors, P., Makinson, D.: On the logic of theory change: Partial meet contraction and revision functions. J. of Symbolic Logic 50(2), 510–530 (1985)
3. Katsuno, H., Mendelzon, A.: On the difference between updating a knowledge base and revising it. In: Proc. of KR 1991, pp. 387–394 (1991)
4. Gärdenfors, P.: Knowledge in Flux: Modeling the Dynamics of Epistemic States, vol. 103. MIT Press (1988)
5. Hansson, S.O.: A Textbook of Belief Dynamics. Theory Change and Database Updating. Kluwer Academic (1999)
6. Konieczny, S., Pino Pérez, R.: On the logic of merging. In: Proc. of KR 1998, pp. 488–498 (1998)
7. Konieczny, S., Grégoire, É.: Logic-based information fusion in artificial intelligence. Information Fusion 7(1), 4–18 (2006)
8. Doyle, J.: A truth maintenance system. Artificial Intelligence 12, 231–272 (1979)
9. Dalal, M.: Investigations into a theory of knowledge base revision (preliminary report). In: Proc. of AAAI 1988, vol. 2, pp. 475–479 (1988)
10. Revesz, P.Z.: On the semantics of theory change: Arbitration between old and new information. In: Proc. of PODS 1993, pp. 71–82 (1993)
11. Subrahmanian, V.S.: Amalgamating knowledge bases. ACM Transactions on Database Systems 19, 291–331 (1994)
12. Fagin, R., Ullman, J.D., Vardi, M.Y.: On the semantics of updates in databases. In: Proc. of PODS 1983, pp. 352–365 (1983)
13. Besnard, P., Grégoire, É., Ramon, S.: Logic-based fusion of legal knowledge. In: Proc. of Fusion 2012, pp. 587–592. IEEE Press, Singapor (2012)
14. Besnard, P., Grégoire, É., Ramon, S.: Enforcing Logically Weaker Knowledge in Classical Logic. In: Xiong, H., Lee, W.B. (eds.) KSEM 2011. LNCS, vol. 7091, pp. 44–55. Springer, Heidelberg (2011)
15. Besnard, P., Grégoire, É., Ramon, S.: Preemption operators. In: Proc. of ECAI 2012, pp. 893–894 (2012)
16. Besnard, P., Grégoire, É., Ramon, S.: Overriding subsuming rules. In: Liu, W. (ed.) ECSQARU 2011. LNCS, vol. 6717, pp. 532–544. Springer, Heidelberg (2011)
17. Sakallah, K.A., Simon, L. (eds.): SAT 2011. LNCS, vol. 6695. Springer, Heidelberg (2011)
18. Zhuang, Z.Q., Pagnucco, M., Meyer, T.: Implementing iterated belief change via prime implicates. In: Orgun, M.A., Thornton, J. (eds.) AI 2007. LNCS (LNAI), vol. 4830, pp. 507–518. Springer, Heidelberg (2007)
19. Bienvenu, M., Herzig, A., Qi, G.: Prime implicate-based belief revision operators. In: 20th European Conference on Artificial Intelligence (ECAI 2012), pp. 741–742 (2008)
20. Darwiche, A., Marquis, P.: A knowledge compilation map. J. Artif. Intell. Res. (JAIR) 17, 229–264 (2002)
21. Eiter, T., Makino, K.: Generating all abductive explanations for queries on propositional horn theories. In: Baaz, M., Makowsky, J.A. (eds.) CSL 2003. LNCS, vol. 2803, pp. 197–211. Springer, Heidelberg (2003)

Modified Collaborative Filtering Algorithm Based on Multivariate Meta-similarity[*]

Pengyuan Xu and Yanzhong Dang

Institute of Systems Engineering, Dalian University of Technology, Dalian 116024, China

Abstract. This paper further research the recommendation algorithm bases on the meta-similarity. We consider more information about users collect the items , and define the epidemic degree of the item(*EDI*) and user(*EDU*), modify the degree of overlapping of items, and analyze the effect of multivariate similarity in the recommendation system, then we present a modified collaborative filtering algorithm based on multivariate meta-similarity (*MMSCF*). The method reduces the influence of the *EDI* and *EDU*, limited the error to transfer, and enhances the similarity by multivariate meta-similarity. The experiments prove the new recommendation algorithm evaluated by the precision indexes of *ranking score*, *precision* and *recall* have achieved significantly improve.

Keywords: Collaborative Filtering, Recommendation System, Meta-similarity, multivariate meta-similarity.

1   Introduction

Nowadays, the information explosion, which results from the exponential growth of the Internet [1, 2] and Web, has brought us to the information overload era [3].Too much information on the Internet means too much hard work for us to find out the piece which just appears in our mind. Thanks to the search engine, an effective tool for information filtering by which the relevant information could be retrieved by using the keywords, such a problem is relatively being addressed [4, 5]. However, since the search engine relies on the keywords directly, it can not give the users personal information based on the users' interests and habits. Therefore, thus far, various kinds of algorithms have been proposed to provide personal recommendations, including correlation-based methods [6, 7], content-based analysis [8, 9], spectral analysis [10], iteratively self-consistent refinement [11], net-based method [12-15], and so on.

   One of the most successful recommendation algorithms is collaborative filtering (CF) [4, 5] [16, 17]. A number of *CF* methods have been developed in the past decade [18-22]. To make useful predictions about potential interests of a given user, the CF firstly identifies a set of similar users from the past records and then starts to predict based on the weighted combination of those similar users' opinions. There are some other famous collaborative filtering systems [22-25], such as: Tapestry, Ringo.

---

[*] This work is partially supported by the National Natural Science Foundation of China (Grant No. 71031002).

M. Wang (Ed.): KSEM 2013, LNAI 8041, pp. 218–229, 2013.

Recently, some new methods for recommendation systems are proposed including network-based model [12-15] and diffusion-based collaborative filtering [18], which have been demonstrated to be of high accuracy.

In order to improve the accuracy of the standard *CF*, several algorithms [18-22] have been proposed. The paper [26] presents a modified collaborative filtering algorithm base on meta-similarity (*MSCF*), it defines a new similarity: meta-similarity, instead of the traditional measure to calculate the similarity between two users. In accordance to the previous research, the similarity between user *i* and *j* is usually estimated by the Pearson correlation coefficient of their rates for all items, which only considers the initial similarity of the user by the collection of the items. The meta-similarity between *i* and *j* is the Pearson correlation coefficient of their similarity for all the users. And it gets good results on the experiments.

After further research, we find there are some shortcomings of the *MSCF*. One is it ignores the initial information of the items and users, reduce the accurate of the recommend result; other is the meta-similarity is compute by the similarity of the user, if the initial similarity has error, *MSCF* will expand the error; at last, the meta-similarity is a binary similarity, whether it can be extended to multivariate similarity should to be researched further. To solve these problems, this paper modifies the *MSCF* base on multivariate correction meta-similarity.

2   A Modified Collaborative Filtering Algorithm Based on Meta-Similarity

2.1   Collaborative Filtering Algorithm

The aim of a collaborative filtering algorithm is to suggest new items or to predict the utility of a certain item for a particular user based on the user's previous likings and the opinions of other like-minded users. Denoting the item set as $O = \{O_1, O2, ..., O_M\}$, and the user set as $U=\{u_1, u_2, ..., u_N\}$. If only consider the user whether collect the items, the recommender system, consisting of N users and M items can be fully described by an $N\times M$ rating matrix R, with $r_{i\alpha}=1$ the rating user i have already collected items α, and $r_{i\alpha}$ has been set as zero otherwise. In this way, the proposal of a recommender system is to predict personal opinions on those items before the users collected. Another more complicated case is the voting system, where each user can give ratings to objects, and the aim of the recommendation algorithm is to estimate unknown ratings for items. These two problems are closely related, however, in this article we focus on the former case.

In terms of the *CF* algorithm, the similarity $sim(i, j)$ between users i and j has the Pearson-like form, which is given by

$$sim(i, j) = \frac{\sum_{\alpha \in O} r_{i,\alpha} \bullet r_{j,\alpha}}{\min(k(i), k(j))} \tag{1}$$

Where $k(i)$ is the degrees of the user i. O denotes the set of items which has been evaluated by both users i and j. $sim(i, j) \to 1$ means the users i and j are very similar, while $sim(i, j) \to 0$ means the opposite case.

In the most widely applied similarity-based algorithm, the predicted rating is calculated by using a weighted average, as follow:

$$R'_{i,\alpha} = \frac{\sum_{\beta} sim(\alpha, \beta) \bullet r_{i,\beta}}{\sum_{\beta} |sim(\alpha, \beta)|} \tag{2}$$

From Eq (2), we can find that the larger $sim(\alpha, \beta)$ and $r_{i,\beta}$, the bigger $sim(\alpha, \beta) \cdot r_{i,\beta}$. In other words, the conclusion that user i likes item α can be reached through the fact that user i likes more similar item β. For any user u_i all predicted score $R'_{i,\alpha}$ are sorted in descending order, and those objects on the top are recommended.

2.2   Collaborative Filtering Algorithm Based on Meta-similarity

Although the *CF* already has a good algorithmic accuracy, this linear equation (1) only considers the common collections, which is not effective for the users with no common selections before. In [26], a new concept meta-similarity is proposed. It is a high order similarity.

The initial similarity $Isim(i, j)$ between users i and j is:

$$Isim(i, j) = \frac{\sum_{\alpha \in O} r_{i,\alpha} \bullet r_{j,\alpha}}{\min(k(i), k(j))} \tag{3}$$

To use Pearson Correlation, the equation as:

$$metasim(i, j) = \frac{\sum_{\alpha \in O} sim'_{i,\alpha} \bullet sim'_{j,\alpha}}{\sqrt{\sum_{\alpha \in O} (sim'_{i,\alpha})^2} \bullet \sqrt{\sum_{\alpha \in O} (sim'_{j,\alpha})^2}} \tag{4}$$

Where $sim'_{i,\alpha} = sim_{i,\alpha} - <sim_i>$, $sim_{i,\alpha}$ is the similarity of the user i with the user α, $<sim_i>$ is the average similarity of the user i with other users, and α runs over one user sets. Meta-similarity measures the similarity between i and j based on the normal similarity. The sparsity of data set makes the direct similarity calculation less accurate, and therefore a new measure of similarity is expected. Properly integrating meta-similarity correlations may perform better. Motivated by this idea, we propose a linear form to combine similarity and meta-similarity as:

$$newsim = \lambda \times Isim + (1 - \lambda)metasim \tag{5}$$

Where *newsim* is the new similarity matrix and *Isim* is the initial similarity matrix defined by Eq (3), *metasim* is the meta-similarity matrix defined as Eq (4). And λ is a tunable parameter. Experiments show that when $\lambda=0.5$ get good results.

The predicted rating could be obtained by:

$$R_{i,\alpha} = \frac{\sum_{\beta} newsim(\alpha, \beta) \bullet r_{i,\beta}}{\sum_{\beta} \left| newsim(\alpha, \beta) \right|} \tag{6}$$

Through Eq (5), $newsim(\alpha, \beta)$ is the element of matrix *newsim* defined by Eq (4). It is easy to understand that user i likes item α can be got from the fact user i likes more similar item β. It is easy to understand that if user i likes item β, which is similar to item α, the user probably like item α. For any user u_i all predicted score $R'_{i,\alpha}$ are sorted in descending order, and those objects in the top are recommended.

3   Modified Collaborative Filtering Algorithm Based on Multivariate Meta-similarity

3.1   Analyze the *MSCF*

The *MSCF* solves the problems of the data sparsity by compute the Pearson correlation coefficient of their similarity for all the users, and showed good recommendation efficiency. But it also has some problems:

(1) The Eq (3) and Eq (4) are too simple, without considering the distribution of the initial resources. In Eq(3), the user for all the items is equal, some items may be collected by many users, the similarity of these items will be has strong ability of recommendation than the items be collected by few users. The *MSCF* without considering the items amount which are collected by the user, some users may be collect many items, the meta-similarity of this users will affect the recommendation result. To solve the problem, two concepts are proposed, the number of the users collect the item is the epidemic degree of item (*EDI*), the item is collected by more users the *EDI* is bigger, and the number of the items which are collected by the user is the epidemic degree of user (*EDU*), the user collect more items the *EDU* is bigger.

(2) In Eq (4), the meta-similarity is get by Pearson-like form base on Eq (3), the meta-similarity is between the users have the same interesting of the item. Sometime the user collect the item random, the error of the item collected mistake by user, will escalate by Eq(3) and Eq(4). So the meta-similarity should be computed base on the initial data to avoid the bigger error.

(3) The *MSCF* only consider the recommendation of the binary similarity without the multivariate similarity.

3.2   The Meta-similarity Base on the Multivariate Meta-Similarity

This paper modify the meta-similarity by consider the effect of recommendation bye *EDI*, *EDU* and correct overlap, then through the multivariate meta-similarity to modify it farther. The results show the recommendation in the important precision index: *ranking score*, *precision* and *recall* have achieved significantly increase.

The Modified *EDI* and *EDU*

For problem (1), the *EDI* and *EDU* effect the recommendation, we can use the *EDI* factor ω^{δ}_a and the *EDU* factor φ^{γ}_i to modify the effect of them. ω_a is the epidemic degree of item and φ_i is the epidemic degree of user, we can modify the effect of the *EDI* and *EDU*, by the coefficient δ and γ, $\omega^{\delta}_a > 1$ ($\varphi^{\gamma}_i > 1$) represents enhance the effect, while $\omega^{\delta}_a < 1(\varphi^{\gamma}_i < 1)$ represents reduce the effect, and $\omega^{\delta}_a = 1(\varphi^{\gamma}_i = 1)$ represents ignored the effect.

The Modified Degree of Overlapping

For problem (2), we use the degree of overlapping [21,22] for item, it means to limit the meta-similarity by the number of the items which are common collected, and modify the degree of overlapping by the variance of user collection.

$$Modified\_overlapping = \frac{Crossover}{\text{var}_i + \text{var}_j} \tag{7}$$

Modified_overlapping is to compute the similarity of the users by the degree of overlapping and the fluctuation of the user collection. *Crossover* is the number of the common collected items, var_i and var_j are the variance of the collections by user i and j, it reflect the fluctuation of the user collection, if the variance is bigger, it means the user collection is more random, the user is instability. If the *Crossover* is bigger, the degree of overlapping is bigger, it means the user has the more common items and the user is the more credible. The value of $var_i + var_j$ is smaller, the degree of overlapping is bigger, the user has the smaller variance and it is more credible.

This paper limits the meta-similarity by *Modified_overlapping*, the bigger *Modified_overlapping* means the more common items, the users are more credible, and the user collection less random, the meta-similarity is bigger. So we can avoid the bigger error in binary similarity by *Modified_overlapping*, as discussed in problem (2).

Multivariate Similarity

For problem (3), the ternary similarity is:

$$cubesim(i, j) = \frac{\sum\limits_{\alpha \in O} metasim'_{i,\alpha} \bullet metasim'_{j,\alpha}}{\sqrt{\sum\limits_{\alpha \in O} (metasim'_{i,\alpha})^2} \bullet \sqrt{\sum\limits_{\alpha \in O} (metasim'_{j,\alpha})^2}} \tag{8}$$

Where $metasim'_{i,a}=metasim_{i,a}-<metasim_i>$, $<metasim_i>$ is the average similarity of the user i and other users, O is the set of the object user which the meta-similarity are more than 0.

The error of the ternary similarity is more serious, so it can't to recommend as the main similarity, although the noise by the error of the ternary similarity is obvious, it doesn't cover the user has the bigger similarity, we can enhance the similarity of these users by ternary similarity, it can improve the result of the recommendation.

The Meta-Similarity Base on the Multivariate Similarity

We have defined the epidemic degree of the item and user, the degree of overlapping, the multivariate similarity, here define the meta-similarity base on the multivariate similarity as follow steps:

(1) The initial similarity: in Eq(3), consider the EDI and EDU, the initial similarity is :

$$Modified\_Isim(i,j) = \frac{\sum_{\alpha \in O} r_{i,\alpha} \bullet r_{j,\alpha} \bullet w_{\alpha}^{\delta} \bullet \psi_i^{\gamma} \bullet \psi_j^{\gamma}}{min(k(i),k(j))} \qquad (9)$$

(2) The meta-similarity: in Eq(4), consider the EDU and the degree of overlapping, the meta-similarity is:

$$Modified\_metasim(i,j) = metasim(i,j) \bullet \psi_i^{\gamma} \bullet \psi_j^{\gamma} \bullet Modified\_Overlap \qquad (10)$$

Where $metasim(i,j)$ is the similarity get by Eq(7),

$$metasim(i,j) = \frac{\sum_{\alpha \in O} Modified\_Isim'_{i,\alpha} \bullet Modified\_Isim'_{j,\alpha}}{\sqrt{\sum_{\alpha \in O}(Modified\_Isim'_{i,\alpha})^2} \bullet \sqrt{\sum_{\alpha \in O}(Modified\_Isim'_{j,\alpha})^2}} \qquad (11)$$

$$Modified\_Isim'_{i,\alpha} = Modified\_Isim_{i,\alpha} - < Modified\_Isim_i > \qquad (12)$$

Here, $<Modified\_Isim_i>$ is the average similarity of the user i with other users.

Then combine the Eq (8) and Eq (9), the similarity defined as:

$$Combine\_sim(i,j) = \lambda \bullet Modified\_Isim(i,j) + (1-\lambda) \bullet Modified\_metasim(i,j) \quad (13)$$

$\lambda=1$ shows only consider the initial similarity, it is the same as the CF, $\lambda=0$ shows only consider the meta-similarity. $0<\lambda<1$ shows combined consider the initial similarity and meta-similarity.

(3) The multivariate similarity:

$$cubesim(i,j) = \frac{\sum_{\alpha \in O} Modified\_metasim'_{i,\alpha} \bullet Modified\_metasim'_{j,\alpha}}{\sqrt{\sum_{\alpha \in O}(Modified\_metasim'_{i,\alpha})^2} \bullet \sqrt{\sum_{\alpha \in O}(Modified\_metasim'_{j,\alpha})^2}} \qquad (14)$$

$$Modified \_ metasim'_{i,\alpha} =$$
$$Modified \_ metasim_{i,\alpha} - < Modified \_ metasim_i > \tag{15}$$

Here, $<Modified\_metasim_i>$ is the average modified meta-similarity of the user i with other users.

(4) Final the similarity is defined as:

$$MMSCF \_ sim(i, j) =$$
$$(\lambda \bullet Modified \_ Isim(i, j) + (1 - \lambda) \bullet Modified \_ metasim(i, j)) \bullet e^{cubesim(i,j)} \tag{16}$$

Because the big error of the multivariate similarity, we use the exponential factor e to enhance the meta-similarity of the users with higher initial similarity.

To sum up presents a modify meta-similarity, use the Modified_overlapping to avoid the error spreading in binary similarity. The bigger value of Modified_overlapping shows the more possibly of the user i accredits user j, or reduce the meta-similarity of i and j, it is a pre decision.

φ'_i, φ'_j is the epidemic degree of the user i and j respective. When φ'_i, $\varphi'_j > 1$, enhance the effect of the user has higher epidemic degree, φ'_i, $\varphi'_j < 1$, reduce the effect of the user has lower epidemic degree, φ'_i, $\varphi'_j = 1$, ignore the effect of the epidemic degree.

When $\lambda = 0.5$, the $Modified\_metasim(i, j)$ as follow:

$$Modified \_ metasim(i, j) = newsim(i, j) \bullet \psi_i^\gamma \bullet \psi_j^\gamma \bullet Modified \_ overlapping \tag{17}$$

Consider the effect of the multivariate similarity, the similarity is as follow:

$$cubesim(i, j) = \frac{\sum_{\alpha \in O} metasim'_{i,\alpha} \bullet metasim'_{j,\alpha}}{\sqrt{\sum_{\alpha \in O} (metasim'_{i,\alpha})^2} \bullet \sqrt{\sum_{\alpha \in O} (metasim'_{j,\alpha})^2}} \tag{18}$$

$$metasim'_{i,\alpha} = metasim_{i,\alpha} - < metasim_i > \tag{19}$$

Here, $<metasim_i>$ is the average similarity of the user i with other users. The final similarity is:

$$Fital \_ sim(i, j) = Modified \_ metasim(i, j) \bullet e^{cubesim(i,j)} \tag{20}$$

The predicted rating could be obtained by:

$$R_{i,\alpha} = \frac{\sum_j MMSCF \_ sim(i, j) \bullet r_{j,\alpha}}{\sum_j |MMSCF \_ sim(i, j)|} \tag{21}$$

Here $MMSCF\_sim(i, j)$ is the similarity get by Eq(20). Eq(21) shows that when the item α is collected by more user which are similar with the user u_i, the user u_i

collected α with a greater possibility. In this method, all the items haven't be collected by u_i should be valued, then recommend the best item.

4   Experiment

4.1   Data set

The data set in experiment is from MovieLens, it contain one million information of 6040 users vote to 3952 items, the value of the vote are five numbers 1-5. Here we select 200000 information of 1228 users from the data set randomly. To test the new method, divided the set into two parts randomly, one has 80% data as training set, the other is the test set, use the training set to predict the data of test set. First pre treat the data, if the user evaluation value of the item more than 2, we determinate the user collects the item.

4.2   Evaluation Standard

In predicted processing, we can get a list of the not evaluate items which are sorted by the recommendation method. The results can be evaluated by many precision indexes. Conclude *mean absolute error (MAE), precision, recall, F-measure, ranking score* and *hitting rate*. In these indexes, select *ranking score, precision, recall* to evaluate the new algorithm.

According to the *ranking score* method, we measure the position of an item in the ordered queue. For example, if there are 1000 uncollected items for user i and item j is the 260th from the top, we say the position of j is 260/1000, denoted by r=0.26. Since the probe entries are actually collected by users, a good algorithm is expected to give high recommendations to them, thus leading to a smaller r. The smaller the ranking score, the higher the algorithmic accuracy, and vice versa.

4.3   Results

Parameter Setting
In [26], it has the best result when λ=0.5, if the δ of *EDI* and γ of *EDU* are 0, means ignored the effect of the *EDI* and *EDU*, the mean value of *rank* is 0.101. It shows that only consider the degree of overlapping and multivariate similarity, the result of *rank*

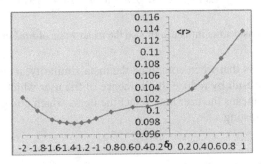

Fig. 1. γ=0, the influence of the change of δ on the mean value of *rank*

get by *MMSCF* is lower than by *CF*(0.120), it is reduced by 15.8%, compared with the result get by *MSCF*(0.106), it is reduced by 4.7%. The Fig. 1 shows the change of δ and γ influence the mean value of *rank*.

In Fig.1, $\delta>0$ shows enhance the influence of the *EDI*, then the value of the rank is large, as while, $\delta<0$, the value is small. It means, the user interest reflect in the unpopular items, $\delta=-1.3$, rank get the minimum, *rank*=0.098, then adjust γ, consider the influence of γ on the mean value of rank in Eq(9) and Eq(10) respectively.

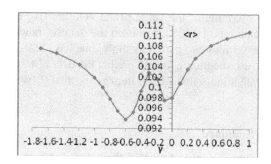

Fig. 2. Consider the influence of γ on the mean value of *rank* only in Eq(9)

Fig. 2 shows that, reduced the *EDU* to a certain degree, the mean value of rank is improved greatly, it means when computed the initial similarity, it can be get an improve recommendation result by reduce the influence of the user who are broad collection on other users. The user may like the item which is collected by the user who has the same interesting with he.

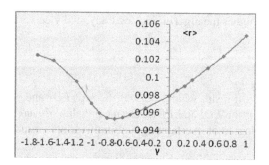

Fig. 3. Consider the influence of γ on the mean value of *rank* only in Eq(10)

The Fig. 3 shows that when computed the meta-similarity, it can be get an improve recommendation result by reduce the influence of the user which are broad collection on other users. It means the user may select the items which collected by the user who has the same interesting.

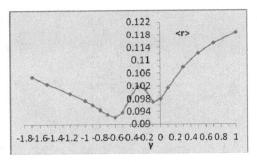

Fig. 4. Consider the influence of γ on the mean value of *rank* both in Eq(9) and Eq(10)

In Fig. 4, $\gamma>0$ shows enhance the influence of the *EDU*, then the value of the rank is large, as while, $\gamma<0.4$, the value is small. It means we should reduce the *EDU* in certain degree, $\gamma=-0.6$, *rank* get the minimum, *rank*=0.092.

Compare the Results with *CF* and *MSCF*

(1) The *CF* computes the initial similarity by Eq(1), and use it to predict.
(2) The *MSCF*: in [26], it gets the initial similarity and meat similarity by Eq(3) and Eq(4), then uses a new similarity get by Eq(5) to predict.
(3) The *MMSCF*: first we get the initial similarity by Eq(9) and the meta-similarity by Eq(10), then computed the multivariate similarity by Eq(14), and used the final similarity get by Eq(20) to predict.

When $\lambda=0.5$, $\delta=-1.3$, $\gamma=-0.8$, the experiment get the best result, the mean value of rank is 0.092, the result of rank get by *MMSCF* is lower than by *CF*(0.120), it is reduced by 23.3%, compared with the result get by *MSCF*(0.106), it is reduced by 13.2%.

In real life, the user usually interest the items in front of the recommendation results, so a good recommendation lie on the front of the results are whether accurate. Here we use *precision* to measure the algorithm. For a user u_i, u_i select the items (*fracR$_i$*) which are in the front of the recommendation list L, *precision$_i$* (L) is the ratio of *fracR$_i$* and L, the smaller *precision$_i$* (L) means the user is interested in more items, the effect of the recommendation is better.

$$precision_i(L) = fracR_i(L) / L \tag{22}$$

To measure the algorithm is not enough by precision only, in the test set, different users collect different items, and the *precision* can't to measure the method when the collection of the user is less. So we select recall to measure the method. *Recall$_i$(L)* is the ratio of *fracR$_i$* and M, M is the total number of the items which are collected by u_i in the test set.

$$recall_i(L) = fracR_i(L) / M$$

In Fig. 5, it compares the results of *precision* and *recall* by the new algorithm. The results show that the *MMSCF* is superior to the *CF* and *MSCF* obviously.

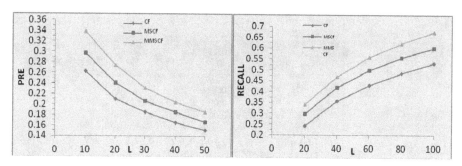

Fig. 5. Descriptions of the *precision* and *recall* get by *CF*, *MSCF*, *MMSCF* with the change length of the recommendation list

5   Conclusion

In this paper, based on meta-similarity, consider more information about users collect the items, *EDI* is the epidemic degree of the item, it reflect the number of the users collect one item; *EDU* is the epidemic degree of the user, it reflect the number of the items collected by one user; *Modified_overlapping* is to compute the similarity of the users by the degree of overlapping, it reflect the number of the items which are common collected, and the fluctuation of the user collection; multivariate similarity can enhance the similarity of these users to improve the result of the recommendation. The final similarity gets by them. Compared the experiments on the evaluation *rank score*, *recall*, *precision* by *MMSCF*, *CF*, *MSCF*, the results indicate the *MMSCF* is better then them.

References

1. Adomavicius, G., Tuzhilin, A.: Toward the next generation of recommender systems: A survey of the state-of-the-art and possible extensions. IEEE Trans. Know. Data Eng. 17, 734 (2005)
2. Brin, S., Page, L.: The Anatomy of a Large-Scale Hyper-textual Web Search Engine. Computer Networks and ISDN Systems 30, 107 (1998)
3. Broder, A., Kumar, R., Maghoul, F.: Graph structure in the Web. Comput. Netw. 33, 309 (2000)
4. Kleinberg, J.M.: Authoritative sources in a hyperlinked environment. ACM 46, 604 (1999)
5. Herlocker, J., Konstan, J., Riedl, J.: Explaining Collaborative Filtering Recommendations. In: Proceedings of CSCW 2000, pp. 241–250 (2000)
6. Konstan, J.A., Miller, B.N., Maltz, D.: Grouplens: applying collaborative filtering to use net news. Commun. ACM 40, 77–87 (1997)
7. Balabanovic, M., Shoham, Y.: Fab: Content-based, collaborative recommendation. Commun. ACM 40, 66–72 (1997)
8. Pazzani, M.J.: A framework for collaborative, content-based and demo-graphic filtering. Artif. Intell. Rev. 13, 393–408 (1999)
9. Ren, J., Zhou, T., Zhang, Y.: Information filtering via self-consistent refinement. Europhys. Lett. 82, 58007 (2008)

10. Zhang, Y., Blattner, M., Yu, Y.-K.: Heat Conduction Process on Community Networks as a Recommendation Model. Phys. Rev. Lett. 99, 154–301 (2007)
11. Goldberg, K., Roeder, T., Gupta, D.: Eigentaste: A constant time collaborative filtering algorithm. Inform. Ret. 4, 133–151 (2001)
12. Zhang, Y., Medo, M., Ren, J., et al.: Recommendation model based on opinion diffusion. Europhys. Lett. 80(6), 68003, 1–5 (2007)
13. Zhou, T., Ren, J., Medo, M., Zhang, Y.: Bipartite network projection and personal recommendation. Phys. Rev. E 76, 046115 (2007)
14. Zhou, T., Jiang, L.-L., Su, R.-Q., Zhang, Y.-C.: Effect of initial configuration on network-based recommendation. Europhys. Lett. 81, 58004 (2008)
15. Herlocker, J.L., Konstan, J.A., Terveen, L.G., Riedl, J.T.: Evaluating Collaborative Filtering Recommender Systems. ACM Transactions on Information Systems 22(1), 5–53 (2004)
16. Liu, J.-G., Wang, B.-H., Guo, Q.: Improved collaborative filtering algorithm via information transformation. Int. J. Mod. Phys. C 20 (2009)
17. Liu, J.-G., Dang, Y.-Z., Wang, Z.-T., Zhou, T.: Relationship between the in-degree and out-degree of WWW. Physica A 371, 861–869 (2006)
18. Liu, J.-G., Xuan, Z.-G., Dang, Y.-Z.: Highly accurate recommendation algorithm based on high-order similarities. Physica A 377, 302 (2007)
19. Liu, Jia, C.-X., Zhou, T., Sun, D., Wang, B.-H.: Personal Recommendation via Modified Collaborative Filtering. Physica A 388, 462–468 (2009)
20. Liu, J.-G., Zhou, T., Wang, B.-H., et al.: Highly accurate recommendation algorithm based on high-order similarities. Physica A 389, 881–886 (2008)
21. Herlocker, J.L., Konstan, J.A., Terveen, K.: Evaluating Collaborative Filtering Recommender Systems. ACM Trans. Inform. Syst. 22, 5 (2004)
22. Zhao, B.Y., Kubiatowicz, J.D., Joseph, A.D.: Tapestry: An infrastruc-ture for fault-tolerant wide - area location and routing. Tech. Rep. CSD- 01- 1141, U. C. Berkeley (April 2001)
23. Hildrum, K., Kubiatowicz, J.D., Rao, S., et al.: Distributed object location in a dynamic network. In: Proceedings of SPAA, vol. 15, pp. 41–52 (2002)
24. Shardanand, U., Maes, P.: Social information filtering: Algorithms for automating "word of mouth". In: Conference on Human Factors in Computing Systems, CHI 1995. Denver (May 1995)
25. Zhou, T., Su, R.-Q.: Accurate and diverse recommendation via eliminating redundant correlations. New Journal of Physics 11 (2009)
26. Xu, P.-Y., Dang, Y.-Z.: A Modified Collaborative Filtering Algorithm Based on Meta-similarity. Application Research of Computers 28(10) (2011)

Representation and Reasoning on RBAC: A Nonmonotonic Approach

Wei Zhang and Zuoquan Lin

School of Mathematical Sciences, Peking University
{zw,lzq}@is.pku.edu.cn

Abstract. Role-Based Access Control (RBAC) is recognized as the predominant model for access control nowadays. However, the ANSI RBAC model provides no mechanism for various rules and policies. To address this issue, a formal logical foundation of RBAC is urgently needed. In this paper, we present an ASP-based nonmonotonic approach to formalize ANIS RBAC model. The proposed formalization provides a proper expression for RBAC components, and an efficient reasoning mechanism for authorization decisions. We show that the formalism can capture RBAC models well and accomplish specific nonmonotonic reasoning tasks flexibly.

Keywords: RBAC, ASP, Formalization.

1   Introduction

RBAC is used to restrict system access to authorized users. In RBAC, permissions are assigned to roles rather than users. A user can acquire some permissions only if he can activate a role assigned to him that corresponds to the permissions. Compared with traditional access control models, RBAC obviously reduces the security administration cost and improves the system efficiency. Sandhu *et al.* present the famous RBAC96 model in [1], which divides RBAC into four conceptual models. Based on the RBAC96 model, an INCIT 359-2004 standard for RBAC is accredited by ANSI (American National Standards Institute) in 2004 [2].

RBAC has achieved great success in the enterprise. However, the ANSI RBAC reference model provides no mechanism for representing and reasoning access control policies. Simple rules can only be coded as a part of the application systems, which complicates the policy setting and its management.

To address this problem, a formal analysis of RBAC is required. Researchers have developed many policies languages for access control. XACML is an extensible access control markup language which formats the access control request and reply. OASIA (Organization for the Advancement of Structured Information Standards) proposes a specification for core RBAC and hierarchical RBAC in [3]. However, due to the limited expression abilities, XACML cannot support RBAC constraints and other complex policies well. Other works can be divided into two main categories: graph-based approaches [4,5,6] and logic-based approaches. The former approaches

M. Wang (Ed.): KSEM 2013, LNAI 8041, pp. 230–240, 2013.

employ graph transformations to represent the state change of access control, which are intuitive. However, the logic-based approaches provide a more precise mathematical foundation for reasoning, and better support for computation.

In this paper, we propose a nonmonotonic approach to formalize the ANSI RBAC reference model. The information in access control systems is changing all the time. In order to support the nonmonotonic features in access control well, we employ answer set programming (ASP) as the policy language in our approach. As a form of declarative programming, ASP roots in nonmonotonic logic, deductive database and logic programming with negation as failure. We will specify how to represent the RBAC components and do flexible reasoning tasks via ASP in our approach. Moreover, ASP is limited by abandoning function, recursion and resolution principle to keep its solvers highly efficient. The development of existing answer set solvers, such as DLV [7], SMODELS [8], ensures the stability and efficiency of our approach in the enterprise.

The rest of the paper is organized as follows. In Section 2 we provide relevant background notions. Then the representation on RBAC is introduced in Section 3. In Section 4 we discuss the nonmonotonic features in RBAC and formalize them respectively. We also illustrate how to accomplish reasoning tasks through our approach. Finally, we conclude the paper with related work in Section 5 and conclusions in Section 6.

2   Backgrounds

2.1   ANSI RBAC Model

ANSI RBAC reference model consists of four model components: core RBAC, hierarchical RBAC, static separation of duty relations (SSD), and dynamic separation of duty relations (DSD).

Core RBAC defines the minimum collection of RBAC elements, including element sets, assignment relations and mapping functions. It is required in any RBAC systems. Hierarchical RBAC adds a partial order on role sets. If role R_1 is senior to R_2, then all users of R_1 are also users of R_2, and all permissions of R_2 are also permissions of R_1. Separation of duty introduces exclusivity into RBAC. SSD defines that no user can be assigned to n or more than n SSD roles, while DSD states that no user can activate n or more than n DSD roles at the same time.

2.2   ASP

ASP [9] is a form of declarative programming for difficult combinatorial search problems. The basic idea of ASP is to represent a given problem by a logic program whose answer sets correspond to solutions, and then use an answer set solver to compute the answer sets.

A program Π [10] is a set of rules that can be represented as:

$$A \leftarrow B_1, \ldots, B_m, not\ C_1, \ldots, not\ C_n.$$

For any set of literals X, the reduct of Π relative to X (written as Π^X) is obtained by:

1. deleting each rule that contains *not* C_i and $C_i \in X$, and
2. replacing other rules $A \leftarrow B_1, ..., B_m, not\ C_1, ..., not\ C_n$ by $A \leftarrow B_1, ..., B_m$.

All the rules in Π^X can be represented as $A \leftarrow B_1, ..., B_m$. X is closed in Π^X if $A \in X$ whenever $B_1, ..., B_m \in X$. X is an answer set of a program Π, if X is the minimal closed set in Π^X.

3   Representation on RBAC with ASP

We will describe how to conceptualize the basic components in ANSI RBAC model and construct an answer set program in this section. Separation of duty constraints are discussed in the following section.

3.1   Representation on Core RBAC

Core RBAC consists of the basic elements in RBAC. According to the core RBAC specifications in [2], they can be divided into three categories: sets, relations and functions. We will discuss them respectively in the following.

Sets in RBAC are represented as predicates in ASP. For example, we define predicate *user(U)* to represent the user set *USERS*, where U is a variable. Specifically, suppose in an RBAC system, *USERS*={*Tom, Alice*}. We can add the following facts into the program to represent it.

user(tom).

user(alice).

Sets in core RBAC can be represented as shown in Table 1.

Table 1. Representation on sets

Sets	ASP rules	Examples
user sets *USERS*	*user(U).*	*user(tom).*
role sets *ROLES*	*role(R).*	*role(manager).*
operation sets *OPS*	*op(OP).*	*op(modify).*
object sets *OBS*	*ob(OB).*	*ob(codes).*
permission sets *PRMS*	*prm(OP,OB).*	*prm(modify,codes).*
session sets *SESSIONS*	*session(S).*	*session(s₁).*

There are two kinds of relations in core RBAC: user-role assignment relation and permission-role assignment relation. We employ predicate *ur(U,R)*, *rp(R,OP,OB)* to represent the relations respectively, where U, R, OP, OB are all variables. For example, *ur(tom, manager)* represents that role *manager* is assigned to user *Tom*.

Functions in core RBAC can be regarded as a special relation. For example, *assigned_users(manager)* stands for the users that the *manager* role is assigned to. According to the schematic program semantic [10], the fact *ur(U,manager)* can be used to represent the relation between users sets *assigned_users(manager)* and the role *manager*.

Relations and Functions in core RBAC can be represented as shown in Table 2.

Table 2. Representation on relations and fuctions

Relations and functions	ASP rules	Examples
user-role *UA*	*ur(U,R).*	*ur(tom,manager).*
permission-role *PA*	*rp(R,OP,OB).*	*rp(coder,modify,codes).*
assigned_users(R)	*ur(U,R).*	*ur(U,manager).*
assigned_permissions(R)	*rp(R,OP,OB).*	*rp(coder,OP,OB).*
session_users(S)	*us(U,S).*	*us(U,s_1).*
session_roles(S)	*sr(S,R).*	*sr(s_1,R).*
avail_session_perms(S)	*sp(S,OP,OB).*	*sp(s_1,OP,OB).*

3.2 Representation on Hierarchical RBAC

Hierarchical RBAC introduces role hierarchies into core RBAC, which is a partial order on the role sets. We employ the predicate $rh(R_1,R_2)$ to represent the partial order $R_1 \geqslant R_2$ in hierarchical RBAC.

The partial order on roles is transitive, which can be defined as the following rule.

Definition 1 (Role partial order rule)

$$rh(R_1,R_2) \leftarrow rh(R_1,R_3), rh(R_3,R_2). \tag{1}$$

where R_1, R_2, R_3 are all variables.

Table 3 shows the representation of functions in hierarchical RBAC.

Table 3. Representation on hierarchical fuctions

Hierarchical functions	ASP rules	Examples
authorized_users(R)	*uhr(U,R).*	*uhr(U,manager).*
authorized_permissions(R)	*rhp(R,OP,OB).*	*rhp(coder,OP,OB).*

In the hierarchical RBAC, a user *U* is authorized to role *R*, if he is directly assigned to role *R*, or he is assigned to a role R_0 that is senior to *R*. These rules can be represented by ASP as follows.

Definition 2 (Hierarchical user-role assignment rules)

$$uhr(U,R) \leftarrow ur(U,R). \tag{2}$$

$$uhr(U,R) \leftarrow ur(U,R_0), rh(R_0,R). \tag{3}$$

where U, R, R_0 are all variables.

Similarly we can define the hierarchical role-permission assignment rules.

Definition 3 (Hierarchical role-permission assignment rules)

$$rhp(R,OP,OB) \leftarrow rp(R,OP,OB). \tag{4}$$

$$rhp(R,OP,OB) \leftarrow rp(R_0,OP,OB), rh(R,R_0). \tag{5}$$

where R, R_0, OP, OB are all variables.

The above rules mean a role R is authorized to permission (OP,OB), if it is directly assigned to permission (OP,OB),   or R_0 is assigned to permission (OP,OB) and R is senior to R_0.

Above all, the basic components and hierarchical rules have been represented with ASP.

4   Nonmonotonic Reasoning on RBAC with ASP

Programs introduce *not* to support negation as failure. For example, the following rule means that if *Polly* is a bird, and it cannot be proved to be a penguin, then *Polly* can fly.

$$fly(polly) \leftarrow bird(polly), not\ penguin(polly).$$

In this section, we specify the nonmonotonic features of RBAC and do some reasoning tasks on them. We demonstrate that negation as failure, as a nonmonotonic inference rule in logic programming, can support RBAC policies and authorization decisions well.

4.1   Role Activation

The ANSI RBAC model develops the role activation mechanism. A user U can acquire some permission (OP,OB) only if he successfully activates a role R that corresponds to the permission. First we present the user-role activation rule.

Definition 4 (User-role activation rule)

$$uar(U,R) \leftarrow us(U,S), sr(S,R). \tag{6}$$

where U, R, S are all variables.

This rule means that user U activates role R if U and R are in the same session S. Then the user-permission authorization rule can be achieved as follows.

Definition 5 (User-permission authorization rule)

$$uap(U,OP,OB) \leftarrow uar(U,R), rp(R,OP,OB). \tag{7}$$

where U, R, OP, OB are all variables.

According to definition 4 and 5, we can perform some reasoning tasks for authorization requests. We employ the predicate $requp(U,OP,OB)$ to represent the request of user U for permission (OP,OB). Then the following rules are added into our program.

Definition 5 (Authorization rules)

$$sr(S,R) \leftarrow us(U,S), uhr(U,R), rp(R,OP,OB), requp(U,OP,OB). \qquad (8)$$

$$permit(U,OP,OB) \leftarrow requp(U,OP,OB), uap(U,OP,OB). \qquad (9)$$

$$prohibit(U,OP,OB) \leftarrow requp(U,OP,OB), not\ permit(U,OP,OB). \qquad (10)$$

where U, S, R, OP, OB are all variables.

For any authorization request $requp(U,OP,OB)$, we add it as a fact into our program and compute the answer set, which corresponds to the solutions. If $permit(U,OP,OB)$ is in the answer set, the authorization access succeeds; if $prohibit(U,OP,OB)$ is in the answer set, the authorization access fails.

The role activation authorization process is nommonotonic. Suppose, for example, that in a corporate intranet, user *Tom* is assigned to role *tester*, which has the permission to test codes. There is another role *coder* in the system, with the permission to modify codes. In a session s, *Tom* request to modify codes. First, we add the following facts into our program:

$$ur(tom,tester).\ rp(tester,test,codes).\ rp(coder,modify,codes).$$

$$us(tom,s).\ requp(tom,modify,codes).$$

Then we compute the answer set via answer set solvers and get the following results:

$$\{ ur(tom,tester).\ rp(tester,test,codes).\ rp(coder,modify,codes).$$

$$us(tom,s).\ requp(tom,modify,codes).\ prohibit(tom,modify,codes). \}$$

Thus *Tom* cannot acquire the requested permission. However, if the security administrator assigns the role *coder* to *Tom* to satisfy some temporarily need and adds the additional fact $ur(tom,coder)$ into the program. The following answer set can be reached:

$$\{ ur(tom,tester).\ ur(tom,coder).\ rp(tester,test,codes).\ rp(coder,modify,codes).$$

$$us(tom,s).\ requp(tom,modify,codes).\ sr(s,coder).\ uar(tom,coder).$$

$$uap(tom,modify,codes).\ permit(tom,modify,codes).\}$$

Thus *Tom* can get the permission to modify codes.

4.2   Permission Revoke

The revoke mechanism is an important nonmonotonic feature in RBAC. Revoking permission (OP,OB) from role R prevents R from acquiring (OP,OB) anymore, represented as $revoke(R,OP,OB)$ in ASP.

To extend our approach for permission revoke, we introduce the concept of role-permission assignment trend. There is a role-permission assignment trend if permission (OP,OB) is assigned to role R, written as $rpt(R,OP,OB)$ in ASP. This concept takes no consideration on permission revoke. Then we propose the role-permission assignment rule with permission revoke:

Definition 6 (Role-permission assignment rule with permission revoke)

$$rp(R,OP,OB) \leftarrow rpt(R,OP,OB), not\ revoke(R,OP,OB). \tag{11}$$

where R, OP, OB are all variables.

This rule means that role R can be assigned to permission (OP,OB) if there is a role-permission assignment trend $rpt(R,OP,OB)$ and the corresponding permission revoke does not exist.

4.3   Separation of Duties

As a key security principle in RBAC, separation of duty is used to ensure that no one individual can complete a task by himself. Constrained RBAC supports both SSD and DSD. We will discuss them respectively in the following.

Suppose U is a variable, $rs=\{R_1,...,R_m\}$ is an SSD role set, n is a natural number, then no user in U can be assigned to n or more than n roles in rs, written as:

$$\forall(rs,n) \in SSD, \forall t \in rs: |t| \geq n \Rightarrow \bigcap_{r \in t} assigned\_users(r)=\emptyset.$$

ASP provides expressions for constraints and cardinality [9]. Thus we define the SSD rules as follows.

Definition 7 (SSD rules)

$$\{ur(U,R_1), ..., ur(U,R_m)\}^c. \tag{12}$$

$$\leftarrow not\ \{ur(U,R_1), ..., ur(U,R_m)\} < n. \tag{13}$$

where U is a variable and $R_1,..., R_m$ are constants.

These rules restrict that there are less than n elements that are both in $\{ur(U,R_1), ..., ur(U,R_m)\}$ and the answer set. Specially, if $m=2$, then R_1 and R_2 are mutually exclusive roles. No user can be assigned to both R_1 and R_2. We can simplify the SSD rule as follows.

Definition 8 (Mutually exclusive roles rule)

$$\leftarrow ur(U,R_1), ur(U,R_2). \tag{14}$$

where U is a variable and R_1, R_2 are constants.

Then considering hierarchical RBAC, we modify definition 7 to support role hierarchies.

Definition 9 (Role hierarchical SSD rules)

$$\{uhr(U,R_1), \ldots, uhr(U,R_m)\}^c . \tag{15}$$

$$\leftarrow not \{uhr(U,R_1), \ldots, uhr(U,R_m)\} < n. \tag{16}$$

where U is a variable and $R_1, \ldots, R_m$ are constants.

DSD ensures that no user can activate n or more than n roles in $\{R_1,\ldots,R_m\}$ in a session. Similar to SSD, the DSD rules are defined as follows.

Definition 10 (DSD rules)

$$\{uar(U,R_1), \ldots, uar(U,R_m)\}^c . \tag{17}$$

$$\leftarrow not \{uar(U,R_1), \ldots, uar(U,R_m)\} < n. \tag{18}$$

where U is a variable and $R_1, \ldots, R_m$ are constants.

We add definition 7,9,10 into the program, and then get new answer sets that satisfy both SSD and DSD rules. Suppose, for example, that there are two roles *accountant* and *cashier* in a company. A DSD rule prohibits any user from activating both roles in a session. Then we can add the following rule into the program to perform this rule.

$$\leftarrow uar(U,accountant), uar(U,cashier).$$

4.4   Complex Policies

A policy is a set of rules that can guide decisions and achieve expected outcomes. As for the limited expression abilities, ANSI RBAC model only supports simple policies such as SSD and DSD. There is no mechanism to support complex policies and policy combinations, which can be processed in our approach.

For example, the deny-overrides open policy states that if a user U is not be explicitly forbidden from acquiring permission (OP,OB), then he is permitted to get the permission. In order to represent this policy, we introduce the concept of negative authorization first.

In a negative authorization, a user U is forbidden from activating role R anymore, written as $nuar(U,R)$ in ASP. Then the following rule can be reached.

Definition 11 (Negative authorization rule)

$$nuap(U,OP,OB) \leftarrow nuar(U,R), rp(R,OP,OB). \tag{19}$$

where U, R, OP, OB are all variables.

This rule means that user U is forbidden from acquiring permission (OP,OB), if there is a negative authorization between U and R, and (OP,OB) is assigned to role R. Next, we can program the deny-overrides open rules as follows.

Definition 12 (Deny-overrides open rules)

$$permit(U,OP,OB) \leftarrow not\ nuap(U,OP,OB). \qquad (20)$$

$$prohibit(U,OP,OB) \leftarrow nuap(U,OP,OB). \qquad (21)$$

where U, OP, OB are all variables.

Rule (19)-(21) constitute the deny-overrides open policy. More detailed discussion on other complex policies is included in [11].

5   Related Work

Researchers have spent a few decades focusing on the formalization of access control rules and policies. Woo and Lam [12] propose a logic language based on default logic for access authorization. This language can represent various kinds of policies well. However, their approach needs to compute extension, which is NP-complete even for propositional default logic. Jajodia *et al.* [13] present a stratified logic approach for access control policies. However, they only focus on different kinds of policies, rather than an integrated formalization framework. Besides, their work lacks support for a specific access control model. An approach based on C-Datalog is put forward in [14]. Yet this approach concentrates on the comparison and analysis of different access control models, without detailed discussion on policies. Barker and Stucky [15] employ constraint logic programming to formalize policies. Nonetheless they cannot support the constraints in RBAC well.

Some researchers employ description logic as the policy language [16,17,18,19,20].However, as a decidable subset of first-order logic, description logic cannot fit constraints in RBAC well. Moreover, the description logic approaches provide no support for negation as failure and other nonmonotonic features in RBAC. Other works include [21,22,23], which are based on first-order logic or its extension. Thus an efficient reasoning mechanism lacks.

Our approach employs ASP to formalize RBAC. We show that ASP can represent the RBAC components and policies well. Besides, the highly expressive and efficient answer set solvers ensure the flexible nonmonotonic reasoning in our approach. Ahn *et al.* propose an ASP-based approach for XACML policies [24]. However, no specific formalization framework on RBAC is given in their approach. Moreover, they focus on XACML policies, rather than RBAC administrative policies and complex policies.

6   Conclusion and Future Work

In this paper, we propose an ASP-based nonmonotonic approach to formalize ANSI RBAC model. First, we translate ANSI RBAC components into answer set rules. Then we show how to support various access control rules and policies, and accomplish nonmonotonic reasoning tasks through our approach. We aim to demonstrate that ASP is well suited for the representation and reasoning on RBAC. In the further, we hope to extend our approach to build a universal access control formalization framework.

References

1. Sandhu, R., Coynek, E.J., Feinsteink, H.L., Youmank, C.E.: Role-based access control models. Computer 29(2), 38–47 (1996)
2. ANSI INCITS: INCITS 359-2004, American national standard for information technology, role based access control (2004)
3. Anderson, A.: Core and hierarchical role based access control (RBAC) profile of XACML version 2.0. OASIS XACML-TC, Committee Draft (2004)
4. Koch, M., Mancini, L.V., Parisi-Presicce, F.: A graph-based formalism for RBAC. ACM Transactions on Information and System Security (TISSEC) 5(3), 332–365 (2002)
5. Sandhu, R.: A perspective on graphs and access control models. In: Ehrig, H., Engels, G., Parisi-Presicce, F., Rozenberg, G. (eds.) ICGT 2004. LNCS, vol. 3256, pp. 2–12. Springer, Heidelberg (2004)
6. Ding, G., Chen, J., Lax, R.F., Chen, P.P.: Graph-theoretic method for merging security system specifications. Information Sciences 177(10), 2152–2166 (2007)
7. Eiter, T., Faber, W., Leone, N., Pfeifer, G.: Declarative problem-solving using the DLV system. In: Logic-based Artificial Intelligence, pp. 79–103. Springer US (2000)
8. Niemelä, I., Simons, P.: Smodels—an implementation of the stable model and well-founded semantics for normal logic programs. In: Fuhrbach, U., Dix, J., Nerode, A. (eds.) LPNMR 1997. LNCS, vol. 1265, pp. 420–429. Springer, Heidelberg (1997)
9. Lifschitz, V.: What is answer set programming. In: Proceedings of the AAAI Conference on Artificial Intelligence, pp. 1594–1597 (2008)
10. Lloyd, J.W.: Foundations of logic programming, 2nd edn. Springer, Berlin (1984)
11. Zhang, W., Lin, Z.: A Logic-based RBAC Framework for Flexible Policies. In: 2012 Eighth International Conference on Semantics, Knowledge and Grids (SKG), pp. 279–282. IEEE (2012)
12. Woo, T.Y., Lam, S.S.: A semantic model for authentication protocols. In: Proceedings of the 1993 IEEE Computer Society Symposium on Research in Security and Privacy, pp. 178–194. IEEE (1993)
13. Jajodia, S., Samarati, P., Sapino, M.L., Subrahmanian, V.S.: Flexible support for multiple access control policies. ACM Transactions on Database Systems (TODS) 26(2), 214–260 (2001)
14. Bertino, E., Catania, B., Ferrari, E., Perlasca, P.: A logical framework for reasoning about access control models. ACM Transactions on Information and System Security (TISSEC) 6(1), 71–127 (2003)
15. Barker, S., Stuckey, P.J.: Flexible access control policy specification with constraint logic programming. ACM Transactions on Information and System Security (TISSEC) 6(4), 501–546 (2003)
16. Zhao, C., Heilili, N., Liu, S., Lin, Z.: Representation and reasoning on RBAC: A description logic approach. In: Van Hung, D., Wirsing, M. (eds.) ICTAC 2005. LNCS, vol. 3722, pp. 381–393. Springer, Heidelberg (2005)
17. Chae, J.H., Shiri, N.: Formalization of RBAC policy with object class hierarchy. In: Dawson, E., Wong, D.S. (eds.) ISPEC 2007. LNCS, vol. 4464, pp. 162–176. Springer, Heidelberg (2007)

18. Knechtel, M., Hladik, J., Dau, F.: Using OWL DL Reasoning to decide about authorization in RBAC. In: OWLED 2008: Proceedings of the OWLED 2008 Workshop on OWL: Experiences and Directions (2008)
19. Kolovski, V., Hendler, J., Parsia, B.: Analyzing web access control policies. In: Proceedings of the 16th International Conference on World Wide Web, pp. 677–686. ACM (2007)
20. Ferrini, R., Bertino, E.: Supporting rbac with xacml+ owl. In: Proceedings of the 14th ACM Symposium on Access Control Models and Technologies, pp. 145–154. ACM (2009)
21. Massacci, F.: Reasoning about security: a logic and a decision method for role-based access control. In: Nonnengart, A., Kruse, R., Ohlbach, H.J., Gabbay, D.M. (eds.) FAPR 1997 and ECSQARU 1997. LNCS, vol. 1244, pp. 421–435. Springer, Heidelberg (1997)
22. Mossakowski, T., Drouineaud, M., Sohr, K.: A temporal-logic extension of role-based access control covering dynamic separation of duties. In: Proceedings of the 10th International Symposium on Temporal Representation and Reasoning and Fourth International Conference on Temporal Logic, pp. 83–90. IEEE (2003)
23. Crescini, V.F., Zhang, Y.: A logic based approach for dynamic access control. In: Webb, G.I., Yu, X. (eds.) AI 2004. LNCS (LNAI), vol. 3339, pp. 623–635. Springer, Heidelberg (2004)
24. Ahn, G.J., Hu, H., Lee, J., Meng, Y.: Reasoning about xacml policy descriptions in answer set programming (preliminary report). In: 13th International Workshop on Nonmonotonic Reasoning, NMR 2010 (2010)

System Dynamics Modeling of Diffusion of Alternative Fuel Vehicles

Feng Shen and Tieju Ma

School of Business, East China University of Science and Technology, Shanghai, China
{fshen,tjma}@ecust.edu.cn

Abstract. With the intensification of global energy crisis and the growing importance of environmental protection, the promotion of alternative fuel vehicles becomes an inevitable trend. The promotion of alternative fuel vehicles lies in the development of alternative fuel vehicle technology. By means of patent data analysis, this paper explores the change of alternative fuel vehicle technology, studies the stage of technology life cycle which alternative fuel vehicle technology is in, and thus to predict the market holdings of alternative fuel vehicles in future. By considering the factors which affect the diffusion of alternative fuel vehicles, such as technology, government grants, construction of fuel-filling stations, this article uses system dynamics to build a causality diagram and a flow chart diagram of diffusion of alternative fuel vehicles. On this basis, the diffusion process of alternative fuel vehicles is discussed.

Keywords: alternative fuel vehicle, technology diffusion, patent analysis, system dynamics.

1 Introduction

At present, the world is facing a situation of oil and other fossil energy increasing depletion, traditional internal combustion engine (ICE) automobiles are dependent on fossil energy, which speeds up the rate of consumption of the fossil energy on Earth. In addition, due to the geopolitical instability, oil price is soaring. Correspondingly, price of fuel is very high and bring a huge economic cost to the people's life.

Traditional internal combustion engine (ICE) vehicles use fuel as a power of travel, its exhaust emissions have become a major source of pollution in many cities in the world. The protection of environment is becoming more important, environmental issues brought about by the ICE is gradually get the attention of the various countries. Development of alternative fuel vehicles, can reach the goal of energy-saving and environmental protection, has become an important measure to solve the above problems. Developed countries, such as the United States, Europe and Japan, developing countries, such as Brazil, are active in the development of alternative fuel automotive. On the one hand, these countries have issued a series of policy for the development of alternative fuel vehicles, which covers cash subsidies, tax incentives and other measures. On the other hand, they formulate rules and regulations, which

M. Wang (Ed.): KSEM 2013, LNAI 8041, pp. 241–251, 2013.
© Springer-Verlag Berlin Heidelberg 2013

propose mandatory requirements of the development of alternative fuel vehicles to auto companies. In order to form a globally competitive auto industry in the future, in October 2010, China identified alternative fuel automotive industry as one of the strategic emerging industries [1]. On this basis, China deployed R&D and industrialization of alternative fuel vehicles.

Major world auto giants such as Toyota, GM, etc., have strengthened the research of alternative fuel vehicles. The number of alternative fuel vehicles patents they obtained has occupied a fairly large proportion of the related fields. These companies will play a dominant role in the future development of alternative fuel vehicle technology. Patent means the emergence of technology innovation [2-3], which will promote alternative fuel vehicles to replace the traditional ICE vehicle in future. The replacement process is a course of technology diffusion. This study will pay attention to the diffusion of alternative fuel vehicle technology and be useful to promote the application of alternative fuel and realize the goal of energy-saving and emission reduction.

The diffusion of alternative fuel vehicle technology is effected by multiple factors. Nuno Bento analyzed this topic and found that the diffusion rate is affected by the double impact of the income and the level of investment [4]. Chi chunjie, etc. analyzed the case of the diffusion of natural gas vehicles in Shanghai [5]. The study found that the initial distribution fuel-filling stations directly affect the effectiveness of the diffusion of natural gas vehicles, and the strong support of the government is quite critical to market penetration of natural gas vehicles. Therefore, the factors, such as cost, government subsidies, facilities, are all directly or indirectly related to the successful promotion of alternative fuel vehicles.

Current researches about the diffusion of alternative fuel vehicles are often from a variety of local perspectives. The study, in which the diffusion of alternative fuel vehicles is considered as a complete system, is relatively rare. System Dynamics is a systematic approach to deal with complex issues, which uses causal relationship to describe the system dynamic complexity [6-7]. Therefore, this paper proposes using system dynamics to build a framework of diffusion of alternative fuel vehicles, which is used to describe the mechanism of the diffusion process.

2   Model Assumptions and Variable Design

2.1   Model Basic Assumptions

Assumption 1. With more attention to alternative fuel vehicles, related supports of government are increasing. The acquisition cost of alternative fuel vehicles is gradually reduced, and can be easily accepted by most people.

Assumption 2. In order to develop alternative fuel vehicles, the state focused on promoting infrastructure construction, such as building more fuel-filling stations. With the increase in the number of alternative fuel vehicles, the corresponding increase will appear in the number of fuel-filling stations. This situation will reduce people's level of worry about fuel-filling for alternative fuel vehicles.

Assumption 3. The patents of the alternative fuel vehicles are able to predict the development prospects of alternative fuel vehicle technology. The rapid increase in the number of patents indicates that alternative fuel vehicle technology is entering a stage of rapid development. In this situation, we should increase investment, speed up the layout of patents to lay the foundation of winning the market in future. By means of the study of cumulative patent application and its changes, we can evaluate which stage it is in the technological life cycle (emerging, growth, maturity, or saturation) [8]. At different stages, production of alternative fuel vehicles, and their average life span are all different.

Assumption 4. With lower acquisition cost and more convenient fuel-filling of alternative fuel vehicles, the number of user of alternative fuel vehicles will increase. As a result, more and more ICE vehicles in the market will be replaced by alternative fuel vehicles.

2.2   The Main Variables of the Model

Stock Variables
In this paper, four stock variables are considered. They are shown in Table 1.

Table 1. Stock variables in the diffusion model of alternative fuel vehicles

Stock variables
(1) alternative fuel vehicle ownership
(2) cumulative production of alternative fuel vehicles
(3) fuel-filling station quantity
(4) cumulative patents of alternative fuel vehicles

Flow Variables
In this study, six flow variables are designed. They are shown in Table 2.

Table 2. Flow variables in the diffusion model of alternative fuel vehicles

Flow variables
(1) alternative fuel vehicle sales
(2) amount of alternative fuel vehicles scrapped
(3) alternative fuel vehicle production
(4) amount of fuel-filling station construction
(5) amount of fuel-filling station demolition
(6) alternative fuel vehicle patent applications

Auxiliary Variables

Thirty-seven auxiliary variables are discussed in this article. They are shown in Table 3. Some of auxiliary variables are interpreted as followed.

Government grants represent using central financial funds to increase investment on industrial development of alternative fuel vehicles. Government subsidy measures include four main aspects: consumer subsidies, manufacturer subsidies, fuel-filling station subsidies and researcher subsidies.

Pay coefficient of government grants is an effective means of regulation. It controls the amount of government subsidies through sources of grant funds doubled or times less.

Operating income threshold of a fuel-filling station is a reasonable value, which is estimated by means of statistical research and scientific calculation. It can be regarded as an important reference for fuel-filling station construction or demolition.

Potential cumulative patents of alternative fuel vehicles refer to the total number of patents accumulated when alternative fuel vehicle technology is saturated. It is a valuation by predicting.

Maturity coefficient of alternative fuel vehicle technology is used to describe the stage of development of alternative fuel vehicle technology quantitatively. Higher the coefficient, more the production and longer the life span of alternative fuel vehicles.

Learning rate of alternative fuel vehicle technology is regarded as the cost reduction ratio of manufacturing alternative fuel vehicles, with the increasing of cumulative production of alternative fuel vehicles.

Diffusion rate of alternative fuel vehicles refers to a measure of the rate of market diffusion of alternative fuel vehicles. Higher the diffusion rate, larger the market acceptance and better the sales of alternative fuel vehicles.

Through using alternative fuel vehicles, benefits of fuel-saving and environmental protection can be gained. These benefits can be transformed into social welfare, which can be shared for the whole society. The government, as a representative of the common people, should make payments to the related objects, such as the users of alternative fuel vehicles. The form of payment is all kinds of subsidies.

Table 3. Auxiliary variables in the diffusion model of alternative fuel vehicles

Auxiliary variables
(1) government grants
(2) pay coefficient of government grants
(3) start-up capital of government grants
(4) consumer subsidies coefficient
(5) manufacturer subsidies coefficient
(6) fuel-filling station subsidies coefficient
(7) researcher subsidies coefficient
(8) fuel-filling cost of a alternative fuel vehicle

(9) average operating income of a fuel-filling station

(10) operating income threshold of a fuel-filling station

(11) total amount of R & D funding

(12) own R & D funding

(13) average R & D cost of a patent

(14) potential cumulative patents of alternative fuel vehicles

(15) maturity coefficient of alternative fuel vehicle technology

(16) average life span of alternative fuel vehicle

(17) initial production of alternative fuel vehicle

(18) initial average production cost of alternative fuel vehicle

(19) initial production cost of alternative fuel vehicle

(20) learning rate of alternative fuel vehicle technology

(21) production cost of alternative fuel vehicles

(22) average production cost of alternative fuel vehicle

(23) average acquisition cost of alternative fuel vehicle

(24) total funds of consumers

(25) per capita disposable income

(26) total population in one region

(27) potential size of the market of alternative fuel vehicles

(28) diffusion rate of alternative fuel vehicles

(29) external influence coefficient

(30) internal influence coefficient

(31) fuel savings

(32) emission reductions

(33) fuel-saving benefits conversion coefficient

(34) environmental protection benefits conversion coefficient

(35) fuel saving and environmental protection benefits of a alternative fuel vehicle

(36) fuel saving and environmental protection benefits

(37) social welfare

3   Diffusion Model of Alternative Fuel Vehicles

3.1   Causality Diagram of Diffusion of Alternative Fuel Vehicles

There are seven feedback loops in the causality diagram of diffusion of alternative fuel vehicles.

(1) Government subsidies to fuel-filling stations can cause increase in the number of stations. The growth will reduce consumers' worries about fuel-filling for alternative fuel vehicles, and promote the rate of diffusion of alternative fuel vehicles. As a result, the ownership of alternative fuel vehicles will be enhanced, the goal of energy saving and environmental protection will be achieved. Therefore, the level of society welfare can be improved. This is the first positive feedback loop.

(2) Government subsidies to consumers can cause reduce in the average acquisition cost of alternative fuel vehicles. The reduction will promote the sales of alternative fuel vehicles. This situation is consistent with the concept of fuel saving and environmental protection which government advocates, will make the whole society benefit. This is the second positive feedback loop.

(3) Government subsidies to manufacturers can cause reduce in the average production cost of alternative fuel vehicles. The reduction will urge automotive manufacturers to increase investment in alternative fuel vehicles production capacity. Consequently, the production and sales of alternative fuel vehicles will be increased dramatically. As a result, the ownership of alternative fuel vehicles in the market will be enhanced. More fuel saving and less pollution, better social life. This is the third positive feedback loop.

(4) Government subsidies to researchers can cause increase in the number of patents of alternative fuel vehicles. The rise indicates that the alternative fuel vehicle technology is more mature. It will cause two consequences. One is the average production costs will reduce, another is the average life span will increase.

Firstly, with reduction in average production cost of alternative fuel vehicles, a lot of consumers can buy cheaper alternative fuel vehicles. It is useful to promote the sales of alternative fuel vehicles.

Secondly, with increase in average life span of alternative fuel vehicles, they can be used for longer.

Above both will help increase the quantity of alternative fuel vehicles on the market. Furthermore, it will promote the goal of fuel saving and environmental protection. These are the fourth and fifth positive feedback loop.

(5) Ownership of alternative fuel vehicles will affect the rate of diffusion of alternative fuel vehicles, and then affect the sales of alternative fuel vehicles. The sales rise is bound to further increase the ownership of alternative fuel vehicles. This is the sixth positive feedback loop.

(6) The relationship between the diffusion of alternative fuel vehicles and the construction of the corresponding fuel-filling stations is just a "chicken and egg" problem. The one hand, increase in the number of fuel-filling station can reduce consumers' worries about fuel-filling. And then, it can promote the production and sales of alternative fuel vehicles. On the other hand, increase in the amount of alternative fuel car ownership can lead to more demand of fuel-filling, which means more profit can be made. There is no doubt that this is very attractive to fuel-filling stations. And then, more fuel-filling station will be built. This is the seventh positive feedback loop.

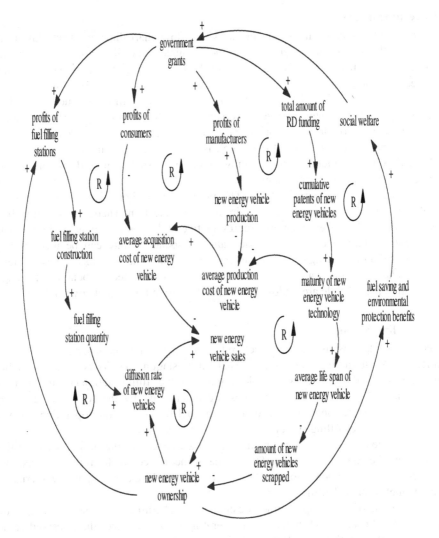

Fig. 1. The causality diagram of diffusion of alternative fuel vehicles

3.2   Flow Chart Diagram of Diffusion of Alternative Fuel Vehicles

According to the causality diagram and the explanation of main variables, this paper can offer the flow chart diagram of diffusion of alternative fuel vehicles.

In the flow chart diagram, there are four stock variables, six flow variables and thirty-seven auxiliary variables. They have been mentioned in previous chapter.

The flow chart diagram can be divided into four parts, which are government grants, fuel-filling stations, patents of alternative fuel vehicles, and alternative fuel vehicles. They are described in the next sections.

Government Grants

Government grants are important in the whole flow chart diagram. The relationship between the diffusion of alternative fuel vehicles and the construction of the corresponding fuel-filling stations is just a "chicken and egg" problem, which actually makes technological learning of alternative fuel vehicles extremely difficult to start spontaneously. 'At this situation, in order to make the diffusion successful, it is necessary for government to give some support, such as offer subsidies, or establish some demonstration projects. Sources of government grants include two aspects:

One aspect is the enabling fund which is established to promote the diffusion of alternative fuel vehicles. The enabling funds are usually offered by the central government and local governments.

Another aspect is social welfare. The promotion of alternative fuel vehicles can realize the goal of fuel-saving and environmental protection. Therefore, social welfare is increasing. The government, as a representative of the common people, should make payments to the related objects. The value of social welfare can be measured by benefits from fuel-saving and environmental protection. The benefits can be calculated easily. The method is that fuel-saving and environmental protection benefits of a single alternative fuel vehicle are multiplied by the alternative fuel vehicle ownership.

Fuel-Filling Stations

Fuel-filling stations, as the infrastructure to offer fuel-filling service, are necessary for diffusion of alternative fuel vehicles. Fuel-filling station quantity is a stock variable, which is affected by two flow variables (amount of fuel-filling station construction and demolition). These two flow variables are determined by the following auxiliary variables: average operating income of a fuel-filling station, operating income threshold of a fuel-filling station.

In order to calculate average operating income of a fuel-filling station, we must first determine the source of income. The source of income comes from two aspects. One aspect is the government subsidies. Another aspect is the expenses paid by alternative fuel vehicles for fuel-filling.

The setting of operating income threshold of a fuel-filling station should refer to the level of local economy. When average operating income exceeds the threshold value, new fuel-filling stations should be built. On the contrary, some of the extant fuel-filling stations should be removed out.

Patents of Alternative Fuel Vehicles

Alternative fuel vehicle technology development is inseparable from the support of patents of alternative fuel vehicles. Some literature data shows that patent is usually used as the predictor of the technology. By analyzing changes of the patent, it is clear grasp of current technology, such as the stage in the life cycle of technology, as well as future trends. Accordingly, businesses can choose different investment decisions in different stages, which ensure the largest gains. For example, for the high-growth patents, the technology is promising. The business should increase the investment.

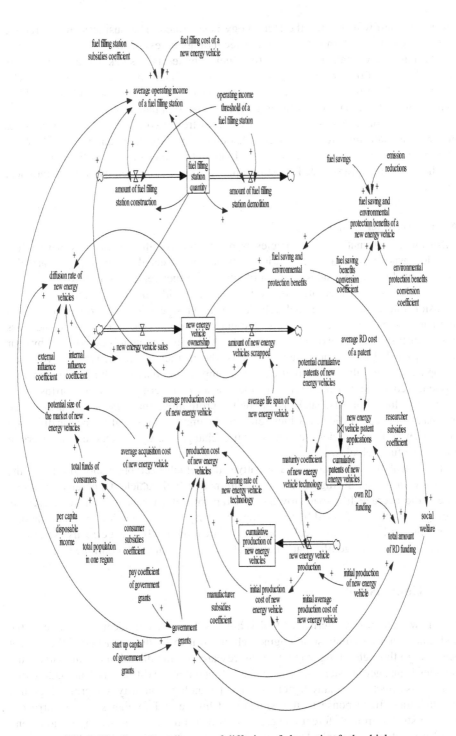

Fig. 2. The flow chart diagram of diffusion of alternative fuel vehicles

For the low-growth patents, the technology is saturated. The business should reduce their investment and turn to the next new technology investment.

The cumulative patent of alternative fuel vehicles is a stock variable, which is affected by a flow variable (alternative fuel vehicle patent applications). The calculation about this flow variable should refer to the following auxiliary variables: total amount of R & D funding, average R & D cost of a patent.

The source of R & D funding usually comes from two aspects. One aspect is the government subsidies to researchers. Another aspect is researchers' own R & D funding, which can be raised from research institutes, enterprises or individuals.

The value of average R & D cost of a patent can be estimated by way of statistical research.

Alternative Fuel Vehicles

Diffusion of alternative fuel vehicles is main focus of this study. Alternative fuel vehicle ownership is a stock variable, which is affected by two flow variables (alternative fuel vehicle sales, amount of alternative fuel vehicles scrapped).

The calculation of alternative fuel vehicle sales should refer to the auxiliary variable: diffusion rate of alternative fuel vehicles. In order to calculate the diffusion rate of alternative fuel vehicles, we can use the classic Bass model. According to the Bass model, diffusion quantity of alternative fuel vehicles in a moment is determined by four factors [9-10]. First is the potential size of the market of alternative fuel vehicles. Second is current alternative fuel car ownership on the market. Third is the external influence coefficient, which refers to the impact of external factors on diffusion of alternative fuel vehicles, such as publicity, promotion, mass media. Fourth is the internal influence coefficient, which refers to the impact of internal factors on diffusion of alternative fuel vehicles, such as verbal exchanges between users and nonusers of alternative fuel vehicles.

The calculation of amount of alternative fuel vehicles scrapped should refer to the auxiliary variables: average life span of alternative fuel vehicle. Generally speaking, longer the average life span, lower the scrap rate of alternative fuel vehicles. The average life span is affected by technology maturity. If the technology is more mature, problems similar to a rechargeable battery life are more likely to get effective solutions. As a result, the average life span of alternative fuel vehicles will continue to grow.

4   Conclusions

As a new technology, alternative fuel vehicle is treated as an ideal alternative to the traditional internal combustion engine vehicle. But to be the mainstay of the market needs to go through a long stage. Current research on the diffusion of alternative fuel vehicles is just getting started. Scholars have found that the diffusion of alternative fuel vehicles is affected by many factors, such as technology maturity, government grants, and facilities. In response to this situation, Chinese and foreign scholars carry out detailed studies from different angles, such as distribution of fuel-filling stations, and regulation of government. These studies have achieved fruitful results. However, research on the diffusion of alternative fuel vehicles in this stage is lack of a clear and unified model.

This paper argues that future research can take full advantage of the existing research results. Based on the method of system dynamics, this study deeply analyzes the relationship between the influencing factors of diffusion of alternative fuel vehicles. And then, a complete alternative fuel vehicles diffusion model is built.

At this stage, we start research from consideration of the main factors affecting diffusion of alternative fuel vehicles. The scope of the study will gradually be expanded in future. As a result, diffusion process of alternative fuel vehicles can be described more realistically. In subsequent studies, a case study will also be introduced to validate the proposed system dynamics model. In addition, it is necessary to constantly explore some theories and methods of other disciplines. On this basis, a mature model of diffusion of alternative fuel vehicles will be created.

Acknowledgement. This research was sponsored by NSFC (No.71125002).

References

1. The State Intellectual Property Office of Planning and Development Division. Alternative fuel vehicles industry patent trend analysis report. Patent Statistics Briefing (18) (2011) (in Chinese)
2. Yin, Y., Liao, C., Zhao, D.: An empirical research and analysis of a new low-carbon technology diffusion in China. Science & Technology Progress and Policy 27(23), 20–24 (2010) (in Chinese)
3. Wang, K., Zhang, Q.: Technological innovation diffusion and its barriers: an analysis at micro-level. Studies in Science of Science 23(1), 139–143 (2005) (in Chinese)
4. Bento, N.: Dynamic competition between plug-in hybrid and hydrogen fuel cell vehicles for personal transportation. International Journal of Hydrogen Energy 35, 11271–11283 (2010)
5. Chi, C., Ma, T., Ning, F.: Technology/infrastructure diffusion of natural gas vehicles: the case of Shanghai. International Journal of Energy Sector Management 6, 33–49 (2012)
6. Liang, D., Xu, C., Ma, D.: The research on the model and stability of CCS based on the system dynamics. Journal of Management Sciences in China 15(7), 36–49 (2012) (in Chinese)
7. Wang, Q.: System dynamics, pp. 10–35. Tsinghua University Press, Beijing (1994) (in Chinese)
8. Ernst, H.: The Use of Patent Data for Technological Forecasting: The Diffusion of CNC-Technology in the Machine Tool Industry. Small Business Economics 9, 361–381 (1997)
9. Bass, F.M.: A new product growth model for consumer durables. Management Science 15, 215–227 (1969)
10. Dong, H., Jie, J.: Estimation of private passenger vehicles in China through Bass Model. Journal of Beijing Technology and Business University (Natural Science Edition) 25(4), 63–66 (2007) (in Chinese)

Discovering the Overlapping and Hierarchical Community Structure in a Social Network*

Jiangtao Qiu and Yixiao Hu

School of Information, Southwestern University of Finance and Economics, China
jiangtaoqiu@gmail.com

Abstract. Social networks often show a hierarchical organization, with communities embedded within other communities; moreover, nodes can be shared between different communities. Discovering the overlapping and hierarchical community structure of a social network can provide researchers a deeper understanding of the social network. In this paper, we define the overlapping and hierarchical community as a hierarchy presenting overlapping communities of a social network at different levels of granularity. We propose an algorithm DOHACS to derive overlapping and hierarchical communities from a social network which learn Gaussian mixture models from the social network at various granularities, and then organizing the overlapping communities into a hierarchy. The experiments conducted on synthetic and real dataset demonstrate the feasibility and applicability of the proposed algorithm.

Keywords: Social Network Analysis, Hierarchy, Overlapping Community.

1 Introduction

Community discovery is an important problem in social network analysis, where the goal is to identify related groups of members such that intra-community associations are denser than inter-communities associations. Researchers have developed various methods [1, 6, 9, 10, 14] to discover communities from the SN. Most of methods assume that communities are disjoint, assigning each node only one label of community. In real world, however, one member may be interested in several topics in forum; researchers may belong to more than one research groups. The overlapping community assigning one node to more than one label of community better represents properties of social network. Figure 1 illustrates the overlapping communities discovered from a social network. From the figure, we can observe node 7 in level 1 is assigned into both communities $K1A$ and $K1B$. i.e., node 7 is a hub of $K1A$ and $K1B$. It can be concluded the overlapping community provides a deeper insight into revealing relations among members in a social network than do non-overlapping community.

Hierarchical community presents the communities of a SN at different levels of granularity. Finer granularity tends to generate more and smaller communities from a

---

* This work was supported by the Humanities and Social Science Foundation for the Youth Scholars of Ministry of Education of China (No. 09YJCZH101).

M. Wang (Ed.): KSEM 2013, LNAI 8041, pp. 252–262, 2013.

SN while coarser granularity may obtain a macro-view of communities. For example, it is possible to learn sectors of an enterprise by detecting communities of a SN of the enterprise at a coarser granularity; groups in the sector may be detected at a finer granularity. Level 1 in Figure 1 is the communities detected at a coarse granularity while Level 2 presents community in a finer granularity.

We believe that combining hierarchical community and overlapping community can do a better job on understandings of the social network than do single either overlapping or hierarchical community. From the overlapping and hierarchical community structure illustrated in Figure 1, it can be observed both communities *K2B* and *K2C* in level 2 belong to community *K1B* in level 1. Exploring the SN from a hierarchical community's perspective, we can conclude node 7 and node 10 are far in relation in community *K1B*. Discovering the overlapping and hierarchical community is an interesting topic. There are a little researchers working on this field [5, 7]. However, their works either investigate the overlapping community and the hierarchical community separately [7], or can be only applied in unweighted undirected graph [5].

Our contributions in this paper include develop a general approach for discovering the overlapping and hierarchical community structure in a social network.

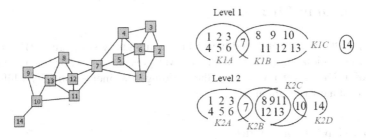

Fig. 1. An example for the overlapping and hierarchical community structure

2   Related Works

Many researches on overlapping community use the technique of discovering cliques in the SN. The first attempts to unveil the overlapping community structure of a network appear in [4]. The algorithm in [4] retrieves all k-cliques (complete sub-graphs of size k) of the graph. Further, a k-clique-community as a union of all k-cliques can be reached from each other through a series of adjacency k-cliques (where adjacency means sharing k−1 nodes). Shen et. al. [12] propose a metric Qc to quantify the overlapping community structure. Based on the betweenness centrality measure, Gregory developed a overlapping community discovery algorithm CONGA[8] and a improved algorithm Peacock [13] by extending Girvan and Newman's well-known algorithm[10]. However, these algorithms, capable of detecting overlapping communities, perform well only when the extent of community overlap is kept to modest levels. Many earlier efforts employ hierarchical clustering method [6] to derive hierarchical structure of a SN. They may generate a dendrogram of the communities which present community dividing at different levels. Researchers also developed some methods based on either probability model [11] or structure clustering [2].

In recent years, researchers [5, 7] have believed that combining the overlapping community and the hierarchical community can capture much richer information about the social network. Lancichinetti [7] presents an algorithm that can find both overlapping communities and the hierarchical structure. The method is based on the local optimization of a fitness function. Community structure is revealed by peaks in the fitness histogram. The resolution can be tuned by a parameter enabling to investigate different hierarchical levels of organization. Shen et. al. [5] propose an algorithm to detect both the overlapping and hierarchical properties of complex community structure together. This algorithm deals with the set of maximal cliques and adopts an agglomerative framework. However, only un-weighted and undirected networks are considered in this algorithm. In this paper, we shall show that a hierarchy of overlapping community can naturally capture the relationships between overlapping groups at different levels of granularity.

3   Using GMM to Discover the Overlapping and Hierarchical Community Structure

In this paper, the overlapping and hierarchical community structure (OHCS in short) is defined as a hierarchy in which every level presents an overlapping community dividing of a SN. Lower levels show the dividing at a finer granularity while higher levels do so at a coarser granularity.

3.1   Gaussian Mixture Model

The GMM is a superposition of K Gaussian densities in the form

$$p(\mathbf{x}) = \sum_{k=1}^{K} \pi_k N(\mathbf{x}|\mu_k, \Sigma_k)$$

Each Gaussian density $N(\mathbf{x}|\mu_k, \Sigma_k)$ is called a component of the mixture and has its own mean μ_k and covariance Σ_k. The parameters π_k are called mixing coefficients satisfying $0 \leq \pi_k \leq 1$, $\sum_{k=1}^{K} \pi_k = 1$. GMM has been proven to be effective on clustering tasks. In a GMM, learned from a dataset, each component represents a cluster. The posterior probability $p(k|x)$ indicates the probability of data x belong to cluster k. they satisfy $\sum_{k=1}^{K} p(k|x) = 1$. Learning GMM from a social network, we can calculate the probability of one node belong to one community, and then further discover overlapping community. Adjusting the number of components K of GMM, we shall discover the overlapping community at different levels of granularity. In this way, an overlapping and hierarchical community structure may be obtained.

3.2   Transition Probability Matrix

Employing GMM to discover OHCS, we need to represent a social network in a matrix where each row vector denotes a node. For an unweighted undirected graph, element of matrix M_{ij} indicates linkage between node i and node j. If there exists an edge between both nodes, let $M_{ij}=1$. Otherwise, $M_{ij}=0$. M_{ii} is set to 1 by default. In the context of the unweighted undirected graph, matrix M is an adjacency matrix. For a weighted directed graph, if there is one edge from node i to j, let M_{ij} be a weight of the edge. Otherwise, $M_{ij}=0$. In this way, our method may overcome limitation of the types of graph. We shall only consider unweighted undirected graph at following examples.

At the beginning, we use an adjacency matrix to represent social network. Figure 2 gives an example of an adjacency matrix. In practice, however, learning a GMM from adjacency matrix cannot achieve good enough performance as we expect. We found element of adjacency matrix is in discrete 0 or 1 while GMM is a model for the distribution of continuous variable. Therefore, we seek to use continuous variable to represent nodes of social network.

Random walk on graph [3] calculates the transition probability for each pair of nodes in a SN. One-step transition probability from node j to l $P_{t+1|t}(l|j)$ may be calculated in the form of $P_{t+1|t}(l|j) = \begin{cases} (1-p)M_{jl}/\sum_i M_{ji} & \forall l \neq j \\ p & l = j \end{cases}$ where M is an adjacency matrix; M_{ij} indicates that there is an edge between node i and node j; p denotes the self-transition probability. We organize the transition probability as a matrix A where $A[j,l]$ is a probability from node j to l $P_{t+1|t}(l|j)$. After obtaining one-step transition probability matrix A, we further calculate t-step transition probability matrix $A^t=A(A...(A))$ and normalize A^t according to $A^t= Z^{-1}*A^t$, where Z is a diagonal matrix. Figure 2 gives an example of one-step transition probability matrix where parameter $p=0.6$.

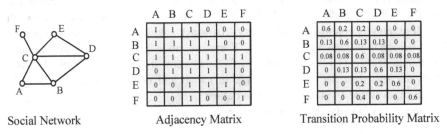

| Social Network | Adjacency Matrix | Transition Probability Matrix |

Fig. 2. Representing a social network with adjacency matrix and transition probability matrix

3.3   Discovering the Overlapping and Hierarchical Community Structure

We employ GMM to calculate poster probability of one node belonging to one community p(k|x), and further discover overlapping community. However, in practice a problem is encountered that one node is often assigned to one community in a very high probability, along with assigned to others community in a low probability. For example, in Figure 1, node 7 belongs to community {8, 9, 10, 11, 12, 13, 14} in a probability close to 1, belonging to community {1,2,3,4,5,6} at a probability near to 0.

By observation, the mean vector of GMM reveals rich information about community. Table 1 lists the mean vectors of GMM learned from the social network in Figure 1. Elements of the mean vectors may be regarded as indictors, showing what extent a node belongs to one community. For example, in table 1, values of node 7 is 4.97 and 5.01 in μ_1 and μ_2 respectively, having large values compared to others nodes in both vectors. This means that node 7 belong to the both community (each component of GMM or row vector is regarded as one community) in a large power. Node 14 has a less value than other nodes in both vectors μ_1 and μ_2. It is reasonable for regarding node 14 as an outlier, not belonging to any community. These analyses can be verified by observing Figure 1.

Organizing K mean vectors of GMM as a matrix, we, to discover the overlapping community by analyzing the matrix, need to set two thresholds in percentage rth and cth. rth is a threshold in row vector. When value of one node is less than the product of max value in the row vector and rth, the node does not belong to the corresponding community. We further explore column vector. If the value of one node is less than the product of max value in the column vector and cth, the node also will not belong to the corresponding community.

Table 1. Mean vector of GMM with 2 components

node	1	2	3	4	5	6	7	8	9	10	11	12	13	14
μ_1	1.33	0.42	0.42	1.33	1.48	0.90	4.97	5.37	4.64	5.97	2.74	3.97	5.91	1.86
μ_2	6.09	5.42	5.42	6.09	5.90	8.43	5.01	0.97	0.21	0.28	0.76	0.91	0.37	0.03

In table 1, max (μ_1)=5.97, max (μ_2)=6.09 and max(7)=5.01. If we set rth=50% and cth=50%, value of node 7 in μ_1, 4.97, is larger than max $(\mu_1) \times r$th and max(7) $\times c$th. Therefore, node 7 belongs to community 1 (row μ_1). Likewise, we can learn that node 7 also belong to community 2 (row μ_2). Therefore, node 7 is a hub of both communities. For node 14, its value is 1.86 and 0.03 in vector μ_1 and μ_2, respectively. Both values are less than the product of max $(\mu_1) \times r$th and max $(\mu_2) \times r$th, respectively. Therefore, the node 14 is regarded as an outlier.

In this paper, Akaike Information Criterion

$$\underset{rth,cth}{\arg\min} \quad [\ln(\sum RSS(rth,cth)) + \lambda D/(2 \times \ln(N))]$$

is employed to determine the optimal rth and cth where $\sum RSS(rth,cth)$ is the sum of residual sum of squares of all communities detected with parameters rth and cth; D is the number of the communities (including outliers); N is the number of nodes in the network; λ is a weighting factor. A large value of λ favors solutions with few communities.

Setting the number of components of GMM to n on learning a GMM from a social network, we can generate $k<=n$ overlapping community. Let $n^{\text{new}} = \begin{cases} k & k < n \\ n-1 & k = n \end{cases}$, we then use n^{new} as the number of components of GMM in next iteration. Repeating this step until $n^{\text{new}}=2$ and organizing the overlapping community discovered in each of iterations into a hierarchy, finally, we can obtain an OHCS. Experiential, initial value of n may be one third or a quarter of the number of nodes in a SN. An algorithm for discovering OHCS is described as follows.

Algorithm 1. Discovering the **O**verlapping and **H**ier**A**rchical **C**ommunity **S**tructure: *DOHACS*

Input: Social network G

Output: The overlapping and Hierarchical Community Structure *OHCS*

Parameter: steps t, self-transition probability p

1. $A\leftarrow$generate one-step transition probability matrix from G with parameter p

2. $Z_{ij}\leftarrow\Sigma_i[A^t]_{ij}$

3. $A^t\leftarrow A^t\times Z^{-1}$ // obtain normalized t-step transition probability matrix A^t

4. $n\leftarrow$ the number of nodes in $G/3$

5. **while** $(n>1)$

6. $GMM\leftarrow$ learning a GMM with n components from A^t

7. $\mu\leftarrow$*get centroids of n components from GMM*

8. $[rth, cth]\leftarrow argmin_{rth,cth}[ln(\Sigma RSS\ (rth,cth))+ \lambda D/(2*ln(N)]$

9.   **For** each element of μ_{ij}

10. $r_{ij}\leftarrow$if $\mu_{ij}>max(\mu_i)\times rth$ then 1 else 0

11. $c_{ij}\leftarrow$if $\mu_{ij}>max(node_j)\times cth$ then 1 else 0

12. $\mu_{ij}\leftarrow r_{ij}$ & c_{ij} // logic 'And' operation in matrix r and c

13.   **End**

14. $OHCS[n]\leftarrow$let nodes whose element is 1 in row vector μ_i belong to community i, obtain the overlapping community at level n

15     $n\leftarrow min$(the number of community in $OHCS[n]$, n-1)

16. **loop**

---

4   Experiments

In this section, algorithm DOHACS is evaluated using both synthetic and real datasets. We firstly explore what parameters settings enable the algorithm to achieve optimal performance. In the section 4.2, we conduct experiments on three real datasets to discover their OHCSs. Also, we compare DOHACS to k-clique [4] and EAGLE [5] on the overlapping community discovery.

4.1   Experiment for Exploring the Parameters of Algorithm

DOHACS algorithm can be used to find single-label community (one node only belong to one community) when the parameter rth and cth are set on 0% and 100%, respectively. To evaluate parameters self-transition probability p and steps of random walk t, in this experiment, we apply algorithm DOHACS to discover single-label community on synthetic dataset that may be generated based on method proposed in [10]. Modularity measure proposed in [10] can evaluate the discovered communities from the SN. The greater the value of the modularity, the better the result. Our experiments employed the modularity measure as a metric for evaluating the discovered communities. On generating a synthetic network, some parameters should be chosen such as the number of vertices v, the number of community c, the average number of edges for each vertex z, and the average number of edges linking one vertex outside communities $zout$.

We set the parameters for generating synthetic network as $v=128$, $c=4$, $z=16$, $zout=\{0, 2, 5\}$. Such parameters settings mean that each synthetic network has 128 nodes, 4 communities. parameter settings $z=16$, and $zout=8$ indicates that there are an average of 16 edges at every node with eight of them linked to nodes in other communities, implying that the generated network does not contains a community structure. Parameter settings $z=16$, and $zout=0$ indicates that there does not exist edges among communities. Therefore, as the value of $zout$ decreases, the generated networks will present an increasingly strong community structure.

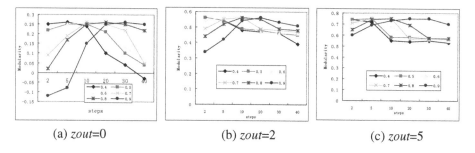

(a) $zout=0$ (b) $zout=2$ (c) $zout=5$

Fig. 3. Evaluating the self-transition probability and random walk steps

Let the number of components of GMM be 4, self-transition probability $p=\{0.4, 0.5, 0.6, 0.7, 0.8, 0.9\}$ and steps of random walk $t=\{2,5,10,20,30, 40, 50\}$. We explore how parameters p and t impact the performance of DOHACS. Generating 100 synthetic networks under each of $zout=\{0,2,5\}$, we run DOHACS in each of combination of three parameters $<p, t, zout>$. The average modularity of discovered communities on the 100 network are illustrated in Figure 3 where each of curves corresponds to one p value. We can observe from the figure that each curve can achieve almost best performance along with the modularity measures for the same. The optimal parameters pair for $<p,t>$ are $\{0.4, 2\}$, $\{0.5, 5\}$, $\{0.6, 5\}$, $\{0.7, 10\}$, $\{0.8, 10\}$, $\{0.9, 20\}$.

4.2   Experiments for Discovering *OHCS*

In this section, we employ DOHACS to discover OHCS on real social networks, and then compare it to k-clique and EAGLE on the overlapping community discovery. All of following experiments use parameter settings $p=0.7$ and $t=10$.

SN1 in Figure 4 (a) is a social network shown in [1], which includes an outlier and a hub. Many works use the network to examine algorithm's capability on finding the outlier and hub. Figure 4 (b) illustrated the discovered OHCS where K indicate the number of components of GMM. When $K=2$, we can find that node 14 is an outlier and node 7 is a hub. Community $K2A$ contains nodes $\{1, 2, 3, 4, 5, 6, 7\}$. Exploring the SN at a finer granularity $K=6$, we can find community $K6A$ is a denser association group in $K6B$. $K2A$ has the same set of nodes with $K6B$. This means the hierarchical community can provide views of community at different levels of granularity. We further find that community $K6E$ is composed of node 14 and 10, and node 10 is a hub of community $K6E$ and $K6D$. Communities $K6C$ and $K6D$ show node 7 and node 10 are far in relation in community $K2B$. DOHACS can discover not only low overlapping communities, such as $K2A$ and $K2B$, but also high overlapping communities, such as $K6C$ and $K6D$.

SN2 in Figure 5 (a) is a social network shown in [1]. We apply algorithm DOHACS to SN2. Figure 5 (b) is an OHCS obtained from SN2. When $K=2$, two communities $K2A$ and $K2B$ and one outlier are found from SN2. At a finer granularity, $K=5$, nodes $\{8, 11\}$ are hubs of both community $K5A$ and $K5B$; nodes $\{19, 20\}$ are hubs of community $K5B$ and $K5C$. Also, we find that there exists a denser group $K7B$ in $K7A$. This shows that communities at a lower level of OHCS reveal inner information of communities at a higher level.

Karate club dataset [15] in Figure 6 (a) represents a relationship network at a university karate club which consisted of 34 members. Researches use the dataset to evaluate their method of community discovery. There are two communities in the social network in real world that are told using different colors in Figure 6 (a). We also employ the dataset to evaluate our algorithm on discovering OHCS. When $K=2$, we find the nodes $\{3, 31, 9\}$ are hubs of communities $K2A$ and $K2B$. When $K=4$, $K4A$ community is an inner group of $K2A$. This means that nodes in $K4A$ have closer relations than other nodes in $K2A$. It also can be seen that node 1 is a hub of community $K4A$ and $K4B$; $K4C$ is an inner group in $K3C$; node 3 is a hub of $K4C$ and $K4B$; nodes $\{33, 34\}$ are hubs of Community $K4C$ and $K4D$. Three OHCSs above illustrate that changing K may obtain overlapping community of social network at different levels of granularity. OHCS can reveal the relations among nodes, which provides researchers deeper understandings of the SN.

Algorithm DOHACS may find some interesting information that other methods cannot provide. For example, Figure 5 (b) presents that there exists a denser group $K7B$ in community $K7A$; Figure 4 shows, from a view presented on $K=6$, node 7 and 10 in $K2B$ community are in a far relation. The experiments demonstrate that discovering OHCS is an interesting work.

k-clique [4] is a well known algorithm for detecting overlapping communities in social network. A Python language software package, NetworkX, implements the *k*-clique algorithm in it's latest version. To compare DOHACS with k-clique, we employ the NetworkX to detect the communities of three datasets SN1, SN2 and karate club. The results are illustrated in Figure 4, Figure 5 and Figure 6, respectively.

Figure 4 (c) shows that the overlapping communities detected by *k*-clique with the size of clique *k*=3 are identical to the communities in *K*=2 Level of OHCS of SN1 shown in Figure 4 (b). It also can be observed that communities in Figure 5 (c) is identical to *K*=5 level of OHCS of SN2 in Figure 4 (b). Although there is no a ground truth, we can confirm the result detected by both *k*-clique and DOHACS is reasonable by analyzing the both social networks illustrated in Figure 4 (a) and Figure 5 (a). Further observing the results of *k*-clique algorithm running in karate club dataset, it can be seen that the overlapping community on the setting of size of clique *k*=3 contains a large mount outliers (outlier are not illustrated in the figure). On the setting of *k*=4, the detected result contain only two outliers, node 10 and node 12. Compare to *K*=2 Level of OHCS of karate club, however, we can find the result detected by *k*-clique is less reasonable. In real world [15] node 1 and 2 are in one community different from that of node 33 and 34. *k*-clique fail to divide them into two communities.

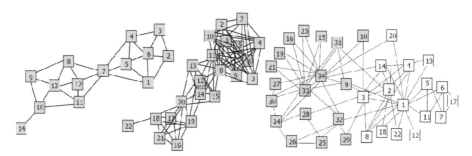

Fig. 4. (a) dataset SN1   **Fig. 5. (a)** dataset SN2   **Fig. 6. (a)** Karate club dataset

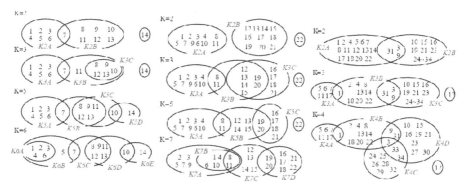

Fig. 4. (b) OHCS of SN1   **Fig. 5. (b)** OHCS of SN2   **Fig. 6. (b)** OHCS of Karate club

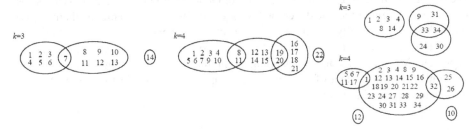

Fig. 4. (c) Communities of SN1 detected using k-clique algorithm

Fig. 5. (c) Communities of SN2 detected using k-clique algorithm

Fig. 6. (c) Communities of Karate club detected using k-clique algorithm

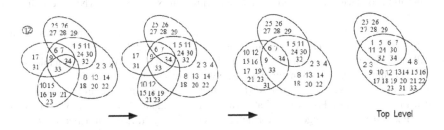

Fig. 7. Communities of Karate club detected using EAGLE algorithm

Also, we implement EAGLE and compare it to DOHACS. The part of dendrogram derived by EAGLE from karat club dataset shows in Figure 7. On detecting Hierarchical and overlapping communities, EAGLE obtains maximal cliques of the network and then adopts an agglomerative framework to generate a dendrogram. In practice, however, it generally obtains so much maximal cliques that the dendrogram is far too complicated to show a clear view about the communities just like it shows in the figure. Also, we observed that in top level the derived two communities fail to reflect ones in real world.

From the experiment, we can conclude that DOHACS is capable of effectively detecting the overlapping community from the social network. Even it may outperforms k-clique and EAGLE in some settings.

5   Conclusions

Discovering the overlapping and hierarchical community structure may obtain deeper understandings of a social network than does single either overlapping community or hierarchical community. For example, it can do a good job on exploring inner relationship in an organization; finding researchers across multiple academic fields; finding important role in a community. In this paper, we combine random walk on graph and Gaussian mixture model to obtain the overlapping community by analyzing mean vectors of GMM. Changing the number of components of GMM, we can discover the overlapping communities at different levels of granularity. Organizing them into a hierarchy, we can finally obtain OHCS of a SN. The experiments conducted on synthetic and a real dataset demonstrate the feasibility and applicability of the proposed approach. Most of existing overlapping community discovery algorithms perform well only when the extent of community overlap is kept to modest levels. DOHACS can overcome this limitation.

References

1. Xu, X., Yuruk, N., Feng, Z., Schweiger, T.A.J.: SCAN: A Structural Clustering Algorithm for Networks. In: The Proceeding of SIGKDD 2007 (2007)
2. Huang, J., Sun, H., Han, J., Deng, H., Sun, Y., Liu, Y.: SHRINK: A Structural Clustering Algorithm for Detecting Hierarchical Communities in Networks. In: The Proceedings of CIKM 2010 (2010)
3. Craswell, N., Szummer, M.: Random walks on the click graph. In: The Proceedings of the 30th Annual International ACM SIGIR Conference, pp. 239–246 (2007)
4. Palla, G., Derényi, I., Farkas, I., Vicsek, T.: Uncovering the overlapping community structure of complex networks in nature and society. Nature 435, 814–818 (2005)
5. Shen, H., Cheng, X., Cai, K., Hu, M.-B.: Detect overlapping and hierarchical community structure in networks. Physica A: Statistical Mechanics and its Applications 388(8) (2009)
6. Newman, M.E.J.: Fast algorithm for detecting community structure in networks. Physical Review E 69, 066133 (2004)
7. Lancichinetti, A., Fortunato, S., Kertész, J.: Detecting the overlapping and hierarchical community structure in complex networks. New Journal of Physics 11(3) (2009)
8. Gregory, S.: Finding Overlapping Communities Using Disjoint Community Detection Algorithms. In: Fortunato, S., Mangioni, G., Menezes, R. (eds.) Complex Networks: CompleNet 2009. SCI, vol. 207, pp. 47–61. Springer, Heidelberg (2009)
9. Leicht, E.A., Clarkson, G., Shedden, K., Newman, M.E.J.: Large-scale structure of time evolving citation networks. The European Physical Journal B 59(1) (2007)
10. Newman, M.E.J., Girvan, M.: Finding and evaluating community structure in networks. Physical Review E 69, 026113, 1–15 (2004)
11. Clauset, A., Moore, C., Newman, M.E.J.: Hierarchical structure and the prediction of missing links in networks. Nature 453, 98–101 (2008)
12. Shen, H.-W., Cheng, X.-Q., Guo, J.-F.: Quantifying and identifying the overlapping community structure in networks. Journal of Statistical Mechanics: Theory and Experiment (7) (2009)
13. Gregory, S.: Fuzzy overlapping communities in networks. Journal of Statistical Mechanics: Theory and Experiment 2011(2) (2011)
14. Chekuri, C., Goldberg, A., Karger, D., Levin, M., Stein, C.: Experimentao study of minimum cut algorithms. In: Proc. 8th ACM-SAIM Symposium on Discreet Algorithm, pp. 324–333 (1997)
15. Zachary, W.W.: An information flow model for conflict and fission in small groups. Journal of Anthropological Research 33, 452–473 (1977)

A Knowledge Based System of Principled Negotiation for Complex Business Contract

Xudong Luo[1], Kwang Mong Sim[2], and Minghua He[3]

[1] Institute of Logic and Cognition, Sun Yat-sen University, Guangzhou, China
[2] School of Computing, University of Kent, Kent, UK
[3] School of Engineering and Applied Science, Aston University, Birmingham, UK

Abstract. Automated negotiation systems can do better than human being in many aspects, and thus are applied into many domains ranging from business to computer science. However, little work about automating negotiation of complex business contract has been done so far although it is a kind of the most important negotiation in business. In order to address this issue, in this paper we developed an automated system for this kind of negotiation. This system is based on the principled negotiation theory, which is the most effective method of negotiation in the domain of business. The system is developed as a knowledge-based one because a negotiating agent in business has to be economically intelligent and capable of making effective decisions based on business experiences and knowledge. Finally, the validity of the developed system is shown in a real negotiation scenario where on behalf of human users, the system successfully performed a negotiation of a complex business contract between a wholesaler and a retailer.

Keywords: automated negotiation, agents, knowledge based system, knowledge engineering, e-business.

1 Introduction

Negotiation is a communication process among a group of agents, which aims to reach an agreement on some matter [1]. It often occurs in our daily lives. For example, a teenager discusses, with the parents, over what food to have for dinner. Nevertheless, negotiation is often not an easy task for human beings [2]. Thus, recently there has been a surge of interest in automated negotiation systems that are populated with artificial agents [3,2]. In fact, a wide range of models for automated negotiation have been developed, for example, auctions [4,5,6], direct one-to-one negotiations [1,7,8], and argumentation [9].

However, few researchers work on automated negotiation systems for complex business contract in real life although this kind of negotiation is very important. In order to tackle the problem, in this paper we will develop such an automated negotiation system based on principled negotiation theory [10]. The system has the theory as its base for a number of reasons. First, if we want our system to be widely adopted in real business, it has to be built upon the solid research of business science. It is not difficult to imagine how difficult an automated negotiation

M. Wang (Ed.): KSEM 2013, LNAI 8041, pp. 263–279, 2013.

system based on computer scientists' heuristics would be accepted and adopted in real business. Principled negotiation is one of the most influential approaches if it is not the most influential approach in current manual negotiation theory [11]. This approach has been widely embraced and often cited both in academia and among industry professionals dealing with the problem of negotiation. Second, it can make both parties win, rather than sticking to an assumed position where there must be a winner and a loser. Basically, both parties in the negotiation want to maximise their interests. So, the negotiation result should be fair for both. That means both should be the winners. In other words, an automated negotiation system should be able to produce win-win outcomes. The principled negotiation approach can guarantee win-win outcome.

Our system is developed as a knowledge based one according to knowledge engineering methodology KEMNAD that is proposed in [2] for facilitating negotiating agent development. This is because an automated negotiation system in a specific business domain has to be economically intelligent and capable of making effective decisions based on business experiences and knowledge. Our system is made of four templates from KEMNAD: classification, diagnosis, deduction and utility. The first two are used to carry out the task of negotiation situation analysis; and the last two are for the task of decision making during the course of a negotiation. The developed system works in this way: firstly classify input/responses, analyse such responses, and make deductions based on the result of response diagnosis and measure utility. It stores up a history of negotiations that will be fed into new cycles. On the other hand, this developed system for automated principled negotiation shows further the validity of KEMNAD for developing negotiating agents. This is because this hybrid model accommodates main human negotiation methods: making pie bigger, tradeoff, argumentation and concession. In other words, it shows that the KEMNAD methodology is effective for major negotiation approaches. On the other hand, the model is made of four templates, while previously the BDI-based negotiation model that we used to show the validity is made of only two templates: classification and deduction.

The rest of this section is organised as follows. Section 2 presents the main task model of our negotiation system, while its subtask task models are given in Section 3. Section 4 illustrates the validity of the proposed system by a complex business negotiation between a wholesaler and a retailer. Section 5 discusses the related work. Finally, Section 6 summarises the paper and points out the direction of the future work.

2   Main Task Model

This section will develop the main task model of our negotiation system.

The key points of the principled negotiation approach [10] are: (i) The interests behind the conflicting positions of the negotiating parties could not be in conflict. (ii) There are some options that can guarantee the interests of both parties and even increase the gain of both parties. (iii) When conflicts are unavoidable, the argumentation proceeds according to objective criteria. Accordingly, the basic

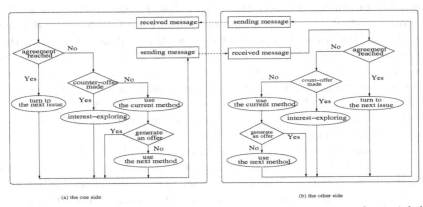

Fig. 1. The basic idea of the negotiation proceeding in our automated principled negotiation system

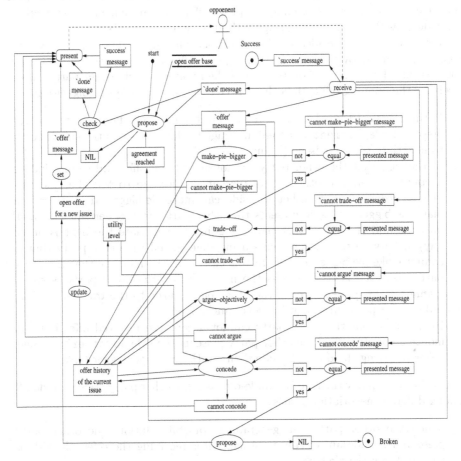

Fig. 2. Graphic specification of task knowledge of an automated principled negotiation

working flow of principled negotiation is as follow. Firstly, try to solve conflict by exploring mutual interests. If it does not work, then to try to create different options by making tradeoff between attributes. If it still does not work, then try to argue and/or concede according to objective criteria. In short, the first component tries to increase mutual gain, the second one tries to ensure mutual gain is non-decreasing, and the third tries to keep one party's gain non-decreasing if possible; otherwise both parties concede if possible. Moreover, if the both sides of a negotiation employ the principled negotiation approach, before they try the next method, they should see what the other side can do.

That is, as shown in Figure 1, there are three cases: (i) If the other side accepts its offer about the current negotiation issue, they turn to the next issue. (ii) If the other side cannot accept its offer nor make any counter offer, this side tries the current method to generate a new offer. If it can do, it sends the offer to the other side and sequentially the other side tries the interest exploring method to handle the offer; otherwise this side tries the next method to generate a new offer. (iii) If the other side cannot accept its offer but can make a counter offer, this side tries the interest exploring method to handle the offer.

More specifically, the idea of the main task knowledge model (as shown in Figure 2)[1] is as follows. At the beginning, one side of a negotiation proposes an opening offer for a negotiation issue, and then presents to its other party. After receiving the opening offer, the other party turns to the make-pie-bigger task.

- In the case that the other party can accept the received offer, the other party presents an 'agreement reached' message to the side. Then the side is going to negotiate the next issue (if any).
- In the case that the other party cannot accept the offer and cannot induce a counter offer, the other party presents a message to tell the side this fact in order to see whether the side can make-pie-bigger. If the side can make-pie-bigger, the side presents the other party an 'agreement reached' message and then the other party is going to negotiate the next issue (if any); otherwise this side presents a message to tell the other party that it cannot make-pie-bigger, either. In this case (*i.e.*, the message that the other party presented is the same as the message that the other party received: 'cannot make-pie-bigger'), the other party turns to the trade-off task in order to keep its current utility level unchanged.
- In the case that the other party cannot accept the received offer but can induce a counter offer, the other party presents the side an 'offer' message. After receiving the offer, the side turns to the make-pie-bigger task.

When the other party turns to the trade-off task, the other party tries to trade-off among different negotiation issues.

- When the other party can generate a tradeoff solution, the other party presents this side an 'offer' message. After receiving the offer, the side is turn to the make-pie-bigger task.

---

[1] Notice both sides of a negotiations employ the same task model.

- When the other party cannot **generate** a tradeoff solution, the other party presents a message to tell the side this fact in order to see whether the side can **trade-off**. If the side can **generate** a trade-off solution, the side **presents** an 'offer' message to the other party and after receiving the offer the other party turns to the make-pie-bigger task; otherwise this side **presents** to tell the other party that this side cannot **trade-off**, either. In this case (*i.e.*, the message that other party **presented** is the same as the message the other party received: 'cannot trade'), the other party turns to the argue-objectively task.
- When the other party can accept the received offer, the other party **presents** the side an 'agreement reached' message. After receiving the message, the side is going to negotiate next issue.

When the other party turns to the argue-objectively task, the other party tries to argue with this side according to the relevant axioms.

- When the other party can find a counter offer with arguments[2], the other party **presents** the side an 'offer' message. After receiving the offer, this side turns to the make-pie-bigger task.
- When the other party cannot find a counter offer with arguments, the other party **presents** this side a message to tell the side this fact in order to see whether the side can **argue-objectively**. If in the argue-objective task the side can make a counter offer, then the side **presents** an 'offer' message to the other party and after receiving the offer the other party turns to the make-pie-bigger task; otherwise this side **presents** a message to tell the other party that this side cannot argue-objectively, either. In this case (*i.e.*, the message that the other party **presented** is the same as the message that the other party received: 'cannot argue'), this other party turns to the concede task.
- When the other party can accept the received offer, the other party **presents** the side an 'agreement reached' message. After receiving the message, the side is going to negotiate next issue.

When the other party turns to the concede task, the other party will consider whether it could make some concession.

- When the other party can make some concession, that is, the other party can **generate** a counter offer, the other party **presents** the side an 'offer' message. After receiving the offer, this side turns to the make-pie-bigger task.
- When the other party cannot make any more concession, the other party presents a message to tell the side this fact in order to see whether this side can **concede**. If this side can **generate** an alternative by conceding, the side **presents** the other party an 'offer message and after receiving the offer the other party turns to the make-pie-bigger task; otherwise this side **presents** a message to tell the other party that this side cannot **concede**, either. In this case, the other party left the issue as unsolved one[3] and is going to negotiate the next issue.

---

[2] We view the same offer but with different arguments as different offer.

[3] Maybe later on when solving other negotiation issue, the negotiating parties can solve this issue, for example, by making a tradeoff between the two issues.

– When the other party can accept the received offer, the other party presents the side an 'agreement reached' message. After receiving the message, the side is going to negotiate next issue.

When the agent is going to negotiate the next issue, if the agent can propose the opening offer of the next issue. Then the new round of negotiation starts like the above procedure. In the case where no more issues are left, the agent first checks whether the previous received message is a 'done' message (meaning that for the other party no more issues that need to be negotiated). If it is a 'done' message, both sides have no more issues that need to be negotiated and so the negotiation can be terminated, and accordingly the agent presents a 'success' to the other party agent and the agent's negotiation procedure is terminated successfully; otherwise, the agent presents a 'done' message to the other party, meaning that the agent has no more issues that need to be negotiated. In addition, if the agent receives a 'success' message, the agent's negotiation procedure is terminated.

3   Subtask Models

From the the previous section, we can see the main task knowledge model of automated principled negotiation is made of four task knowledge models: trade-off, make-pie-bigger, argue, and concede. In this section, we will discuss their details.

Actually, each of these tasks can be viewed as a relatively independent negotiation task, and so can be constructed from the templates and the generic main task model of KEMNAD [2]. That is, the whole procedure to develop these task knowledge models includes using templates to build situation analysis and decision making component, then substituting them into the generic main task model to obtain an integrated main task model, and finally, if necessary, simplifying the integrated main task model to obtain the integrated, simplified main task models. Although we indeed construct the make-pie-bigger, trade-off, argue and concede tasks step by step according to the procedure, for the sake of space we omit the procedure and just present the final results. Also for the sake of space, we present their graphical specification but omit their textual ones.

3.1   The Subtask of make-pie-bigger

The graphical specification of the make-pie-bigger task is shown in Figure 3. It is made of three templates [2]: the classification method, the deduction method, and the diagnosis method. The classification method is used to classify the received message ('offer' or 'cannot make-pie-bigger') into the three decision categories as follows: (i) When it is unnecessary to explore the interest behind the other party's offer, directly induce a counter offer or a decision to accept the other party's offer or nothing. In the first two cases, set the response as an 'offer' message or an 'agreement reached' message, respectively. And in the case of a count offer induced successfully, update the offer history (of the

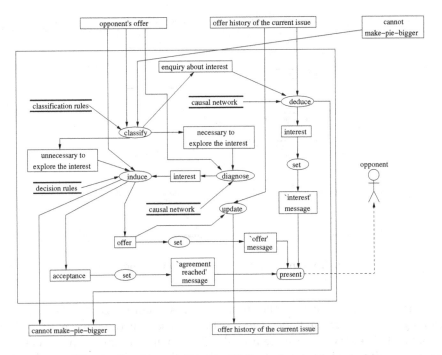

Fig. 3. Graphic specification of the make-pie-bigger task

issue being negotiated) by adding the counter offer in. The third case means that the agent cannot make-pie-bigger and so quits the make-pie-bigger task (and presents a 'cannot make-pie-bigger' message to the other party). (ii) When it is necessary to explore the interest behind the other party's offer, the diagnosis method [2] is used to explore the other party's interest. After the interest is known, the deduction method is used to make a decision and accordingly set the response message. (iii) When it is an enquiry about the interest behind the agent's offer, try to deduce the interest. If the interest could be revealed, then present the other party an 'interest' message; otherwise it means that the agent cannot make-pie-bigger and so quits the make-pie-bigger task (and presents a 'cannot make-pie-bigger' message to the other party).

Finally, present the response message to the other party. Notice that the situation analysis subtask of the make-pie-bigger task is actually comprised of two templates [2]: classification and diagnosis, and its decision making subtask is actually comprised of the deduction method [2] that is simply made of inference functions deduce and induce.

3.2   The Subtask of trade-off

The graphical specification of the trade-off task is shown in Figure 4. Its situation analysis subtask uses the classification method to classify the received message ('offer' or 'cannot make-pie-bigger' or 'cannot trade-off') into the three

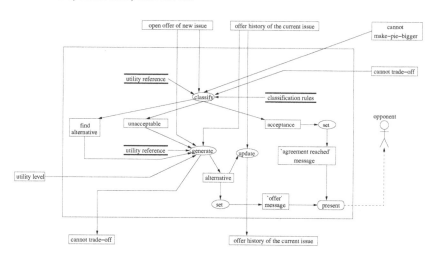

Fig. 4. Graphic specification of the trade-off task

cases: (i) In the case that the other party's offer is acceptable, set the response as an 'agreement reached' message. (ii) In the case that the other party's offer is unacceptable, try to generate a tradeoff solution that can keep the current expected utility level unchanged. If an alternative cannot be generated, then quit the trade-off task (and present a 'cannot trade-off' message to the other party); otherwise update the offer history by adding the alternative in, and set the response as an 'offer' message. (iii) In the case of finding alternative (*i.e.*, the received message is 'cannot make-pie-bigger' or 'cannot trade-off'), try to generate an alternative tradeoff solution that can keep the current expected utility level unchanged. If an alternative cannot be generated, it means that the agent cannot trade-off and so quits the trade-off task (and present a 'cannot trade-off' message to the other party); otherwise update the offer history by adding the alternative in, and set the response as an 'offer' message.

Finally, present the response message to the other party. Notice that the decision making subtask of the trade-off task is simply inference function generate.

3.3   The Subtask of argue

The graphical specification of the argue task is shown in Figure 5. Its situation analysis subtask uses the classification method to classify the received message ('offer' or 'cannot trade-off' or 'cannot argue') into the following two cases: (i) When the received message is 'cannot trade-off' or 'cannot argue', try to find an argument for its offer. If the agent can really find one, then update the offer history by adding the offer with the argument in, and set the response message as an 'offer' message; otherwise it means that the agent cannot argue for its offer and so quits the argue-objectively task (and present the other party a 'cannot argue' message). (ii) When the received message is argumentation, try to prove whether the received offer (with argument) is correct or not according

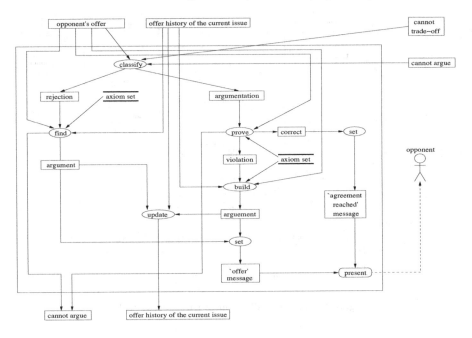

Fig. 5. Graphic specification of the argue task

to the axiom set. (a) If it is correct, then set the response as an 'agreement reached' message. (b) If it is incorrect, then build a counter argument and set the response as an 'offer' message. (c) If it cannot be proved to either correct or incorrect, this means that the agent cannot argue and so quits the argue-objective task (and presents the other party a 'cannot argue' message).

Finally, present the response message to the other party.

3.4   The Subtask of concede

The graphical specification of the concede subtask is shown in Figure 6. Its situation analysis subtask uses the classification method [2] to classify the received message ('offer' or 'cannot argue' or 'cannot concede') into the three cases: (i) When the current utility level is less than the concession threshold, this means that the agent cannot concede any more and so quits the concede task (and presents the other party a 'cannot concede' message). (ii) When the current utility level is not less than the concession threshold, reduce the current utility level. Further, try to generate a solution at the current utility level. If the solution can be generated, then set the response as an 'offer' message and update the offer history (of the issue being negotiated) by adding the new solution in; otherwise although the utility level is not less than the threshold, the agent is still not able to concede and so quits the concede task (and presents the other party a 'cannot concede' message). (iii) When the other party's offer is acceptable, set the response as an 'agreement reached' message.

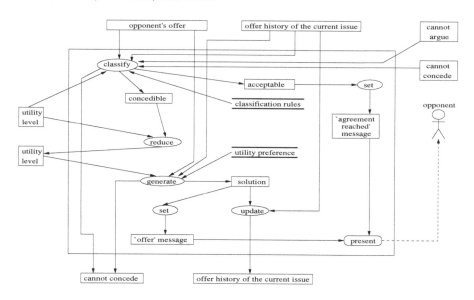

Fig. 6. Graphic specification of the concede task

Finally, present the response message to the other party. Notice that the decision making subtask of the concede task is actually comprised of inference functions reduce and generate.

4 Computer Wholesale Scenario

In this section, we first give a simplified (without losing generality) scenario of computer wholesale negotiation,[4] then describe the operation of our automated principled negotiation model in realising the negotiation.

Suppose Chris Jennings acting on behalf of Hewlett Packard company tries to sell, by bulk, some of HP Pavilion DV6 15.6 Inch 750GB 2.3GHz i3 Laptop to Michael Luo. Michael Luo acts on behalf of a computer retailer that retails computers to university students in Singapore.

The above negotiation is summarised in Table 1. For the issue of *delivery time*, the make-pie-bigger method is used. Initially the buyer wants the computers are delivered at the end of July or the beginning of August. The seller thinks it is too early. Then they explore the interest behind the position of the buyer and find it is because the buyer wants to sell the computers to university students from 1 September (the first date of a new academic year). Thus, they find an alternative position that can also meet the interest: the seller can deliver within

---

[4] In the field of English language learning, there are some text books (e.g., [12]), which give out many examples of *real* negotiation dialog. Although our example here is adapted from these examples of real negotiation, it does lose generality in practical applications.

Table 1. The Summary of the negotiation issues, the initial positions, the final agreement, and the method in the scenario of laptop wholesale

Issues	Seller	Buyer	Agreement	Approaches
Delivery time	Too early	End of July / Beginning of August	End of August, late fine S$20 per day per unit, 30% of saved storage fee to seller	Make-pie-bigger
Price	S$1,500	30% discount	25% discount	Trade-off & concession
Payment Terms	Immediately	1/3 in three installments	1/3 in three installments	Trade-off
Quantity	20	100	100	Trade-off
Insurance	Only WPA covered	WPA & TPND (covered by price)	Only WPA covered	Argue objectively

the last week of August but will be fined S$20 per unit per day if it is delayed, and meanwhile the seller can get the 30% of the storage fee the buyer saves. So, the buyer's interest is guaranteed and it saves 70% of storage fee. The agreement makes both the sides at a position better than their initial positions (*i.e.*, the pie has been made bigger). The knowledge that the buyer uses is as follow:

(i) *If sell the computers from 1 September and there are uncertain factors for the delivery time then the delivery time should be the end of July or the begin of August.*

(ii) *If sell the computers from 1 September and delayed delivery is fined S$20 per unit per day then it is accepted to deliver the computer within the last week of August.*

The knowledge that the seller uses is as follow:

(i) *If the delivery time is just before the time the buyer starts to sell the computers then the buyer can save some storage fee.*

(ii) *If the buyer can save some storage fee and the delayed delivery will be fined then ask for 30% of the storage fee saved.*

For the issues of *price, payment terms*, and *quantity*, the tradeoff method and the concession method are used. First, the seller likes to offer S$1,500 per unit for 20 sets. By making tradeoff, since the buyer can buy 100 sets that is much more than 20 sets, it wants 30% discount. The seller makes a trade-off as well: 30% discount for 100 sets is fine but the buyer needs to pay all immediately. The buyer cannot accept this and thus proposes a plan of three installments. The seller responds with a tradeoff of 20% discount for the installment plan. Finally, both concede 5% and reach the agreement: the unit price is $1500 \times (1 - 25\%)$, the quantity is 100 sets, and the payment term is a plan of three installments (each is 1/3 of total). The knowledge that the buyer uses for making a tradeoff is as follow:

If the seller wants S\$1,500 per unit for 20 sets then increase the quantity to 100 sets but asks for 30% discount and buy on a three-installments plan.

The knowledge that the seller uses for making a tradeoff is as follow:

(i) *If the buyer can increase the quantity to 100 sets but asks for 30% discount then 30% discount is fine but the buyer needs to pay all immediately.*

(ii) *If the buyer buys 100 sets on the three-installments plan (each is 1/3 of total) then only 20% discount is fine.*

And both the sides' strategy of concession is each time to concede 1/4 of the difference.

Finally, for the issue of insurance, the agreement is reached by arguing according to the axiom: WPA can be covered on CIF but TPND cannot.

5   Related Work

Negotiation is an important topic in multiple disciplines. In game theory, variouse negotiation models are investigated theoretically [13,14]. However, the theoretic work is based on an assumption that is hardly to be true in real world, and so the outcome of game theory is intractable for practical applications [15,16]. In fact, in real negotiation, it is difficult to guess what the opponents are thinking, and also difficult to assume the opponents are rational and will not make any mistakes. Thus, in business and management science, for manual negotiation in real life, a set of theories and methods (e.g., [15,10]), which are independent of game theory, have been developed. In recent years, great effort has been made in on automating such negotiations (see [17,18,19,20] for surveys). This is because automated systems can negotiate faster even with hundreds or thousands of participants at the same time, but need not be co-located in space and time; and the automated systems can perform in a professional and patient manner, without imparting the emotions or frustrations of a human negotiator [2]. In order to facilitate the applications of automated negotiation in a wide range of domains, [2] also proposes a methodology of knowledge-based software engineering for developing negotiating agents efficiently.

An automated negotiation system can be viewed as a point in multi-dimension space. These dimensions include: (1) The number of issues/attributes to be negotiated (single-issue/attribute or multi-issue/attribute). For single-issue negotiation systems, the focus is on how to design concession strategies. For multi-issue negotiation, the main challenges are how to make a tradeoff between different issues [21,22,23,24,25] and how to find win-win agreements. Some examples of multi-issue negotiation are agenda-based negotiation [26,27], mediator based negotiation [28,29], and learning-based approaches [30,31]. (2) The number of parties involved in the negotiation (bilateral or multilateral). For multilateral negotiation systems, some examples of research topics are negotiation partner selection and how the relationship among partners impact their negotiation behavior. (3) The nature of the negotiation environment (static or dynamic). In a static environment, the number of negotiating agents are fixed, while in a dynamic environment, negotiating agents can enter into or leave off at any point during the course of negotiation. The dynamic nature can also by referred to random factors that can impact the issues to be negotiated. (4) The relation of negotiating parties (competitive, semi-competitive, or cooperative). (5) The negotiation style (position or principled approaches). (6) The technologies used in the negotiation (fuzzy reasoning, utility, constraint, case based reasoning, machine learning, searching, and so on). For example, in [32,33,4,34,35,36,37], the methods of fuzzy reasoning is employed to decide how to make proper concession in dynamic environment of negotiation. (7) The application domains such as service-oriented computing [38,39], the Grid [40,41], peer-to-peer systems [42,43], pervasive computing [44,45] and e-business [17,46]. Actually, it has been argued that such a negotiation is the standard mode of interaction in all computational systems that could be composed of autonomous agents [47].

However, few automated negotiation models are based on principled negotiation approaches [10] although there is still much work about how human beings should apply principled negotiation (e.g., [48,49]). A well-known example is an accommodation renting problem of negotiation in which the buyer's requirements and preferences with respect to the desired accommodation are represented as fuzzy constraints [1]. However, the system does not used the *make-pie-bigger* method, its argumentation function is just implemented by the *reward* method, and it is designed for retailers; while our work in the current paper employs the *make-pie-bigger* method, its argumentation function is based on more complicated reasoning, and it is designed for more complicated wholesalers. Other examples include principled negotiation model for air traffic management [50], but it is not for business contract negotiation. Recently, some researchers develop a decision support system for principled negotiation [51], but it is not an automated negotiation system.

6   Conclusions

The motivation of this paper is two-fold. On the one hand, we aim at developing an automated model for the valuable manual negotiation approach. On the other hand, we use this to show that knowledge engineering methodology KEMNAD [2] can work well in building complex negotiation systems. In particular, the important aim of identifying templates (inference library) in Sections 3 and 4 in [2] is to make it possible to build hybrid negotiation models; the reason why the principled negotiation model is put into the paper is just for proving that our methodology has really achieved this aim. Instead, the principled negotiation model can be viewed as a hybrid one since it is made of four templates: classification, diagnosis, deduction and utility, which are identified in KEMNAD [2]. In the future, we can build more various negotiation systems according to KEMNAD.

Acknowledgements. The authors would like to thank the anonymous referees for their comments, which helped us to improve the paper a lot. Moreover, this paper is partially supported by National Natural Science Foundation of China (No. 61173019), Bairen Plan of Sun Yat-sen University, Fundamental Research Funds for the Central Universities in China, and Major Projects of the Ministry of Education (No. 10JZD0006) China. Finally, Kwang Mong Sim gratefully acknowledges financial support from the School of Computing at the University of Kent for supporting the visit of Xudong Luo to carry out this research from 15 October 2012 through 14 December 2012.

References

1. Luo, X., Jennings, N.R., Shadbolt, N., Leung, H.F., Lee, J.H.M.: A fuzzy constraint based model for bilateral, multi-issue negotiation in semi-competitive environments. Artificial Intelligence 148(1-2), 53–102 (2003)
2. Luo, X., Miao, C., Jennings, N.R., He, M., Shen, Z., Zhang, M.: KEMNAD: A knowledge engineering methodology for negotiating agent development. Computational Intelligence 28(1), 51–105 (2012)
3. Jennings, N.R., Faratin, P., Lomuscio, A.R., Parsons, S., Sierra, C., Wooldridge, M.: Automated negotiation: Prospects, methods and challenges. International Journal of Group Decision and Negotiation 10(2), 199–215 (2001)
4. He, M., Rogers, A., Luo, X., Jennings, N.R.: Designing a successful trading agent for supply chain management. In: Proceedings of the Fifth International Conference on Autonomous Agents and Multi-Agent Systems, Hakodate, Japan, pp. 61–62 (2006)
5. Chang, M., He, M., Luo, X.: Designing a successful adaptive agent for TAC Ad auction. In: Proceedings of the Nineteen European Conference on Artificial Intelligence, pp. 587–592 (2010)
6. Chang, M., He, M., Luo, X.: AstonCAT-plus: an efficient specialist for the tac market design tournament. In: Proceedings of the Twenty-Second International Joint Conference on Artificial Intelligence, vol. 1, pp. 146–151. AAAI Press (2011)
7. Pan, L., Luo, X., Meng, M., Maio, C., He, M., Guo, X.: A two-stage win-win multiattribute negotiation model: Optimization and then concession. Computational Intelligence 29(2) (2013)
8. de la Hoz, E., López-Carmona, M.A., Marsá-Maestre, I.: Trends in multiagent negotiation: From bilateral bargaining to consensus policies. In: Agreement Technologies, pp. 405–415. Springer (2013)
9. Bench-Capon, T.J.M., Dunne, P.E.: Argumentation in artificial intelligence. Artificial Intelligence 171(10-15), 619–641 (2007)
10. Fisher, R., Ury, W., Patton, B.: Getting to yes: Negotiating an agreement without giving in. Penguin Books (1991), This is the revised 2nd edn. The 1st edn. unrevised, is published by Houghton Mifflin (1981)
11. Turel, O., Yuan, Y.: Online dispute resolution services: Justice, concepts and challenges. In: Kilgour, D.M., Eden, C. (eds.) Handbook of Group Decision and Negotiation. Advances in Group Decision and Negotiation, vol. 4, pp. 425–436. Springer Netherlands (2010)
12. Feng, Q.: Interactive multimedia: Examples of business negotiations English Encyclopaedia. China Aerospace Press (2009)
13. Nash, J.: The bargaining problem. Econometrica 18(2), 155–162 (1950)
14. Rubinstein, A.: Perfect equilibrium in a bargaining model. Econometrica 50(1), 97–109 (1982)

15. Raiffa, H.: The Art and Science of Negotiation. Harvard University Press, Cambridge (1982), sixteenth printing (2002)
16. Lai, G., Li, C., Sycara, K., Giampapa, J.: Literature review on multi-attribute negotiations. Carnegie Mellon University, Robotics Institute, Technical Report CMU-RI-TR-04-66 (2004)
17. He, M., Jennings, N.R., Leung, H.F.: On agent-mediated electronic commerce. IEEE Transactions on Knowledge and Data Engineering 15(4), 985–1003 (2003)
18. Lomuscio, A.R., Wooldridge, M., Jennings, N.R.: A classification scheme for negotiation in electronic commerce. International Journal of Decision and Negotiation 12(1), 31–56 (2003)
19. Rahwan, I., Ramchurn, S.D., Jennings, N.R., McBurney, P., Parsons, S., Sonenberg, L.: Argumentation-based negotiation. The Knowledge Engineering Review 18(4), 343–375 (2004)
20. Wellman, M., Greenwald, A., Stone, P.: Autonomous Bidding Agents: Strategies and Lessons from the Trading Agent Competition. MIT Press, Cambridge (2007)
21. Faratin, P., Sierra, C., Jennings, N.R.: Using similarity criteria to make issue trade-offs in automated negotiations. Artificial Intelligence 142(2), 205–237 (2002)
22. Luo, X., Jennings, N.R., Shadbolt, N.: Knowledge-based acquisition of trade-off preferences for negotiating agents. In: Proceedings of the Fifth International Conference on Electronic Commerce, Pittsburgh, USA, pp. 138–144 (2003)
23. Luo, X., Jennings, N.R., Shadbolt, N.R.: Acquiring tradeoff preferences for automated negotiations: A case study. In: Faratin, P., Parkes, D.C., Rodríguez-Aguilar, J.-A., Walsh, W.E. (eds.) AMEC 2003. LNCS (LNAI), vol. 3048, pp. 37–55. Springer, Heidelberg (2004)
24. Luo, X., Jennings, N.R., Shadbolt, N.: Acquiring user tradeoff strategies and preferences for negotiating agents: A default-then-adjust method. International Journal of Human Computer Studies 64(4), 304–321 (2006)
25. Luo, X.: The evaluation of a knowledge based acquisition system of fuzzy tradeoff strategies for negotiating agents. In: Proceedings of the 14th Annual International Conference on Electronic Commerce, pp. 157–158. ACM (2012)
26. Fatima, S., Wooldridge, M., Jennings, N.R.: An agenda-based framework for multi-issue negotiation. Artificial Intelligence 152(1), 1–45 (2004)
27. Fatima, S.S., Wooldridge, M., Jennings, N.R.: On optimal agendas for package deal negotiation. In: Proceedings of the Tenth International Conference on Autonomous Agents and Multiagent Systems. International Foundation for Autonomous Agents and Multiagent Systems, vol. 3, pp. 1083–1084 (2011)
28. Lai, G., Li, C., Sycara, K.: A general model for Pareto optimal multi-attribute negotiations. In: Ito, T., Hattori, H., Zhang, M., Matsuo, T. (eds.) Proceedings of the Second International Workshop on Rational, Robust, and Secure Negotiations in Multi-Agent Systems. SCI, vol. 89, pp. 59–80. Springer, Heidelberg (2006)
29. Chalamish, M., Kraus, S.: Automed: An automated mediator for multi-issue bilateral negotiations. Autonomous Agents and Multi-Agent Systems 24(3), 536–564 (2012)
30. Hindriks, K., Tykhonov, D.: Opponent modelling in automated multi-issue negotiation using Bayesian learning. In: Proceedings of the Seventh International Conference on Autonomous Agents and Multi-Agent Systems, pp. 331–338 (2008)
31. Yu, C., Ren, F., Zhang, M.: An adaptive bilateral negotiation model based on Bayesian learning. In: Ito, T., Zhang, M., Robu, V., Matsuo, T. (eds.) Complex Automated Negotiations. SCI, vol. 435, pp. 75–93. Springer, Heidelberg (2013)

32. He, M., Leung, H.F., Jennings, N.R.: A fuzzy logic based bidding strategy in continuous double auctions. IEEE Transactions on Knowledge and Data Engineering 15(6) (2003)

33. He, M., Jennings, N.R.: Designing a successful trading agent: A fuzzy set approach. IEEE Transactions on Fuzzy Systems 12(3), 389–410 (2004)

34. Fu, L., Feng, T.: A fuzzy reasoning based bidding strategy for continuous double auctions. In: Proceedings of 2007 IEEE International Conferences on Control and Automation, pp. 1769–1774 (2007)

35. Ma, J., Goyal, M.L.: A fuzzy bidding strategy in automated auctions using agent's perspective. In: Proceedings of 2008 International Conferences on Computational Intelligence for Modelling, Control and Automation, pp. 907–911 (2008)

36. Jain, V., Deshmukh, S.: Dynamic supply chain modeling using a new fuzzy hybrid negotiation mechanism. International Journal of Production Economics 122(1), 319–328 (2009)

37. Lin, C.-C., Chen, S.-C., Chu, Y.-M.: Automatic price negotiation on the web: An agent-based web application using fuzzy expert system. Expert Systems with Applications 38(5), 5090–5100 (2010)

38. Cappiello, C., Comuzzi, M., Plebani, P.: On automated generation of web service level agreements. In: Krogstie, J., Opdahl, A.L., Sindre, G. (eds.) CAiSE 2007 and WES 2007. LNCS, vol. 4495, pp. 264–278. Springer, Heidelberg (2007)

39. Koumoutsos, G., Thramboulidis, K.: Towards a knowledge-base framework for complex, proactive and service-oriented e-negotiation systems. Electronic Commerce Research 9(4), 317–349 (2009)

40. Chao, K.-M., Younas, M., Godwin, N., Sun, P.-C.: Using automated negotiation for grid services. International Journal of Wireless Information Networks 13(2), 141–150 (2006)

41. Guan, S., Dong, X., Mei, Y., Wu, W., Xue, Z.: Towards automated trust negotiation for grids. In: Proceedings of the 2008 IEEE International Conference on Networking, Sensing and Control, pp. 154–159 (2008)

42. Koulouris, T., Spanoudakis, G., Tsigkritis, T.: Towards a framework for dynamic verification of peer-to-peer systems. In: Proceedings of the Second International Conference on Internet and Web Applications and Services, pp. 2–12 (2007)

43. Ragone, A., Noia, T., Sciascio, E., Donini, F.: Logic-based automated multi-issue bilateral negotiation in peer-to-peer e-marketplaces. Autonomous Agents and Multi-Agent Systems 16(3), 249–270 (2008)

44. Alcalde Bagüés, S., Mitic, J., Zeidler, A., Tejada, M., Matias, I.R., Fernandez Valdivielso, C.: Obligations: Building a bridge between personal and enterprise privacy in pervasive computing. In: Furnell, S.M., Katsikas, S.K., Lioy, A. (eds.) TrustBus 2008. LNCS, vol. 5185, pp. 173–184. Springer, Heidelberg (2008)

45. Park, S., Yang, S.-B.: An efficient multilateral negotiation system for pervasive computing environments. Engineering Applications of Artificial Intelligence 21(8), 633–643 (2008)

46. Loutaa, M., Roussakib, I., Pechlivanos, L.: An intelligent agent negotiation strategy in the electronic marketplace environment. European Journal of Operational Research 187(3), 1327–1345 (2008)

47. Jennings, N.R.: An agent-based approach for building complex software systems. Comms. of the ACM 44(4), 35–41 (2001)

48. Zwier, P.J.: Principled Negotiation and Mediation in the International Arena: Talking with Evil. Cambridge University Press (2013)

49. Nomura, Y.: Rethinking the method of principled negotiation–1981-2011. Osaka School of International Public Policy, Osaka University, Tech. Rep. (2012)
50. Wangermann, J., Stengel, R.: Principled negotiation between intelligent agents: A model for air traffic management. Artificial Intelligence in Engineering 12, 177–187 (1998)
51. Carneiro, D., Novais, P., Andrade, F., Zeleznikow, J., Neves, J.: Using case-based reasoning and principled negotiation to provide decision support for dispute resolution. Knowledge and Information Systems, 1–38 (2012)

Evolving Model of Emergency Planning System in China

Jing Wang and Lili Rong

Institute of Systems Engineering, Dalian University of Technology, Dalian, China

Abstract. The emergency planning system in China is an emergency plan network which consists of the different levels and different types of plans. In this paper, in order to improve the emergency planning system and seek its best evolving mode, according to the establishment of emergency plans, the ordered mode based on the administrative ranking and the disordered mode based on local demand that drive the establishment of the emergency plan are proposed. Considering the reference relationship between emergency plans, the ordered and disordered evolving network models of emergency planning system are proposed based on complex network. The simulation results indicate that the hierarchical tree structure of the emergency planning system is scale-free. Compared to the ordered evolving model, the disordered evolving model which according to the actual needs have smaller hierarchy, and the operation of the emergency planning system is affected by most emergency plans, not only depend on the high-level emergency plans. The disordered evolving model of the emergency planning system is more stability and long-term effectively.

Keywords: emergency planning system, establishment of the emergency plan, complex network, evolving model.

1 Introduction

In recent years, more and more paroxysmal events have been happening frequently all over the world. So the ability of emergency management has attracted much social attention. Especially in response to disaster, the necessity and importance of emergency plans are reflected fully in the disaster response[1, 2].From 2006, China has promulgated a series of emergency plans, including the national plans, the provincial plans, municipal plans, the district level plans and overall plans, specific plans, department plans. Until 2006, there have 24 thousand emergency plans in China which basically cover all types of public emergencies. At the end of 2007, the sum of emergency plans in China is more than 1.5 million[3].The emergency planning system is formed basically, which reduces the emergency hazards effectively[4].Although the emergency plan system has been formed, the system is not yet completed[5, 6].The plans for some disaster such as earthquake are relatively complete, but the plans for general emergencies are seldom. Therefore, the emergency plan system need to be improved further.

M. Wang (Ed.): KSEM 2013, LNAI 8041, pp. 280–288, 2013.

Most of the previous works about emergency planning system are the descriptions and definitions of emergency plan system and some qualitative analysis and evaluation[7, 8, 9, 10].The research in the view of model to analysis the emergency plan system is seldom. In this paper, according to the establishment of emergency plans, the process of the emergency plans join into the system actually is a process of dynamic evolution of emergency plan system. Thousands of emergency plans form a complex system. Recently, the study of complex networks has attracted increasing attention and become a common focus of many branches of science. Many efforts have been made to understand the evolution of networks, the relations between topologies and functions, and the network characteristics. The study of complex networks provides a new view and method to analysis the evolving of emergency plan system.

Recently, there is a lot of research about the evolving models of complex network. Many real complex systems can be described by complex network models such as social, biological and information systems[11].Among these evaluation models, there are a large number of classical models. Watts proposes the Small World model and Barabsi proposes the BA model[12, 13].Li proposes a novel evolving network model with the new concept of local-world connectivity[14].Tao takes into account the competition between structure and nodes in the specific networks (energy network, transportation network), use competition factor to improve BA model.[15].Banavar et al models the transportation network to minimize transportation costs.[16]. These evolving models of complex network can help people to understand the microscopic formation mechanism of a variety of networks[17].

In this paper, we propose two evolving model of emergency plan system based on complex network. Furthermore, we explore the form mode, structural characteristics and the dynamic evolution of China's emergency plan system to improve the emergency planning system and seek its best evolving mode.

2   Evolution Strategies

There are many works about emergency plan system[3, 18]. It is found that the emergency plan system in China is an interwoven network. In vertical, it includes at least four levels which are nation, province, city and country town. In the transverse, there are also many levels which can be divided into comprehensive plans, special plans, site plans and individual plans in function[19]. This system contains many sub-systems for events, such as emergency plan system for earthquake, for geological disaster[20], for disaster relief[21].It also includes emergency plan system for different administrative areas, such as emergency plan system for flood control and drought in Liaoning Province[22],emergency plan system for flood in Beijing [23].Here we focus on the emergency plan system of the nation, province and city three levels.

In this paper, we model the emergency plan system in vertical (level) and in transverse(type), shown in Fig.1. In vertical, it includes three levels. In transverse, the public emergency is divided into four categories which are natural disaster, accidents, public health events and social security events. Each category is divided into 19 sub-categories.

Different levels and categories of emergency plans need to be normalized and should be fully collected and refer to the existing plans to ensure coordination and consistency. Therefore, consider each emergency plan as the node, and the relationship of cross-reference between plans as the edge, which constitute a network model of emergency plan system. Based on the actual situation of es-

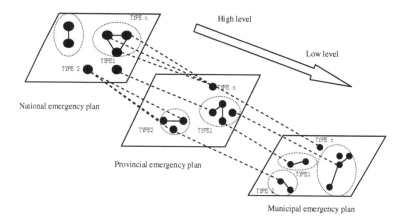

Fig. 1. Model of the emergency planning system

tablishment, the ordered model based on the administrative ranking and the disordered mode based on local demand are proposed. The ordered mode means the lower level plans will be established based on the higher level plans in the same type,if not, the lower level plans will not established. In the same category, provincial emergency plans always are established after the national emergency plans and before municipal emergency plans [7].

On the other hand, the need of different regions always is different so that the establishment of plans is not all driven by the level of administrative. For example, if it is possible to occur a terrorist attacks in a certain area, but there isnt any response plan for terrorist attacks. In order to response such emergency effectively, the plan should be established. The establishment of emergency only depends on the actual demand, no matter there are higher level plan or not.

3   Model and Method

3.1   The Ordered Model

According to the ordered evolution strategy, the ordered evolving model of emergency plan system is:

1. Initial condition: network starts with 21 fully connected nodes which are 21 national plans. Define the property of the emergency plan as $\eta = (\varepsilon, \alpha, \beta).\varepsilon$ is the level of the emergency plan,national plans is $1.\alpha = 1$ 19 means the type of the plans. β means the administrative area.

2. In each time interval, set the priority probability $p1$ and $p2.r \in [0,1]$.If $r < p1$ skip to 3.If $r > p2$,skip to 5. If $p1 \leq r \leq p2$, skip to 4.

3. If the new plan is national plan, the property vector of the node is $(\varepsilon, \alpha, \beta)$, $\varepsilon = 1$, $\beta = 0$, $\alpha \in [1,19]$. Consider the type of the new node i is α_i, the type of existing node j is α_j, if $\alpha_i = \alpha_j$, connect node i and j.

4. If the new node is provincial plan, $\varepsilon = 2$, $\alpha \in [1,19]$.There are 34 provinces in China, so we set β as a random integer between 1 to 34 represents the province of the new plan. The type of the new node i is α_i, the province is β_i, G is the set of all national emergency plans, $G(\alpha_i)$ is the set of nodes with the same typeα_i in G.If $G(\alpha_i) = \emptyset$, regenerate the new node, else connect node i with all the nodes in $G(\alpha_i)$.Suggest P is the set of all provincial plans.P_i is the province of plan i. $P_i(\alpha_i)$ is the set of nodes with the same type α_i in P_i. If $P_i(\alpha_i) \neq \emptyset$, connect node i with all the nodes in $P_i(\alpha_i)$. $P(\alpha_i)$ is the set of nodes with the same type α_i in P. If $P(\alpha_i) \neq \emptyset$, connect node i with the node of largest degree in $P(\alpha_i)$.

5. If the new node is municipal plan.$\varepsilon = 3$, $\alpha \in [1,19]$. β is the city that the plan belong. There are about 333 municipal administrative units. Set β as the decimal and $\beta \in [1,34]$. Set $N1$ is an integer, the new node i belong to the city β_i if $N1-1 < \beta_i < N1$, β_i belongs to the province $\beta = N1$. Suggest the type of i is α_i, and belongs to the city β_i, β_i belongs to the province P_i. $P_i(\alpha_i)$ is the set of nodes with the same type α_i in P_i. If $P_i(\alpha_i) \neq \emptyset$, connect node i with all the nodes in $P_i(\alpha_i)$. Suggest C is the set of municipal emergency plans. The city of i is C_i. $C_i(\alpha_i)$ is the set of nodes with the same type α_i in C_i. If $C_i(\alpha_i) \neq \emptyset$, connect node i with all the nodes in $C_i(\alpha_i)$. $C(\alpha_i)$ is the set of nodes with the same type α_i in C. If $C(\alpha_i) \neq \emptyset$, connect node i with the node of largest degree in $C(\alpha_i)$.

3.2   The Disordered Model

According to the disordered evolution strategy, the ordered evolution model of emergency plan system is:

1. Same as step 1 in the ordered model.
2. Same as step 2 in the ordered model.
3. Same as step 3 in the ordered model.
4. If the new node is provincial plan,$\varepsilon = 2$, $\alpha \in [1,19]$. Set β as a random integer between 1 to 34 represents the province of the new plan. The type of the new node i is α_i,the province is β_i, G is the set of all national emergency plans, $G(\alpha_i)$ is the set of nodes with the same typeα_i in G. If $G(\alpha_i) \neq \emptyset$ connect node i with all the nodes in $G(\alpha_i)$. Suggest P is the set of all provincial plans. P_i is the province of plan i. $P_i(\alpha_i)$ is the set of nodes with the same type α_i in P_i. If $P_i(\alpha_i) \neq \emptyset$, connect node i with all the nodes in $P_i(\alpha_i)$.

$P(\alpha_i)$ is the set of nodes with the same type α_i in P. If $P(\alpha_i) \neq \emptyset$, connect node i with the node of largest degree in $P(\alpha_i)$.

5. If the new node is municipal plan. $\varepsilon = 3, \alpha \in [1, 19]$. β is the city that the plan belong. Set β as the decimal and $\beta \in [1, 34]$. Suggest the type of i is α_i, and belongs to the city β_i, β_i belongs to the province P_i. $P_i(\alpha_i)$ is the set of nodes with the same type α_i in P_i.If $P_i(\alpha_i) \neq \emptyset$, connect node i with all the nodes in $P_i(\alpha_i)$. Suggest C is the set of municipal emergency plans. The city of i is C_i. $C_i(\alpha_i)$ is the set of nodes with the same type α_i in C_i. If $C_i(\alpha_i) \neq \emptyset$, connect node i with all the nodes in $C_i(\alpha_i)$. $C(\alpha_i)$ is the set of nodes with the same type α_i in C.If $C(\alpha_i) \neq \emptyset$, connect node i with the node of largest degree in $C(\alpha_i)$.

4   Experiments and Results

in this section we will use tools of complex network to analyze the structural characteristics of Chinas emergency plan system, in order to seek its best evolution mode and make the emergency planning system more stable and more effective. Degree can represent the influence and importance of the emergency plan. The greater the degree of the plan is, the greater influence of the plan is. The degree distribution can show the characteristics of the whole system. Cluster- ing coefficient is used to measure the aggregation of the system. The larger the clustering coefficient is, the closer the plans related and more stable the system is.

4.1   Degree Distribution

Set $p1 = 0.003, p2 = 0.3, N = 5000$. In simulation, it is found that the probability of nodes with degree zero is large. Therefore, we use the method in[24].

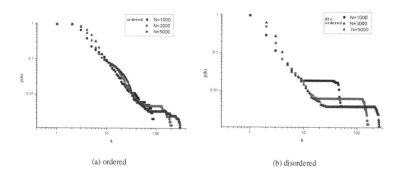

(a) ordered   (b) disordered

Fig. 2. Degree distribution in the two evolving model

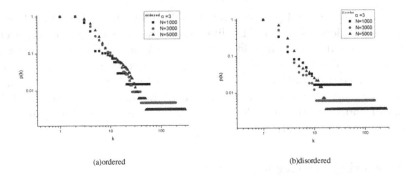

(a)ordered   (b)disordered

Fig. 3. Degree distribution in the two sub-nets of the same type emergency plans

Fig 2 and Fig 3 show that degree distribution of the emergency plan system is approximately power-law distribution[25, 26].It means that the status of each plan is different. Minority of the node is authoritative which have great impact to the emergency plan system this is consistent with the reality. In the ordered model, the long-tail appears when $k = 100$,but it is small in the disordered model. So, compared with ordered model, the number of nodes with large degree reduces, and the number of nodes with small degree increases in the disordered model. The heterogeneity of network will gradually reduce which means not only the national plan but also some lower level plans are very important in the system.

In vertical, it includes three levels. Here we analysis the degree distribution of the plans in each level.

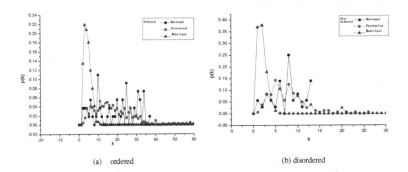

(a)   ordered   (b) disordered

Fig. 4. Degree distribution of the network in different levels

Fig 4 shows the degree distribution of the network in different levels of two models. It is found that degree distribution of the network in each level is Poisson distribution. That means the emergency plans in the same level are equal and the influence is almost the same. It can be seen in the ordered model, most of the degree of the national plan nodes is $k = 20 - 30$, but the provincial level is $k \approx 15$, the city level is $k \approx 5$. The difference between each level is large. Once one of the national plans is invalid, it will seriously affect the whole emergency plan system. In disordered model, the difference between each level is not so large, the degree distribution almost is $k = 2 - 10$ it means that the important plans are not only national plans but also contain the plans in each level. The majority of plans affected the effectiveness of the emergency plan system, that making the system more stable.

4.2   Clustering Coefficient

Clustering coefficient C is an important parameter to measure the collectivize level in the networks. Consider C_i is the Clustering coefficient of node i. Suggest the degree of i is k_i,it means there are k_i nodes is the neighbor node of i. $E_i$is the number of edges between the k_i nodes.Then the clustering coefficient of node i is:

$$C_i = \frac{2E_i}{k_i(k_i - 1)} \tag{1}$$

The clustering coefficient of network is:

$$C = \frac{1}{N} \sum_{i=1}^{n} C_i \tag{2}$$

Fig 5 shows clustering coefficient of the network in two models. It is found that the clustering coefficient increases with the increasing of the number of nodes. The coefficient increases of two models are both large, but the disordered model is smaller than ordered model in the same size. Therefore, when a plan fails, the disordered model is more stable.

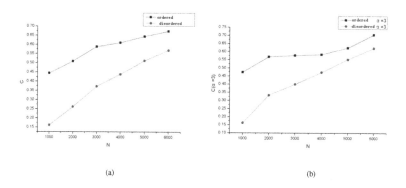

(a)   (b)

Fig. 5. Clustering coefficient of the network.(a) is the Clustering coefficient of whole network.(b) is Clustering coefficient of the subnet of $\alpha = 3$.

5   Conclusions

In this letter, we analysis the emergency plan system and the establish mechanism of emergency plan. The attributes of emergency plans are defined by level, type and administrative area. The method of complex network is used in the research. The ordered mode based on the administrative ranking and the disordered mode based on local demand that drive the establishment of the emergency plan are proposed. Considering the reference relationship between emergency plans, the ordered and disordered evolving network models of emergency planning system are proposed. The simulation results indicate that the hierarchical tree structure of the emergency planning system is scale-free property and the network of the same level plans is like random network. Compared to the ordered evolving model, the disordered evolving model which according to the actual needs have smaller hierarchy and smaller clustering coefficient, the operation of the emergency planning system is affected by most emergency plans, not only depend on the high-level emergency plans. The disordered evolving model of the emergency planning system is more stability and more long-term effectively. Therefore, the emergency plans must be established according to the need of each region, not only dependent on the level of administrative. The evolving of the emergency planning system remains to be further research to improve the emergency planning system and seek its best evolving mode.

Acknowledgments. This work is partially supported by the National Natural Science Foundation of China(Grant No. 91024003 and 91024031), the Major Program of the National Natural Science Foundation of China(Grant No. 71031002)and the National Science & Technology Pillar Program(Grant No. 2011BAH30B01-03).

References

[1] Wang, F., Xu, Z.: Study on compiling technique for emergency rescue protocol of accident in enterprise. China Safety Science Journal 15, 101–105 (2005)

[2] Perry, R.W., Lindell, M.K.: Preparedness for emergency response: guidelines for the emergency planning process. Disasters 27, 336–350 (2003)

[3] Li, H.: How to improve the effectiveness of emergency plan. Modern Occupational Safety 80, 70–73 (2008)

[4] Zan, C.: Theory and method to improve the emergency management system in China. Journal of Political Science 3, 92–98 (2009)

[5] Shan, C., Zhou, L.: From SARS to snowstorm:reflection and experience of the development of the emergency management system in China. Gansu Social Sciences 5, 40–43 (2008)

[6] Zhong, K., Zhang, J.: The preparation and management of emergency plan. Gansu Social Sciences 3, 240–243 (2006)

[7] Jiang, Q.: Study on earthquake emergency predetermined plan system in our country (2008)

[8] Du, J., Wu, Z.: Present condition analysis of earthquake predetermined plan system in China. Journal of Institute of Disaster Prevention 13, 71–74 (2011)

[9] Dong, C., Wang, W., Yang, P.: Emergency plan system ontology and its application. Computer Engineering and Applications 463, 235–238 (2010)

[10] Xie, Y., Zhu, C., Zhou, G., Wang, L.: Study on emergency response plan system. Journal of Safety Science and Technology 6, 214–218 (2010)

[11] Newman, M.E.J.: The structure and function of complex networks. SIAM Review 45, 167–256 (2003)

[12] Watts, D.J., Strogatz, S.H.: Collective dynamics of small world networks. Nature 393, 440–442 (1998)

[13] Barabási, A.L., Albert, R.: Emergence of scaling in random networks. Science 286, 509–512 (1999)

[14] Li, X., Chen, G.: A local-world evolving network model. Phys. A 328, 274–286 (2003)

[15] Tao, S., Guo, C.: BA Extended Model Based on the Competition Faetors. In: Proeeedings of the 2008 Workshop on Power Electronies and Intelligent Transportation System, pp. 561–567 (2008)

[16] Banavar, J.R., Maritan, A., Rinaldo, A.: Size and form in efficient transportation network. Nature 399, 130–132 (1999)

[17] Barabási, A.L.: Scale-Free Networks: A decade and beyond. Science 325, 412–413 (2009)

[18] Shan, C.: Strengthen emergency plan system and improve the ability to respond to risk. Modern Occupational Safety 70, 72–75 (2007)

[19] Liu, G., Liu, T.: Study on administrative levels and structure of emergency planning for Unban accidence. In: The First Annual Meeting of the Occupational Safety and Health Association Occupational Safety and Health Forum, pp. 284–287 (2004)

[20] Liao, S., Jiang, H.: Discussion on management of emergency plan for paroxysmal geological hazards. The Chinese Journal of Geological Hazard and Control 20, 109–112 (2009)

[21] Yang, W., Yang, Z.: Thinking of the establishment of mechanism for a major disaster emergency. Disaster Reduction in China 8, 20–23 (2009)

[22] Kang, J., Xiong, J., Bao, J.: The build and practice of flood plan system of Liaoning province. Water Resources and Hydropower of Northeast China 26, 51–53 (2008)

[23] Wu, F.: The preparation of flood emergency response plan in Beijing. China Flood and Drought Management 5, 47–51 (2007)

[24] Li, L., Alderson, D., Doyle, J.C., Willinger, W.: Towards a theory of scale-free graphs: definitions, properties, and implications. Internet Math. 2, 431–523 (2005)

[25] Amaral, L.A.N., Scala, A., Barthélémy, M., Stanley, H.E.: Classes of small-world networks. PNAS 97, 11149–11152 (2000)

[26] Liu, Z.H., Lai, Y.C., Ye, N., Dasgupta, P.: Connectivity distribution and attack tolerance of general networks with both preferential and random attachments. Physics Letters A 303, 337–344 (2002)

Navigability of Hierarchical-Tag-Based Folksonomies

Tao Wang, Huiyu Liu, and Haoxiang Xia

Institute of Systems Engineering
Dalian University of Technology, Dalian, 116024, P.R.C
{herbawong,hyliu}@mail.dlut.edu.cn,
hxxia@dlut.edu.cn

Abstract. Although the effectiveness of social-tagging-systems has been illustrated in information retrieval and organization, there still are some urgent requirements including reducing redundancy and saving semantic relations for them. A natural solution is to organize tags in a hierarchical means. In this paper, we propose the idea of hierarchical-tag or HTag for short, and verify the improvements in the navigability of the HTag-base folksonomy by computational stimulations on the basis of a delicious.com dataset. Our findings may shed light on designing social-tagging-systems to improve the navigability.

Keywords: Folksonomy, Hierarchical Tag, Social Tagging System, Navigability.

1 Introduction

In social tagging systems, users organize information by using tags to annotate various resources such as URLs on Del.icio.us, photos on Flickr, and academic references on CiteULike. Users can also socially share their annotations to generate a "folksonmy", or the folk taxonomy created by massive users. Folksonomies are represented as tripartite graphs $F \subseteq U \times T \times R$ [1], where U, T, and R are three finite sets, i.e. the user set, the tag set and the resource set, respectively.

In the last few years, such social tagging systems have been widespread in all sorts of Web applications. However, today's mainstream tagging systems, which are based on flat tags, may encounter the difficulty of tag management with scaling up of the tagging system. To cope with this problem, we in our previous work have introduced the idea of hierarchical tagging and developed a reliable method to merge this kind of meta-structures or hierarchical tags into a folksonomy [2]. That work paved the way to utilize hierarchical-tags (HTag) in actual social tagging systems. But the applicability of the proposed HTag method has not been fully examined. In particular, more theoretical and pragmatic investigations are deserved to test the abilities of the hierarchical-tag-based-folksonomies in the tasks of navigation and search.

In order to test the navigability of the proposed HTag-system, we use an evaluation framework, which is based on Helic et al.'s method [3], to simulate a user's navigation guided by the two different folksonomies that are respectively constructed by using

M. Wang (Ed.): KSEM 2013, LNAI 8041, pp. 289–297, 2013.

Heymann's hierarchical clustering algorithm [4] from the flat-tags and hierarchical-tags, respectively. As a work in progress report, in this paper we are to narrate our experiments and results, which primitively illustrate the advantage of the suggested hierarchical-tagging method.

2   HTag-Based Folksonomy and Its Navigability Problem

In [2], we suggested replacing flat-tags with hierarchical-tags (HTag for short) to improve the information organization in the tagging system. A hierarchical tag can be expressed as: *ParentTag*→*{ChildTag1, ChildTag2}*.

The benefits for information organization with hierarchical tags are apparent. Assume that a user A defines a hierarchical tag, a parent-to-child relation *{ParentTag1, ChildTag1}* and a sibling relation *{ChildTag1, ChildTag2}* can then be introduced and these two semantic structures are incorporated into the user's tag system. This tag system is significantly different from the classical flat-tag-system, from which only the co-occurrence relations can be extracted. The reserved semantic relations are useful for guiding information access. However, the introduction of the hierarchical tags would on the other hand bring problems for the management of tags at the collective level. Two problems are particularly critical for the applicability of hierarchical tagging in actual systems. First, structural and semantic conflicts are inevitable, when millions of users produce huge amount of HTags. It is therefore a critical problem to generate a consistent folksonomy from those inconsistent tags. This problem has been partly resolved in [2]. Second, there is a need to test the navigability of the suggested hierarchical tagging system, i.e. to examine the effectiveness and efficiency to retrieve resources through a navigation process that is directed by the folksonomy generated from the hierarchical tags of massive users.

Navigation is the common mode for a user to use the folksonomies to retrieve information. As illustrated in Fig.1, the user wanders along the tag network to seek for resources of interest. The tags that are accessed during this navigational process constitute the trail or path from a source tag to a sink tag that directly indexes the desired resource, e.g. the path from tag 13 to 1 to 21 to 31 and finally to 33 as demonstrated in Fig. 1. Consequently, a key criterion to evaluate a folksonomy is its performance in guiding such type of tag-to-tag wandering or navigation to reach the desired resource.

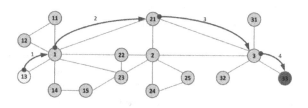

Fig. 1. Illustration of Navigation in A Folksonomy (Source: [3])

The navigability of a folksonomy can then be evaluated by examining whether the folksonomy provides background knowledge to facilitate the prior navigation, i.e. to find a short path from an entrance tag to the destination tag. In this paper, we are to give an evaluation for the suggested hierarchical tagging by comparing the navigability of the folksonomy constructed by the hierarchical-tagging method and that of the folksonomy constructed by the traditional flat-tagging method, by adopting the evaluation framework suggested by Helic et al. [3].

3   Model and Experiments

In accordance with the prior discussion, we use Helic et al.'s [3] evaluation framework to test the navigability of the HTag-based folksonomy. We first generate two folksonomies that are respectively based on the traditional flat-tagging system and the hierarchical-tagging system. Then computer simulations are conducted to mimic a real-world user who is surfing a social tagging system to seek for a desired or destination tag by using a decentralized search strategy. Following Helic et al.'s framework, the adopted search strategy uses an evaluation metric to measure the "relevance" or "semantic similarity" between the destination tag and the neighboring tags of the current tag. The virtual information-seeker (i.e. the simulation program) thus executes tag-to-tag hops, moving from the current tag to one of the neighboring tags that is most relevant to the destination tag according to the given relevance metric. The generated folksonomies are used to provide the tag network and the background knowledge to guide the local search.

With the previous simulation model, the navigability of a folksonomy can then be measured by the success rate in complete a plenty number of randomly-assigned navigation tasks, and the average path length to complete the navigation tasks.

3.1   Measuring the Semantic Relevance between Tags

As previously stated, the suggested navigation process relies on the folksonomy that is generated from the social tags to determine the neighborhood structure of the tags and the background knowledge that measures the semantic relevance between two tags. We use Heymann et al.'s [4] algorithm to generate folksonomy from the social tags. This algorithm is also based on the tag-relevance measure. Thus, we first define the relevance measure between two HTags.

In order to simplify experiments, we in this paper just consider 1-to-1 HTags, indicating that an HTag contains only one child tag. We define the relevance between two tags T_1 and T_2 as by Eq. (1).

$$rel(T_1, T_2) = (1 - \lambda) \cdot cooccurel(T_1, T_2) + \lambda \cdot structrel(T_1, T_2) \tag{1}$$

As shown in Equation (1), the relevance measure is comprised of two parts that are leveraged by a weight parameter λ. The "$cooccurel$" part refers to the relevance between the two tags with respect to their co-occurrence in tagging resources.

Quantitatively it equals the cosine similarity as proposed by Heymann et al. [4]. The second part ("*structrel*") is the structural relevance between two tags connected by the parent-child relation. This structural relevance is further defined by Eq. (2).

$$
structrel(T_i, T_j) = \begin{cases} 1 \\ \dfrac{1 + Num(T_i \to T_j)/Num(T_i \to T_l)}{2}, l \neq j \\ \dfrac{\max(Num(T_i \to T_j), Num(T_j \to T_i))}{Num(T_i \to T_j) + Num(T_j \to T_i)} \end{cases} \tag{2}
$$

Three situations should be taken into account for calculating the structural relevance.

1. If there is no structural conflict between two tags in the dataset, the structural relevance is set to1 and this corresponds to the situation that all the users vote to support the potential semantic relevance between tag i and tag j.
2. If there is another tag l being as the child tag of tag i .This corresponds to a polysemy situation and some disagreements on the relevance among the users are present.
3. If conflicts exist, we calculate the amount of relevant HTags to measure the relevance, As shown in the third part of Eq. (2), if the HTag {twitter→social-media} appeared totally 80 times and {social-media→twitter} 20 times, we should say the structure relevance between the tag i and tag j are 0.8.

3.2   Folksonomy Construction

Based on the previously-defined relevance measure, we use Heymann et al.'s [4] algorithm to construct the folksonomies as background knowledge to steer navigation.

As shown in Eq. (1), when λ=0, the relevance between two tags is reduced to the co-occurrence measure; and the relevance between to flat-tags can be calculated by this co-occurrence measure. When λ>1, the second part of Eq. (1) is taken into account, reflecting the situation of hierarchical tagging as suggested.

In folksonomy construction, we give a simple modification for Heymann et al.'s algorithm. A fixed "root" tag is used to be the absolute root of all the tags, so as to ensure the traversal to the full folksonomy in navigation tasks. The similarity measure suggested by Heymann et al. is also replaced by the HTag relevance mentioned above. Through adjustments of the value of parameter λ, different kinds of folksonomies can be conducted from the same dataset. Correspondingly, the distance between any two tags in a constructed folksonomy can be calculated by using the method suggested in [5].

3.3   Dataset

For our experiments, we extract a dataset from Delicious.com, by acquiring the del.icio.us tags in the period between July 2009 and November 2009. The data source is with about 1.25 million U-T-R triple structures or posts. We defined a post $A= \{URL, tag1, tag2, ...,tagk\}$, where k is a value with a significant probability (0.90) less than or equal to 3. Averagely, a post contains about 3 tags. We lowercase all the tags and remove the typos and the useless tags which either occur only once or can not be identified as a word in dictionaries. Then we extract the tag-to-tag networks by extracting largest connected component from original tag co-occurrence network. We use this dataset as the test data for flat tags.

To compare the performance between flat-tags and hierarchical-tags, we also need an HTag dataset. However, such data is lacking as the idea of hierarchical tags has not widely been adopted in the actual social tagging systems. To overcome this, we artificially generate a set of hierarchical tags from the previous del.ious.us dataset. The tactic to generate the hierarchical tags is to let the more frequently-used tags as the parent tag in a co-occurred tag-pair and the second tag as the child tag, which is with lower frequency to be used in indexing resources.

With the above method, we generate two sets of tags from the same data source of delicious.com, respectively representing flat-tag and hierarchical-tag datasets. Thus, we conduct our simulations on two different folksonomies constructed from the same dataset. The basic statistics of the generated dataset are shown in Table 1.

Table 1. Statistical properties of the Dataset derived from Delicious.com

HTags	14783
Tags	19264
Links	49898
Average Shortest Path Length	3

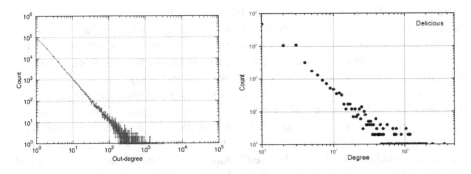

Fig. 2. Degree distributions of a navigable tag-network (Left, Source [5]) and the tag network generated from the delicious.com dataset (Right)

The effective diameter of the tag network is also in a reasonable value area according to [3], indicating that the generated datasets can support navigational tasks.

Fig. 2 shows the degree distributions of the hierarchies depicting the existence of hub nodes. The left part of the figure plots the out-degree distribution of an actual tag-network depicted in [5], while the right part plots the degree distribution of the tag-network that is generated from the decilious.com dataset. The two distributions are highly similar. According to [5], the generated tag network can support the navigational tasks.

Accordingly, the folksonomy generated from the hierarchical tags is with a structure as illustrated in Fig. 3, which shows the tag hierarchy of the top 10% frequent HTags. This hierarchy shows a broad and shallow folksonomy that contains 5 hierarchical levels in depth.

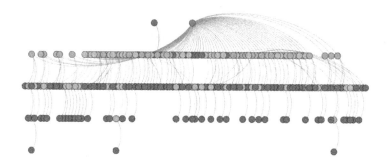

Fig. 3. The folksonomy generated from the top 10% frequent HTags in the dataset

3.4   Simulation Setting

Using the prior datasets, we construct two folksonomies, which respectively represent a flat-tag-based folksonomy and a hierarchical-tag-based folksonomy. Then navigational tasks are tested upon these two folksonomies in order to compare the navigability. To simulate a navigational process under the guidance of a folksonomy, we arbitrarily select a pair of tags $[m, n]$, where m is the source tag and n is the destination tag. Both the source and the destination are set to be low-degree tags to avoid the trivial navigation process between two high degree tags.

With such simulation setting, we run the simulations on those search pairs. Two indicators are used to measure the effectiveness of the navigations in our simulations, i.e. the success rates s and the stretch τ which is the ratio between the simulator's actual hops and the global shortest path (h/l) that connects the source and destination tags. In the simulations, a task is declared failure if its destination tag can't be reached within 20 hops. Any occurrence of loop also leads to a failure.

4   Result Analysis

Based on the previously-described simulation procedure, we design a simulator to test the navigational performance under the two folksonomies generated from the same del.icio.us dataset. 10,000 pairs of tags are selected as the navigational tasks; and the shortest paths between the two tags are recorded.

We compare the navigational efficiency of HTag-based folksonomy with that of the flat-tag-based folksonomy by comparing the success rate s and the ratio h/l. According to Eq. (1), when setting the parameter $\lambda =0$, the folksonomy generated from the dataset reflects the situation of flat social-tagging; when $\lambda >0$, the structural relevance between two tags are taken into account, and the generated folksonomy reflect the situation of hierarchical-tagging. Thus, we compare the navigational performance under the conditions of $\lambda =0$, and $\lambda >0$.

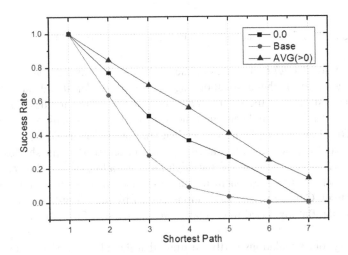

Fig. 4. The success rates of navigations guided by different folksonomies

The comparison of success rate is shown in Fig. 4. The success rate line of "AVG" in Fig. 4 reflects the case of HTag-based folksonomy. This line depicts the mean success rates under $\lambda =0.2$, 0.4, 0.6, 0.8 and 1.0 in different shortest paths (the horizontal axis is the shortest paths between the source and destination tags). The line marked "0.0" reflects the case of navigation guided by the flat-tag-based folksonomy. Finally, the line marked "BASE" plots the success rates of navigation by using the random-walk strategy, without being guided by the background knowledge of the folksonomy. The simulation results show that the folksonomies are useful for directing navigation since the success rates under the random-walk strategy are the lowest. Furthermore, the adoption of the HTag-based folksonomy improves the success rate of navigation.

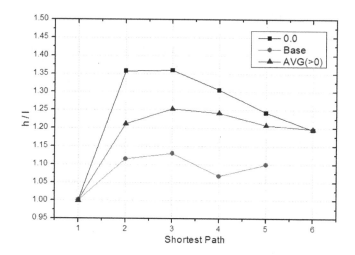

Fig. 5. The *h/l* ratio values of navigations guided by different folksonomies

Fig. 5 illustrates the *h/l* ratio values of successful navigations. This *h/l* ratio depicts the efficiency to reach the destination tag. The case of *h/l*=0 indicates that the simulator successfully find the shortest path between the source and destination tags. Higher values of this ratio indicate more hops to reach the destination and therefore the lower efficiency in completing the navigational tasks. The results show that the navigations with the random-walk strategy are most efficient when the shortest path is less than 5. However, because of the high-rate of failure under this navigation strategy, we cannot draw the conclusion that the random-walk strategy is the most efficient strategy. Comparing the navigations guided by the HTag-based folksonomy with those by the flat-tag-based folksonomy, we can see that former is more efficient than the latter.

In summary, our simulations partly illustrate that the adoption of hierarchical tags may improve information-seeking via tag-to-tag navigation, comparing with the traditional flat tags.

5   Conclusions

This work presents work in progress report for incorporating the idea of hierarchical tags into social tagging systems, following the authors' previous efforts in [2]. The main contribution of this work is to partly illustrate the effectiveness and efficiency of information retrieval by using the suggested hierarchical tagging method. However, our current findings are limited to a single dataset. We are on the way to check the navigability of the suggested hierarchical-tagging system by testing more dataset from different types of actual tagging datasets like Flickr and CiteULike. What's more, a future work is to develop a proof-of-concept prototype of hierarchical tagging system to verify its applicability.

Acknowledgments. This work is supported by Natural Science Foundation of China under Grants 70871016, 71031002, respectively. The authors are grateful for the suggestions from the anonymous referees.

References

1. Mika, P.: Ontologies are us: A unified model of social networks and semantics. Proceedings of the Web Semantics: Science, Services and Agents on the World Wide Web 5(1), 5–15 (2007)
2. Wang, T., Xia, H., Liu, H.: Folksonomy Construction through User-Defined Hierarchical Tags. In: Proceedings of the 4th China National Conference on Social Computing, Beijing (2012)
3. Helic, D., Strohmaier, M., Trattner, C., Muhr, M., Lerman, K.: Pragmatic evaluation of folksonomies. In: Proceedings of the 20th International Conference on World Wide Web, pp. 417–426. ACM Press, New York (2011)
4. Heymann, P., Garcia-Molina, H.: Collaborative Creation of Communal Hierarchical Taxonomies in Social Tagging Systems. Technical Report of InfoLab of Stanford University (2006)
5. Helic, D., Körner, C., Granitzer, M., Strohmaier, M., Trattner, C.: Navigational efficiency of broad vs. narrow folksonomies. In: Proceedings of the 23rd ACM Conference on Hypertext and Social Media, pp. 63–72. ACM Press, New York (2012)

TYG: A Tag-as-You-Go Online Annotation Tool for Web Browsing and Navigation

He Hu and Xiaoyong Du

School of Information, Renmin University of China, Beijing 100872, China
Key Laboratory of Data Engineering and Knowledge Engineering, MOE, Beijing 100872, China
{hehu,duyong}@ruc.edu.cn

Abstract. We present TYG (Tag-as-You-Go) in this paper, a chrome browser extension for personal knowledge annotation on standard web pages. We investigate an approach to combine a *K-Medoid*-style clustering algorithm with the user input to achieve semi-automatic web page annotation. The annotation process supports user-defined tagging schema and comprises an automatic mechanism that is built upon clustering techniques, which can group similar html nodes into clusters corresponding to the user specification. TYG is a prototype system illustrating the proposed approach. Experiments with TYG show that our approach can achieve both efficiency and effectiveness in real world annotation scenarios.

1 Introduction

With the exponential growth of web contents being generated by classical Web, Web 2.0 and Semantic Web services, web users are in dire need of tools that can help them manage the massive and often dynamic web data. However, most web pages are designed for human viewing only and may not be suitable for other purposes. Reusing web content generally requires explicit identification of structural and semantic roles of various content components.

Extracting informative content from web pages has attracted many research efforts in recent years [1,2,3,4,5]. The identification of titles [4], authors [5], and the main article [1,2,3] is studied in the literature. In this paper, we focus on identifying general semantic groups from data-intensive web pages with a hybrid approach which combines user guidance and unsupervised clustering techniques. In particular, we adapt the *K-Medoid* clustering and utilize the users input as the initial centroids. The approach is implemented in a prototype system TYG (Tag-as-You-Go), which is a browser extension to support users to do online annotations during the course of web browsing and navigation.

TYG (Tag-as-You-Go) is a browser based tool for the display, manipulation and annotation of web contents. In addition to classical manual tagging functions, TYG offers novel ways of automatically clustering similar html nodes according to user specifications. Tagging schemas can be interactively manipulated and edited with

M. Wang (Ed.): KSEM 2013, LNAI 8041, pp. 298–309, 2013.

user-defined vocabulary in TYG tool. We present the architecture of the tool, and elaborate the main algorithms it uses in the automatic annotation process.

The original contributions of the presented work are manifold: (1) We propose the "tag as you go" style annotation which seamlessly combines user input with automatic clustering algorithms; (2) We adapt a path based measure with *TagPathDistance* algorithm for comparing the distances among web elements; (3) We describe the efficient *K-MedoidUserInput* algorithm which eliminates the need to calculate cluster centroids; (4) We run extensive experiments showing high efficiency and effectiveness of our approach on 20 different web sites of different categories.

The paper is structured as follows. Section 2 gives a brief overview of related work in the literature. In section 3, we present our approach to support online annotation combining user input with clustering techniques. We illustrate the algorithms behind our approach and discuss their characteristics for the hybrid annotation. Section 4 describes TYG system, including its architecture, user interface, and main functions. Section 5 studies the performance of TYG in real world application scenarios. Both efficiency and effectiveness are examined in this section. We conclude the paper in the final section.

2   Related Work

In this section, we review a number of existing works in the areas of information extraction, semantic annotation, and browser based tools implementation.

Extracting informative content from web pages has attracted many research efforts in the literature. [1,2,3] have studied main article extraction on the Web; [4] focuses on the identification of titles in web pages, [5] investigates authors extraction of web document. Earlier, [6] has proposed a rule learning algorithm LP2, and implemented an automatic annotation module used by some semantic annotation systems [9]. These approaches process on web sites and rely on multiple web pages to induct the general template structures. We focus on in-page annotation in this paper. Our method need only the current web page and does not require information from other web pages.

Our work is also related to semantic annotation [7] or semantic tagging [8] which is a research topic in Semantic Web community Many annotation systems have been developed in recent years [9,10,11]; they are all desktop applications and do not support integration with modern web browsers. Web browsers are probably used more often than desktop applications. We implement our approach with browser based tool to help users to do online annotation while browsing and navigating the Web.

Nowadays all major browsers (IE, Firefox, Chrome etc.) are providing support for extensions (plug-ins) development. Browser extensions are third-party software components that extend the functionality of web content by calling into browser application programming interfaces. Many browser based tools have been implemented in recent years. [12] has developed a plug-in tool for Internet Explorer, providing ink annotations on the web pages. [13] has realized an extension for the Google Chrome web browser that allows a user to generate a WARC file from the current webpage. [14] has implemented a Firefox plug-in to support URL recommendation. There are no related browser extensions to do semi-automatic online annotation. Our system is the first browser extension to help user to tag the web contents semi-automatically.

Automatic annotation has been formulated as a *K-Medoid*-style clustering problem in this paper. Previous studies on *K-Medoid* clustering have focus on the performance issues and comparison between the *K-Means* and *K-Medoid* algorithms. [15] has studied to improve *K-Medoid*'s performance by storing the distance matrix. [16] compares *K-Means* and *K-Medoid* algorithms for normal and uniform distributed data points. The most common realization of *K-Medoid* clustering is the Partitioning Around Medoids (PAM) algorithm. In this paper, we propose a related algorithm *K-MedoidUserInput* which has O(n) complexity.

3   Method

Our method for semi-automatic annotation relies on three algorithms. The **Walk-DOMTree** builds features from HTML documents; the **TagPathDistance** calculates the distances between different tag paths for HTML elements; the **K-MedoidUserInput** uses user input as centroids to cluster other HTML elements.

The **WalkDOMTree** algorithm walks the DOM tree in recursion, from the root node (document element) down to the leaves (tag elements or texts). In traversing, the algorithm records every node's tag path by calling **get_tagpath** function, which is a simple function calculating the tag path from the current node to the root node.

Algorithm WalkDOMTree

Input: A HTML document root element *e*
Output: A list of *tag nodes N* containing each DOM node with various features

01:   *children* ← all children of element *e*
02:   **if** *children*.length = 0 **then**
03:       create new result object *obj*
04:       obj.tagpath ← **get_tagpath**(*e*)
05:       obj.xpath ← **get_xpath**(*e*)
06:       obj.depth ← **get_depth**(*e*)
07:       obj.spatial ← **get_spatial**(*e*)
08:       obj.value ← *e.nodeValue*
09:       obj.type ← *e.nodeType*
10:       *N*.**push(obj)**
11:   **else**
12:       **for each** *child* in *children* **do**
13:               **WalkDOMTree**(*child*)

The distances between different tag paths are calculated with **TagPathDistance** algorithm, which is adapted from *Levenshtein distance*. Levenshtein distance (or *edit distance*) is a string metric for measuring the difference between two string sequences. The Levenshtein distance between two words is the minimum number of single-character edits (insertion, deletion, substitution) required to change one word into the other. When it comes to tag paths, the original Levenshtein distance is not applicable to list structures. We modified the original Levenshtein distance to use lists as the data structure instead of strings. The adapted algorithm **TagPathDistance** can be used to caculate the minimum number of node edits (insertion, deletion or substitution) in the tag path list, so **TagPathDistance** can also be referred to as *node distance* algorithm.

Algorithm TagPathDistance

Input: A pair of Tag Paths $p1$ and $p2$, each path p is a list: [tag1,tag2,tag3, …]
Output: A Distance metric d representing the difference of two elements

01:   **if** $p1$.length = 0 **then**
02:       **return** $p2$.length
03:   **if** $p2$.length = 0   **then**
04:       **return** $p1$.length
05:   $new1 \leftarrow p1;   new2 \leftarrow p2$
06:   $n1 \leftarrow   new1$.shift()
07:   $n2 \leftarrow   new2$.shift()
08:   $d1 \leftarrow$ **TagPathDistance**($new1, p2$) + 1
09:   $d2 \leftarrow$ **TagPathDistance**($p1, new2$) + 1
10:   $d3 \leftarrow$ **TagPathDistance**($new1, new2$) + ($n1$!= $n2$)?1:0
11:   $d = \min(d1,d2,d3)$
12:   **return** d

K-Means and *K-Medoid* are two widely used unsupervised clustering algorithms. Compared to K-Means algorithm, *K-Medoid* algorithm is computationally expensive especially for huge datasets. In K-Means, centroid point can be calculated by simply averaging subcluster points, while in *K-Medoid*, averaging calculation is often not supported. As for our case, we cannot do averaging either on tag paths or on xpaths. *K-Medoid* picks existing data points as centroid points, and it has to calculate every possible pairs of data points to find the centroid points. Park et. al. [15] proposed a new algorithm for *K-Medoids* clustering which runs not worse than the K-Means. The algorithm calculates the distance matrix once and uses it for finding new medoids at every iterative step.

Our **K-MedoidUserInput** algorithm uses user input as the centroids points, and clusters other data points according to these centroids. The user supplied *xpath* is extracted from the HTML DOM tree, so the **05** line of **K-MedoidUserInput** algorithm is garenteed to succeed once and only once in the loop. The *tp* variable is the corresponding tag path of the element whose xpath is *xpath*. For performance, we can keep the results of distance matrix caculated with **TagPathDistance** and use it repeatedly in the algorithm, as in [15], then the complexity of **K-MedoidUserInput** algorithm is O(n), where n is the number of nodes. This result indicates the algorithm will be efficient in complex web pages.

Algorithm K-MedoidUserInput

Input: A list of Tag Nodes N, number of clusters c, user input I, a support threshold s;
      I is of the form: [xpath1:i1, xpath2:i2, …]; i1, i2 … take values from [1, c]
Output: N with each node n's cluster property assigned   with a value from [0, c]

```
01:   for each n in N do
02:      n.cluster ← 0
03:   for each (xpath, i) in I do
04:      for each n in N do
05:         if n.xpath = xpath then
06:            tp ← n.tagpath
07:      for each n in N do
08:         if TagPathDistance(n.xpath, tp) < s then
09:            n.cluster ← i
10:   return N
```

4   TYG System

TYG is an extension for the Google Chrome web browser that allows a user to annotate content from web pages. In addition to creating annotations by hand, the extension provides options that automatically cluster similar html DOM nodes and tag
these nodes according to existing annotations.

4.1   Architecture of TYG

Our goal with the annotation tool is to extend a browser with a user interface to provide the necessary interaction operations. We choose Chrome browser because it's
fast among various browsers and it provides built-in developer tools which supports
source viewing, debugging and console editing utilities; so as a proof of concept we
concentrated our efforts on Chrome browser. TYG Chrome extension separates into
two main components: the extension scripts and the content scripts.

Since TYG content scripts run in the context of a web page and not the extension,
they often need some way of communicating with the rest of the TYG extension. In
general, a content script exchanges messages with its parent extension, so that the
parent extension can perform cross-site requests on behalf of the content script.
Communication between TYG extension and its content scripts works by using message passing. Either side can listen for messages sent from the other end, and respond
on the same channel. A message can contain any valid JSON (JavaScript Object Notation) object. Figure 1 illustrates the architecture of TYG tool. The dashed lines
show the communication relationships of different components of TYG.

To create an annotation from the current webpage, the user clicks on the browser
extension's icon in the address bar (see Figure 2). The browser extension injects content scripts into the current page which create an *iframe* element and open a side

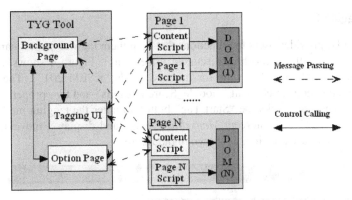

Fig. 1. Architecture of TYG tool

window within the current page. The icon 🖉 indicates that the TYG tool is disabled, while the icon 🖉 indicates the extension tool is functioning. Figure 2 illustrates the Chrome toolbar after installation of TYG.

Fig. 2. TYG's icon on the browser toolbar

4.2   Schema UI

TYG supports user to use self-defined vocabulary to define a tagging schema for the current browsing page. Figure 3 shows the schema definition UI of TYG. Tree-shaped schemes are very popular in annotation. Figure 3 illustrates such a schema; the schema contains a root "ConceptName" which has 3 properties: "PropertyName1", "PropertyName2" and "PropertyName3". The background colors of each property will be used to tag the elements of web pages. The names in the figure are automatically generated and can be manipulated by the user.

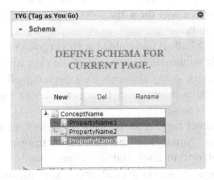

Fig. 3. TYG's UI for defining tree-structured schemes

4.3   Tagging UI

The tagging UI provides users the functions that help them tag web pages manually or automatically with the algorithms discussed in section 3. Figure 4 shows a manual tagging with Google search results for the key words "KSEM 2013". The tagging schema contains the root element "Google_Search_results" and two properties "title" and "links". The user clicks the "Start Tag" button to enter the tagging mode. In the tagging mode, TYG will intercept *MouseMove* event and display a movable box to track the user's mouse position, as shown in the figure.

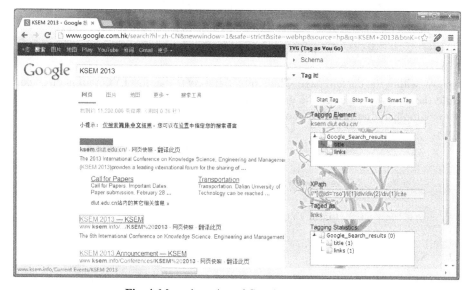

Fig. 4. Manual tagging of Google search results

The user can move the mouse to the web page element he/she wants to tag and click the mouse. TYG intercepts *MouseClick* event and will show the selected element's text on the side window ("ksem.dlut.edu.cn/" in this case). The user can then click the colored property name to specify that the element should be tagged as the property ("links" in this case). Once user clicked the colored property name, the current element of the web page will show the color of the property (yellow color in this case) as its background color.

The tagged element's XPath is determined with *get_xpath* function (see section 3) and is shown on the side window.   Below area of the side window shows the current statistics of tagging, i.e. how many elements have been tagged to their corresponding property tags. The statistics information is organized into the schema tree and is shown with the tree node's name.

Figure 4 shows a manual tagging scenario of TYG tool, as for the automatic tagging, TYG provides the "Smart Tag" button; which triggers **WalkDOMTree** and **K-MedoidUserInput** algorithm discussed in the previous section. We will examine the automatic tagging performance in the following experiments.

5   Experiments

5.1   Experimental Setup

We carry out experiments on 20 different web sites drawn from roughly 4 categories. The 4 categories are: Search engines; News Media; Science Index / famous science sites and some major publishers for research journals and conferences. For each category, we select 5 popular sites as representatives. Note that our approach and TYG system can be used on any web pages and are not confined to the pages discussed here. The experiments are carried out online and operate at web page level, which demonstrates the real world application setting for web users.

All the experiments are performed on an Intel Pentium P6000 Notebook with a 1.87GHz CPU and 2G of RAM. TYG can be used by causal web users, and we can expect better performance if users use a more powerful computing device. Figure 5 and Figure 6 shows the automatic tagging results on web pages of Google and CNN news site respectively.

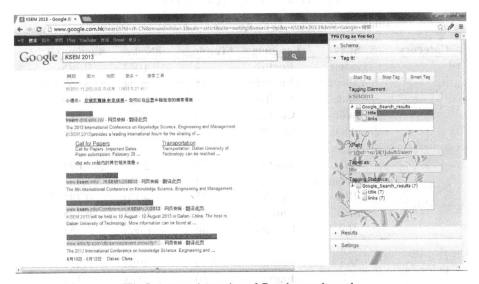

Fig. 5. Automatic tagging of Google search results

5.2   Efficiency Evaluation

The runtime of TYG tool is mainly affected by two aspects: the number of user inputs and the number of nodes in a HTML DOM tree. We observe that the variance of the number of nodes in web pages is not very big in different sites. The number of user inputs is usually a small number ranging from 1 to 10. We test our approach in pages from different sources. TYG is applied on 20 different web sites (see Table 1).

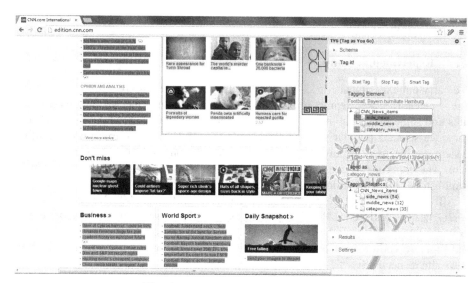

Fig. 6. Automatic tagging of CNN news

Table 1. Efficiency evaluation results of TYG on 20 web sites

Category	Sites	Language	#Taggings	#Nodes	Runtime (s)
Search Engine	Google	English	14	1044	0.285
	Yahoo!	English	18	3144	0.557
	Bing	English	20	1322	0.588
	Baidu	Chinese	20	672	0.340
	Sogou	Chinese	20	839	0.448
News Media	CNN	English	101	2427	1.697
	Yahoo News	English	45	3543	1.421
	BBC	English	55	4070	0.781
	Sina	Chinese	62	9345	6.475
	163	Chinese	46	13926	5.752
Index database/ Nature/ Science	SCI	English	50	3010	2.677
	EI	English	50	3173	3.323
	CPCI/ISTP	English	20	3004	3.244
	Nature	English	16	1533	1.524
	Science	English	18	1095	0.650
Journal/ Conf	Elsevier	English	16	829	0.738
	Springer	English	30	1010	0.669
	IEEE	English	75	5376	4.615
	Kluwer	English	21	401	0.603
	CNKI	Chinese	80	2214	1.160
Average			39	3099	2.027 (s)

5.3   Effectiveness Evaluation

We also record the correct and wrong tagging in the experiments. Precision and recall statistics can be drawn from these data to evaluate the effectiveness of TYG tool. We manually label all data fields in these tagging by 5 invited editors. To evaluate the precision of our approach, they label the tagging results of the pages in each site to get true positives and false positives. True positives are the set of annotations correctly processed by TYG, and false positives are the set of annotations processed incorrectly. Precision and recall metrics for each site can be calculated from the results. The total experimental results of TYG tool are shown in Table 2. As is shown in Table 2, TYG achieves very high precision and good recall on most sites. On average, it achieves over 95% F1 score in the experiments.

Table 2. Effectiveness evaluation results of TYG on 20 web sites

Sites	Actual	Correct	Wrong	Missed	Precision	Recall	F1
Google	14	14	0	0	1.00	1.00	1.00
Yahoo!	18	18	0	0	1.00	1.00	1.00
Bing	20	19	0	1	1.00	0.95	0.97
Baidu	20	20	0	0	1.00	1.00	1.00
Sogou	20	20	0	0	1.00	1.00	1.00
CNN	101	83	18	0	0.82	1.00	0.90
Yahoo News	45	45	0	0	1.00	1.00	1.00
BBC	55	45	10	0	0.82	1.00	0.90
Sina	62	62	3	1	0.95	0.98	0.97
163	46	44	2	0	0.96	1.00	0.98
SCI	50	50	0	0	1.00	1.00	1.00
EI	50	50	0	0	1.00	1.00	1.00
CPCI/ISTP	20	20	0	0	1.00	1.00	1.00
Nature	16	16	0	0	1.00	1.00	1.00
Science	18	18	0	0	1.00	1.00	1.00
Elsevier	16	16	0	0	1.00	1.00	1.00
Springer	30	30	0	0	1.00	1.00	1.00
IEEE	75	50	25	0	0.67	1.00	0.80
Kluwer	21	14	7	0	0.67	1.00	0.80
CNKI	80	80	0	0	1.00	1.00	1.00
Average	39	36	3.25	0.11	0.917	0.997	0.955

6   Conclusion

In this paper, we investigate a hybrid approach which combines user input with automatic clustering techniques. By using this approach, Users can specify the interested elements on the web pages with self-defined property tags; subsequent *k-medoid* clustering can help users gather similar elements automatically. This user-guided approach provides more control over the annotation process and incurs less noise than batch-mode approaches. Moreover, errors can be easily detected during the annotation process, so users can prevent errors propagating in the knowledge base at an early stage.

TYG is a prototype system we developed to show the efficiency and effectiveness of the proposed approach. It can be used to both static and dynamic web pages and gain high performance annotation results. We discuss the main architectural components and techniques constituting TYG tool. The tool is evaluated through real world application scenarios. Experimental results indicate that the system achieves good performance for online tagging.

Acknowledgement. The research is supported by the Fundamental Research Funds for the Central Universities, and the Research Funds of Renmin University of China (11XNJ002).

References

1. Pasternack, J., Roth, D.: Extracting article text from the web with maximum subsequence segmentation. In: WWW, Madrid, Spain, pp. 971–980 (2009)
2. Wang, J., et al.: Can we learn a template-independent wrapper for news article extraction from a single training site? In: SIGKDD, Paris, France, pp. 1345–1353 (2009)
3. Luo, P., et al.: Web Article Extraction for Web Printing: a DOM+Visual based Approach. In: DocEng, Munich, Germany, pp. 66–69 (2009)
4. Fan, J., Luo, P., Joshi, P.: Title identification of web article pages using HTML and visual features. In: Proc. of SPIE-IS&T Electronic Imaging, vol. 7879 (2011)
5. Liu, J., Song, X., Jiang, J., Lin, C.: An unsupervised method for author extraction from web pages containing user-generated content. In: Proceedings of CIKM, Maui Hawaii (2012)
6. Ciravegna, F.: (LP)2, an adaptive algorithm for information extraction from web-related texts. In: Proceedings of the IJCAI 2001 Workshop on Adaptive Text Extraction and Mining held in conjunction with 17th IJCAI 2001, Seattle, USA, pp. 1251–1256 (2001)
7. Kiryakov, A., Popov, B., Terziev, I., Manov, D., Ognyanoff, D.: Semantic Annotation, Indexing, and Retrieval. Journal of Web Semantics 2(1) (2005)
8. Dill, S., Eiron, N., Gibson, D., Gruhl, D., Guha, R., Jhingran, A., Kanungo, T., Rajagopalan, S., Tomkins, A., Tomlin, J.A., Zien, J.Y.: SemTag and Seeker: bootstrapping the semantic web via automated semantic annotation. In: Proceedings of the 12th International Conference on WorldWideWeb (WWW 2003), Budapest, Hungary (2003)
9. Handschuh, S., Staab, S., Ciravegna, F.: S-CREAM – semi-automatic CREAtion of metadata. In: Gómez-Pérez, A., Benjamins, V.R. (eds.) EKAW 2002. LNCS (LNAI), vol. 2473, pp. 358–372. Springer, Heidelberg (2002)
10. Popov, B., Kiryakov, A., Kirilov, A., Manov, D., Ognyanoff, D., Goranov, M.: KIM – semantic annotation platform. In: Fensel, D., Sycara, K., Mylopoulos, J. (eds.) ISWC 2003. LNCS, vol. 2870, pp. 834–849. Springer, Heidelberg (2003)
11. Tang, J., Li, J., Lu, H., Liang, B., Huang, X., Wang, K.: iASA: Learning to annotate the semantic web. In: Spaccapietra, S. (ed.) Journal on Data Semantics IV. LNCS, vol. 3730, pp. 110–145. Springer, Heidelberg (2005)

12. Plimmer, B., et al.: iAnnotate: Exploring Multi-User Ink Annotation in Web Browsers. In: Proc. of the 9th Australasian Conf. on User Interface, pp. 52–60. Australian Comp. Soc. (2010)
13. Kelly, M., Weigle, M.C.: WARCreate: create wayback-consumable WARC files from any webpage. In: Proceedings of the 12th ACM/IEEE-CS Joint Conference on Digital Libraries, Washington, DC, USA, pp. 437–438 (2012)
14. Muhlestein, D., Lim, S.: Online learning with social computing based interest sharing. Knowledge and Information Systems 2(1), 31–58 (2011)
15. Park, H.S., Jun, C.H.: A simple and fast algorithm for K-medoids clustering. Expert Systems with Applications 36(2), 3336–3341 (2009)
16. Velmurugan, T., Santhanam, T.: Computational Complexity between K-Means and K-Medoids Clustering Algorithms for Normal and Uniform Distributions of Data Points. Journal of Computer Science 6(3), 363–368 (2010)

On the Use of a Mixed Binary Join Tree for Exact Inference in Dynamic Directed Evidential Networks with Conditional Belief Functions

Wafa Laâmari[1], Boutheina Ben Yaghlane[1], and Christophe Simon[2]

[1] LARODEC Laboratory - Institut Supérieur de Gestion de Tunis, Tunisia
[2] CRAN - Université de Lorraine - CNRS, UMR 7039, France

Abstract. Dynamic directed evidential network with conditional belief functions (DDEVN) is a framework for reasoning under uncertainty over systems evolving in time. Based on the theory of belief function, the DDEVN allows to faithfully represent various forms of uncertainty.

In this paper, we propose a new algorithm for inference in DDEVNs. We especially present a computational structure, namely the mixed binary join tree, which is appropriate for the exact inference in these networks.

Keywords: Dynamic evidential models, belief function theory, changes over time, reasoning under uncertainty, exact inference.

1 Introduction

In the recent few years, there has been a growing interest in reasoning under uncertainty over time-evolving systems. Several popular techniques in artificial intelligence have been typically introduced in real-world domains for modeling the uncertainty in the existing knowledge and handling changes over time, including dynamic Bayesian networks [5], dynamic possibilistic networks [2] and dynamic evidential networks [3,12] which are respectively designed for probabilistic [6], possibilistic and evidential reasoning [7].

In this paper, we are interested in reasoning over temporal changes when the uncertainty in knowledge is expressed using the belief function formalism [7] which is known as a general mathematical framework extending the probability and the possibility theories. The expressive power of the belief function theory, also referred to as evidence theory or Dempster-Shafer (DS)'s theory, comes from its ability to represent in a very flexible way the full knowledge, the partial ignorance and the total ignorance.

In the literature, two principal evidential graphical models have been developed for handling the temporal dimension with the belief function formalism: dynamic evidential networks (DENs) [12] which extend static evidential networks (ENs) [10] and dynamic directed evidential networks with conditional belief functions (DDEVNs) [3,4] which are an extension of directed evidential networks with conditional belief functions (DEVNs) [1].

M. Wang (Ed.): KSEM 2013, LNAI 8041, pp. 310–324, 2013.

Based on an extension of the Bayes' theorem to the representation of the DS's theory, DENs exploit rules similar to the ones used in Bayesian networks [12]. Unlike DENs, DDEVNs are based solely on rules of the evidence theory [3].

The aim of this paper is to present a new algorithm for making an accurate inference in DDEVN. The remainder of the paper is as follows: In section 2, we recall some basics about the belief function framework, then we give in section 3 a short review of the binary join tree (BJT) [8] and the modified binary join tree (MBJT) [1] which are the architectures used for exact inference in the evidential networks. In section 4, we briefly present the DDEVN and some of its theoretical concepts. Section 5 introduces our new algorithm for exact inference in DDEVNs and gives a modified version for constructing the BJT leading to a new computational structure that we call mixed binary join tree. Section 6 is devoted to a short example illustrating the evidential reasoning in the DDEVNs using the proposed algorithm. We present our conclusions in section 7.

2   Dempster-Shafer's Theory Preliminaries

In DS's belief function theory [7], a frame of discernment $\Theta = \{\theta_1, ..., \theta_n\}$, refers to the finite set of all the propositions θ_i of interest. A basic belief assignment (bba), also called a mass function, is a mapping m from the power set 2^Θ to the interval [0,1] and verifies: $\sum_{B \in 2^\Theta} m(B) = 1$ and $m(\emptyset) = 0$.

Let V $=\{X_1, X_2...X_n\}$ be a finite set of variables, where each variable X_i in V is associated with a frame of discernment Θ_{X_i}. Let U and T be two disjoint subsets of V. Their frames Θ_U and Θ_T are the Cartesian product of the frames for the variables they include. The space Θ_U and the product space $\Theta_U \times \Theta_T$ are shortly denoted by U and UT respectively.

2.1   Marginalization

Let ψ_1 be a potential[1] defined on the domain UT with an associated bba m^{UT}. The marginalization of ψ_1 to U produces a potential ψ_2 with a bba m^U as follows:

$$m^{UT \downarrow U}(A) = \sum_{B \subseteq (UT), B^{\downarrow U} = A} m^{UT}(B) . \tag{1}$$

where $B^{\downarrow U}$ is the projection of $B \subseteq UT$ to U by dropping extra coordinates in each element of B.

2.2   Ballooning Extension

Let $m^U[t_i](u)$ be a conditional mass distribution defined on U for $t_i \in T$. The ballooning extension of $m^U[t_i](u)$ denoted $m^{U \uparrow UT}$ is defined on UT as follows:

$$m^{U \uparrow UT}(\omega) = \begin{cases} m^U[t_i](u) & \text{if } \omega = (u, t_i) \cup (U, \overline{t_i}), \\ 0 & \text{otherwise} \end{cases} \tag{2}$$

---

[1] In the evidence theory, beliefs are represented by mass functions called potentials.

2.3   Vacuous Extension

Let ψ_1 be a potential with the associated bba m^U. The vacuous extension of ψ_1 to UT produces a potential ψ_2 with the following associated bba $m^{U \uparrow UT}$:

$$m^{U \uparrow UT}(A) = \begin{cases} m^U(B) & \text{if } A = B \times T, B \subseteq U \\ 0 & \text{otherwise} \end{cases} \qquad (3)$$

2.4   Combination

Given two potentials ψ_1 and ψ_2 defined on U and T respectively with their associated bba's m^U and m^T, the combination of these two potentials produces a new potential $\psi_{12} = \psi_1 \otimes \psi_2$ with an unnormalized bba m^{UT} as follows:

$$m^{UT}(A) = \sum_{B \subseteq U, C \subseteq T, B \cap C = A} m^{U \uparrow UT}(B).m^{T \uparrow UT}(C) \ . \qquad (4)$$

where $m^{U \uparrow UT}$ and $m^{T \uparrow UT}$ are obtained using equation (3).

3   Binary Join Tree and Modified Binary Join Tree

BJTs and MBJTs are frameworks for local computations on evidential networks.

3.1   Binary Join Tree

The BJT structure was introduced by Shenoy in [8] as an improvement of the Shenoy-Shafer architecture [9]. The BJT algorithm compiles a directed acyclic graph (DAG) G into a tree whose nodes N_i are subsets of variables in G.

Binary Join Tree Construction. To get a BJT, Shenoy's binary fusion algorithm (FA) [8] which is considered as a recursive variable elimination is applied using an elimination sequence (ES) S specifying the order in which the variables must be eliminated and a hypergraph H containing subsets of variables in G.

Each subset Sub_i in H contains a node X_i in G and its parent nodes. Let $V = \{X_1, X_2...X_n\}$ denote the set of nodes in G. The hypergraph H is formally defined by the following equation:

$$H = \{\{X_1 \cup Pa(X_1)\}, \{X_2 \cup Pa(X_2)\}...\{X_n \cup Pa(X_n)\}\} \ . \qquad (5)$$

where $Pa(X_i)$ are the parent nodes of X_i.

Let $\Psi = \{\psi_1, \psi_2...\psi_n\}$ be a set of given potentials. There exists one to one mapping between subsets in H and potentials in Ψ. This means that every potential ψ_i in Ψ is defined exactly on a domain $D(\psi_i) = Sub_i = \{X_i \cup Pa(X_i)\}$, and represents the conditional belief for X_i given $Pa(X_i)$.

Based on the hypergraph H and a specified ES $S = \vartheta_1 \vartheta_2...\vartheta_n : \vartheta_i \in V$, the BJT is constructed step-by-step by eliminating one variable after another.

At each step k of the FA, we particular eliminate the kth variable ϑ_k in S. To eliminate ϑ_k, the FA combines by equation (4) all the potentials which contain this variable in their domains on a binary basis [8], i.e. at each time, two potentials are selected to be combined. Then it marginalizes the variable out of the combination using equation (1). Each elimination of a variable ϑ_k generates a subtree of the BJT. This procedure is recursively repeated till we eliminate the last variable ϑ_n in S and obtain the final BJT.

Example 1. Let us consider the DAG G: A→ C ← B and let us assume that ψ_1, ψ_2 and ψ_3 are three potentials with the corresponding domains $D(\psi_1) =$ {A}, $D(\psi_2) =$ {B} and $D(\psi_3) =$ {A,C,B}. If we apply the FA according to the ES ACB then BAC, we obtain the two corresponding BJTs shown respectively in Fig. 1 (a) and Fig. 1 (b). Clearly, the BJT structure depends strongly on the ES: Using different ESs leads to different BJTs with different joint nodes. For instance, the first BJT contains a joint node C,B while the second one does not. This is simply explained by the fact that the combination operations performed when applying the FA do not give the same intermediate results when changing the order in which the variables are eliminated.

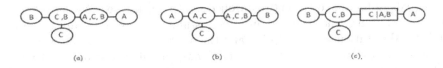

(a)   (b)   (c).

Fig. 1. Binary join trees (a,b) and modified binary join tree (c)

3.2   Modified Binary Join Tree

MBJT [1] was proposed to avoid a problem arising when constructing the BJT, which is the loss of information about the (in)dependencies among the variables.

Modified Binary Join Tree Construction. The MBJT construction process is an adaptation of that of the BJT with some modifications used for showing explicitly the conditional relationships between variables in the model. To obtain a MBJT, a BJT is first generated using the FA. Then every joint node (i.e circular node) in the BJT containing variables among which there is a conditional relationship in the model is converted to a conditional node (i.e. a rectangular node). This conditional node emphases this relationship, contrary to the corresponding joint node used in the BJT and which contains just the set of the involved variables. Using conditional nodes in the MBJT for representing qualitatively the conditional (in)dependencies is also computationally more efficient than using joint nodes in the corresponding BJT [1]. Associated with conditional belief functions, these conditional nodes, allow to reduce the amount of computations during the propagation process since representing belief functions using the conditional form allows to avoid the computation of joint belief functions on the product space by performing efficient computations using the disjunctive rule of combination and the generalized Bayesian theorem [11]. About how to efficiently perform computations in the conditional nodes using these two rules, please refer to [1], we do not give the details here.

Example 2. Considering the same DAG of the previous example, the MBJT obtained according to the ES ACB is shown in Fig. 1 (c). The conditional node C|A,B in this MBJT shows explicitly that A and B are the parent nodes of C in the model. While the joint node A,C,B in the corresponding BJT constructed using the same ES (Fig. 1 (a)) shows only these variables without maintaining the relationship between them (this is the information loss problem of the BJT).

4   Dynamic Directed Evidential Network with Conditional Belief Functions

Dynamic directed evidential networks with conditional belief functions [3], abbreviated to DDEVNs, are dynamic graphical models based on the DS's theory.

DDEVNs allow to compactly model systems or processes that evolve over time. Unfortunately, the number of time slices to be represented for these systems can not be known beforehand, and as a consequence the number of variables is potentially infinite and the entire unrolled model quickly becomes intractably large. To deal with this intractability, two assumptions are typically made in DDEVNs: the stationary assumption and the Markov assumption [3]. The first assumption means that the transition-belief mass distributions[2] do not depend on the time slice. Whereas the second one means that every node in the model at a particular time slice t depends only on its parent nodes which can be either in the same time slice t or in the immediately previous time slice $t-1$.

Under these two assumptions, systems can be simply modeled with a DDEVN unrolled slice by slice. This DDEVN is defined in a concise way by two evidential networks D_0 and D_t, where:

1. D_0 is an evidential network that represents the joint mass distribution over variables of the initial time slice $t = 0$.
2. D_t is an evidential network made up of only two consecutive time indexed slices $t-1$ and t, namely a 2-time slices DEVN (2-TDEVN).
   If the state of a variable X_{t-1} at the time slice $t-1$ directly affects the state of a variable Y_t at the time slice t, then this temporal dependency between the two variables is qualitatively represented by a directed arc connecting them. This arc denotes a transition-belief mass distribution which quantifies how Y_t depends on X_{t-1}.

Definition 1. The incoming interface in-I_{t-1} is the set of nodes of the last time slice $t-1$ which have one child node or more in the current time slice t. The outgoing interface out-I_t is the set of nodes of the current time slice t which have one child node or more in the next time slice $t+1$.

More formally, the in-I_{t-1} and the out-I_t are defined in the DDEVN as follows:

$$\text{in-}I_{t-1} = \{Z_{t-1} \in N_{t-1} | (Z_{t-1}, X_t) \in E(t-1, t) \text{ and } X_t \in N_t\} . \tag{6}$$

$$\text{out-}I_t = \{X_t \in N_t | (X_t, Y_{t+1}) \in E(t, t+1) \text{ and } Y_{t+1} \in N_{t+1}\} . \tag{7}$$

---

[2] The transition-belief mass distributions represent the temporal dependencies between variables at time slice t and variables at time slice $t+1$.

where N_{t-1}, N_t and N_{t+1} are the set of nodes modeling respectively the time slices $t - 1$, t, and $t + 1$ and $E(t, t+1)$ is the set of edges between two successive time slices t and $t + 1$.

Definition 2. A 1.5 DDEVN unrolled for time slices $(t - 1, t)$ is a graph which consists of all the nodes of the current time slice t and all the incoming interface nodes from the preceding time slice $t-1$, i.e. the in-I_{t-1}. It results from a 2-TDEVN by the elimination of nodes which belong to the time slice $t - 1$ but not to the incoming interface in-I_{t-1}.

According to *Definition 1* and *Definition 2*, the 1.5 DDEVN unrolled for time slices $(t-1, t)$ emphasizes the in-I_{t-1} and the out-I_t interfaces. The in-I_{t-1} allows to take into account the effect of the time slice $t - 1$ on the current time slice t, and the out-I_t allows to summarize the information about variables in the current time slice t influencing the future time slice $t + 1$.

4.1   Inference Algorithm in DDEVN

In front of an observed situation at the time slice $t = 0$ and based on transition distributions illustrating the temporal dependencies between two consecutive time slices, the aim of the inference algorithm in a DDEVN is to compute the belief mass distribution of each variable X in the system at any time step $t = T$.

The Markov assumption of DDEVNs [3] justifies the fact that the future time slice is conditionally independent of the past ones given the present time slice.

Therefore, the out-I_t interface which forms the set of nodes of the current time slice t with outgoing arcs to the time-slice $t + 1$ d-separates the past from the future since it encapsulates the information about previous time slices (time slices before the time t) which is the only necessary information to perform propagation in next time slices (time slice $t + 1$ and later) [3].

The propagation algorithm, proposed in [3] for inference in DDEVN, creates a MBJT M_0 for the initial time slice and another one, called M_t, for the 1.5 DDEVN unrolled for time slices $(t - 1, t)$. This algorithm approximates the joint mass distribution α over the interface by representing it as a product of the marginals α_i of the involved variables. More precisely, both the outgoing interface and the incoming interface are represented in the MBJTs by many nodes. Therefore, to advance propagation from time slice t to $t + 1$, marginals α_i of variables in the out-I_t are computed in nodes representing the out-I_t in the 1.5 DDEVN unrolled for time slices $(t - 1, t)$, then they are multiplied onto the nodes representing the in-I_t in the 1.5 DDEVN unrolled for time slices $(t, t + 1)$.

5   New Inference Algorithm in Dynamic Directed Evidential Network with Conditional Belief Functions

The accuracy of the inference algorithm in dynamic models depends on the number of nodes used to represent the interfaces. The exact interface algorithms proposed for reasoning in dynamic probabilistic and dynamic possibilistic networks require the integration of a single node for the outgoing interface and a single node for the incoming interface in the computational structures [5,2]. Using more than one node sacrifice intuitively some accuracy for speed.

The algorithm proposed in [3] for inference in DDEVNs uses several nodes for representing each interface in the MBJT which is the computational structure used in DDEVN. Therefore, it corresponds to approximate inference algorithm since the requirement that all variables in the interface need to be in a single node is dropped. Imposing a new restriction $\mathbf{R}_1$ that all the incoming interface nodes (respectively all the outgoing interface nodes) in the DDEVNs belong to one node in the computational structure, will allow the application of the exact interface algorithm in these networks.

Following this line of reasoning and in order to accurately perform the propagation process in DDEVNs, we propose in this section a new algorithm for making inference in these networks. The main principle underlying this algorithm is the use of the BJT structure with necessary modifications to perform the exact inference. These modifications guarantee to integrate in the obtained structure two nodes representing the incoming and the outgoing interfaces. We refer to the new tree structure as mixed binary join tree (MixBJT). Like the BJT algorithm, the MixBJT algorithm builds using the FA a tree structure for performing local computations but with the restriction $\mathbf{R}_1$.

The basic idea of our algorithm is :

✓ to create for the initial time slice $t = 0$ of the DDEVN a MixBJT Mix_0 in which variables of the out-I_0 belong to a single node that we call N_0.
N_0 encapsulates the sufficient information needed for propagation in the next time slice $t = 1$. Mix_0 is used only to run belief propagation in the DDEVN at the time slice $t = 0$.

✓ to create for the 1.5 DDEVN unrolled for time slices $(t - 1, t)$ a MixBJT structure Mix_t in which there are two nodes representing the in-I_{t-1} and the out-I_t. We respectively refer to these nodes as N_{t-1} and N_t. As explained previously, when advancing the time in the DDEVN from $t - 1$ to t, N_{t-1} allows to take into account the effect of the time slice $t - 1$ on the current time slice t, whereas N_t encapsulates the sufficient information affecting the next time slice $t + 1$. Mix_t is used for inference in the DDEVN at each time slice $t \geq 1$.

5.1   From Dynamic Directed Evidential Networks with Conditional Belief Functions to Mixed Binary Join Trees

To generate MixBJTs Mix_0 and Mix_t for a dynamic evidential network, we adapt the BJT construction process presented in section 3.1 but with some modifications to satisfy the new requirement $\mathbf{R}_1$.

Example 1 has shown that based on two different ESs, the FA creates for the same model two different BJTs with nodes having different labels or domains. If our aim were to construct for this model a BJT containing a node having as a label C, B, then we could not know in advance that BAC is not the appropriate ES and that it leads to a BJT which does not contain this node (see Fig. 1 (b)).

Thus, given a DDEVN and a chosen ES, generally one can not successfully determine beforehand whether or not this ES would guarantee to make the out-I_0 interface a node in the tree structure Mix_0. In addition, neither the incoming interface in-I_{t-1} nor the outgoing interface out-I_t are guaranteed to be two nodes in Mix_t. Nevertheless, it is far from trivial to generate all the possible ESs and to build the corresponding BJTs in order to verify which best suits our need.

To overcome this difficulty, we propose to add the interfaces for which nodes must exist in the resulting tree to the hypergraph since we are sure that whatever the ES is, there is always in the final BJT one node for each subset of the hypergraph used to construct it. More precisely, if we add the out-I_0 as a subset in the hypergraph used to construct Mix_0, this guarantees to obtain in the resulting structure the node N_0 whose label is the domain of out-I_0. Considering also the in-I_{t-1} and the out-I_t of the 1.5 DDEVN as subsets in the hypergraph used to build Mix_t, enables the FA to generate a tree containing obligatory the two nodes N_{t-1} and N_t corresponding respectively to the in-I_{t-1} and the out-I_t. Therefore, to construct for a given DDEVN, the tree structures Mix_0 and Mix_t satisfying the requirement $\mathbf{R}_1$, we define new hypergraphs H_0 and H_t by the following equations:

$$H_0 = \{\{X_0^1 \cup Pa(X_0^1)\}, \{X_0^2 \cup Pa(X_0^2)\}, ..., \{X_0^n \cup Pa(X_0^n)\}\} \cup \text{out-}I_0 \ . \quad (8)$$

$$H_t = \{\{X_t^1 \cup Pa(X_t^1)\}, ..., \{X_t^n \cup Pa(X_t^n)\}\} \cup \{\text{in-}I_{t-1}, \text{out-}I_t\} \ . \quad (9)$$

where X_t^i refers to the i-th variable of the time slice t in the 2-TDEVN.

In order to maintain the (in)dependence relationships between variables in the produced tree structures Mix_0 and Mix_t, we suggest to exploit the structure of the DDEVN and to use the conditional nodes for showing explicitly the conditional relationships. Like in MBJTs [1], these conditional nodes in MixBJTs will avoid the computation of joint belief functions on the product space and reduce the dimensionality of the belief functions involved in the computation.

Naturally, a question crosses the mind: If conditional nodes which represent the transformations essentially made on the produced BJTs to obtain the MixB-JTs Mix_0 and Mix_t, are the same transformations made on BJTs to obtain the MBJTs, why are they called "mixed binary join trees" ? The following example attempts to answer this question by showing that the tree structure proposed in this paper for inference in DDEVN is neither a purely BJT nor a purely MBJT.

Example 3. Let us consider the 2-TDEVN and the corresponding 1.5 DDEVN shown respectively in Fig. 2 (a) and Fig. 2 (b). The out-I_0 interface for the initial time slice is represented by the subset $\{A_0, C_0\}$.

To generate Mix_0 that represents the time slice $t = 0$, we use the hypergraph H_0 obtained by equation (8): $H_0 = \{\{A_0\}, \{B_0, A_0\}, \{C_0, A_0\}, \{E_0, B_0\}\}$.

Using this hypergraph with any ES enforces the FA to satisfy $\mathbf{R}_1$ by making the out-I_0 a node N_0 in the BJT Mix_0. For instance, considering H_0 with the ES $C_0 A_0 B_0 E_0$, we obtain the BJT structure shown in Fig. 2 (c) in which the joint node $N_0 = A_0, C_0$ represents the out-I_0 interface. Once the inference for time slice 0 has been accomplished in Mix_0, this node encapsulates the exact quantity relative to the joint distribution over the out-I_0 that we will use to perform inference in the time slice $t = 1$.

Now, if we exploit the DDEVN structure and use conditional nodes instead of joint nodes to represent explicitly all the conditional dependencies in Mix$_0$, we obtain the structure depicted in Fig. 2 (d) which integrates three conditional nodes $B_0|A_0$, $E_0|B_0$ and $C_0|A_0$. Clearly, after making the modifications, the obtained structure Mix$_0$ coincides with the MBJT structure. The node N_0 which was initially a joint node C_0, A_0 in Fig. 2 (c) has become a conditional one in Fig. 2 (d) since C_0 is the child node of A_0 in the model.

It is noteworthy to note that during the inference through a MBJT [1], conditional nodes are asked neither to send nor to receive messages because they are regarded as bridges between joint nodes. They do not computations and hold just a decisive role allowing to determine whether the message sent from a joint node to another joint node is a parent message or a child message. At the end of the belief propagation process, distributions of variables in MBJT are computed only in the joint nodes and one can not compute them in the conditional nodes.

Thus, despite satisfying the requirement $\mathbf{R}_1$ in Mix$_0$ by making the out-I_0 interface a node N_0 in this tree, this node does not yield after the propagation process the joint distribution α over the involved variables since it is a conditional node. This problem arises here because the DDEVN structure presents a conditional relationship between variables of the interface. To deal with this problem, our solution is based on the following simple idea: Since the MixBJT algorithm takes advantage of the DDEVN structure to reduce the amount of computations by converting each joint node having as a label $X_t^i \cup Pa(X_t^i)$ to a conditional node $X_t^i|Pa(X_t^i)$, we should pay attention when making these modifications to the fact that node N_0 in Mix$_0$ as well as nodes N_{t-1} and N_t in Mix$_t$ must be maintained as joint nodes even if they represent implicitly an (in)dependence relationship over the involved variables (requirement $\mathbf{R}_2$). Imposing this new requirement guarantees to have at the end of the propagation process the joint distribution needed to advance the inference from time slice to another.

When maintaining nodes representing interfaces as joint nodes, we must distinguish between them and other joint nodes (which do not represent implicitly conditional relationships over the involved variables). For this reason, we propose to qualitatively represent them by rectangles containing a joint relation between variables (mixed nodes). This representation reveals explicitly that there is a conditional dependence between the variables of the mixed node in question, but according to the requirement $\mathbf{R}_2$, this node is maintained a joint node when making modifications in an attempt to have the joint distribution over the corresponding interface.

Considering the new requirement $\mathbf{R}_2$ during the construction process of Mix$_0$ leads to the resulting tree structure shown in Fig 2.(e). The node $N_0 = C_0, A_0$ that we kept a joint node will provide at the end of the propagation process in Mix$_0$ a joint distribution over variables A_0 and C_0 (which form the out-I_0).

The obtained tree can not be considered as a MBJT since the algorithm for constructing a MBJT imposes the requirement to convert without exception each joint node showing implicitly an (in)dependence relationship between the involved variables to a conditional node and this requirement is dropped in our algorithm by R_2. We can not also consider the obtained structure as a BJT because it may contain both conditional and joint nodes contrary to the BJT which contains only joint nodes. Since this structure may contain joint nodes (like BJT), conditional nodes (like MBJT) and mixed nodes to which particular attention must be paid during the initialization phase, we call it MixBJT.

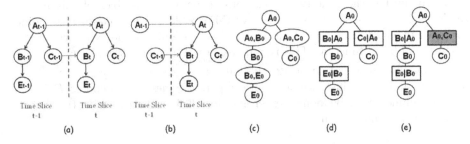

Fig. 2. A 2-TDEVN, a 1.5 DDEVN, the BJT_0, the $MBJT_0$ and the Mix_0 for the time slice $t = 0$

Mixed Binary Join Tree Mix_0 Construction. Given a 2-TDEVN and an elimination sequence S_0, the general construction process of the MixBJT Mix_0 is described by the following steps:

1. Identify the out-I_0 interface in the 2-TDEVN using equation (7).
2. Remove each node belonging to the time slice 1 from the 2-TDEVN (obtain the evidential network D_0 which represents only the time slice $t = 0$).
3. Determine the hypergrah H_0 for the resulting structure D_0 by equation (8).
4. Apply the FA to construct a BJT B_0 for D_0 using S_0 and H_0.
5. Make each joint node in B_0 (except N_0 which represents the out-I_0) representing an (in)dependence relationship between the involved variables a conditional node.

Mixed Binary Join Tree Mix_t Construction. Given a 2-TDEVN and an elimination sequence S_t, the general construction process of the MixBJT Mix_t is described by the following steps:

1. Identify the in-I_{t-1} and the out-I_t in the 2-TDEVN by equations (6) and (7).
2. Remove each node belonging to the time slice $t - 1$ and not to the in-I_{t-1} from the 2-TDEVN (obtain a 1.5 DDEVN unrolled for time slices$(t - 1, t)$).
3. Determine the hypergrah H_t for the resulting structure by equation (9).
4. Apply the FA to construct a BJT B_t for the 1.5 DDEVN using S_t and H_t.
5. Make each joint node in B_t (except N_{t-1} and N_t which represent the in-I_{t-1} and the out-I_t interfaces) representing an (in)dependence relationship between the involved variables a conditional node.

Mixed Binary Join Trees Initialization. The MixBJT Mix_0 is initialized by potentials relative to time slice $t = 0$. Whereas the MixBJT Mix_t is initialized by potentials relative only to time slice t. Composed of both joint and conditional nodes, a MixBJT is initialized, like a MBJT, by means of joint belief functions for not conditional nodes (i.e. joint nodes) and conditional belief functions for conditional ones. Mixed nodes which represent conditional relationships between the involved variables $X_t^i \cup Pa(X_t^i)$ and which we maintain as joint nodes ($\mathbf{R_2}$) should receive a specific attention when initializing the MixBJTs. Naturally a mixed node must be initialized by a joint belief function $m^{X_t^i \times Pa(X_t^i)}$ since it is a joint node. However we have the conditional belief function $m[Pa(X_t^i)](X_t^i)$ which quantifies the conditional relationship between variables of this mixed node in the DDEVN. One way to solve this problem is to construct the joint belief distribution from the conditional one, i.e. to compute the joint form of the conditional belief function. This is done using the following two steps:

a. Compute using equation (2) the ballooning extension $m[p(X_i)]^{\uparrow X_i \times P(X_i)}$ of $m[p(X_i)]$ for each instance $p(X_i)$ of the parent nodes of X_i denoted $P(X_i)$.

b. Combine all the ballooning extensions for all the instances using the Dempster's rule of combination.

5.2   Inference Algorithm in Mixed Binary Join Trees

Once Mix_0 and Mix_t are constructed and initialized, we can compute the marginals of variables at any time step $t = T$ by recursively performing the bidirectional message-passing scheme in Mix_t using the following algorithm:

Algorithm 1: Exact inference in DDEVNs
Data: Mix_0, Mix_t
Result: Marginal Distributions in the time slice $t = T$
begin
  |1. Performing the propagation in Mix_0 [a];
  |2. distr=Joint_distribution(out-I_0);
  |3. **if** $T > 0$ **then**
  | **for** $i=1$ to T **do**
  | Associate distr to node N_{i-1} representing the in-I_{i-1} in Mix_t;
  | Performing the propagation in Mix_t ;
  | distr= Joint_distribution(out-I_i);
  | **end**
  |
  | **end**
  |4. **if** $T=0$ **then**
  | | Compute_marginals(Mix_0)[b];
  | **else**
  | | Compute_marginals(Mix_t);
  | **end**
  end

---

[a] The inference is performed in the MixBJT as in the MBJT since both of them are composed of joint and conditional nodes. For details, the reader is referred to [1]
[b] To compute the marginal for a node, we combine its own initial potential with the messages received from all the neighbors during the propagation process

6   Illustrative Case Study

For the sake of illustration, let us consider again the evolving-time system with four variables modeled by the 2-TDEVN displayed in Fig. 2(a). Each node X_t in the 2-TDEVN represents the variable X at the time slice t which has two states ($\{x\},\{\bar{x}\}$). Fig. 2(b) shows the 1.5 DDEVN created from the 2-TDEVN given in Fig. 2(a) by removing all nodes in the time slice $t - 1$ not belonging to out-I_{t-1}. Table 1 represents the a priori mass distribution of A at the time step 0. The conditional mass distributions relative to nodes B_t, C_t and E_t are respectively given in Tables 2, 3 and 4. The belief mass distribution of variables A_t and B_t at the time slice t which depend respectively on the distributions of A_{t-1} and C_{t-1} at the time step $t - 1$ are respectively represented in Tables 5 and 6.

Table 1. A priori mass tables $M(A_0)$

a	$\bar{a}$	$a \cup \bar{a}$
0.7	0.2	0.1

Tables 2. 3 and 4. Conditional mass tables $M[A_t](B_t)$, $M[A_t](C_t)$ and $M[B_t](E_t)$

$B_t|A_t$	a	$\bar{a}$	$a \cup \bar{a}$	$C_t|A_t$	a	$\bar{a}$	$E_t|B_t$	b	$\bar{b}$	$b \cup \bar{b}$
b	0.6	0.7	0.5	c	0.7	0.5	e	0.5	0.6	0.7
$\bar{b}$	0.2	0.2	0.4	$\bar{c}$	0.2	0.3	$\bar{e}$	0.2	0.3	0.1
$b \cup \bar{b}$	0.2	0.1	0.1	$c \cup \bar{c}$	0.1	0.2	$e \cup \bar{e}$	0.3	0.1	0.2

Tables 5 and 6. Conditional mass tables $M[A_{t-1}](A_t)$ and $M[C_{t-1}](B_t)$

$A_t|A_{t-1}$	a	$\bar{a}$	$a \cup \bar{a}$	$B_t|C_{t-1}$	c	$\bar{c}$	$c \cup \bar{c}$
a	0.4	0.5	0.7	b	0.5	0.6	0.7
$\bar{a}$	0.4	0.1	0.2	$\bar{b}$	0.2	0.3	0.1
$a \cup \bar{a}$	0.2	0.4	0.1	$b \cup \bar{b}$	0.3	0.1	0.2

Construction and Initialization of Mix₀ and Mix_t. To construct Mix_0 and Mix_t, we first define using equations (8) and (9) the hypergraphs H_0 and H_t that respectively characterize the subsets of variables that will be used by the fusion algorithm: $H_0 = \{\{A_0\}, \{B_0, A_0\}, \{C_0, A_0\}, \{E_0, B_0\}\}$ and $H_t = \{\{A_t, A_{t-1}\}, \{B_t, A_t\}, \{B_t, C_{t-1}\}, \{C_t, A_t\}, \{E_t, B_t\}, \{A_{t-1}, C_{t-1}\}\}$. Given these two hypergraphs, the 2-TDEVN in Fig.2 (a) and the two elimination sequences $S_0 = C_0 A_0 B_0 E_0$ and $S_t = A_{t-1} C_{t-1} A_t C_t B_t E_t$, we obtain the MixBJTs Mix_0 and Mix_t depicted respectively in Fig.2 (e) and Fig.3.

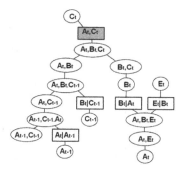

Fig. 3. The mixed binary join tree Mix$_t$ of the DDEVN in Fig.2 (a)

Using the a priori and the conditional mass tables, Mix$_0$ and Mix$_t$ are initialized by assigning each belief function distribution to the corresponding node.

For the mixed node $N_0 = C_t, A_t$, since we have the conditional belief function $M[A_t](C_t)$, we must compute the joint form of this belief function over the joint space C_0, A_0 to initialize the joint rectangular nodes $N_0 = \{A_0, C_0\}$ in Mix$_0$ and $N_t = \{A_t, C_t\}$ in Mix$_t$. The first step of computing this joint form consists in computing the ballooning extensions of $m[a]^{C_t \uparrow C_t \times A_t}$ and $m[\bar{a}]^{C_t \uparrow C_t \times A_t}$ using equation (2) as shown in Table 7.

Table 7. Ballooning extension of the conditional belief function $m[A_t]^{C_t \uparrow C_t \times A_t}$

a	$m(\{c,a\}, \{c,\bar{a}\}, \{\bar{c},\bar{a}\}) = m[a](c) = 0.7$
	$m(\{\bar{c},a\}, \{c,\bar{a}\}, \{\bar{c},\bar{a}\}) = m[a](\bar{c}) = 0.2$
	$m(\{C_t \times A_t\}) = m[a](\{c,\bar{c}\}) = 0.1$
$\bar{a}$	$m(\{c,\bar{a}\}, \{c,a\}, \{\bar{c},a\}) = m[\bar{a}](c) = 0.5$
	$m(\{\bar{c},\bar{a}\}, \{c,a\}, \{\bar{c},a\}) = m[\bar{a}](\bar{c}) = 0.3$
	$m(\{C_t \times A_t\}) = m[\bar{a}](\{c,\bar{c}\}) = 0.2$

Now that we have computed the ballooning extension of the conditional belief function $M[A_t](C_t)$, we compute the joint form of this conditional belief function by combining the ballooning extensions obtained in Table 7 using Dempster's rule of combination. The resulting joint form of $M[A_t](C_t)$ which will be used to initialize nodes N_0 and N_t is depicted in Table 8.

Table 8. Joint Form of $M[A_t](C_t)$

	$m(\{c,a\}, \{c,\bar{a}\}, \{\bar{c},\bar{a}\})$ $= 0.7$	$m(\{\bar{c},a\}, \{c,\bar{a}\}, \{\bar{c},\bar{a}\})$ $= 0.2$	$m(\{C_t \times A_t\})$ $= 0.1$
$m(\{c,\bar{a}\}, \{c,a\}, \{\bar{c},a\})$ $= 0.5$	$m(\{c,\bar{a}\}, \{c,a\})$ 0.35	$m(\{c,\bar{a}\}, \{\bar{c},a\})$ $= 0.1$	$m(\{c,\bar{a}\}, \{c,a\}, \{\bar{c},a\})$ $= 0.05$
$m(\{\bar{c},\bar{a}\}, \{c,a\}, \{\bar{c},a\})$ $= 0.3$	$m(\{\bar{c},\bar{a}\}, \{c,a\})$ $= 0.21$	$m(\{\bar{c},\bar{a}\}, \{\bar{c},a\})$ $= 0.06$	$m(\{\bar{c},\bar{a}\}, \{c,a\}, \{\bar{c},a\})$ $= 0.03$
$m(\{C_t \times A_t\})$ $= 0.2$	$m(\{c,a\}, \{c,\bar{a}\}, \{\bar{c},\bar{a}\})$ $= 0.14$	$m(\{\bar{c},a\}, \{c,\bar{a}\}, \{\bar{c},\bar{a}\})$ $= 0.04$	$m(\{C_t \times A_t\})$ $= 0.02$

Once Mix_0 and Mix_t are constructed and initialized, we can compute the marginals of variables at any time step $t = T$ by recursively performing the bidirectional message-passing scheme in Mix_t.

Performing the Propagation Process in the DDEVN. Suppose now that we wish to compute the marginal of the variable E_t at time step $t = 1000$. We first perform the inference process in Mix_0 and we compute the joint distribution over nodes in the outgoing interface out-$I_0 = \{A_0, C_0\}$ (the first and the second steps of algorithm 1).

The joint distribution over the outgoing interface out-I_0's nodes will be used when performing the second propagation in the MixBJT Mix_t in the next time slice ($t = 1$). It will be introduced in the in-I_0 of Mix_t. After performing the inference algorithm, Mix_t yields the joint distribution of the out-I_1 (forming the outgoing interface out-I_1) which is the sufficient information needed to continue the propagation in the following time slice $t = 2$.

After carrying out the inference process in the MixBJT Mix_t recursively for 1000 time slices, we obtain the following distribution for node E_{1000}: $M(E_{1000}) = [m(\emptyset) = 0 \ \ m(\{e\}) = 0.00218 \ \ m(\{\bar{e}\}) = 0.8525 \ \ m(\{e, \bar{e}\}) = 0.14525]$.

7 Conclusion

We have presented in this paper an algorithm for inference in Dynamic directed evidential network with conditional belief functions based on a computational structure, called the mixed binary join tree, which is proposed for making the exact inference in these networks. In future work, the development of new algorithms to perform faster propagation in the DDEVN will be of a great interest, since exact propagation still a NP-Hard problem.

References

1. Ben Yaghlane, B., Mellouli, K.: Inference in Directed Evidential Networks Based on the Transferable Belief Model. Int. J. Approx. Reasoning 48(2), 399–418 (2008)
2. Heni, A., Ben Amor, N., Benferhat, S., Alimi, A.: Dynamic Possibilistic Networks: Representation and Exact Inference. In: IEEE International Conference on Computational Intelligence for Measurement Systems and Applications (CIMSA 2007), Ostuni, Italy, pp. 1–8 (2007)
3. Laâmari, W., Ben Yaghlane, B., Simon, C.: Dynamic Directed Evidential Networks with Conditional Belief Functions: Application to System Reliability. In: Greco, S., Bouchon-Meunier, B., Coletti, G., Fedrizzi, M., Matarazzo, B., Yager, R.R. (eds.) IPMU 2012, Part III. CCIS, vol. 299, pp. 481–490. Springer, Heidelberg (2012)
4. Laâmari, W., Ben Yaghlane, B., Simon, C.: On the Complexity of the Graphical Representation and the Belief Inference in the Dynamic Directed Evidential Networks with Conditional Belief Functions. In: Hüllermeier, E., Link, S., Fober, T., Seeger, B. (eds.) SUM 2012. LNCS, vol. 7520, pp. 206–218. Springer, Heidelberg (2012)
5. Murphy, K.: Dynamic Bayesian Networks: Representation, Inference and Learning. PhD thesis, Dept. Computer Science, UC, Berkeley (2002)

6. Pearl, J.: Probabilistic Reasoning in Intelligent Systems: Networks of Plausible Inference. Morgan Kaufmann (1988)
7. Shafer, G.: A Mathematical Theory of Evidence. Princeton University Press, Princeton (1976)
8. Shenoy, P.P.: Binary Join Trees for Computing Marginals in the Shenoy-Shafer Architecture. Int. J. Approx. Reasoning 17(2-3), 239–263 (1997)
9. Shenoy, P.P., Shafer, G.: Axioms for probability and belief functions propagation. Uncertainty in Artificial Intelligence 4, 159–198 (1990)
10. Simon, C., Weber, P., Evsukoff, A.: Bayesian networks inference algorithm to implement Dempster Shafer theory in reliability analysis. Reliability Engineering and System Safety 93, 950–963 (2008)
11. Smets, P.: Belief Function: the Disjunctive Rule of Combination and the Generalized Bayesian Theorem. Int. J. Approx. Reasoning 9, 1–35 (1993)
12. Weber, P., Simon, C.: Dynamic Evidential Networks in System Reliability Analysis: A Dempster Shafer Approach. In: 16th Mediterranean Conference on Control and Automation, France, vol. 93, pp. 262–267 (2008)

Multiobjective-integer-programming-based Sensitive Frequent Itemsets Hiding

Mingzheng Wang, Yue He, and Donghua Pan

No.2, Linggong Road, Ganjingzi District, Dalian 116024, P.R. China
{mzhwang,gyise}@dlut.edu.cn, yue_he@mail.dlut.edu.cn

Abstract. Due to substantial commercial benefits in the discovered frequent patterns from large databases, frequent itemsets mining has become one of the most meaningful studies in data mining. However, it also increases the risk of disclosing some sensitive patterns through the data mining process. In this paper, a multi-objective integer programming, considering both data accuracy and information loss, is proposed to solve the problem for hiding sensitive frequent itemsets. Further, we solve this optimization model by a two-phased procedure, where in the first procedure the sanitized transactions can be pinpointed and in the second procedure the sanitized items can be pinpointed. Finally, we conduct some extensive tests on publicly available real data. These experiments' results illustrate that our approach is very effective.

Keywords: privacy preserving data mining, association rule, sensitive knowledge protection, multi-objective integer programming.

1 Introduction

The widespread use of Internet and rapid advances in software and hardware technologies enable users to collect huge amounts of data. At the same time, high-speed computation has made it possible to extract valuable information contained therein by analyzing these data. This so-called *data mining* plays an important role in many applications such as database marketing, Web usage analysis and fraud detection. However, it also causes some privacy disclosure risks. For example (see, Menon et al.[1]), there is a retailer who has implemented shelf-space-allocation strategies that are currently resulting in some very profitable market baskets being purchased at the retailer's outlets. By providing the data to any external party, the retailer risks disclosing these market baskets that could result in their competitors replicating this strategy. Therefore, the retailer must hide the sensitive patterns through some specific privacy policies before sending the data to the cooperator, which results in that the retailer will only share the modified database with his cooperator. In general, the data owner and the cooperator intend to create a win-win situation. So it is important for the data owner to keep original database's utility and the sensitive patterns are not discovered by the cooperator.

M. Wang (Ed.): KSEM 2013, LNAI 8041, pp. 325–335, 2013.

There are many researchers to study sensitive frequent itemsets hiding issues, see [2][3][4][5]. These studies employed different heuristics to preserve the quality of the database. Unfortunately, heuristic algorithms may cause undesirable side-effects to nonsensitive rules, which is due to the fact that heuristics always aim at getting locally optimal solutions in terms of hiding of sensitive frequent itemsets. Both Menon and Sarkar[6] and Menon et al. [1] presented model-based methods for hiding sensitive frequent itemsets. Menon et al.[1]'s paper focuses on frequent item sets, the identification of which forms a critical initial step in a variety of data-mining tasks. It presents an optimal approach for hiding sensitive item sets, while keeping the number of modified transactions to a minimum. Menon et al. [1] formulated an integer programming model to hide sensitive frequent patterns with maximizing the data accuracy, while Menon and Sarkar[6] formulated the model for concealing the sensitive information based on minimizing information loss. In fact, we find that it is necessary to make trade-offs between the transaction accuracy and the revealing non-sensitive frequent itemsets (i.e. information loss), because they are not negatively correlated with each other in some cases. It is possible that two modified databases have different information losses, containing the same transaction accuracy. Guo [7] presented a reconstruction-based method to sanitize the database for hiding sensitive association rules, while Divanis and Verkios [8] proposed a database extension method for that purpose. Unfortunately, it causes some new generated spurious rules after their sanitization process. In this paper, we present a item-based model to surmount that problem and propose the sanitization strategy about items by deleting some items to avoid bringing the fake frequent itemsets in the published database.

Our main contributions are two-fold: (1) multi-objective optimization-based privacy preserving model is proposed in order to address frequent itemsets hiding problem when considering both transactions accuracy and information utility. (2) a item-based hiding strategy instead of itemset-based one can be obtained from the result of the presented model. Results from extensive tests conducted on publicly available real data indicate that the approach presented is effective in both the transaction accuracy and itemsets utility.

The remainder of the paper is organized as follows. Section 2 describes the problem, along with useful related concepts. Section 3 introduces the inter-programming formulation for the frequent itemset hiding problem. Section 4 describes an example illustrating the entire sanitization process. Section 5 discusses the data used in conducting computational tests and analyzes the results of these tests. Finally section 6 concludes the paper.

2   Problem Description

The problem can be defined as follow. Let $I = \{i_1, i_2, ..., i_m\}$ be the set of items. Any subset $I_k \subseteq I$ is an itemset. Let D be the database of transactions. Any transaction T_i actually is an itemset. It is said that a transaction T_i includes an itemset I_j if $I_j \subseteq I_k$. The support of I_j in D is defined as the number of transactions of D, which is denoted by $s(I_j)$. An itemset is called frequent itemset if at least σ_{min} (a predefined minimum number) transactions support it. Define F the set of frequent itemsets. Let $S \subseteq F$ be the set of frequent itemsets. The owner of the database would like to conceal these itemsets (i.e. the sensitive itemsets). Correspondingly, $N = F \setminus S$ is defined as the set of nonsensitive frequent itemsets. The frequent itemset hiding(FIH) problem aims at minimally impacting the original database and potential patterns contained by frequent itemsets can still be mined after all the sensitive frequent itemsets are concealed.

3   Model Formulation

In this paper, the frequent itemset hiding process is transferred into sanitizing the transactions. More specifically, a transaction i is said to be sanitized if it is altered in such a way that it no longer supports sensitive itemset S_j. The item included in the sensitive itemset is called sensitive item. If we delete the sensitive item in the sanitized transaction, the support of the sensitive itemset will decrease. It means that hiding an itemset j can be achieved through reducing its support below σ_{min}. After that the non-frequent itemsets can not be mined (i.e. can not be discovered) in the new database. For the nonsensitive itemsets, they may have the common item(sensitive item) with the sensitive ones. The support of the nonsensitive itemset decrease if this common item is deleted from the sensitive itemset. The support of the nonsensitive itemset will also decrease to σ_{min} after sanitizing the transactions. Hence, it can not be mined in the modified database, which leads to the negative influence on the data availability. Here, it is regarded as the information loss if the nonsensitive itemset is lost. We wish the less information loss the better during the sanitization process. In this section, we propose a multi-objective integer programming model to maximize the transaction accuracy and to minimize the information loss for the frequent itemset hiding problem. The definition of parameters and variables used in the remainder of this paper is as follows.

Sets:

F : the set of frequent itemsets
S : the set of sensitive frequent itemsets
N : the set of non-sensitive frequent itemsets
J_j : the set of items which are included in the itemset j.

Parameters:

b_{kj} : be one if the itemset j supports item k, and zero otherwise.
s_j : the support of itemset j
σ_{min} : mining threshold given by the data owner, the frequent itemsets are the ones whose support are over σ_{min}
M : a large enough positive coefficient to ensure that no feasible solution has been cut off, especially M=σ_{min}.

Decision Variables:

$$q_{ij} = \begin{cases} 1, & \text{if the itemset } j \text{ loses the support of transaction } i; \\ 0, & \text{otherwise.} \end{cases}$$

$$p_{ik} = \begin{cases} 1, & \text{if the item } k \text{ is removed from transaction } i; \\ 0, & \text{otherwise.} \end{cases}$$

$$z_i = \begin{cases} 1, & \text{if transaction } i \text{ is not sanitized;} \\ 0, & \text{otherwise.} \end{cases}$$

$$y_j = \begin{cases} 1, & \text{if the itemset } j \text{ gets hidden;} \\ 0, & \text{otherwise.} \end{cases}$$

Obviously we know that $y_j = 1$ if and only if $\sum_i q_{ij} \geq s_j - \sigma_{min} + 1$.

Our problem is then formulated as the following multiobjective integer programming which describes trade-offs between the transaction accuracy and information loss:

$$\max \sum_{i \in D} z_i \qquad (a)$$

$$\min \sum_{j \in N} y_j \qquad (b)$$

$$\min \sum_{j \in S} \sum_{i \in D} q_{ij} \qquad (c)$$

(FIH)

$$\text{s.t.} \quad \sum_i q_{ij} \geq s_j - \sigma_{min} + 1 \qquad \forall j \in S$$

$$\sum_i q_{ij} \leq s_j - \sigma_{min} + M \times y_j \qquad \forall j \in N$$

$$z_i = \prod_k (1 - p_{ik}) \qquad \forall i \in D$$

$$q_{ij} = 1 - \prod_{k \in J_j} (1 - p_{ik}) \qquad \forall i \in D \ \ \forall j \in F$$

$$q_{ij}, p_{ik}, y_j, z_i \in \{0, 1\}$$

Although we can easily protect privacy by sanitizing the transaction, it is still important for data owner to guarantee the transaction integrality (transaction accuracy). In this paper we define the transaction accuracy by the proportion of transactions that are unchanged in the sanitized database. Hence, objective (a) is used to measure the transaction accuracy. In FIH problem, the number of hidden nonsensitive itemsets should be as less as possible. So the side-effect is measured by the number of concealing nonsensitive itemsets, that is, objective (b). The objective (c) measures over-hiding for sensitive itemset. Actually it is feasible to decrease all the sensitive itemsets' support to 0 according to the target of hiding the sensitive frequent itemsets, but it will make a huge effect on the utility of data. In practice, the object of protecting privacy information can be achieved only if the support of sensitive itemset is just below the mining threshold. So we only require that the support of sensitive itemset is just below the mining threshold.

In order to solve the model simply, we transform the above model into a single objective model as follows:

$$\max \sum_{i \in D} z_i - \sum_{j \in S} \sum_{i \in D} q_{ij} - \sum_{j \in N} y_j$$

$$\text{s.t.} \quad \sum_i q_{ij} \geq s_j - \sigma_{min} + 1 \quad \forall j \in S \tag{1}$$

(FIH − 0)
$$\sum_i q_{ij} \leq s_j - \sigma_{min} + M \times y_j \quad \forall j \in N \tag{2}$$

$$z_i = \prod_k (1 - p_{ik}) \quad \forall i \in D \tag{3}$$

$$q_{ij} = 1 - \prod_{k \in J_j} (1 - p_{ik}) \quad \forall i \in D \ \forall j \in F \tag{4}$$

$$q_{ij}, p_{ik}, y_j, z_i \in \{0, 1\} \tag{5}$$

Constraint (1) ensures that the sensitive itemsets are all concealed, while constraint (2) ensures that nonsensitive itemset is lost if its support drops below σ_{min}. The parameter M in constraint (2) is a positive coefficient who is large enough to ensure that no feasible solution has been cut off, especially $M = \sigma_{min}$. Constraint (3) indicates the relationship between z_i and p_{ik}. If one or more items are marked for removal from a specific transaction, then the new transaction obtained is completely different from the original one. If the transaction is sanitized, then at least one item is marked for removal from that transaction. Constraint (4) indicates the relationship between q_{ij} and p_{ik}. If one or more items are marked for removal from a particular itemset supported by a specific transaction, that itemset loses the support of that transaction. If an itemset is marked as having lost the support of a transaction, at least one item is marked for removal from that itemset for that transaction. Constraint (5) says that all decision variables are binary.

The FIH-0 model can be viewed as a generalization of the set-covering problem, which is known to be a difficulty in general. In order to simplify the primal model, FIH-0 model is implemented in two phases, FIH-1 and FIH-2, in this paper. The first phase, as shown in Section 3.1, is maximizing the transaction accuracy. In the second phase, as described in Section 3.2, it makes sure that the information loss of the altered database is minimum based on the first phase.

3.1 Maximizing Accuracy Process

In FIH-1 model, the transaction accuracy is maximized under the constraints where all sensitive itemsets are hidden, while it does not concern with nonsensitive itemsets. The reason is that the primary target of FIH is to hide all the sensitive itemsets while it is inevitable to influence some nonsensitive itemsets during the sanitization process. Therefore, the outcome of the sanitization for sensitive itemsets is considered solely in the first phase. Maximizing accuracy process is detailed as follows:

$$\max \sum_{i \in D} (z_i - \sum_{j \in S} q_{ij})$$

(FIH − 1) $s.t. \sum_{i \in D} q_{ij} \geq s_j - \sigma_{min} + 1 \qquad \forall j \in S$ (6)

$$z_i = \prod_{j \in S}(1 - q_{ij}) \qquad \forall i \in D$$ (7)

$$z_i, q_{ij} \in \{0, 1\}$$ (8)

The objective is maximizing the data accuracy which includes maximizing the transaction accuracy and avoiding sensitive itemsets over-hiding. Constraint (6) ensures that more than $(s_j - \sigma_{min})$ transactions supporting each sensitive itemset j are sanitized (i.e. the sensitive itemsets are all concealed). Constraint (7) indicates the relationship between z_i and q_{ij}. If the transaction i is sanitized, then at least one sensitive itemset in the transaction i will lose its support. And if one or more sensitive itemset lose their support in the transaction i, then that transaction should be sanitized. By using FIH-1 model, it can be found which transactions should be sanitized and how the supports of sensitive itemsets decrease.

3.2   Minimizing Information Loss Process

We define a new database D', which consists of sanitized transactions (i.e. $z_i=0$ in the result of FIH-1 model). The nonsensitive frequent itemsets, $\tilde{N}$, involved in the second process should satisfy the following two conditions: (1) at least one sensitive item is included; (2) the support of nonsensitive frequent itemset is not less than σ_{min} in the set of unsanitized transactions (i.e. $D \setminus D'$) should be excluded. Based on the FIH-1 model, the following model FIH-2 aims to minimize the number of concealing nonsensitive itemset. Minimizing information loss process is detailed as follows:

$$\min \sum_{j \in \tilde{N}} y_j$$

(FIH − 2) $s.t. \sum_{k} b_{kj} \times p_{ik} \geq 1 \quad \forall i \in D', \forall j \in S : q_{ij} = 1$ (9)

$$\sum_{i \in D'} (1 - \prod_{k \in J_j}(1 - p_{ik})) \leq s_j - \sigma_{min} + M \times y_j \quad \forall j \in \tilde{N}$$ (10)

$$y_j, p_{ik}, b_{kj} \in \{0, 1\}$$ (11)

In this paper, none of new items or itemsets are added into the transactions, because the sanitization process tries to erase the original items in the database. From the above analysis, we can know that the sole resource of the information loss is the hidden nonsensitive frequent itemsets. Hence, the objective of this model is to minimize the number of the lost nonsensitive frequent itemsets. Constraint (9) states that at least one item has to be removed from each of the sensitive itemsets, which are supported by the sanitized transactions. In the other words, all the sensitive itemsets are hidden in the FIH-2 model. Constraint (10) indicates that if more than $(s_j - \sigma_{min})$ transactions supporting a nonsensitive itemset j, are marked for sanitizing, then itemset j is regarded as the lost one. If $y_j = 1$, then nonsensitive frequent itemset j will be lost after the sanitization process. If $p_{ik} = 1$, then the item k in the transaction i will be deleted during the sanitization process. From the result of FIH-2 model, we can find how to sanitize exactly the item in the sanitized transactions in order to hide the sensitive frequent itemsets. And the lost nonsensitive itemsets can be obtained from the results of solving the FIH-2 model.

4 An Illustrative Example

In this section, we illustrate the entire process using the database D in Table 1. Database D is a 7-transaction database. We assume $\sigma_{min} = 3$, so we have the frequent (non-singleton) itemsets listed in Table 2. Let three of these itemsets, $\{i_4, i_5\}, \{i_6, i_8\}$ and $\{i_2, i_3, i_4\}$, be the sensitive itemsets (denoted in bold) and others are non-sensitive itemsets.

Table 1. Database D

TID	items
t1	$i_1, i_2, i_3, i_4, i_5, i_6, i_8$
t2	i_1, i_2, i_6, i_7, i_8
t3	i_1, i_2, i_5
t4	i_1, i_2, i_3, i_4, i_5
t5	i_2, i_3, i_4
t6	i_6, i_8
t7	i_4, i_5, i_6, i_8

Table 2. Frequent (Non-singleton) Itemsets for D $\sigma_{min} = 3$

ID	Itemsets	σ_j	ID	Itemsets	σ_j
1	$\{i_1, i_2\}$	4	6	$\{i_3, i_4\}$	3
2	$\{i_1, i_5\}$	3	7	$\{i_4, i_5\}$	3
3	$\{i_2, i_3\}$	3	8	$\{i_6, i_8\}$	4
4	$\{i_2, i_4\}$	3	9	$\{i_2, i_3, i_4\}$	3
5	$\{i_2, i_5\}$	3	10	$\{i_1, i_2, i_5\}$	3

Based on the approach described in Section 3.1, we formulate the maximizing accuracy model in the following:

$$max \quad z_1 + z_2 + z_3 + z_4 + z_5 + z_6 + z_7 - (q_{17} + q_{18} + q_{19}) - q_{28} - (q_{47} + q_{49}) - q_{59} - q_{68} - (q_{77} + q_{78})$$

$$s.t. \quad q_{17} + q_{47} + q_{77} \geq 3 - 3 + 1 \qquad (\{i_4, i_5\})$$

$$q_{18} + q_{28} + q_{68} + q_{78} \geq 4 - 3 + 1 \qquad (\{i_6, i_8\})$$

$$q_{19} + q_{49} + q_{59} \geq 3 - 3 + 1 \qquad (\{i_2, i_3, i_4\})$$

$$z_1 = (1 - q_{17})(1 - q_{18})(1 - q_{19})$$

$$z_2 = 1 - q_{28}$$

$$z_4 = (1 - q_{47})(1 - q_{49})$$

$$z_5 = 1 - q_{59}$$

$$z_6 = 1 - q_{68}$$

$$z_7 = (1 - q_{77})(1 - q_{78})$$

$$z_1, z_2, z_3, z_4, z_5, z_6, z_7, q_{17}, q_{18}, q_{19}, q_{28}, q_{47}, q_{49}, q_{59}, q_{68}, q_{77}, q_{78} \in \{0, 1\}$$

One optimal solution is $z_2 = z_3 = z_4 = z_5 = z_6 = 1, q_{17} = q_{18} = q_{19} = q_{78} = 1$. This implies that transaction $t1$ and $t7$ are dealt and the accuracy of the transaction is $5/7 = 71.4\%$. The transaction accuracy is measured by the proportion of transactions that are accurate in the sanitized database(i.e. left un-sanitized). The solution also implies that the itemsets $\{i_4, i_5\}, \{i_6, i_8\}$ and $\{i_2, i_3, i_4\}$ in transaction $t1$ as well as $\{i_4, i_5\}$ in transaction $t7$ should be dealt. Then we formulate the minimizing information loss model through above results. So the new database D' consists of transaction $t1$ and $t7$. Furthermore $\widetilde{N} = \{\{i_1, i_2\}\{i_1, i_5\}\{i_2, i_3\}\{i_2, i_4\}\{i_2, i_5\}\{i_3, i_4\}\{i_1, i_2, i_5\}\}$. So we have the minimizing information loss model as follow:

$$min \quad y_2 + y_3 + y_4 + y_5 + y_6 + y_{10}$$

$$s.t. \quad p_{1i_4} + p_{1i_5} \geq 1 \qquad (q_{17} = 1)$$

$$p_{1i_6} + p_{1i_8} \geq 1 \qquad (q_{18} = 1)$$

$$p_{1i_2} + p_{1i_3} + p_{1i_4} \geq 1 \qquad (q_{19} = 1)$$

$$p_{7i_6} + p_{7i_8} \geq 1 \qquad (q_{78} = 1)$$

$$1 - (1 - p_{1i_5}) \leq 3 - 3 + 3y_2 \qquad (y_2)$$

$$1 - (1 - p_{1i_2})(1 - p_{1i_3}) \leq 3 - 3 + 3y_3 \qquad (y_3)$$

$$1 - (1 - p_{1i_2})(1 - p_{1i_4}) \leq 3 - 3 + 3y_4 \qquad (y_4)$$

$$1 - (1 - p_{1i_2})(1 - p_{1i_5}) \leq 3 - 3 + 3y_5 \qquad (y_5)$$

$$1 - (1 - p_{1i_3})(1 - p_{1i_4}) \leq 3 - 3 + 3y_6 \qquad (y_6)$$

$$1 - (1 - p_{1i_2})(1 - p_{1i_5}) \leq 3 - 3 + 3y_{10} \qquad (y_{10})$$

$$y_2, y_3, y_4, y_5, y_6, y_{10}, p_{1i_2}, p_{1i_3}, p_{1i_4}, p_{1i_5}, p_{1i_6}, p_{1i_8}, p_{7i_6}, p_{7i_8} \in \{0, 1\}$$

One optimal solution is $y_4 = y_6 = 1, p_{1i_4} = p_{1i_6} = p_{7i_6} = 1$. This implies that we should remove the item i_4 and i_6 from transaction $t1$ and remove the item i_6 from transaction $t7$. As a result, we lose two itemsets, $\{i_2, i_4\}$ and $\{i_3, i_4\}$, and the information loss is $2/7=28.6\%$. Consequently, we should remove three items to hide the sensitive itemsets and the accuracy of transaction is 71.4%, meanwhile the information loss is 28.6%.

5   Computational Experiment and Results

All the datasets that used in this section are publicly available through the FIMI repository located at http://fimi.cs.helsinki.fi/. The following datasets, CHESS and MUSHROOM, were prepared by Roberto Bayardo from the UCI datasets and PUMSB. CHESS has 3196 transactions and there are 75 items in the transactions. There are 609 non-singleton frequent itemsets when we set σ_{min} to be 2877. MUSHROOM has 8124 transactions and there are 119 items in the transactions. There are 544 non-singleton frequent itemsets when we set σ_{min} to be 3250. Average transaction length (the number of items in each transaction in average.) of CHESS is 37 while one of MUSHROOM is 23.

Table 3. Results from Solving Model FIH-1

HS	chess			mushroom		
	# of sanitized Tran.	# of sanitized itemset	time (s)	# of sanitized Tran.	# of sanitized itemset	time (s)
(3,3)	118	160	0.064	579	581	0.074
(7,3)	186	247	0.095	1523	2033	0.078
(3,5)	107	115	0.061	458	485	0.068
(7,5)	190	385	0.162	1025	1981	0.125

In this paper, the sensitive itemsets are selected randomly by "Hiding Scenarios". The "Hiding Scenarios"(HS) indicates the nature and the number of sensitive itemsets which are selected for hiding. More specifically, we make use of the notion (a, b) to illustrate by a the number of sensitive itemsets that are

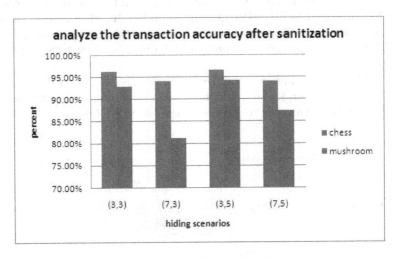

Fig. 1. Analyze the Transaction Accuracy

Table 4. Sanitization in Different Hiding Scenarios

	CHESS		MUSHROOM	
HS	# of sanitized Tran.	# of deleting item	# of sanitized Tran.	# of deleting item
(3,3)	118	118	579	582
(7,3)	186	200	1523	1574
(3,5)	107	107	458	511
(7,5)	190	222	1025	1031

hidden each time, as well as, by b the average length of the sensitive itemset (i.e. the number of the item which the itemset supports). In particular, we assume that all the sensitive itemsets are the same length.

The code was implemented in Matlab on a PC running Window XP on an Intel Core 2, 2.93GHz processor, with the integer programming being solved using CPLEX 12.1.

Table 3 shows that sanitized transaction number is always less than or equal to altered itemset number under different HS. That satisfies our purpose, protecting the transaction integrality, so it alters itemsets in the same transaction as many as possible. As MUSHROOM is a sparse database, its sanitization transaction number is more than the sanitization transaction number in CHESS. Sanitized transactions are more disperse owning to the disperse itemset distribution in sparser database. The σ_{min} of MUSHROOM is lower than that of CHESS, which also leads to the more process itemsets. The reason is that hiding sensitive itemsets is achieved by transforming them into non-frequent itemsets and whether to be the non-frequent ones is concerned with σ_{min}.

Figure 1 shows the transaction accuracy in different databases and HS. The transaction accuracy is measured by the proportion of transactions that are accurate in the sanitized database. In CHESS database the transaction accuracy is all higher than that in MUSHROOM under all HS. It indicates that the sanitization influence to the dense database is less than that to the sparse database. From Figure 1, the transaction accuracy of CHESS is at least 94% and that of MUSHROOM is above 80%.

From table 4, the number of both sanitized transaction and deleting item become larger when parameter a becomes larger if parameter b is the same as before. This trend is much more obvious in the MUSHROOM.

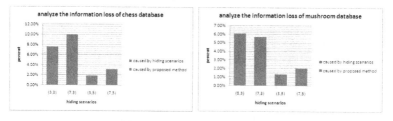

Fig. 2. Analyze Information Loss

The information loss is measured by the proportion of nonsensitive itemsets that are concealed by mistake. As can be seen from Figure 2, the loss rate is within 10%. In both CHESS and MUSHROOM, the longer the sensitive itemset, the less the information loss is.

6   Conclusion

This paper solves the problem of hiding sensitive patterns in the field of association rule mining. Differing from the previous methods, we present a novel exact approach to address related problems when considering the tradeoff between transaction accuracy and information loss. As the problem is very difficult to solve, we propose a two-phase approach in order to reduce the complexity of the problem. The experiments' results illustrate that our approach is both effective and efficient. With the development of society and economy, the distributed databases are applied in nowadays. The association rule hidden problem in the distributed mining will be an important venue for the future research.

Acknowledgements.   This   work   is   partially   supported   by   NSFC (No.71171027), the key project of NSFC(No.71031002), NCET-12-0081, LNET(No.LJQ2012004) and the Fundamental Research Funds for the Central Universities(No. DUT11SX11 and DUT12ZD208) and MOST (No.2011BAH30B01-03).

References

1. Menon, S., Sarkar, S., Mukherjee, S.: Maximizing accuracy of shared databases when concealing sensitive patterns. Information Systems Research 16(3), 256–270 (2005)
2. Oliveira, S.R.M., Zaïane, O.R.: Privacy preserving frequent itemset mining. In: Proc. of the 2002 IEEE International Conference on Privacy, Security and Data Mining (CRPIT 2002), pp. 43–54 (2002)
3. Oliveira, S.R.M., Zaïane, O.R.: Protecting sensitive knowledge by data sanitization. In: Proceeding of the Third IEEE International Conference on Data Mining (ICDM 2003), pp. 211–218 (2003)
4. Amiri, A.: Dare to share:Protecting sensitive knowledge with data sanitization. Decision Support Systems 43(1), 181–191 (2007)
5. Sun, X., Yu, P.S.: A border-based approach for hiding sensitive frequent itemsets. In: ICDM 2005: Proceedings of the Fifth IEEE International Conference on Data Mining, pp. 426–433 (2005)
6. Menon, S., Sarkar, S.: Minimizing information loss and preserving privacy. Management Science 53(1), 101–116 (2007)
7. Guo, Y.: Reconstruction-Based Association Rule Hiding. In: Proceedings of SIG-MOD 2007 Ph.D.Workshop on Innovative Database Research 2007 (IDAR 2007), June 10 (2007)
8. Gkoulalas-Divanis, A., Verykios, V.S.: Exact Knowledge Hiding through Database Extension. IEEE Transactions on Knowlege and Data Engineering 21(5), 699–713 (2009)

Measuring Formula-Variable Relevance in Four-Valued Logic by Forgetting

Xin Liang and Zuoquan Lin

School of Mathematical Sciences, Peking University, Beijing 100871, China
{xliang,lz}@pku.edu.cn

Abstract. This paper discusses an approach to evaluate the relevance between a formula and a variable in it under the four-valued semantics. In the existing literature, for the classical two-valued propositional logic, there are definitions of whether a variable is independent to a certain formula. These definitions are based on forgetting, which is an operation to maintain and manage knowledge bases. Forgetting has its semantic connections with relevance. In the previous work in which the authors of this paper participated, an approach of quantitatively defining formula-variable relevance in two-valued propositional logic was proposed, which not only gave the judgement of relevant or not, but also gave a measurement of how relevant they are (i.e., the degree of relevance), also using the tool of variable forgetting. In this paper, we adapt the method to four-valued logic. Because forgetting has not been defined in four-valued logic yet, our first step is to define forgetting in four-valued logic. Then we will define formula-variable relevance quantitatively in four-valued logic. It will be a tool for the management of the knowledge bases under four-valued logic.

Keywords: knowledge representation, four-valued logic, relevance, forgetting.

1 Introduction

Belnap's four-valued logic [1,2] is a typical kind of multivalued logics [3], introducing *paraconsistency* into the classical inference, thus it is a kind of paraconsistent logics [4], which give a solution to the problem of *ex falso quodlibet* (or principle of explosion). If there exists a formula ϕ and its negation $\neg\phi$ simultaneously in the knowledge base, then we call the knowledge base *inconsistent*. In classical logics, such as the two-valued propositional logic and first-order logic, inconsistency causes every well-formed formula to become a theorem in the knowledge base, and this makes the knowledge base trivial, and useless. This is the principle of explosion. There are methods for handling inconsistencies [5]. Paraconsistent logics are a set of logics in which inconsistencies are not eliminated, but tolerated during inference, by rejecting the principle of explosion, which means that, if there exist both ϕ and $\neg\phi$ in the knowledge base, there are still formulas which can not be deduced from it, thus this avoids the triviality of the knowledge base, and some

M. Wang (Ed.): KSEM 2013, LNAI 8041, pp. 336–348, 2013.

meaningful inferences can be done in the knowledge base. Four-valued logic is a kind of paraconsistent logics, by introducing two new truth values "⊤" and "⊥". The paraconsistency lies in the new truth values.

Forgetting [6] is a tool for managing knowledge bases. By forgetting the irrelevant, outdated, incorrect or inconsistent information, we may improve the efficiency of inference by focusing on the most important issues, as well as resolve inconsistencies by forgetting the inconsistent information. Thus forgetting is a way to keep the knowledge base well-maintained. Forgetting is not simply an operation rudely "deleting" information. Actually, it is a process weakening knowledge bases, and the process itself has its own semantic meanings. Lin and Reiter firstly introduced forgetting into the field of artificial intelligence [6], and discussed forgetting a fact (a ground atom in first-order logic, or a variable when it comes to the situation of propositional logic), and forgetting a relation (a predicate in first-order logic). Lang, Liberatore, and Marquis discussed literal (a variable or a negation of a variable) forgetting [7]. Xu and Lin extended the concept to formula forgetting [8].

Forgetting has its connections with relevance. Take formula-variable relevance for an example. Forgetting is an operation regarding a formula and a variable in it. If the forgetting process does not change the original theory too much, then it may be a sign showing that the variable itself is not so relevant to the formula before forgetting. If forgetting does change the theory a lot, it means that the variable may be strongly relevant to the original formula. Forgetting could be used as a tool to define relevance between formulas and variables. In fact, in the previous literature [6,7], such concepts of relevance, or independence, were defined. However, these definitions are "binary", only showing two possible conditions: relevant or not, and lacking the quantitative measurement of the degree of relevance between a formula and a variable. Thus, in the recent work that the authors of this paper participated in [9], a method was proposed to define how relevant a formula and a variable are. The key idea is to evaluate the degree of change of the formula before and after forgetting the variable.

Defining formula-variable relevance is also an interesting issue in four-valued logic. It is important to know whether a variable is relevant or not, and how relevant, with respect to a certain formula when we maintain and manage the four-valued knowledge base. This gives us a better understanding of the current status of the various elements (variables) of the knowledge base. Also, when we manipulate the knowledge base according to some preferential relations among variables, the quantitative measurement of formula-variable relevance could be used as a criteria for defining preference. In the case of classical two-valued propositional knowledge base, forgetting could be used as a tool to resolve inconsistencies [10,11], and during the process, a preference relation among sets of variable sets is needed. Formula-variable relevance could be used to define such preference relation [9]. In the case of four-valued knowledge base, there should be other circumstances that preference relations among variables are needed, and in these cases, defining formula-variable relevance for four-valued logic is useful, not only for the theoretical need, but also for the practical one.

In this paper, we will use the method by which formula-variable relevance was defined in the classical two-valued propositional logic [9]. We apply this method to define the degree of relevance between a formula and a variable in it in four-valued logic. This approach uses variable forgetting as a tool. However, by now, we have seen forgetting defined in description logics [12], logic programming [13], modal logics [14], etc., but not defined in four-valued logic yet. So the first part of our work would be introducing variable forgetting into four-valued logic, applying the key ideas and techniques of forgetting to four-valued logic. Then we use this tool to define formula-variable relevance in four-valued logic.

The main contributions of this paper could be summarized as follows:

1. Defining variable forgetting in four-valued logic, and giving some properties of it.
2. Using the tool of variable forgetting to define formula-variable relevance in four-valued logic, and discussing some of its properties.

The next sections of this paper are organized as follows: In Section 2, we talk about some background knowledge and notations, including four-valued logics, forgetting, and formula-variable relevance. In Section 3, we define forgetting in four-valued logic, with properties and examples. In Section 4, we use the tool of forgetting to define formula-variable relevance in four-valued logic, also with properties and examples. In Section 5, we conclude our work, discuss related work and future work.

2 Preliminaries

This paper is based on propositional four-valued logic. In this section, we introduce some background knowledge and notations.

2.1 Four-Valued Logic

Here is a brief introduction of four-valued logic. For the details, the readers may refer to relevant papers [1,2].

The significant feature of four-valued logic is that it has four truth values, instead of two. The truth values are $t, f, \top, \bot$, where t, f have their meanings just as in classical logic, and $\top$ means "too much knowledge" (i.e., inconsistent), while $\bot$ means "lack of knowledge". The relations of the four truth values are shown in Figure 1.

The x-axis of this figure stands for the "degree" of truth, and the y-axis of this figure stands for the "degree" of knowledge. There are two partial orders on the four truth values, namely $\leq_t$ and $\leq_k$. A line between two truth values stands for that they have relations with respect to the partial orders, and the orders are shown just as the direction of the two axises indicated in the figure. For example, $f \leq_t \top, f \leq_k \top, \top \leq_t t, t \leq_k \top$, and by transitivity, we have $f \leq_t t$. However, neither $f \leq_k t$, nor $t \leq_k f$. The two truth values are not comparable with respect to $\leq_k$. Two binary operators on the truth values, $\wedge$ and $\vee$, are

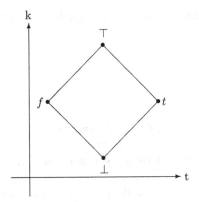

Fig. 1. (From Arieli and Avron's paper [2].) The relations of the four truth values

defined respectively as the greatest lower bound and the least upper bound with respect to $\leq_t$. For example, $f \wedge t = f$, $\top \wedge \bot = f$, $f \vee t = t$. It is easy to notice that they are extensions of the corresponding operators in classical two-valued semantics. Other two binary operators, $\otimes$ and $\oplus$, are defined respectively as the greatest lower bound and the least upper bound with respect to $\leq_k$. There is also an unary operator $\neg$, which has the same role as it does in two-valued logics, i.e., $\neg f = t$ and $\neg t = f$. However, it has no effects on $\top$ or $\bot$, i.e., $\neg \top = \top$ and $\neg \bot = \bot$.

Let PS be a set of propositional variables that has the form of $\{p, q, r, \ldots\}$ or $\{p_1, p_2, \ldots\}$. Given PS, we may have formulas defined on it. A valuation is a mapping from PS to $\{t, f, \top, \bot\}$, assigning every variable a truth value. By the semantics of the connectives, given a formula and a valuation, we may get the truth value of the entire formula. There is a set of designated elements $\mathcal{D} = \{t, \top\}$. We say that a valuation ν satisfies a formula ϕ, denoted by $\nu \models_4 \phi$, if ϕ's truth value is in $\mathcal{D}$ under valuation ν. Given a formula ϕ and a valuation ν, we say that ν is a model of ϕ, if ν satisfies ϕ. The set of all the models of ϕ is denoted as $\mathrm{Mod}_4(\phi)$. Given two formulas ϕ and ψ, we say that ϕ implies ψ, denoted by $\phi \models_4 \psi$, if all the models of ϕ are the models of ψ, i.e., $\mathrm{Mod}_4(\phi) \subseteq \mathrm{Mod}_4(\psi)$. In this paper, we say $\phi \equiv_4 \psi$, if ϕ and ψ have the same set of models, i.e., $\mathrm{Mod}_4(\phi) = \mathrm{Mod}_4(\psi)$ [1]. The language of four-valued logic consists of PS, and the logical connectives $\{\vee, \wedge, \oplus, \otimes, \supset, t, f, \bot, \top\}$. Here t (or f, $\bot$, $\top$) represents a formula whose truth value is always t (or f, $\bot$, $\top$) under all possible valuations. The value of $p \supset q$ is the value of q if the value of p is in $\mathcal{D}$, and otherwise the value of $p \supset q$ is t.

---

[1] Note that because there are two designated elements in the set $\mathcal{D}$, so although if two formulas have the same model set, it does not mean that they have the same truth value under all valuations. There may be a case that under one model, one formula is valued t while the other valued $\top$, and there may be a case that under one valuation which is not a model, one formula is valued f while the other valued $\bot$. So having the same model set is a sense of "weakly equivalent", which is slightly different from the case in two-valued logics.

Example 1. There is a famous dilemma on logic-based knowledge representation. It says:

- All birds can fly.
- Penguins are birds.
- Penguins cannot fly.
- Tweety is a penguin.

Then we get both "Tweety can fly" and "Tweety cannot fly", which is inconsistent. In classical two-valued logics, this yields the "explosion" of the knowledge base, which means that every formula is a theorem of the knowledge base, thus the knowledge base becomes trivial. In four-valued semantics, such inconsistencies are tolerated. Here we introduce the formulation in Arieli and Avron's paper [2], which is using propositional four-valued logic, and only concerning the situation of Tweety, instead of other birds or penguins.

- $TB \mapsto TF$.
- $TP \supset TB$.
- $TP \supset \neg TF$.
- TB.
- TP.

The set of formulas is denoted by Γ, and $p \mapsto q$ means $\neg p \vee q$. TB means Tweety is a bird; TF means Tweety can fly; TP means Tweety is a penguin. Γ has six models in four-valued semantics, listed in Table 1.

Table 1. $\mathrm{Mod}_4(\Gamma)$

No.	TB	TF	TP
1	$\top$	$\top$	$\top$
2	$\top$	$\top$	t
3	$\top$	f	$\top$
4	$\top$	f	t
5	t	$\top$	$\top$
6	t	$\top$	t

In four-valued semantics, Γ *does* have models, and this makes it not necessarily imply everything. Triviality is removed. Actually, in four-valued semantics, $p \wedge \neg p$ has a model (when p is assigned as $\top$). From this the readers may see that the paraconsistency of four-valued logic comes from the new truth values, the selection of designated elements $\mathcal{D} = \{t, \top\}$, and the semantics of logical connectives as well.

2.2 Forgetting

In this paper, we only consider the case of forgetting a variable from a formula in propositional case. Now we introduce forgetting in two-valued propositional logic. The readers may refer to relevant papers for details [6,7]. Notice that the concepts and symbols used in this subsection and the next subsection are all based on two-valued propositional logic.

The idea of forgetting is to expand the current set of models of a certain formula in a certain knowledge base. If we want to forget a variable p in a formula ϕ, then we find all the models of ϕ. For each model of ϕ, we change the valuation of p to the opposite one, and get a new model (denoted by ω'). If $\omega' \in \text{Mod}(\phi)$, then it is fine and we do not do anything. If not, then we add ω' as a new model. For each original model of ϕ, we do this, and finally we get a new model set, which is actually the model set of the formula after forgetting p in ϕ. From this new model set we may give a formula corresponding to it. Actually, forgetting is an operation to add new models into the formula's model set, thus it is a process weakening formulas, because the more models we have in a formula, the less information it contains. (One extreme condition is that a formula does not have any models, then it implies everything, and has the strongest inference ability, although it may be trivial. Another extreme condition is that every possible valuations are models of a certain formula, then this formula is a tautology, and could not imply anything except other tautologies, so it has the weakest inference ability.) Let us come back to the case of forgetting p in ϕ. After forgetting p, we may get pairs of models in the new model set. In each pair, the valuation of all the variables except p are the same, and for p, in one model it is assigned as t, and in the other it is assigned as f. In this way, we weaken the formula by "erasing" information about the truth or falsity of the variable p (we cannot judge the truth or falsity of p in this new model set), and our aim of forgetting p in ϕ is reached.

Formally, let ϕ be a formula, and p a variable in it. The result of forgetting p in ϕ is denoted as ForgetVar(ϕ, p). Its model set is described as follows:

$$\text{Mod}(\text{ForgetVar}(\phi, p)) = \text{Mod}(\phi) \cup \{\text{Switch}(\omega, p) \mid \omega \models \phi\} , \qquad (1)$$

where $\text{Switch}(\omega, p)$ means to change the valuation of p in ω to its opposite one.

Above is the semantic definition of forgetting. There is a syntactic counterpart, which is stated as:

$$\text{ForgetVar}(\phi, p) \equiv \phi_{p \leftarrow t} \vee \phi_{p \leftarrow f} , \qquad (2)$$

where $\phi_{p \leftarrow t}$ (or $\phi_{p \leftarrow f}$) means to change every occurance of p in ϕ to tautology t (or contradiction f). Note that a tautology is a formula which is always true under any possible valuations, and a contradiction is a formula which is always false under any possible valuations.

The semantic and syntactic definitions of forgetting are equivalent, in the sense that

$$\text{Mod}(\phi_{p \leftarrow t} \vee \phi_{p \leftarrow f}) = \text{Mod}(\phi) \cup \{\text{Switch}(\omega, p) \mid \omega \models \phi\} . \qquad (3)$$

Example 2. Let ϕ_1 be $p \wedge q$, and ϕ_2 be $p \vee q$, then ForgetVar$(\phi_1, p) \equiv q$, and ForgetVar(ϕ_2) is tautology.

2.3   Formula-Variable Relevance

Forgetting could be used to define the relevance between a formula and a variable in it. For the "binary" definitions (i.e., relevant or not, without "degrees"), the readers may refer to the existing literature in forgetting [6,7]. For quantitatively measuring the degree of formula-variable relevance in two-valued propositional logic, the readers may refer to the previous work before this paper [9]. Here we briefly introduce the method for evaluating formula-variable relevance quantitatively.

The idea of measuring formula-variable relevance is firstly using forgetting as a tool. Given a formula ϕ and a variable p, we forget p in ϕ. This procedure is not really forgetting it in the knowledge base, but using forgetting as a tool to measure the relevance. We compare the change of model set before and after the forgetting process. The intuition is that the more the model set changes during forgetting, the more relevant ϕ and p are. If the model set does not change a lot, it means that the original formula ϕ is not so relevant to p (so forgetting p does not change a lot). If the model set changes a lot, it means that ϕ is strongly relevant to p (so the forgetting process brings a big change). The formal definition is: Let ϕ be a formula defined on PS, p a variable in it, then the degree of formula-variable relevance, denoted by $R_{\mathrm{FV}}(\phi, p)$, is

$$R_{\mathrm{FV}}(\phi, p) = \frac{|\mathrm{Mod}(\mathrm{ForgetVar}(\phi, p)) \setminus \mathrm{Mod}(\phi)|}{2^{|PS|-1}} . \tag{4}$$

The value ranges from 0 to 1.

3   Forgetting in Four-Valued Logic

In this section, we give a formal definition of forgetting in four-valued logic, applying the ideas of forgetting in classical logics. This could be seen as extending the concept of forgetting to four-valued logic, and also could be seen as a tool ready for defining formula-variable relevance in four-valued logic in Section 4.

3.1   The Definition

Here we consider the problem from a semantic aspect of view. Because forgetting is a process introducing new models to the existing model set, and the new models are obtained by switching the valuation of the variable which is to be forgotten, and keeping other variables' valuations. This idea could also be applied when we define forgetting in four-valued logic.

Definition 1. *Let PS be a variable set, ϕ a formula on PS in four-valued logic, and p a variable in it. The result of forgetting p in ϕ is denoted by* $\mathrm{ForgetVar}_4(\phi, p)$. *The model set of it is defined as follows:*

$$\mathrm{Mod}_4(\mathrm{ForgetVar}_4(\phi, p)) = \bigcup_{v \in \{t, f, \top, \bot\}} \{\mathrm{Assign}(\omega, p, v) \mid \omega \models_4 \phi\} , \tag{5}$$

where ω is a valuation, and $\mathrm{Assign}(\omega, p, v)$ means to modify the valuation of p in ω to v.

Next let us come to a simple example showing that this definition is suitable and intuitive.

Example 3. Let $PS = \{p, q\}$, ϕ be $p \wedge q$, then we may get the model set of ϕ in Table 2.

Table 2. The model set of $p \wedge q$ in four-valued semantics

No.	p	q
1	t	t
2	t	$\top$
3	$\top$	t
4	$\top$	$\top$

According to our definition, we could get the model set of ForgetVar$_4(\phi, p)$, shown in Table 3.

Table 3. The model set of ForgetVar$_4(p \wedge q, p)$

No.	p	q
1	t	t
2	t	$\top$
3	$\top$	t
4	$\top$	$\top$
5	f	t
6	$\bot$	t
7	f	$\top$
8	$\bot$	$\top$

There are eight models (four of them are new models) for ForgetVar$_4(\phi, p)$. Notice that ForgetVar$_4(\phi, p) \equiv_4 q$.

From the enlarged new model set obtained after forgetting, we could actually find a corresponding formula just having the new model set. (See the example above.) Notice that formulas sharing the same model set in four-valued logic are only "weakly equivalent" (denoted by $\equiv_4$ in this paper). For this "weak equivalence", the readers may refer to the footnote in Section 2.1.

3.2 Properties

In this subsection, we show that some nice properties of forgetting which hold in the classical two-valued logic (see Lin and Reiter's paper [6]) also hold in four-valued logic.

Firstly, forgetting is a process weakening formulas.

Proposition 1. *Let ϕ be a formula, p a variable in it, then we have*

$$\phi \models_4 \text{ForgetVar}_4(\phi, p) \ . \tag{6}$$

Proof. By the definition, forgetting is a process introducing new models, while keeping the original models. So $\text{Mod}_4(\phi) \subseteq \text{Mod}_4(\text{ForgetVar}_4(\phi, p))$, thus $\phi \models_4$ ForgetVar$_4(\phi, p)$. □

The next proposition shows that forgetting a variable does not affect the inference ability about a formula which does not contain this variable.

Proposition 2. *Let ϕ be a formula on PS, p a variable in it, and ψ a formula on PS, but not mentioning p, then we have*

$$\phi \models_4 \psi \text{ if and only if } \text{ForgetVar}_4(\phi, p) \models_4 \psi . \tag{7}$$

Proof. "$\Longrightarrow$". From $\phi \models_4 \psi$, we may get $\text{Mod}_4(\phi) \subseteq \text{Mod}_4(\psi)$. Notice that ψ does not contain p. Under PS, we may say that ψ have groups of models, and in each group, the valuations of variables except p are the same, and the valuation of p varies from $\{t, f, \top, \bot\}$. In such conditions, consider $\text{Mod}_4(\text{ForgetVar}_4(\phi, p))$. Because the forgetting operation only changes the valuation of p and adds these new models, so whatever the valuation of p it may change to, the modified model is still in $\text{Mod}_4(\psi)$. Thus $\text{Mod}_4(\text{ForgetVar}_4(\phi, p)) \subseteq \text{Mod}_4(\psi)$. So ForgetVar$_4(\phi, p) \models_4 \psi$.

"$\Longleftarrow$". From ForgetVar$_4(\phi, p) \models_4 \psi$, we have $\text{Mod}_4(\text{ForgetVar}_4 (\phi, p)) \subseteq \text{Mod}_4(\psi)$. Because $\text{Mod}_4(\phi) \subseteq \text{Mod}_4(\text{ForgetVar}_4(\phi, p))$, we have $\text{Mod}_4(\phi) \subseteq \text{Mod}_4(\psi)$, and thus $\phi \models_4 \psi$. □

The next theorem shows that just like the case in classical logic, forgetting also has a syntactic representation in four-valued logic. Notice that symbol "$\equiv_4$" in this paper does not mean "exactly equivalent" in four-valued semantics. It has the sense of "weakly equivalent" (see the footnote in Section 2.1).

Theorem 1. *Let ϕ be a formula, and p a variable in it, then we have*

$$\text{ForgetVar}_4(\phi, p) \equiv_4 \phi_{p \leftarrow t} \vee \phi_{p \leftarrow f} \vee \phi_{p \leftarrow \top} \vee \phi_{p \leftarrow \bot} . \tag{8}$$

Proof. We need to prove $\text{Mod}_4(\text{ForgetVar}_4(\phi, p)) = \text{Mod}_4(\phi_{p \leftarrow t}) \cup \text{Mod}_4(\phi_{p \leftarrow f}) \cup \text{Mod}_4(\phi_{p \leftarrow \top}) \cup \text{Mod}_4(\phi_{p \leftarrow \bot})$. Notice that $\text{Mod}_4(\phi)$ can be divided into four parts by the four different valuations of p in the models, denoted by M_t, M_f, $M_\top$, and $M_\bot$, respectively, according to the valuation of p. For each part, take the process (of adding new models) in forgetting, and we get the new model sets M'_t, M'_f, $M'_\top$, and $M'_\bot$, respectively. The union of the four sets is actually the model set of ForgetVar$_4(\phi, p)$. Also notice that M'_t, M'_f, $M'_\top$, and $M'_\bot$ are actually $\text{Mod}_4(\phi_{p \leftarrow t})$, $\text{Mod}_4(\phi_{p \leftarrow f})$, $\text{Mod}_4(\phi_{p \leftarrow \top})$, and $\text{Mod}_4(\phi_{p \leftarrow \bot})$, respectively. Thus, we have proved our desired conclusion. □

By the above theorem, the next conclusion is not hard to prove. It shows that forgetting is irrelevant with order when we have multiple variables to forget (in the sense of "weakly equivalent").

Proposition 3. *Let ϕ be a formula, p_1, p_2 two variables occuring in it. We have*

$$\text{ForgetVar}_4(\text{ForgetVar}_4(\phi, p_1), p_2) \equiv_4 \text{ForgetVar}_4(\text{ForgetVar}_4(\phi, p_2), p_1) . \tag{9}$$

Proof. Use the syntactic representation in the theorem above, and notice that these symbolic transformations are irrelevant with order. □

4   Formula-Variable Relevance in Four-Valued Logic

Now, with the tool of forgetting, we can give a definition of formula-variable relevance in four-valued logic. The key is to measure the change of model set after forgetting, applying the idea of defining formula-variable relevance in classical logics.

Definition 2. *Let ϕ be a formula defined on the proposition set PS, p a variable in ϕ. The degree of relevance between ϕ and p, denoted by $R_{FV4}(\phi, p)$, is defined as:*

$$R_{FV4}(\phi, p) = \frac{|Mod_4(ForgetVar_4(\phi, p)) \setminus Mod_4(\phi)|}{3 \times 4^{|PS|-1}} . \tag{10}$$

The denominator is to restrict the value of relevance to $[0, 1]$, which is shown in the next proposition.

Proposition 4. *For any ϕ defined on a certain PS, and variable p occurring in ϕ,*

$$0 \leq R_{FV4}(\phi, p) \leq 1 , \tag{11}$$

and for any $i \in \{0, 1, 2, \ldots, 3 \times 4^{|PS|-1}\}$, we can find some ϕ and p, such that $R_{FV4}(\phi, p) = i/(3 \times 4^{|PS|-1})$.

Proof. For the first part, it is obvious that $R_{FV4}(\phi, p) \geq 0$. Note that ϕ could have at most $4^{|PS|}$ models. If $|Mod_4(\phi)| \in [0, 4^{|PS|-1}]$, then by adding models during forgetting, we have $|Mod_4(ForgetVar_4(\phi, p)) \setminus Mod_4(\phi)| \leq 3 \times |Mod_4(\phi)| \leq 3 \times 4^{|PS|-1}$. If $Mod_4(\phi) \in [4^{|PS|-1}, 4^{|PS|}]$, then the number of valuations that are not models of ϕ would not be greater than $4^{|PS|} - 4^{|PS|-1} = 3 \times 4^{|PS|-1}$. So the number of increased models after the forgetting process would not be greater than $3 \times 4^{|PS|-1}$. Thus, in all cases, we have $R_{FV4}(\phi, p) \leq 1$.

For the second part, we may construct these kinds of model sets such that $|Mod_4(ForgetVar_4(\phi, p)) \setminus Mod_4(\phi)| = i$, and from the model set, we may get the corresponding formula. $\qquad\square$

Notice that when we have a larger set of variables PS' (such that $PS' \supseteq PS$), and ϕ is a formula on PS, p a variable in ϕ, then $R_{FV4}(\phi, p)$ under PS' is the same as $R_{FV4}(\phi, p)$ under PS.

Next we show an example of formula-variable relevance under the four-valued semantics.

Example 4. Consider the relevance between p and $p \wedge q$. The readers may refer to Table 2 and Table 3 for the model sets. By the definition, we may calculate that $R_{FV4}(p \wedge q, p) = 1/3$.

Notice that in classical two-valued propositional logic, the value of $R_{FV}(p \wedge q, p)$ would be $1/2$. The readers may see that the difference of values is caused by the difference of semantics of the two kinds of logics.

Example 5. $R_{FV4}(p, p) = 2/3$, and $R_{FV4}(p \wedge \neg p, p) = 1$.

Also, notice that in two-valued circumstance, $R_{FV}(p,p) = 1$, while $R_{FV}(p \wedge \neg p, p) = 0$. They are different from their four-valued counterparts. This is due to the difference between two-valued semantics and four-valued semantics.

Then we show some properties of formula-variable relevance in four-valued logic.

Proposition 5. *Let ϕ and ψ be two four-valued formulas on PS, and p a variable in them.*

1. *If $\phi \equiv_4 \psi$, then $R_{FV4}(\phi, p) = R_{FV4}(\psi, p)$.*
2. *$R_{FV4}(t, p) = R_{FV4}(f, p) = R_{FV4}(\top, p) = R_{FV4}(\bot, p) = 0$.*
3. *If $\phi \models_4 p$ or $\phi \models_4 \neg p$, and ϕ has models, then $R_{FV4}(\phi, p) > 0$.*

Proof. 1. In this case, ϕ and ψ have the same model set, so they share the same forgetting process and have the same relevance value with respect to p.
2. In all the four cases, no new models are introduced during the forgetting process, so the relevance values with respect to p are all 0.
3. We know that ϕ has models. If $\phi \models_4 p$, then in all the models of ϕ, p is assigned as t or $\top$. If $\phi \models_4 \neg p$, then in all the models of ϕ, p is assigned as f or $\top$. So there are definitely new models introduced during the forgetting process, thus $R_{FV4}(\phi, p) > 0$.   □

As we have mentioned, defining formula-variable relevance in four-valued logic could be useful in some circumstances of knowledge base management. For example, if there are some limitations on the number of variables in the knowledge base under four-valued logic, and when the number of variables exceeds the limit, we "forget" some existing variables, and by the symbolic transformation way of forgetting (see Theorem 1) , we may get rid of these variables. This may lose some information, but it is probably needed when the number of variables is limited. Formula-variable relevance could be used as a criteria for choosing which variables to forget. We may find the least relevant variables with respect to the knowledge base.

Example 6. Let ϕ be $(p \wedge q) \vee (\neg p \wedge q)$, then we have $R_{FV4}(\phi, p) = 1/6$, and $R_{FV4}(\phi, q) = 1/2$. Thus, ϕ is more relevant with q than with p. If we have to eliminate one of the variables in ϕ due to the limitation of the maximum number of variables, according to our criteria, forgetting p in ϕ (by symbolic transformation) would be a better choice (when there is only one formula ϕ listed as above).

Let us come back to the Tweety example (see Example 1). To calculate the relevance between a *knowledge base* and a variable, we may treat the knowledge base as one formula, which is the conjunction of all the formulas in the knowledge base, then we can calculate the formula-variable relevance for the variable as the desired result.

Example 7. For the Tweety example (see Example 1), by the model set of Γ (see Table 1), we may calculate that $R_{FV4}(\Gamma, TB) = R_{FV4}(\Gamma, TF) = 5/24$, while $R_{FV4}(\Gamma, TP) = 1/8$. So TP is the least relevant variable according to our criteria.

Notice that because the intuition of four-valued semantics is more complex than the classical two-valued semantics, so the intuition of the definition of forgetting and formula-variable relevance in four-valued logic is more complex than the case in two-valued logic. This may need further and more detailed investigations.

5 Related Work and Conclusions

For the definitions of forgetting in different types of logics, the readers may refer to the work of Wang et al. [12] for DL-Lite (a type of description logics), Eiter and Wang's work [13] for answer set programming, as well as Zhang and Zhou's work [14] for modal logic S5, etc. Our work applies the idea of forgetting to four-valued logic. In fact, forgetting can be applied to many types of logics. The key idea is enlarging model sets by changing the valuation of the variable which is to be forgotten, and adding these new models to the model set. However, in different types of logics, forgetting may play different roles, and may have different flavors and specific techniques, due to the different features of logics. In four-valued logic, the main problem comes from the new truth values and the two truth values in the set $\mathcal{D}$ of designated elements, which makes the semantics more complex.

For the formula-variable relevance, in the papers where forgetting was proposed [6] or discussed [7], concepts such as relevance or independence were proposed. However, these concepts are all in a "binary" way. They only give a judgement of "relevant", "independent" or not. In many cases, this is sufficient. But sometimes we need more detailed information of how relevant they are. For example, sometimes we need a ranking function to define preferences on variable sets. Thus, defining formula-variable relevance quantitatively is needed. In the work of Liang, Lin, and Van den Bussche [9], an approach of measuring the degree of formula-variable relevance was proposed in the two-valued propositional setting. This paper here extends the idea of quantitatively evaluating formula-variable relevance in two-valued logic to the case of four-valued logic. To do this, we firstly defined forgetting in four-valued logic. Thus, forgetting, as well as formula-variable relevance, could also be used as tools for knowledge base management under four-valued logic.

In this paper, we defined forgetting in four-valued logic, and showed that some nice properties which hold in two-valued forgetting also hold for four-valued forgetting. Then we gave a definition of formula-variable relevance in four-valued logic, with some properties. For the future work, more novel properties and applications could be proposed. Also, we notice that the two new truth values make the semantics of four-valued logic more complex, so is there a more intuitive approach to define formula-variable relevance, or even forgetting? This is also a topic worth studying.

Acknowledgements. The authors wish to thank Chen Chen, Chaosheng Fan, Zhaocong Jia, Kedian Mu, Jan Van den Bussche, Geng Wang, Dai Xu, Xiaowang Zhang, and other people with whom we have had discussions on this topic. We would also like to thank the anonymous reviewers for the helpful comments. This work is supported by the program of the National Natural Science Foundation of China (NSFC) under grant number 60973003, and the Research Fund for the Doctoral Program of Higher Education of China.

References

1. Belnap Jr., N.D.: A useful four-valued logic. In: Modern Uses of Multiple-Valued Logic, pp. 5–37. Springer (1977)
2. Arieli, O., Avron, A.: The value of the four values. Artificial Intelligence 102(1), 97–141 (1998)
3. Ginsberg, M.L.: Multivalued logics: A uniform approach to reasoning in artificial intelligence. Computational Intelligence 4(3), 265–316 (1988)
4. Priest, G.: Paraconsistent logic. In: Handbook of Philosophical Logic, pp. 287–393. Springer (2002)
5. Bertossi, L., Hunter, A., Schaub, T. (eds.): Inconsistency Tolerance. LNCS, vol. 3300. Springer, Heidelberg (2005)
6. Lin, F., Reiter, R.: Forget it. Working Notes of AAAI Fall Symposium on Relevance, pp. 154–159 (1994)
7. Lang, J., Liberatore, P., Marquis, P.: Propositional independence. Journal of Artificial Intelligence Research 18, 391–443 (2003)
8. Xu, D., Lin, Z.: A prime implicates-based formulae forgetting. In: 2011 IEEE International Conference on Computer Science and Automation Engineering (CSAE), vol. 3, pp. 128–132. IEEE (2011)
9. Liang, X., Lin, Z., Van den Bussche, J.: Quantitatively evaluating formula-variable relevance by forgetting. In: Zaïane, O.R., Zilles, S. (eds.) Canadian AI 2013. LNCS, vol. 7884, pp. 271–277. Springer, Heidelberg (2013)
10. Lang, J., Marquis, P.: Resolving inconsistencies by variable forgetting. In: International Conference on Principles of Knowledge Representation and Reasoning, pp. 239–250. Morgan Kaufmann Publishers (2002)
11. Lang, J., Marquis, P.: Reasoning under inconsistency: A forgetting-based approach. Artificial Intelligence 174(12), 799–823 (2010)
12. Wang, Z., Wang, K., Topor, R., Pan, J.Z.: Forgetting concepts in DL-Lite. In: Bechhofer, S., Hauswirth, M., Hoffmann, J., Koubarakis, M. (eds.) ESWC 2008. LNCS, vol. 5021, pp. 245–257. Springer, Heidelberg (2008)
13. Eiter, T., Wang, K.: Semantic forgetting in answer set programming. Artificial Intelligence 172(14), 1644–1672 (2008)
14. Zhang, Y., Zhou, Y.: Knowledge forgetting: Properties and applications. Artificial Intelligence 173(16), 1525–1537 (2009)

Evolution Analysis
of Modularity-Based Java System Structure

Hongmin Liu[1] and Tieju Ma[2]

[1]School of Business, East China University of Science and Technology,
Meilong Road 130, Shanghai 200237, China
294172877@qq.com
[2] School of Business, East China University of Science and Technology,
Meilong Road 130, Shanghai 200237, China
tjma@ecust.edu.cn

Abstract. In order to study the modularity of system structure and analyze the law of system architecture evolution, we summarised three typical definitions about system modularity from different perspectives in this paper. Then we adopted the quantitative calculation method, established three optimization models, got some methods of module partition and evaluated and analyzed the effects of the different divisions. Finally, we used the Java class diagram (JDK) as an example, analyzed the modular structures of different JDK versions, interpreted the results, drew patterns and trends in the evolution of each version.

Keywords: modularity, module partition, system structure evolution.

1   Introduction

The concept of modularity, whose origin has been so long, has been applied in the manufacturing sector for more than a century as a production principle. Modularity by the virtue of information hiding inside the module, has been widely used in such issues as cluster organizational restructuring, industrial upgrading, technological innovation, and has been nature of the structure of the new industries in information age.

The research on system modularity appears in different disciplines and research areas, such as engineering, manufacturing operations, marketing, and organization and management, but also in psychology, biology, education, mathematics and other disciplines. The understanding of the module and modularity and the focus of the study in the different disciplines are not consistent.

In the study of the evolution of large-scale software systems, Chen Tao et al (Chen Tao, 2008), on the basis of the BA model, put forward the improved model of the software evolution: the evolution based on module. They analyzed the evolution of JDK versions from the perspective of complex network and then simulated it.

In the field of quantitative research on system modularity, how to measure modularity is an open question. System modularity in traditional sense is an extension of the concept of cohesion and coupling in software systems, whose nature is

M. Wang (Ed.): KSEM 2013, LNAI 8041, pp. 349–359, 2013.

maximization of the within-module relations or minimization of the between-module relations. The entire optimization process is: to constantly adjust the position of components in its module as much as possible to increase the within-module relations. However, if we optimize the system by the target to maximize the within-module relations, then it clearly appears such a case as all components are present in a single module, and none member in other modules(Newman,2004). This is a meaningless result. Therefore, the domestic and foreign scholars raised a lot of methods to measure modularity on the basis of the traditional definition. Some of them are such typical as the expected modularity definition of M. E. J. Newman (Newman, 2006) and the balanced modularity definition of Roger Guimera ` & Luı´s A. Nunes Amaral (Roger Guimera ` & Luı´s A. Nunes Amaral , 2005. Hereinafter referred to as GA).

In the past few years, the modularity for Java has been an active topic. From JSR 277 to the JSR 291, Modularity seems the only way which must be passed in the process of evolutionary of Java. Even the future languages based on the JVM, such as Scala, also consider modular. As a general concept, modularity makes applications decomposed into different parts, each of which can be developed and tested separately. In the previous literature, the researches on Java stayed in architecture design and the test of performance, but never carried out the quantitative study of the structural relationship from the perspective of optimization.

In this paper, from the perspective of the association between components in the system, we change the system structure by adjusting the position of module, and explore the changes of the system in terms of modularity. We established three optimization models on the basis of three typical definitions of system modularity, got three methods of module partition respectively and then analyzed and evaluated the optimized system. We used the Java class diagram (JDK) as an example in this paper, analyzed the modular structures of different JDK versions, interpreted the results, drew patterns and trends in the evolution of each version.

2   Three Typical Definitions about Modularity

In general, when we are in the study of the structure of a system, we need to know clearly the relationship between the components of the system (For Java system as an example, there are many kinds of relationship between the class files in JDK class library just like aggregation, implementation, inheritance and so on). The mutual relationship between components constitutes a relation matrix (or adjacency matrix), denoted by A_{nxn} in this article. We make the following conventions on the relationship between the components in this paper:

(1) Ignoring specific type of the relationship. In other words, we don't make a distinction between the various relationships.

(2) Appointing relationship is mutual. Component a had some relationship with b, then b is also regarded as a reverse relationship with a. That's similar to the action and reaction in physics.

(3) Cancelling the accumulation of relationship. Even though there are several relationships between a and b at the same time, the value in the relationship matrix is still 1.

After this appointment, the relationship matrix A is actually a 0-1 symmetric matrix.

After understanding of the interrelationship between the components, another important information is the dependence state of the components, which describes the affiliation between component and module. We denote this affiliation by matrix X, which implys it depends on the binary value of x_{ij} in X whether component i belongs to module j. This is a state matrix, and also the decision variable in our optimization model, which determines the entire structure of the system. We do not consider hierarchical modules. That is to say, all modules are independent of each other, so one component can only exist in one module. Thus, X is also a 0-1 matrix.

Let x_i denote the state row vector of component i in X. If component i belongs to module j, then $x_{ij}=1$,else $x_{ij}=0$.So the value of $x_i*x_j^T$ contains a special meaning:

$$x_i \cdot x_j^T = \begin{cases} 1, & i,j \text{ in the same module} \\ 0, & i,j \text{ in different modules} \end{cases} \tag{1}$$

Let L denote the sum of the relations in system, then we can get:

$$L = \sum_{i>j} a_{ij} \tag{2}$$

where a_{ij} is an element of A(the relation matrix).

The relations of within module can be calculated as follows:

$$L_{within} = \sum_{i>j} a_{ij} * (x_i \cdot x_j^T) \tag{3}$$

Let M denote our modularity.

As to three definition about modularity mentioned in the introduction, we will next introduce them individually.

(1) Traditional definition: the ratio of the sum of within-module relations and the sum of total relations in system, which can be calculated as follows:

$$M = \frac{L_{within}}{L} = \frac{\sum_{i>j} a_{ij} * (x_i \cdot x_j^T)}{\sum_{i>j} a_{ij}} \tag{4}$$

This is just a traditional concept, which had been involved in many papers, but there is never such formal quantitative definition.

(2) Newman expected modularity definition [3]: within the same module, the sum of the values that the actual relations of the two components minus the random expected

relations. He believes that a good module-division method should be better than the expected one generated randomly. Let k_i denote the degree of node i, then we can get the expected number of edges between node i and j if edges are placed at random is $k_i k_j / 2m$. So we can then express the modularity as:

$$M = \frac{1}{2L} \sum_{i,j} (a_{ij} - \frac{k_i k_j}{2L})(x_i \cdot x_j^{T}) \tag{5}$$

where

$$k_i = \sum_j a_{ij} \tag{6}$$

$$\sum_i k_i = \sum_i \sum_j a_{ij} = 2L \tag{7}$$

(3) GA balanced modularity definition [4]: the equation can be expressed as:

$$M = \sum_{s=1}^{N} \left[\frac{l_s}{L} - (\frac{d_s}{2L})^2 \right] \tag{8}$$

where N is the number of modules, l_s is the number of links between nodes in module s, and d_s is the sum of the degrees of the nodes in module s. $\sum_{s=1}^{N} l_s$ and d_s can be calculated respectively as:

$$\sum_{s=1}^{N} l_s = L_{within} = \sum_{i>j} a_{ij}(x_i \cdot x_j^{T}) \tag{9}$$

$$d_s = \sum_{i=1}^{n} \sum_{j=1}^{n} a_{ij} x_{is}, s = 1, 2, ..., N \tag{10}$$

The rationale for this definition of modularity is the following. A good partition of a network into modules must comprise many within-module links and as few as possible between-module links. However, if we just try to minimize the number of between-module links (or, equivalently, maximize the number of within-module links), the optimal partition consists of a single module and no between-module links. GA definition can overcome this difficulty by imposing M=0 if all nodes are in the same module. So when M>0, we can exclude this situation.

3   Module-Based Optimization Models about Structure System

In this paper, three optimization models about system structure were established respectively by adding appropriate restrictions based on three typical definitions about modularity above.

3.1   Traditional Model

In the introduction, we had analyzed the disadvantage of the traditional definition. In order to avoid empty module, we artificially add some restrictions on module capacity, and then maximize modularity. This is our traditional model.

$$\max M = \frac{\sum_{i>j} a_{ij} * (x_i \cdot x_j^T)}{\sum_{i>j} a_{ij}}$$

$$\begin{cases} l_j \le \sum_{i=1}^{n} x_{ij} \le u_j, j = 1, 2, ..., N \\ \sum_{j=1}^{N} x_{ij} = 1, i = 1, 2, ..., n \\ x_{ij} \in \{0, 1\} \end{cases}$$ (traditional model)

where l_j and u_j are the lower and upper limits of the module capacity respectively(We simplified the processing in the actual operation. Let l_j is the minimum value of the initial module capacity, u_j is the maximum value of the initial module capacity).

The objective function in traditional model adopts traditional modularity definition. The meaning of the first limiting condition is that the capacity of each module can't exceed the limit of a given module in the process of optimizing, artificially improving the deficiencies of the traditional definition. The latter two are the limitations of modular relationship mentioned previously.

3.2   Newman Model

Newman's modularity definition evolved from the traditional definition, still remaining the inadequacies of traditional definition .When construct the optimization model, we still have to add the module capacity restrictions.

$$\max M = \frac{1}{2L}\sum_{i,j}(a_{ij} - \frac{k_ik_j}{2L})(x_i \cdot x_j^T)$$

$$\left\{ \begin{array}{l} l_j \le \sum_{i=1}^{n} x_{ij} \le u_j, j = 1,2,...,N \\[2mm] \sum_{j=1}^{N} x_{ij} = 1, i = 1,2,...,n \\[2mm] x_{ij} \in \{0,1\} \end{array} \right. \qquad \text{(Newman model)}$$

3.3   GA model

G&A added a balanced control item on the basis of traditional modularity, trending in balancing each module size of the optimal solution .GA's modularity can be able to avoid the emergence of a single-module division. So, when construct the optimization model, there is no need for imposing restrictions on module capacity, but only restrictions on module relations.

$$\max M = \sum_{s=1}^{N}\left[\frac{l_s}{L} - (\frac{d_s}{2L})^2\right]$$

$$\left\{ \begin{array}{l} \sum_{j=1}^{N} x_{ij} = 1, i = 1,2,...,n \\[2mm] x_{ij} \in \{0,1\} \end{array} \right. \qquad \text{(GA model)}$$

It can be proved: that the theoretical maximum M is 1-1 / N, if and only if, the sum of degrees of within-module node in each module is equal, and the number of between-module relations is zero after a reasonable allocation. This can both achieve the equalization of module capacity, and a modular system.

4   The Principle of System Architecture Analysis and Evaluation

In this paper, we treat the equalization of module capacity and modularity as two indicators of evaluating system. In the no-difference system module, we artificially always want the number of components in module to be relatively balanced in the mass, which may be the largest increase in the degree of utilization of space and also easy to update and maintain system.

We can compare the equalization by calculating the standard deviation of the module capacity, and evaluate and compare the modularity by following the traditional modularity definition as an indicator.

Here we define a parameter which measures the optimization potential of system modularity: modularity increments space. It describes the maximum modularity which the system can continue to increase theoretically (As the path-dependent in evolution of system architecture, that we obtain by heuristic algorithm is local optimal solution). The larger the modularity incremental space is, the larger potential there is still for improvement of the system.

We can also evaluate the overall system from the point of components .The idea is that there are some good descriptions on the degree of importance of the components, the classification of component, and the relationship between components and components, components and modules. This is the direction of our future research.

5   The Analysis and Evaluation of Java System Structure

The data in this article comes from the parsing of JDK foundation class library. Extracted the JDK package source file, parsed and obtained the relationship between the '. java' files. Regarded the sub-folder of JDK as a module to analyze (JDK1.1 ~ 1.3 has 12 modules and JDK1.4 ~ 1.7 has 13 modules).

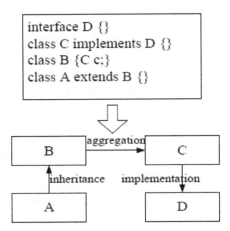

Fig. 1. Java SCG [1]

Some results of our studies are as follows.

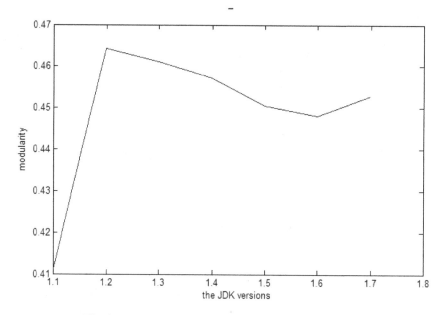

Fig. 2. The trend figure of modularity of the JDK versions

From the figure 2, we can clearly find out the downward trend in modularity of system with the evolution of the JDK versions on the whole. With the development of

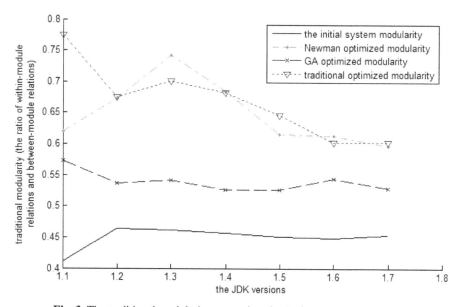

Fig. 3. The traditional modularity comparison in the four system structures

the JDK, the system structure becomes more complex, and the relationship between classes becomes increasingly complex, resulting in the decline of system modularity.

From the figure 3, we can analyze: (1) from the JDK1.1 to JDK1.2 system, the structure has undergone a major change, which can be seen from the fluctuation of the curve in the figure. For the accuracy of the analysis, we can ignore JDK1.1 and analyze from JDK1.2 directly; (2) the trend of Newman optimization and traditional optimization is basically identical: as the complexity of the system increases, the effect of optimization is less and less obvious. Because these two methods are oriented to optimize the traditional modularity, their results are significantly higher than the other two results; (3) with respect to the other two optimization methods, the performance of GA optimization is very gentle, which shows the effect of the GA method in the optimization is very stable. In other words, the performance will not have too large fluctuation as the complexity of the system change.

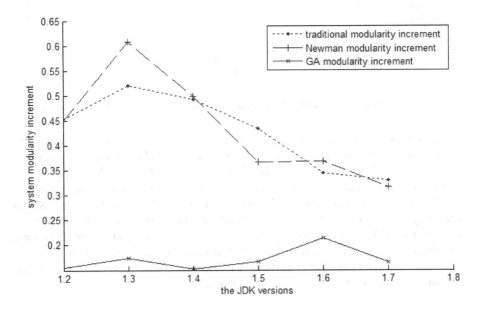

Fig. 4. Comparison of the modularity incremental in three optimized structures

We can find out from the figure 4, GA modular increment space considering module equilibrium is relatively stable, and generally fluctuates in 20%. The other two modular incremental spaces mainly oriented to the modularity have large fluctuation and poor stability, and present a decreasing trend with the increase of system complexity. Overall, however, there is a great degree of improvement in modularity for Java system.

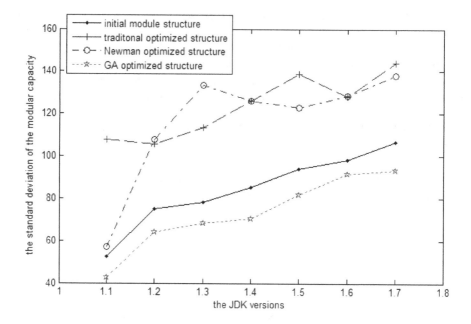

Fig. 5. Comparison of the standard deviation of the four system modular capacity

The standard deviation of module size reflects the degree of the equilibrium of the system. The smaller the standard deviation is, the more balanced the distribution of components is. The figure 5 can be drawn: (1) the standard deviation of GA optimization is the smallest, in other words, the GA optimized system is most balanced; (2) the GA optimized structure and the initial structure has the same trend: that is with increasing system complexity, the standard deviation is greater. And the amplitude is very similar; (3) the standard deviation of Newman optimization and traditional optimization significantly increases. Combined with the previous analysis, it can be drawn: the two optimization methods sacrifice equilibrium of the system and excessively seek the modularity of the system structure.

6   Conclusion and Outlook

In this paper, we established three optimization models on the basis of three typical definitions of system modularity and got three methods of module partition respectively from the perspective of modularity. By the analysis of Java class libraries (JDK), we found that, with the evolution of JDK versions, the complexity of the system was larger and larger, resulting in the downward trend of system modularity. However, after the appropriate adjustment, the JDK system module can have a greater degree of improvement (more than 20%), so it is feasible to consider to increase the modularity of the Java system by adjusting JDK module structure.

Compared three optimization models, we derived modularity defined by Newman was essentially consistent with traditional modularity: they were both methods oriented to maximize the traditional modularity, so it was inevitable that the capacity of the individual modules was too large, while the capacity of the others was too small in the optimization process. GA optimization model, however, on the basis of maintaining its structure substantially the same as the initial one, can guarantee a certain degree of modularity lifting (about 20%), at the same time make the module capacity more balanced.

This article discussed the modularity simply from the point of view of the system relationship. There exist many factors which can impact modularity in the actual situation, such as the hierarchy of system. In the design of system architecture, they are mutual restraint, but indispensable considerations. In the future research we will both consider modularity and hierarchy, and explore their roles in the system structure.

Acknowledgement. This research was sponsored by NSFC(No.71125002).

References

[1] Chen, T., Gu, Q., Wang, S., Chen, X., Chen, D.: Module-based large-scale software evolution based on complex networks. In: 8th IEEE International Conference on Computer and Information Technology, pp. 798–803. IEEE Press, New York (2008)
[2] Newman, M.E.J., Girvan, M.: Finding and evaluating community structure in networks. Phys. Rev. E 69, 026113 (2004)
[3] Newman, M.E.J.: Modularity and community structure in networks. PNAS 103, 8577–8582 (2006)
[4] Roger, G., Luís, A.N.A.: Functional cartography of complex metabolic networks. Nature 433, 895–900 (2005)

Online Knowledge Community: Conceptual Clarification and a CAS View for Its Collective Intelligence

Shuangling Luo[1], Taketoshi Yoshita[2], and Yanzhang Wang[3]

[1] Independent Consultant and Researcher, Dalian, 116023 China
shuanglingluo@gmail.com
[2] School of Knowledge Science, Japan Advanced Institute of Science and Technology,
Nomi, Ishikawa 923-1292 Japan
yoshida@jaist.ac.jp
[3] School of Management Science and Engineering, Dalian University of Technology,
Dalian 116024 China
yzwang@dlut.edu.cn

Abstract. This paper is a position paper to advocate scholarly attention to the online knowledge communities under a uniform framework. The concept of online knowledge community as suggested in this paper is discussed and compared with related conceptions, especially with community of practice and knowledge-building community. Furthermore, focusing on the collective intelligence of the OKC, a viewpoint of complex-adaptive-systems is taken as a basic paradigm to study OKC. From this view, the collective intelligence of the OKC is generated through the co-evolutionary dynamics of the communal knowledge and the communal social-network system. The paper attempts to establish a notional base for further investigations on actual OKCs.

Keywords: Online Knowledge Community, Collective Intelligence, Complex Adaptive System.

1 Introduction

As argued by de Vries et al. [1], human beings have a natural tendency for community and the communities are one of the most essential fabrics of human society. Recently, with the rapid development of modern information and communication technologies (ICT), especially with the growth of Web and Web 2.0 applications, communities have become even more ubiquitous as people can virtually gather in a community via the ICT tools. The rapid growth of the online communities has a profound impact on our society. In particular, the proliferation of the online communities would enormously influence the way people to communicate and to exchange information and knowledge; and the online communities have consequently become a vital platform for knowledge creation, storage, and dissemination. Typical examples include the Wiki-editing communities for collaborative authoring (e.g. the Wikipedian community), the Open Source Software (OSS) communities for programming (e.g. the Linux Kernel developer community), and social tagging communities for folksonomy-building

M. Wang (Ed.): KSEM 2013, LNAI 8041, pp. 360–371, 2013.
© Springer-Verlag Berlin Heidelberg 2013

(e.g. del.icio.us and CiteULike). Such online communities have demonstrated intriguing capabilities for collective knowledge generation and distribution. Correspondingly, the community-based view of knowledge processes has attracted great scholarly attention [2-4]. However, despite all the noteworthy contributions, further investigations on the knowledge activities and processes in these types of online communities are still needed. From the perspective of knowledge management, this research direction is in tune with the shift from the "firm-based model" to the "community-based model" of knowledge generation and distribution [2].

The aforementioned communities are significantly different from one another; but these communities have a common characteristic that they are developed around collective knowledge activities, usually in a specific knowledge domain. In other words, knowledge is the key force to glue the participants into an online community of this type. Furthermore, these communities commonly exhibit the "wisdom of crowds" [5] or "collective intelligence" [6], i.e. the capability of solving complex problems without centralized control or coordination. It may, therefore, be deserved to study these communities under a uniform framework. The first step to establish such a framework is to give a conceptual analysis on these communities, to specify what they are, to clarify how they differ from the other types of communities, and to develop an overall model to understand these communities.

With the argument that the aforementioned online communities can be categorized as "online knowledge communities" (OKC), we in this paper give a position to advocate more scholarly attention to such OKC, as a research field growing in the intersection of online communities, which are mainly studied in the CSCW circle, and knowledge management as a focal topic in the management community. To support this, we give an attempt to conceptualize OKC. The central research question behind OKC is then raised, namely the "collective intelligence" of OKC in solving complex problems and generating and distributing knowledge. A view of complex-adaptive-system (CAS) is accordingly taken to develop an overall understanding of OKC and its collective intelligence, in order to fuel more-concrete investigations.

2   Conceptualizing Online Knowledge Community

2.1   Prior Conceptualizations of OKC

The terms "knowledge community" and "online knowledge community" have been used by other scholars. So first we give a short review on the prior conceptualizations on OKC.

One early effort on OKC was given by de Vries and his colleagues [1]. Defining the OKC as "a group of knowledge workers jointly taking care for a knowledge domain, who meet in frequent social interaction for their professional development by means of an online expertise center", they used the term OKC in a broad sense with the argument that OKC is one of the four major categories of online communities, namely study community, social community, knowledge community and e-commerce community. Then, OKC was further divided into three types, i.e. corporate community, community of practice and social knowledge community. Adopting de Vries et. al.'s conception of OKC, Wu et al. [7] developed a document co-organization system as an attempt for IT support of OKC. To them, the shared knowledge repertoire is central to OKC, and the shared documents serve as key component of the shared knowledge repertoire of OKC.

Different from de Vries' view of OKC, Mueller-Prothmann and Siedentopf [8] noted that "OKCs are viewed as social spaces for creation, exchange, and conservation of knowledge, especially of knowledge that is difficult to externalise and to codify". Accordingly, they regarded the research field of OKC as "a fusion of the preceding concepts of communities of practice and virtual communities, and their adaptation to KM." Based on this "space"-based view of OKC, their work is focused on discussing the software applications to support the communities. This view is basically shared by Erickson and Kellogg [9] in their effort to develop an online environment within which "users can engage socially with one another in the process discover, develop, evolve, and explicate knowledge". Another definition of OKC was given by Lin et al. [10]. In their work, the term "Web-based knowledge community" (WKC) is used instead of OKC. To them, a WKC is a community that allows individuals to seek and share knowledge through a website based on common interests.

Lindkvist [11] regarded a "knowledge community" as an ideal-type notion of epistemology that mirrors a "community of practice" in the real world. In his conception, knowledge communities are almost identical to communities of practice, serving as notional abstraction of communities of practice from the knowledge perspective.

In the prior conceptualizations, some features of OKC have been articulated. However, controversial viewpoints are implicitly contained in the aforementioned contributions. De Vries and his colleagues defined OKC in a rather-loose and somewhat vague manner. There is apparent overlapping between the major categories of communities as they classified, since a uniform criterion of classification seems lacking. In the narratives by Mueller-Prothmann and Siedentopf, Erickson and Kellogg and Lin et al., the ICT platform is focused and they paid less attention to the human-side of an OKC. In this sense, these contributions are limited on establishing a precise understanding for OKC. Lindkvist gave an insightful view to establish a strong connection between knowledge community and community of practice. Nevertheless, his argument may raise a conceptual debate whether knowledge community and community of practice are just identical.

We argue that a concept of OKC is meaningful because of the proliferation of the online communities that are centered on knowledge and knowledge activities, such as the previously-mentioned OSS communities and Wiki-editing communities. In this aspect, the prior conceptualizations seem inadequate to grasp the key characteristics of these communities. A new discussion of the OKC concept may be worthwhile.

2.2   Online Knowledge Community and Its Characteristics

As exemplified in the Introduction section, the key characteristic of the online knowledge communities is that they are formed around the participants' collective knowledge activities in a specific knowledge domain. In this sense, people participate in the community for the sake of creating, distributing, and/or acquiring knowledge. What's more, the participants are interconnected mainly through the knowledge-based interactions; and the community is unlikely to sustain without these knowledge-based interactions. Therefore, we define the online knowledge community as *an online community that consist of people who participate in knowledge work or activities around one specific knowledge domain or in some cases a few interrelated knowledge domains. Within such community, the participants collectively create, transfer, acquire and utilize knowledge, giving rise to the development of the specific knowledge domains as well as promoting the dissemination and utilization of knowledge.*

Comparing the previously-discussed conceptualization, our definition of OKC is more focused on a special type of online communities, which include the Wikipedia community, OSS developer communities, social tagging communities, as well as a number of other communities such as online-resource-sharing and e-print communities (e.g. arxiv.org, ssrn.com), and the online crowdsourcing communities like innocentive.com, yet2.com, and ninesigma.com. The key standpoint behind this definition is to distinguish knowledge communities from knowledge-intensive communities. That is to say, we believe that knowledge communities are basically knowledge intensive; but not all the knowledge-intensive communities are knowledge communities. For instance, an online health-care consultation community is knowledge intensive as professional physicians and health-consultants use knowledge to answer consulters' questions. However, such community can hardly be regarded as a knowledge community as the creation and dissemination of knowledge is insignificant. Thus, under the prior definition of OKC, we emphasize that an OKC is formed and maintained around the activities of knowledge creation and dissemination.

The key characteristics of OKC can, correspondingly, be summarized as follows.

First, the very key characteristic of an OKC lies in that such a community is formed around one or more specific knowledge domains. In other words, the OKC is a community that is enabled by knowledge. For example, people participate in the Wikipedian community because they might have the passion to write a new article or to correct mistakes in an existing one. The participants virtually gather in the community owing to the shared interests in the knowledge items associated with the Wikipedia articles and correspondingly the common practice of collective authoring. The participants even don't have any personal connection other than such indirect expertise-sharing activities. Consequently, the development of the knowledge domain or domains, together with the dissemination and utilization of knowledge in this domain, are the major driving force for the development of the community.

Second, sharing the interests or professions in the specific knowledge domain or domains, the members participate in the community via *knowledge work*. More specifically, here knowledge work essentially means the activities of knowledge creation, dissemination, storage and utilization. Although it is by nature inseparable from the more physical or substantial "practice" in the real-world, such knowledge work is more ideological and information-centric. In other words, the knowledge work that enables a knowledge community is essentially based on more-or-less intangible mind-processes, e.g. thinking, knowing, memorizing, discovering at the individual level, as well as information-passing or communication processes such as articulating, comprehending, and consensus-building at the interpersonal and collective level.

Third, subsequent to this "knowledge work" view of OKC, it can be naturally inferred that knowledge community is generally a *communication-driven community*. In an OKC, the activities of information-passing and expertise-exchange are the most critical activities to make the community animated; in some cases, such communicative activities are even the solely meaningful activities of the community. Furthermore, the computer-mediated communications are the main form of communications that take place in an OKC.

Referring to the "togetherness" of a social community, *community cohesion* is one key aspect of community studies. The fourth key characteristic of OKC is that its cohesion relies heavily on the knowledge domain it involves and the knowledge work that activates it. At the individual level, the *social identity* of an OKC member is primarily based on the personal interests or engagement in the knowledge domain associated with that community. The individual's participation in the community is also centered on the knowledge work around that knowledge domain. Due to this knowledge-work-centered nature of OKC, the community cohesion at the group level would ultimately be established around the joint enterprise in the knowledge work within the specific knowledge domain(s).

Last but not least, one key characteristic of OKC is that it commonly exhibits "collective intelligence" to solve complex problems in a self-organizing mode. This is actually the most intriguing phenomenon of the OKCs. In the case of Wikipedia, thousands of freelance editors can generate an encyclopedia that catches up with Encyclopedia Britannica [12]; and in the case of OSS communities, thousands of voluntary programmers with diverse skill levels can build extremely complex software products like Linux and Apache. However, the deep mechanisms for such amazing capabilities of collective intelligence have not fully been obtained. The exploration and exploitation of such mechanisms would be a key research topic behind OKC.

2.3   Comparison with KBC and CoP

The prior concept of OKC should also be compared with the concepts of learning communities, knowledge-building communities (KBC), and community of practice (CoP), which basically grow from the field of learning and education. We call these types of communities as learning-centered communities.

OKC is conceptually akin to those learning-centered communities. There is apparent conceptual overlapping between online knowledge communities and those learning-centered communities that run on the Internet, i.e. virtual community of practice, online learning community and online knowledge-building community. Some communities can be regarded as OKCs from one aspect, while from another aspect are viewed as communities of practice, learning communities and/or knowledge-building communities. However, the OKC is conceptually distinctive from the learning-centered communities. The OKC is originally conceived from a knowledge-system perspective. In other words, an OKC is formed around one or more specific knowledge domains in which the members are commonly interested or engaged; and the members' collective knowledge work is the key factor to bind the members together. Instead, the learning-centered communities are conceived from the perspective of learning activities. Further comparisons between OKC and the learning-centered communities are given below.

First, let's make a comparison between OKC with knowledge-building community. The knowledge-building community, as coined by Scardamalia & Bereiter [13], is essentially based on the distinction between personal learning and collective knowledge-building, following a "deep constructivism" theory of learning. In this sense, knowledge-building community is essentially a community formed around the learning and knowledge-building activities. In this sense, knowledge-building community can be subsumed by OKC, as the knowledge work in an OKC can contain more activities and processes aside from learning and knowledge-building. For example, the Wikipedian community can be regarded as one knowledge community, and simultaneously a knowledge-building community. However, it would be inappropriate to regard a crowdsourcing community for innovative tasks, which can be viewed as a typical OKC, as a knowledge-building community, since in this community the collective knowledge-building activity is not prominent.

Second, we try to clarify the OKC concept by comparing it with CoP and learning community. As one key logical premise of CoP is the alignment of learning and doing, CoP conceptually subsumes learning community. What's more, the conceptual distinction between OKC and learning community is clearer. Consequently, here we only compare OKC with CoP (i.e. virtual CoP or VCoP).

Essentially, VCoP and OKC are closely-interwoven concepts so that Lindkvist [11] just treated them as almost identical concepts. Especially, in recent years the combination of the ideas of CoP and knowledge management has tightly kneaded "knowledge" and "community of practice" together; and it is even harder to distinguish OKC from VCoP. In fact, quite some actual online communities are OKC and VCoP at the same time. In such communities, Lindkvist's idea is workable that the knowledge community mirrors the CoP from the "common practice" aspect to the knowledge aspect. However, our argument is that such mirroring is not always applicable as these two concepts are actually raised from diverse conceptual bases. CoP is essentially built upon the concept of "situated learning" [14] or learning-by-doing within a specific physical and cultural environment. Instead, the focal attention behind OKC is not learning, but the engagement in knowledge creation, dissemination and utilization.

Moreover, these two concepts are also different from each other in two critical aspects. On one hand, VCoP is the community formed around some "practice" whilst OKC is the community centered on knowledge activities over one or more specific knowledge domains. The "practice" that enables a VCoP is more substantial and tangible than the knowledge work around a knowledge domain that activates a knowledge community. What's more, such conceptual difference would eventually lead to the differences of community boundary, scope, identity and cohesion of these two categories of communities in the real-world. Although quite some real-world communities can be regarded simultaneously as OKCs and VCoPs, these two communities can conceptually be separated. For example, an amateur sports community can basically be considered as a CoP, but it can hardly be identified as a knowledge community because the social force to form such a community is for the members to have fun in sports rather than to create and exchange knowledge.

What's more, communities of practice are "tightly knit" groups that have been practicing together long enough to develop into a cohesive community with relationships of mutuality and shared understanding [11] [15]. In contrast, such tough condition is not a necessity for a knowledge community, which can be formed via weaker social ties. For example, in the Wikipedian community, the Wikipedians don't have to have strong mutual engagement, but their collective editing actions may still cause the continual improvements of the Wikipedia articles. This Wikipedia example also shows the loose-connectivity and openness in an OKC do not necessarily lead to worse efficiency or effectiveness in the development of the associated knowledge domain. Instead, the diversity of the participants sometimes is even beneficial for the advances of knowledge. From this aspect, many of the online knowledge communities such as the OSS communities and the Wikipedian community can hardly regarded as a VCoP, as they are not "tightly-knit".

The following Table summarizes the differences between OKC and VCoP.

Table 1. Comparing OKC with VCoP

OKC	VCoP
OKC is formed around collective knowledge activities in specific knowledge domain(s).	VCoP is focused on learning, subsequent to the concept of "situated learning" [14] or "learning by doing".
OKC is formed around one or more specific knowledge domain; and it is then more "virtual" or intangible.	VCoP is formed around some common practice; thus it is basically substantial or more "real".
OKC must be knowledge centered.	VCoP may be not knowledge-centered.
OKC can be formed via weaker social ties; basically it is the "knowledge" that interconnects people in a KC, e.g. the Wikipedian community can hardly be seen as VCoP, but it is an OKC.	VCoP is usually "tightly-knit" [15] by social ties, characterized by: mutual engagement, joint enterprise, and a shared repertoire [16]

3   From Community Informatics to Community Intelligence: A CAS View

Based on the conceptual clarification of OKC in the previous section, we can give a first glance to the key research questions behind OKC. An OKC is, according to the prior conceptualization, an ICT-supported social system that is centered on knowledge activities and often exhibits collective intelligence. Hence, the researches on OKC should combine three perspectives, namely the ICT perspective, the social-system perspective, and the knowledge perspective. Firstly, from the ICT perspective, the OKC research field can be regarded as a sub-field of "online communities", which generally grows from CSCW and groupware, and manifests the progress of the group-ware technologies from supporting teamwork to supporting collective actions in online communities. Secondly, from the social-system perspective, OKC can be regarded as a sub-field of community studies developed in the disciplines of sociology and social psychology. Thirdly, the studies on OKC should also embrace a knowledge perspective as an expansion of knowledge management studies, since the knowledge activities are the central activities that enable OKCs. In tune with the three perspectives, the researches on OKC should answer the following key questions:

(1) How people interact with each other in the context of OKC, which is a specific type of human communities;

(2) How knowledge is created, transferred and utilized in the OKC; and correspondingly how problems are solved and collective intelligence is developed;

(3) How the social relations and knowledge work in the OKC can be supported by the underlying ICT system.

In the previous three perspectives, the knowledge perspective is critical to understand OKC, as it is the collective knowledge work in specific knowledge domains that distinguishes the OKC from other types of online communities. Accordingly, the explorations from the knowledge perspective also distinguish the OKC research field from the more-general field of online communities, in which the interplay of the social-system perspective and the ICT perspective is centrally concerned.

The prior perspectives and key research questions many bring many research issues. Among them, a particularly important issue is to address the "collective intelligence" of OKC, which reflects the most intriguing phenomenon of the "wisdom of crowds" in different OKCs. The research of such "collective intelligence" exhibits the academic growth from the "community informatics" study for online communities in general to the "community intelligence" study for online knowledge communities in particular.

As a key theme of online community research, the studies on "community informatics" aim at "the complex dynamic relationship between technological innovation and changing social relationships" [17]. More specifically, centered on the ICT use in the (both physical and online) community contexts, community informatics on one hand concerns how the computing technologies influence the community development, regeneration and sustainability, and on the other hand concerns how the social and cultural factors to affect the development and diffusion of the computerized community support systems. In the context of OKC, however, we need to study not only "community informatics", but also "community intelligence" [18], in order to cope with the knowledge processes that are ubiquitous in the OKCs and correspondingly the amazing capabilities for the OKCs to solve the complex problems like software-development and encyclopedia-building by the massive crowds.

The investigations to such "community intelligence", i.e. the collective intelligence of the OKC, should set the focus on the knowledge perspective, since the "community intelligence" is essentially knowledge-centered and the knowledge dynamics in the OKC is the key to establish a good understanding to such collective intelligence. Besides, the social-system perspective and the ICT perspective should also be taken into account. It is the participants of the OKC who conduct the knowledge activities so as to generate "collective intelligence". The study of "collective intelligence" of the OKC should also incorporate the study of the social system of the participants. From the ICT perspective, the ICT system on one hand offers the fundamental channel for the participate to communicate and interact; on the other hand, the ICT system also contributes to the collective intelligence by storing codified knowledge and by providing machine-intelligence capabilities to enhance the intelligence of the human groups. Thus, the key issue from the ICT perspective is to develop computer systems to provide better support for community intelligence.

What's more, a complex-adaptive-system (CAS) point of view should be taken to study the OKC's collective intelligence. The collective intelligence is essentially an emergent property of a complex adaptive social system. The OKC is about a large amount of participants or members who are engaged in some knowledge activities. The members participate in the community with high freedom of join and leave. Their actions are often voluntary, rather than obeying commands and control. However, those loosely-connected groups of independent or semi-independent individuals often exhibit high intelligence at the collective level. Thus, analogous to the swarm intelligence of the ant colonies and the bee swarms, such kind of "community intelligence" essentially emerges from the evolution of the complex adaptive system of the OKC.

Structurally, as discussed in the authors' previous work [18], this complex adaptive system of the OKC can be modeled as a "super-network" of three networks, i.e. the human network of communal participants, the knowledge network that is composed of knowledge items interlinked through semantic and pragmatic relations, and the computer network that support human interactions and knowledge transfer.

In line with this triple-network structure, the OKC system evolves through its internal knowledge activities by individuals and local groups. Such local interactions can, then, give rise to the global consequences simultaneously in two aspects,

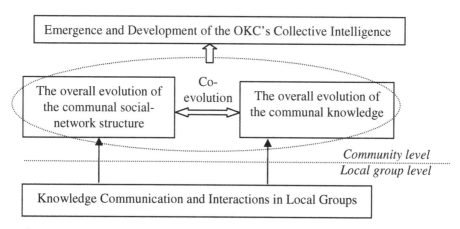

Fig. 1. Community intelligence development as co-evolution of knowledge and social network in the OKC

i.e. the overall process of knowledge evolution on one hand and the overall evolution of the communal social-network on the other hand. Thus, the macroscopic processes of the social system and of the knowledge system are essentially a co-evolutionary process in that the two processes influence each other through the mediation of the local interactions. This co-evolutionary process causes the emergence and development of the community intelligence, as illustrated in Fig. 1.

In [19], the authors have provided some analyses on this co-evolutionary process with an actual case of the Linux Kernel developer community. The driving force of this communal system is the programming activities of the voluntary developers, together with their free-style discussions in the developers' mailing list. Through such activities, the overall evolution of the system is two-fold. On one hand, the knowledge system grows, mainly embodied as the expansion of the Linux Kernel. On the other hand, a "center-periphery" structure of the developer network gradually forms [2]. As the centralized coordination being absent, self-organization is the dominant mechanism that governs the development of the community and its knowledge product, the Linux Kernel.

In all, the previous discussion may give rise to an overall picture for the structure and dynamics of an OKC. From this picture, it can be seen that the study of the co-evolution of knowledge and social-network structure in the OKC would be the key to understand its collective intelligence. This view would shed light on further investigations for OKC at a more-detailed and concrete level.

4   Conclusion and Ongoing Work

In this paper, we try to give an overall conceptualization on the online knowledge communities as a specific type of online communities, which include the wiki-editing communities, OSS communities, social tagging communities, professional-document sharing communities, and crowdsoucing communities, among the others. These communities generally demonstrate high collective intelligence for the solving of complex problems like software-development and encyclopedia-building; therefore, specific academic attention to this community type is deserved. Furthermore, we conceive the key research issues in the field of the OKC, which combine three perspectives, i.e. the knowledge perspective, the social-system perspective, and the ICT perspective. With a CAS point of view is taken, it is argued that the emergence of the OKC's collective intelligence is through the evolution of the complex adaptive system of the OKC, especially through the co-evolutionary processes of the communal knowledge and the communal social-network structure.

This paper is a conceptual extension of the authors' previous endeavors on OKC and its collective intelligence [18-19], with the major purpose being to fuel discussions on the related notions and ideas. Due to the limitation of scope, we do not discuss the concrete research efforts on actual OKCs that support the ideas presented in this paper. Following the present conceptualizations, especially focusing on the co-evolutionary dynamics of knowledge and social-network in OKCs, the more concrete investigations on the actual OKCs are underway.

Acknowledgement.   This work is partly supported by the National Natural Science Foundation of China (NSFC) under Grant No. 91024029, as one key project of NSFC's giant research programs. The authors are grateful for the three anonymous referees for their constructive suggestions.

References

1. De Vries, S., Bloeman, P., Roossink, L.: Online Knowledge communities. In: Proceedings of World Conference on the WWW and Internet 2000, pp. 124–129. AACE, Chesapeake (2000)
2. Lee, G.K., Cole, R.E.: From a Firm-based to a Community-based Model of Knowledge Creation: The Case of Linux Kernel Development. Organization Science 14(6), 633–649 (2003)
3. Kittur, A., Suh, B., Pendleton, B.A., Chi, E.H.: He says, she says: conflict and coordination in Wikipedia. In: Proceedings of the SIGCHI Conference on Human Factors in Computing Systems, pp. 453–462. ACM (2007)
4. Cattuto, C., Loreto, V., Pietronero, L.: Semiotic dynamics and collaborative tagging. Proceedings of the National Academy of Sciences of the United States of America 104(5), 1461–1464 (2007)
5. Surowiecki, J.: The Wisdom of Crowds. DoubleDay, New York (2004)
6. Woolley, A.W., Chabris, C.F., Pentland, A., Hashmi, N., Malone, T.W.: Evidence for a collective intelligence factor in the performance of human groups. Science 330(6004), 686–688 (2010)
7. Wu, H., Gordon, M.D., DeMaagd, K.: Document Co-Organization in an Online Knowledge Community. In: Conference on Human Factors in Computing Systems (CHI 2004), Vienna, Austria, pp. 24–29 (2004)
8. Mueller-Prothmann, T., Siedentopf, C.: Designing Online Knowledge Communities: Developing a Usability Evaluation Criteria Catalogue. In: 3rd European Knowledge Management Summer School, San Sebastian, Spain (2003)
9. Erickson, T., Kellogg, W.A.: Knowledge Communities: Online Environments for Supporting Knowledge Management and Its Social Context. In: Ackerman, M.S., et al. (eds.) Sharing Expertise: Beyond Knowledge Management, pp. 299–325. MIT Press (2003)
10. Lin, H., Fan, W., Wallace, L., Zhang, Z.: An Empirical Study of Web-based Knowledge Community Success. In: Proceedings of the 40th Hawaii International Conference on System Sciences, pp. 1530–1605 (2007)
11. Lindkvist, L.: Knowledge Communities and Knowledge Collectivities: A Typology of Knowledge Work in Groups. Journal of Management Studies 42(6), 1189–1210 (2005)
12. Giles, J.: Internet Encyclopaedias Go Head to Head. Nature 438(7070), 900–901 (2005)
13. Scardamalia, M., Bereiter, C.: Computer Support for Knowledge-Building Communities. Journal of the Learning Sciences 3(3), 265–283 (1994)
14. Lave, J., Wenger, E.: Situated learning: Legitimate peripheral participation. Cambridge University Press (1991)
15. Brown, J.S., Duguid, P.: Organizational Learning and Communities-of-Practice: Toward a Unified View of Working, Learning, and Innovation. Organization Science 2(1), 40–57 (1991)
16. Wenger, E.: Communities of Practice: Learning, Meaning, and Identity. Cambridge University Press, New York (1998)

17. Keeble, L., Loader, B.D. (eds.): Community Informatics: Shaping Computer-Mediated So-
    cial Relations. Routledge, London (2001)
18. Luo, S., Xia, H., Yoshida, T., Wang, Z.: Toward collective intelligence of online commun-
    ities: A primitive conceptual model. Journal of Systems Science and Systems Engineer-
    ing 18(2), 203–221 (2009)
19. Xia, H., Luo, S., Yoshida, T.: Exploring the knowledge creating communities: an analysis
    of the Linux Kernel developer community. In: Chu, S. (ed.) Managing Knowledge for
    Global and Collaborative Innovations. Series of Innovation and Knowledge Management,
    vol. 8, pp. 385–398. World-Scientific Publishing, Singapore (2010)

A New Epistemic Logic Model of Regret Games

Jianying Cui[1], Xudong Luo[1,*], and Kwang Mong Sim[2]

[1] Institute of Logic and Cognition,
Sun Yat-sen University, Guangzhou, China, 510275
[2] School of Computing, University of Kent, Kent, UK, ME4 4AG

Abstract. To many real-life games, the algorithm of Iterated Eliminating Regret-dominated Strategies (IERS) can find solutions that are consistent with experimental observations, which have been proved to be problematic for Nash Equilibrium concept. However, there are a serious problem in characterising the IERS epistemic procedure. That is, the rationality of choosing un-dominated strategies cannot be assumed as the common knowledge among all the players of a game, otherwise the outcome of the IERS cannot be implied. Nevertheless, the common knowledge of rationality among players is an essential premise in game theory. To address these issues, this paper develops a new epistemic logic model to interpret the IERS procedure as a process of dynamic information exchanging by setting the players' rationality as a proper announcement assertion. Finally, we show that under the assumption of rationality common knowledge rather than lower probabilities, our model can successfully solve a well-known traveler dilemma.

Keywords: Epistemic Logic, Game Theory, Elimination Algorithm, Regret, Traveler's Dilemma.

1 Introduction

Now lots of psychological experiments reveal that the outcomes of games in real life often deviate from what game theory predicts [18,19]. To explore the causes of such deviations, Halpern et al. [15,12] proposed an algorithm of Iterated Eliminating Regret dominated Strategy (IERS). They proved the new Game Solutions (GS) obtained by using algorithm IERS exhibit the same behavior as that observed in experiments of many famous games (e.g., Traveler's Dilemma [5]) in real-life, which have been proved to be problematic for the concept Nash Equilibrium (NE).

However, the rationale of an iterated eliminating algorithm needs to be justified, that is, one needs to make sense for the algorithm by providing a reasonable epistemic interpretation for the algorithm [12]. In a game, the cause that each player will not take any dominated strategies of his is that he is rational. The epistemic choosing exactly corresponds to the first round of an iterated eliminating algorithm. Further, after knowing his opponents are rational, a rational

---

* Corresponding author.

M. Wang (Ed.): KSEM 2013, LNAI 8041, pp. 372–386, 2013.

player will make a choice among his non-dominated strategies in a sub-game mode resulting from the first round of deletion. The epistemic interaction of reasoning leads to the second rounds of eliminating in the algorithm. Continuing to reason the interactional knowledge between players, intuitively all algorithms of iterated eliminating, including the rationalisability algorithm [1] and backward induction [2], could be characterised in terms of the common knowledge of choosing un-dominated strategies as rationality.

When Halpern et al. [12] tried to provide a reasonable epistemic interpretation for the game solution obtained using the IERS algorithms, a severe epistemic question arises: the common knowledge of rationality cannot imply the outcomes of IERS. Since the common knowledge of rationality among players is an essential premise in game theory [13]. So, the problem weakens the rationale and the importance of algorithm IERS although it can provide a consistent prediction for the players' behaviors in real-life.

To address this serious issue, in this paper we construct a new epistemic game model to justify algorithm IERS. That is, our model retains that the essential premise of game theory—the rationality should be common knowledge among players. More specifically, in our epistemic model, the IERS procedure is interpreted as a process of dynamic information exchange by defining players' rationality as a proper announcement assertion. Moreover, in Halpern's epistemic model, they had to assign successively lower probabilities to higher orders of rationality, which is difficult to be estimated and understood, while in our model we do not assume any probability estimation. Further, by using the idea behind our model, we can construct a unified framework to analyse and explore rationality in many iterated algorithms (e.g., weak rationality and strong rationality in [16]). Thus, we offer a new method to characterise the algorithms of Iterated Elimination Strictly Dominated strategies (IESD) and the algorithms of rationalisability corresponding to Pearce version [9].

The rest of this paper is organised as follows. Section 2 recaps game theory, the IERS algorithm and public announcement logic. Section 3 defines our epistemic logic model of regret game. Section 4 proves that the IERS procedure can be interpreted as a process of dynamic information exchanging by developing an epistemic regret-game model. To show the merits of our model, Section 5 uses our model to analyse the Traveller's Dilemma. Section 6 discusses how to use the idea behind our model to construct a unified framework for analysing various rationalities, especially weak rationality and strong rationality, in many iterated algorithms. Finally, Section 7 concludes the papers with possible future work.

2 Preliminaries

This section briefs game theory [13], the IERS algorithm[15,12], and public announce logic [4].

Definition 1. *A strategic form game with pure strategies is a 3-tuple of $G = \langle N,$ $\{S_i\}_{i \in N}, \{u_i\}_{i \in N} \rangle$, where*

- N is the set of players in game G,
- S_i is the set of strategies for player i, and
- u_i is a function that assigns a real value to every strategy profile $s = (s_1, \ldots, s_n)$.

Meanwhile, we denote by S_{-i} the set of strategy profiles of the players other than i, that is, $S_{-i} = S_1 \times \ldots \times S_{i-1} \times S_{i+1} \times \ldots \times S_n$. When we want to focus on player i, we denote the strategy profile $s \in S$ by (s_i, s_{-i}) where $s_i \in S_i$ and $s_{-i} \in S_{-i}$.

Thus, for a given game of G, we can define its strategic regret-game as follows:

Definition 2. *For a normal form game of $G = \langle N, \{S_i\}_{i \in N}, \{u_i\}_{i \in N}\rangle$, its strategic regret-game is a 3-tuple of $G' = \langle N, \{S_i\}_{i \in N}, \{re_i\}_{i \in N}\rangle$, where $\{re_i\}_{i \in N}$ stands for player i's expost regret associated with any profile of pure strategies (s_i, s_{-i}), which is calculated as follows:*

$$re_i(s_i, s_{-i}) = \max\{u_i(s'_i, s_{-i}) \mid \forall s'_i \in S_i\} - u_i(s_i, s_{-i}), \tag{1}$$

meaning the regret of choosing s_i for player i when his opponents choose s_{-i}.

Definition 3. *For a given game of $G = \langle N, \{S_i\}_{i \in N}, \{re_i\}_{i \in N}\rangle$, let s_i and s'_i be two strategies of player i, and set $S'_{-i} \subseteq S_{-i}$. Then s_i is regret-dominated by s'_i on S'_{-i} if $Re_i(s'_i) < Re_i(s_i)$, where*

$$Re_i(s_i) = \max\{re_i(s_i, s_{-i}) \mid \forall s_{-i} \in S_{-i}\}. \tag{2}$$

Also, a regret-dominated strategy of s_i is called regrettable for player i. And for $S' \subseteq S$, strategy $s'_i \in S_i$ is unregretted with respect to S'_i if no strategies in S'_i regret-dominate s'_i on S'_{-i}.

According to the above definition, intuitively $Re_i(s_i)$ is the maximal regret value of choosing s_i for player i whichever strategy his opponents choose.

When using algorithm IERS, one needs to delete all regret-dominated strategies of all players in every elimination round at the same time. The elimination process of IERS in [12] can be viewed as the recursive sets of player strategies as follows:

Definition 4. *Given strategic regret-game $G' = \langle N, \{S_i\}_{i \in N}, \{re_i\}_{i \in N}\rangle$, let IUD be the set of iterated regret-undominated strategies of G' recursively defined by:*

$$IUD = \prod_{i \in N} IUD_i, \tag{3}$$

where $IUD_i = \bigcap_{m \geqslant 0} IUD_i^m$ with $IUD_i^0 = S_i$ and $RD_i^0 = \{s_i \mid s_i \in IUD_i^0$ is regrettable with respect to IUD_i^0 in $G'\}$. For $m \geqslant 1$,

$$IUD_i^m = IUD_i^{m-1} \setminus RD_i^{m-1}, \tag{4}$$

where $RD_i^m = \{s_i \mid s_i \in IUD_i^m$ is regrettable in subgame $G'^m\}$.[1]

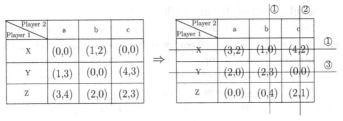

Fig. 1. IERS procedure

By this definition, at each stage all regret-dominated strategies of all the players of a game are simultaneously deleted. Thus, IERS yields a rectangular set of strategy profiles. For example, in Figure 1, the left table denotes game G_1 and the right one is game G_1' (i.e., a regret-game of G). Since $Re_1(X) = 4, Re_1(Y) = 2$ and $Re_1(Z) = 2$, strategy X is a regret-dominated for player 1 by Y and Z. Similarly, strategy b is also regret-dominated by a and c for player 2. So, at the first deletion round, strategies X and c are deleted simultaneously. Then, due to the lack of strategy X and b, $Re_2(a)$ becomes 0 and $Re_2(c)$ becomes 1 respectively, while $Re_1(Y)$ and $Re_1(Z)$ both become 2. So, player 2 needs to delete regret-dominated strategy c for him at the second round. Continuing these steps, finally we can attain a set of strategy profile $IUD = \{(Z, a)\}$. That is,

$$IUD_1^0 = \{X, Y, Z\}, RD_1^0 = \{X\}, \ IUD_2^0 = \{a, b, c\}, \ RD_2^0 = \{b\};$$
$$IUD_1^1 = \{Y, Z\}, \ RD_1^1 = \varnothing, \ IUD_2^1 = \{a, c\}, \ RD_2^1 = \{c\};$$
$$IUD_1^2 = \{Y, Z\}, \ RD_1^2 = \{Y\}, \ IUD_2^2 = \{a\} = IUD_2, \ RD_2^2 = \varnothing;$$
$$IUD_1^3 = \{Z\} = IUD_1.$$

Thus, $IUD = \{(Z, a)\}$. In addition, the following theorem [12] shows that the deletion process converges to IUD. That is, for any nonempty closed strategy set $S = S_1 \times \ldots \times S_n$, IUD is nonempty and is a fixed point of the deletion process.

Theorem 1. *Let $G' = \langle N, \{S_i\}_{i \in N}, \{re_i\}_{i \in N} \rangle$ be a strategic regret-game. If S is a closed, nonempty set of strategies, then IUD is nonempty.*

On the other hand, as a basis for most dynamic epistemic logics [10], Public Announcement Logic (PAL) [4] can deal with the change of information arising from the strategy of public announcement by adding a dynamic modality $[\varphi]$ to standard epistemic logics,[2] i.e., $[\varphi]\psi$ means that after a truthful public announcement of φ, formula ψ holds. Its truth condition is as follows:

$$M, w \vDash [\varphi]\psi \Leftrightarrow M, w \vDash \varphi \text{ implies } M\!\mid_\varphi, w \vDash \psi,$$

---

[1] G'^m is a subgame of G', in which $S_i = IUD_i^m$ and $G'^0 = G'$.

[2] We assume the readers is familiar with the basics concepts of modal logic and epistemic logic. The readers who are not familiar with this area can consult the recent textbook [7] for more details.

where $M \mid_\varphi$ is a submodel of M, in which φ is true.

In PAL, formula $[\varphi]K_j\psi$ means agent j knows that ψ after truthful public announcement of φ, and $[\varphi]C_N\varphi$ expresses after its announcement, φ has become the common knowledge in the group of agents N. PAL can be axiomatised completely. The work of [16] shows that for any model M we can keep announcing φ, retaining only those worlds where φ holds. This yields a sequence of nested decreasing sets, which must stop in finite models, i.e., $\sharp(\varphi, M)$. Formally, we have:

Definition 5. *For any model M and formula φ, the announcement limit $\sharp(\varphi, M)$ is the first submodel in the repeated announcement sequence where announcing φ has no further effect. If $\sharp(\varphi, M)$ is non-empty, we have a model where φ has become the common knowledge. We call such statements self-fulfilling in the given model, and all others self-refutings.*

Public announcements or observations $[\varphi]$ of true propositions φ yield *the information* that changes the current model irrevocably, discarding worlds that fail to satisfy φ; meanwhile agents' interactive knowledge about φ will be increased as models changes. This procedure is similar to an iterated elimination process of many algorithms of this kind in game theory. The reason why dominated strategies are deleted is *rationality*, which is postulated as high-order interactive information among players [13,1]. Therefore, after showing that some rationalities defined by him can be as announcement assertions, van Benthem [16] states the issue of "an announcement limit" has close connections with the equilibria found by using many iterated elimination algorithms.

3 Logic Model of Regret Game

In order to give a dynamic epistemic analysis of IERS, this section defines an epistemic logic model of regret game.

Firstly, we extend public announce logic to define regret game logic as follows:

Definition 6. *Given regret game G', a regret-game logic (G'-logic) is a logic that contains atomic propositions in the following forms:*

- *Pure strategy symbols $s_i, s'_i, \ldots$: the intended interpretation of s_i (or of s'_i) means player i chooses strategy s_i (or s'_i).*
- *Symbol Ra_i^{re} means player i is rational, symbol Br_i^* means the best response of player i and symbol GS means it is a Game Solution with algorithm IERS.*
- *Atomic propositions in the form of $s_i \succeq s'_i$ means strategy s_i is weakly regret-dominant over strategies s'_i for player i, or player i thinks s_i at least as good as strategy s'_i. And $s_i \succ s'_i$ means that for player i, s_i is better than s'_i*

Next, we define a frame for G'-logic as follows:

Definition 7. *Given   regret   game G',   a   frame   of G'-logic   is $\mathfrak{F}'_G = \langle W, \{\sim_i\}_{i \in N}, \{f_i\}_{i \in N} \rangle$, where*

- $W(\neq \emptyset)$ *consists of all players' pure strategy profiles;*
- $\sim_i$ *is an epistemic accessibility relation for player i, which is defined as the equivalence relation of agreement of profiles in the i'th coordinate; and*
- $f_i : W \rightarrow S_i$ *is a pure strategic function, which satisfies the following property: if $w \sim_i v$ then $f_i(w) = f_i(v)$.*

Simply, a frame for G'-logic adds to a classic Kripke S5 frame on agents' knowledge [7] a function that associates with every state w a strategy profile $f(w) = (f_1(w), \ldots, f_n(w)) \in S$, where f_i $(i = 1, \ldots, n)$ means that player i knows his own choice. In other words, if he chooses strategy s_i, then he knows that he chooses s_i. This accords with our intuition. For convenience, we use $R_i(w) = \{v \mid w \sim_i v, w, v \in W\}$ to represent the set of worlds that player i believes possible in world w, and $\|s_i\| = \{w \in W \mid f_i(w) = s_i\}$ to represent the set of the worlds where player i chooses strategy s_i. For example, in the Figure 2, the epistemic regret-game frame $\mathfrak{F}_{G_1'}$ corresponding to the above game G_1, $f_1(w_1) = X$, $f_2(w_1) = b$, and $R_1(w_1) = \{w_0, w_1, w_2\}$, and $R_2(w_1) = \{w_1, w_4, w_7\}$.

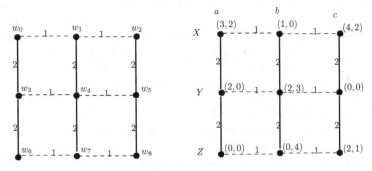

Fig. 2. The epistemic regret-game frame $\mathfrak{F}_{G_1'}$ (the left hand side) corresponding to the above game G_1, where dashed lines for epistemic accessibilities of player 1, and solid lines for player 1. For convenience, the game G_1 with epistemic accessibilities of the players is also illustrated in the right hand side.

Definition 8. *An epistemic game model of $M_{G'}$ over G'-logic is obtained by incorporating the following valuation on $\mathfrak{F}_G'$:*

- $M_{G'}, w \vDash s_i \Leftrightarrow w \in \|s_i\|$;
- $M_{G'}, w \vDash (s_i \succcurlyeq s_i') \Leftrightarrow \exists v \in \|s_i'\|), re_i(s_i, f_{-i}(w)) \leq re_i(s_i', f_{-i}(v))$;
- $M_{G'}, w \vDash (s_i \succ s_i') \Leftrightarrow \forall v \in \|s_i'\|), re_i(s_i, f_{-i}(w)) < re_i(s_i', f_{-i}(v))$; *and*
- $M_{G'}, w \vDash Ra_i^{re} \Leftrightarrow M_{G'}, w \vDash \bigwedge_{a \neq s_i} K_i(f_i(w) \succcurlyeq a)$.

The above definition means that in world w in $M_{G'}$, (i) it is true that player i chooses strategy s_i if and only if the strategy of player i associated with w is s_i (i.e., $f_i(w) = s_i$); (ii) it is true that strategy s_i is weakly regret-dominant over strategies s_i' for player i if and only if there is a world of v in set $\|s_i'\|$ such that s_i in combination with $f_{-i}(w)$ yields a regret value that is not larger than the regret value yielded by s_i' in combination with $f_{-i}(v)$; and (iii) it is true that strategy s_i is strongly regret-dominate over strategy s'_i for player i if and only if for every world v in set $\|s'_i\|$, s_i in combination with $f_{-i}(w)$ yields a regret value that is strictly less than the regret value yielded by s'_i in combination with $f_{-i}(v)$.

More importantly, rationality Ra_i^{re} in world w in $M_{G'}$ is the assertion that for each of available alternative strategies, player i knows that the current one may be at least as good as it, and whereby the truth of interpretation of formula $\neg Ra_i^{re}$ (i.e., an agent is not rational) is that player i still chooses the current strategy although he thinks it is possible there is another strategy of his, which strictly dominants the current strategy $f_i(w)$. Formally, we can express it as

$$M_{G'}, w \vDash \neg Ra_i^{re} \Leftrightarrow M_{G'}, w \vDash \bigvee_{a \neq f_i(w)} (\hat{K}_i(a \succ f_i(w))),$$

where $\hat{K}_i$ is dual of the K_i (i.e., $\hat{K}_i = \neg K_i \neg$), and the interpretation of $\hat{K}_i$ and K_i are as usual in classic epistemic logic (see [7]). Thus, it is easy to verify that Ra_i^{re} fails exactly at the rows or the columns, which is the regret-dominated strategies of player i in an epistemic regret-game model $M_{G'}^*$. For instance, in Figure 2, proposition $(X \succcurlyeq Y)$ holds in world w_1 because of the existence of w_3 so that $re_1(X, f_2(w_1)) \leq re_1(Y, f_2(w_3))$, where $re_1(X, f_2(w_1)) = re_1(X, b) = 1$, and $re_1(Y, f_2(w_3)) = re_1(Y, a) = 2$, but Ra_1^{re} fails at the world w_1 because there exists $w_0 \in R_1(w_1)$ that satified $M_{G'_1}, w_0 \nvDash X \succcurlyeq Y$, thus, $M_{G'_1}, w_1 \nvDash K_i(X \succcurlyeq Y)$. Similarly, Ra_1^{re} fails in worlds w_0 and w_2 as well, and Ra_2^{re} is false in worlds w_1, w_4, and w_7 of the original model $M_{G'_1}$.

As we mentioned previously, the dynamic analysis of iterated elimination algorithms always has to do with changing of an epistemic model. So, in the following, we will call the above epistemic regret-game model $M_{G'}$ a full epistemic regret-game model, and take any submodel of a full epistemic regret-game model $M_{G'}$ as a general epistemic game model $M_{G'}^*$.

4   Property

In this section, we will prove that the rationality notion can be formalised as an assertion of a public announcement (Theorems 2 and 3), and our public announcement procedure of rationality can produce the same result of algorithm IERS (Theorem 4).

It is a general rule in PAL that the assertions, which players publicly announce, must be the statements that they know are true. In the following, Theorems 2 and 3 guarantee that the rationality notion, which we have defined, can be formalised as an assertion of a public announcement.

Theorem 2. *Every finite general epistemic regret-game model has worlds in which Ra^{re} is true, where $Ra^{re} = \bigcap_{i \in N} Ra_i^{re}$.*

Proof. Considering any general game model $M_{G'}^*$, if there are no regret-dominated strategies for all the players in $M_{G'}^*$, then since atomic proposition Ra_i^{re} fails exactly at the rows or columns to which regret-dominated strategies correspond for player i in a general game model, Ra^{re} is true in all the worlds where it is. Thus, the iterated announcement of Ra^{re} cannot change the game model and get stuck in cycles in this situation. If player i has regret-dominated strategy a, then he must have a strategy, say strategy b, which is better than strategy a. Thus, Ra_i^{re} holds in all the worlds that belong to the row or the column corresponding to strategy b. On the other hand, considering player j, if he has no weakly dominated strategy, then also Ra_j^{re} holds in all the worlds. Furthermore, Ra_j^{re} holds in the worlds that belong to the row or the column corresponding to strategy b. So, Ra^{re} holds in the general game model. However, if player j also has a regret-dominated strategy, then he must have a dominant strategy, say strategy Y, and Ra_j^{re} is true in the worlds that belong to the row or the column corresponding to strategy Y. Therefore, Ra^{re} is satisfied in world (Y, b). To sum up the above arguments, every finite general game model has the worlds with true Ra^{re}.   □

The following theorem means that the rationality is epistemically introspective.

Theorem 3. *Formula* $Ra_i^{re} \rightarrow K_i Ra_i^{re}$ *is valid in a general epistemic regret-game model.*

Proof. Consider a general epistemic regret-game model of $M_{G'}^*$ and an arbitrary world of w in $M_{G'}^*$ such that $M_{G'}^*, w \vDash Ra_i^{re}$ but $M_{G'}^*, w \nvDash K_i Ra_i^{re}$. Thus, $\exists v \in R_i(w), M_{G'}^*, v \nvDash Ra_i^{re}$. By Definition 8, $f_i(v)$ is a regret-dominated strategy for player i by some of his strategies. And according to $f_i(w) = f_i(v)$, if and only if $v \in R_i(w)$, we can conclude that $f_i(w)$ is also a regret-dominated strategy of player i. Further, $M_{G'}^*, w \nvDash Ra_i^{re}$, contrary to the precondition, i.e., $M_{G'}^*, w \vDash Ra_i^{re}$. So, formula $Ra_i^{re} \rightarrow K_i Ra_i^{re}$ is valid on a general game model.   □

As a result, by Theorems 2 and 3 we can successively remove the worlds, in which Ra^{re} does not hold in model $M_{G'}^*$ after repeatedly announcing the rationality. Although this scenario of an iterative solution algorithm is virtual, it is significant. This is because we can expect players to announce that they are rational since they would know it. As [16] showed, because the current game model may be changed in light of the information from another player, it makes sense to iterate the process and repeat the assertion Ra^{re} if it is true.

In Figure 3, the left-most model is the strategic regret-game model from Figure 1. The other models are obtained by public announcements of Ra^{re} successively for three times. So, in the last submodel, we have:

$$M_{G'}, (Z, a) \vDash [!Ra^{re}][!Ra^{re}][!Ra^{re}]C_N(GS).$$

It indicates that if the players iteratively announce that they are rational, the process of regret-dominated strategy elimination leads them to a solution that is commonly known to be a game solution (GS). So, we can describe an iterated elimination procedure of algorithm IERS as a process of repeated announcement for the rationality assertion during players deliberate in a one-shot game. Formally, we have:

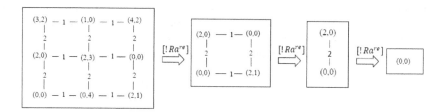

Fig. 3. The public announcement of Ra^{re}

Theorem 4. *Given a full epistemic game model based on finite strategic-form game G' with regret, a general epistemic game model of $M^*_{G'}$, where arbitrary world w is, is stable by repeated announcements of Ra^{re} in $M_{G'}$ for all the players of game G' if and only if $f(w) \in IUD$, i.e., $w \in \sharp(Ra^{re}, M_{G'}) \Leftrightarrow f(w) \in IUD$.*

Proof. (i) From left to right ($\Rightarrow$): If $w \in \sharp(Ra^{re}, M_{G'})$, i.e., $w \in M^*_{G'}$, then $M^*_{G'}, w \vDash Ra^{re}$, i.e., $M^*_{G'}, w \vDash \wedge_{i \in N} Ra^{re}_i$. First we show: $\forall i \in N, f_i(w) \notin RD^0_i$. Suppose not. Then $\exists i \in N$ such that $f_i(w) \in RD^0_i$, that is, $f_i(w)$ of player i is regret-dominated in G' by some other strategy $s'_i \in S_i = IUD^0_i$. It means $Re_i(f_i(w)) > Re_i(s'_i)$. Thus, by formula (2), we have

$$\max\{re_i(f_i(w), s_{-i}) \mid \forall s_{-i} \in S_{-i}\} > \max\{re_i(s'_i, s_{-i}) \mid \forall s_{-i} \in S_{-i}\}. \quad (5)$$

Now set some $s'_{-i} \in S_{-i}$ satisfying $re_i(f_i(w), s'_{-i}) = Re_i(f_i(w))$, and set $s''_{-i} \in S_{-i}$ satisfying $re_i(s'_i, s''_{-i}) = Re_i(s'_i)$. Thus, by inequality (5), we have

$$re_i(f_i(w), s'_{-i}) > re_i(s'_i, s''_{-i}). \quad (6)$$

Furthermore, set $v' \in R_i(w) \cap \|s'_{-i}\|$, where $\|s'_i\| = \{w \in W \mid f_i(w) = s'_i\}$. Then, by inequality (6),

$$re_i(f_i(w), f_{-i}(v')) > re_i(s'_i, s''_{-i}). \quad (7)$$

Thus, considering $\forall v \in \|s'_i\|$, $re_i(s'_i, s''_{-i}) \geq re_i(s'_i, f_{-i}(v))$, from inequality (7) we can derive

$$\forall v \in \|s'_i\|, re_i(f_i(w), f_{-i}(v')) > re_i(s'_i, f_{-i}(v)). \quad (8)$$

According to $f_i(w) = f_i(v')$, which is resulted from $v' \in R_i(w)$ and Definition 7, inequality (7) becomes

$$\forall v \in \|s'_i\|, re_i(f_i(v'), f_{-i}(v')) > re_i(s'_i, f_{-i}(v)). \quad (9)$$

Then, by Definition 8, we have $M^*_{G'}, v' \nvDash f_i(v') \succ s'_i$. Noticing that $f_i(w) = f_i(v')$ again, we have $M^*_{G'}, v' \nvDash f_i(w) \succ s'_i$. By the semantic interpretation of K_i, it follows that $M^*_{G'}, w \nvDash K_i(f_i(w) \succ s'_i)$. Therefore, by Definition 8, we can conclude: $M^*_{G'}, w \nvDash Ra^{re}_i$, which contradicts the hypothesis that $M^*_{G'}, w \vDash \wedge_{i \in N} Ra^{re}_i$. Since $\forall w \in W, f_i(w) \in IUD^0_i = S_i$, it follows that $f_i(w) \in IUD^0_i \setminus RD^0_i = IUD^1_i$.

Next we prove the inductive step. For a given integer $m \geq 1$, suppose that $\forall j \in N$, $f_j(w) \in IUD_j^m$, then we need to show that $f_j(w) \notin RD_j^m$. Suppose not. Then there exists player i who is satisfied with $f_i(w) \in RD_i^m$. That is, $f_i(w)$ is a regret-dominated in G'^m by some other strategy $s_i' \in IUD_i^m$. Then, we have

$$\max\{re_i(f_i(w), s_{-i}) \mid \forall s_{-i} \in IUD_{-i}^m\} > \max\{re_i(s_i', s_{-i}) \mid \forall s_{-i} \in IUD_{-i}^m\}.$$

By the induction hypothesis, $\forall j \in N$, $f_j(w) \in IUD_j^m$ due to the prosperity of $f_i(w)$, that is, $v \in R_i(w) \Rightarrow f_i(w) = f_i(v)$. Thus, we have $\forall v \in R_i(w)$, $f_i(v) \in IUD_i^m$. Further, we have

$$\max\{re_i(f_i(w), f_{-i}(v)) \mid \forall v \in R_i(w)\} > \max\{re_i(f_i(w'), f_{-i}(v)) \mid \forall v \in R_i(w')\},$$

where $w' \in \|s_i'\|$. Thus, similar to the above proof, we can conclude $M_{G'^m}^*, w \nvDash Ra_i^{re}$. Again this is in contradiction with the fact that $M_{G'}^*, w \vDash Ra_i^{re}$ derived from $M_{G'^m}^*$ is a submodel of $M_{G'}^*$ and the prosperities of $M_{G'}^*$. So, $f_i(w) \in IUD_i^m \setminus RD_i^m = IUD_i^{m+1}$.

Thus, by the mathematic induction method, $\forall i \in N$, $f_i(w) \in IUD_i$.

(ii) From right to left ($\Leftarrow$): Let $f(w) \in IUD = \bigcap_{m \geq 0} IUD^m$. Then by Definition 3, $\forall i \in N$, $f_i(w)$ is never regret-dominated in IUD_i^m. It means that after m rounds of public announcement Ra^{re}, $M_{G'^m}^*, w \vDash Ra^{re}$, where $M_{G'^m}^*$ is a general epistemic model related to submodel G'^m. Therefore, due to the arbitrary of m, by Definition 5, we have $w \in \sharp(Ra^{re}, M_{G'})$.   □

5 Analysis on Traveler's Dilemma

This section will show the merits of our model by solving the Travelers' Dilemma [5]. That is, we need not to break a classical premise of game theory—the rationality is the common knowledge among all players; and we need not to assign successively lower probabilities, either.

Table 1. The game of Traveler's Delimma

S_1 \ S_2	100	99	98	$\cdots$	3	2
100	(100,100)	(97,101)	(96,100)	$\cdots$	(1,5)	(0,4)
99	(101,97)	(99,99)	(96,100)	$\cdots$	(1,5)	(0,4)
98	(100,96)	(100,96)	(98,98)	$\cdots$	(1,5)	(0,4)
$\vdots$	$\vdots$	$\vdots$	$\vdots$	$\ddots$	$\vdots$	$\vdots$
3	(5,1)	(5,1)	(5,1)	$\cdots$	(3,3)	(0,4)
2	(4,0)	(4,0)	(4,0)	$\cdots$	(4,0)	(2,2)

The game of Traveler's Dilemma goes as follows. An airline loses two suitcases of two passengers. Both the suitcases and the items contained happen to be the same. An airline manager comes to sort out the problem. He tells the two passengers that the airline only can compensate up to $100 per suitcase if the true value of the items cannot be figured out. Thus, in order to determine a honest appraised value of the suitcases, the manager asks them to write down. separately, the amount of their value in-between $2 and $100. He also tells them that if their numbers are the same, he will think they tell the truth and thus the airline will reimburse them that amount; otherwise, this smaller number will be taken as the true dollar value that the airline should reimburse them, but the one who write down the small number will be rewarded $2, which will be deducted from the one who write down the higher number. Then what should these two do if they do not care the other player's payoff when trying to maximise their own payoffs? The game model is shown in Table 1 (if only integer inputs are taken into account). It is not hard to figure out that the only Nash equilibrium is for both to make the minimum claim of $2.

However, this solution to the game is strange and implausible for human being. In fact, the laboratory findings in [6] shows that the strategy that performed best was $97, instead of playing $2, and the profile ($97, $97) is exactly solution from algorithm IERS as well [12]. Then, what is the reasonable epistemic explanation of the empirical observation or solution of algorithm IERS? Halpern and Pass provided one using Kripke structure in [12]. Nevertheless, it is questionable as they said, in [12], as follows:

> "··· if we make the assumption of common knowledge of rationality and identify rationality with minimizing regret, we seem to run into a serious problem with IERS, which is well illustrated by the Traveler's Dilemma ··· If player 1 knows that player 2 is playing $97, then he should play $96, not $97! That is, among all strategies, $97 is certainly not the strategy minimizes regret with respect to $97."

In order to solve the above problem, they explain RAT_i (i.e., player i is rational in world w) as a best response to strategy sequences $\langle s(B_i^0(w)), s(B_i^1(w)), \ldots \rangle$, where $B_i^0(w)$ consists of the worlds in which player i considers most likely in world w, and the worlds in $B_i^1(w)$ are the ones where player i is less likely, and so on. That is, $M, w \models RAT_i$ if $s_i(w)$ is a best response to the strategy sequence $\langle s(B_i^0(w)), s(B_i^1(w)), \ldots \rangle$. They proved this game solutions resulted from algorithm IERS involves higher and higher levels of belief regarding other players' rationality, but does not involve common knowledge or beliefs of $RAT = \cap_{i \in N} RAT_i$. Consequently, in their epistemic characterisation for algorithm IERS, they have to give up the assumption of common knowledge or common belief of players' rationality in game theory.

However, the necessary epistemic conditions for the outcome is that the players in the dilemma game always tend to be the rational players in the term of our rationality, including the rationality is common knowledge among players. According to the definition of Ra^{re} (see Definition 8), it is clear that Ra^{re} only holds in the worlds where players claim the amount of their suitcases' value not less than $96. This is because the regret value of these strategies are equal to 3, while the regret value of the rest are strictly greater than 3. Thus, after public announcement of Ra^{re} for one time, a submodel showed in Table 2 can

Table 2. The regret-game model of the traveler's dilemma with limitations

S_1 \ S_2	100	99	98	97	96
100	(1,1)	(3,0)	(3,1)	(3,2)	(2,3)
99	(0,3)	(1,1)	(3,0)	(3,1)	(2,2)
98	(1,3)	(0,3)	(1,1)	(3,0)	(2,1)
97	(2,3)	(1,3)	(0,3)	(1,1)	(2,0)
96	(3,2)	(2,2)	(1,2)	(0,2)	(0,0)

be attained. Actually, Table 2 represents a regret-game model under players' strategies in interval [\$96, \$100]. Furthermore, if we assume that both players use a strategy in interval [\$96, \$100], then based on our epistemic regret-game model G'' derived from the above submodel, after public announcement of Ra^{re} for the second time, the only world left in the model exactly corresponds to the strategy of asking for \$97.

At the same time, Theorem 1 ensures that the limit of announcement Ra^{re} (see Definition 5) is not empty. Furthermore, it makes Ra^{re} to be a self-fulfilling proposition. Thus, by Definition 4, there always exist some worlds where Ra^{re} holds in our epistemic regret-game model. Moreover, as the essential prosperities of PAL, after public announcing formula φ, player i can delete the worlds, in his mind, which do not satisfy formula φ. In other words, he never reconsiders these worlds as epistemic possible worlds for him. So, we can retain the classic principal of common knowledge of Ra^{re} in our epistemic regret-game model and avoid the paradox of Halpern and Pass model [12].

In [12], Halpern et al. removed the epistemic paradox by interfering IERS with higher and higher levels of knowledge (or belief) regarding other players' rationality, instead of common knowledge (or belief) of players' rationality, and assigning higher knowledge (or belief) about others' rationality lower probability. However, general speaking, there is little evidence to support such estimations of the probabilities of interactive rationality. In other words, it is difficult to set a criterion to justify what is true probability value for some interactive rationality.

6 Discussion

This section discuss how to use the idea behind our model to analyse and explore various rationalities in many iterated algorithms, especially Weak Rationality (WR) and Strong Rationality (SR) in [16].

Generally speaking, there is a large amount of literatures on the algorithms of iterated elimination in the field of logic, computer science and game theory (e.g., [8,16,11]). In particular, [16] designs different algorithms in game theory by redefining two types of rationality: the weak rationality (WR), which is used to characterise the classic algorithm of Iterated Elimination Strictly Dominated strategies in game theory [13], and the strong rationality (SR) related to algorithm Rationalisability of Pearce [9].

Using the idea behind our epistemic regret-game model, we can re-characterise these concepts of rationality to lay epistemic foundations for these two algorithms as well. In fact, these rationality assertions can be redefined formally as follows:

- $M_{G'}, w \vDash (s_i \succsim s'_i) \Leftrightarrow (re_i(s_i, f_{-i}(w)) \leq re_i(s'_i, f_{-i}(w)))$
- $M_{G'}, w \vDash (s_i > s'_i) \Leftrightarrow (re_i(s_i, f_{-i}(w)) < re_i(s'_i, f_{-i}(w)))$
- $M_{G'}, w \vDash WR_i \Leftrightarrow (M_{G'}, w) \vDash \bigwedge_{a \neq f_i(w)} \hat{K}_i(f_i(w) \succsim a))$
- $M_{G'}, w \vDash SR_i \Leftrightarrow (M_{G'}, w) \vDash \hat{K}_i(\bigwedge_{a \neq f_i(w)} (f_i(w) \succsim a))$

where $\succsim$ is a weak preference relation and $>$ is a strong preference relation. Their intuitive meanings are as follows. In world w in regret-game epistemic model $M_{G'}$, (i) it is true that strategy s_i is at least as **good** as strategy s'_i for player i if and only if the regret value derived from s_i in combination with $f_{-i}(w)$ is not bigger than the regret value yielded by s'_i in combination with $f_{-i}(w)$; (ii) the explanation of strong preference relation $>$ is similar; (iii) weakly rational player i (denoted by WR_i) thinks it is possible that the regret raised by the current strategy is not bigger than that by his other strategies; and (iv) strongly rational player i (denoted by SR_i) thinks it is possible that the current strategy does not make his regret more.

The following theorem reveals the relation between the weak rationality and our rationality:

Theorem 5. *The rationality Ra^{re} is stronger than the rationality WR', i.e., $Ra^{re} \to WR'$, and not vise versa.*

Proof. We can prove this theorem by proving that $\forall i \in N$, formula $\neg WR_i \to \neg Ra_i^{re}$ is valid in a general regret-game model $M'_{G'}$. Suppose $\forall w \in W, (M'_G, w) \vDash \neg WR_i$, i.e.,

$$M_{G'}, w \vDash \neg (\bigwedge_{a \neq f_i(w)} \hat{K}_i(f_i(w) \succsim a)) \Leftrightarrow M'_{G'}, w \vDash \bigvee_{a \in S_i} K_i(a > f_i(w)).$$

Thus, $\exists b \in S_i, M'_G, w \vDash K_i(b > f_i(w))$, and $K_i((b > f_i(w))) \to \hat{K}_i(b > f_i(w))$ is valid in terms of axiom **D**.[3] So, we have $M_{G'}, w \vDash \hat{K}_i(b > f_i(w))$. Furthermore, according to the truth interpretation of Ra_i^{re} in Definiton 8, we have $(M'_G, w) \vDash \neg Ra_i^{re}$. □

However, there are no relations between Ra^{re} and SR'. For example, for game G_3 in Table 3 [17] and its strategic regret-game G'_3 in Table 4, Ra_2^{re} holds in the worlds of $(A, a), (B, a), (C, a), (A, c), (B, c)$, and (C, c), but SR'_2 is true in the worlds of $(A, a), (B, a), (C, a), (A, b), (B, b)$, and (C, b).

7   Conclusions

This paper constructs a new epistemic logic model of regret-game, which provides a dynamic epistemic interpretation for algorithm IERS. Our model does not

---

[3] Axiom **D** refers to $K_i p \to \langle K_i \rangle p$, for arbitrary atomic proposition p in epistemic logic.

Table 3. The game model of G_3

S_1 \ S_2	a	b	c
A	(2,3)	(1,0)	(1,1)
B	(0,0)	(4,2)	(1,1)
C	(3,1)	(1,2)	(2,1)

Table 4. The game model of G_3'

S_1 \ S_2	a	b	c
A	(1,0)	(3,3)	(1,2)
B	(3,2)	(0,0)	(1,1)
C	(0,1)	(3,0)	(0,1)

rely upon complex probability assignments, but retains the classic rule in game theory—the rationality should be common knowledge among players. This is important to consider what each player knows and believes about what the other players know and believe [2] because the reason why players play a game is that they can deduce the outcome from the knowledge that other players are rational. Finally, we use our model to solve Traveler's Dilemma without the disadvantage of Halpern and Pass model [12].

In this paper, we have only considered some cases where games get solved through iterated updating with regret statements. However, many other scenarios can have the same features, including infinite sequences where the approximation behavior itself is the focus of interest. In particular, it is interesting to connect our setting with the learning-theoretic scenarios and extend it using temporal update logics [3,14].

Acknowledgments. This paper is supported partially by National Natural Science Foundation of China (No. 61173019), Major Projects of Ministry of Education (No.10JZD0006), Humanity and Social Science Youth foundation of Ministry of Education (No. 11YJC72040001), Philosophy and Social Science Youth Projects of Guangdong Province (No. GD11YZX03), China Postdoctoral Science Foundation (No. 2011M501370) and Bairen Plan of Sun Yat-sen University. Kwang Mong Sim also gratefully acknowledges financial support from the School of Computing at the University of Kent for supporting the visit of Xudong Luo to carry out this research from 15 October 2012 through 14 December 2012.

References

1. Aumann, R.J.: Rationality and bounded rationality. Games and Economic Behavior 21, 2–14 (1997)
2. Aumann, R.J.: Interactive epistemology I:knowledge. International Journal of Game Theory 28(3), 263–300 (1999)
3. Baltag, A., Gierasimczuk, N., Smets, S.: Belief revision as a truth-tracking process. In: Proceedings of the 12th Conference on Theoretical Aspects of Rationality and Knowledge, pp. 187–190 (2011)
4. Baltag, A., Moss, L.S., Solecki, S.: The logic of public announcements, common knowledge, and private suspicions. In: Proceedings of the 7th Conference on Theoretical Aspects of Rationality and Knowledge, pp. 43–56 (1998)
5. Basu, K.: The Traveler's Dilemma: Paradoxes of rationality in game theory. American Economic Review 84(2), 391–395 (1994)

6. Becker, T., Carter, M., Naeve, J.: Experts playing the traveler's dilemma. Inst. fúr Volkswirtschaftslehre, Univ. (2005)
7. Blackburn, P., van Benthem, J., Wolter, F.: Handbook of Modal Logic. Elsevier Science Inc. (2007)
8. Bonanno, G.: A syntactic approach to rationality in games with ordinal payoffs, vol. 3. Amsterdam University Press (2008)
9. David, P.: Rationalizable strategic behavior and the problem of perfection. Econometrica 52(4), 1029–1050 (1984)
10. Ditmarsch, H., van der Hoek, W., Kooi, B.: Dynamic Epistemic Logic. Springer, Berlin (2007)
11. Halpern, J.Y., Pass, R.: A logical characterization of iterated admissibility. In: Proceedings of the 12th Conference on Theoretical Aspects of Rationality and Knowledge, pp. 146–155. ACM (2009)
12. Halpern, J.Y., Pass, R.: Iterated regret minimization: A new solution concept. Games and Economic Behavior 74(1), 184–207 (2012)
13. Osborne, M., Rubinstein, A.: A Course in Game Theory. The MIT Press (1994)
14. Pacuit, E., Roy, O.: A dynamic analysis of interactive rationality. In: Proceedings of the 3rd International Conference on Logic, Rationality and Interaction, pp. 244–257 (2011)
15. Renou, L., Schlag, K.H.: Implementation in minimax regret equilibrium. Games and Economic Behavior 71(2), 527–533 (2011)
16. van Benthem, J.: Rational dynamics and epistemic logic in games. International Game Theory Review 9(1), 13–45 (2007)
17. van Benthem, J.: Logical Dynamics of Information. Cambridge University Press (2011)
18. Wright, J., Leyton-Brown, K.: Behavioral game-theoretic models: A Bayesian framework for parameter analysis. In: Proceedings of the 11th International Conference on Autonomous Agents and Multiagent Systems, pp. 921–928 (2012)
19. Wunder, M., Kaisers, M., Yaros, J.R., Littman, M.: Using iterated reasoning to predict opponent strategies. In: Proceedings of the 10th International Conference on Autonomous Agents and Multiagent Systems, vol. 2, pp. 593–600 (2011)

A Fuzzy Logic Based Model of a Bargaining Game

Jieyu Zhan[1], Xudong Luo[1,*], Kwang Mong Sim[2], Cong Feng[1], and Youzhi Zhang[1]

[1] Institude of Logic and Cognition, Sun Yat-sen University, Guangzhou, China
[2] School of Computing, University of Kent, Chatham Maritime, Kent, UK

Abstract. This paper proposed a novel bargaining model in which bargainers can change their demand preferences during a bargaining to increase the chance to reach an agreement. In the model, we use fuzzy rules to deal with the degrees to which a bargainer should change his preference. Moreover, we do psychological experiments to show our fuzzy rules reflect well how bargainers' risk attitude, patience and regret degree influence their preference changes during a bargaining. Finally, our model is illustrated by a real life example and analysed by experiments.

Keywords: Game theory, fuzzy logic, bargaining game, preference, agent.

1 Introduction

Since Nash [1] built the first bargaining model numerically, various models of this kind have been proposed [2]. However, it is difficult for people to articulate their preferences by accurate numbers [3,4,5]. Thus, some researchers tried to express bargainers' preferences in an ordinal scale [6,7]. However, the information relevant to the bargainers' risk attitudes, which is a very important factor in bargaining [8,9], is lost [10]. To solve this problem, Zhang [10] introduces a new ordinal bargaining game model and its simultaneous concession solution concept.

However, there still exist some drawbacks in Zhang's model. (i) It is hard to figure out whether a bargainer is risk seeking or averse just through the player's preferences over conflicting demands because the model does not know exactly the original preference ordering of every bargainer. For example, when a couple bargains to purchase fruits, one's preference order is *apple* > *orange* > *banana*,[1] where *orange* and *banana* are conflicting with that of the other. We do not know exactly whether he really does not like *banana* or he is just risk averse for avoiding the bargaining is broken. If we know his original preference order is, for example, *orange* > *apple* > *banana*. Then we know he really does not like *banana* but like *orange*, and so he reverses the order of apples and orange just for preventing breakdown of bargaining. So, we cannot say a bargainer is risk averse even if he puts the conflicting demand at the bottom. (ii) In Zhang's model, there are no functions for bargainers to change their preferences according to dynamic bargaining situations. However, in real life people usually change their preferences when they get more information during a bargaining.

---

* Corresponding author.
[1] Here we just use > as a brief and informal notation of preference order, i.e, $a > b$ means a bargainer prefer a to b. Formal definition will be showed in the next section.

M. Wang (Ed.): KSEM 2013, LNAI 8041, pp. 387–403, 2013.

To address these two issues, this paper develops a new model by adding the dynamic demand preference of bargainers into the model. Thus, a bargainer's risk attitude can be estimated by comparing his original preference with his dynamic one that can be changed during a bargaining. This is very important because the risk attitudes play a key role in strategy changing [8,9]. Thus, we introduce the concept of change degree to which a bargainer's risk attitude and preference need to be changed. Moreover, we introduce a fuzzy logic system to calculate the change degree. Since fuzzy logic can capture well human intuitions [11], with such a system, we can expect an outcome of our model to be more consistent with human intuitions.

Our work advances the state of art in the field of bargaining game in the following aspects. (i) We introduce the concept of dynamic preference into an ordinal bargaining model [10] so that bargainers' risk attitude can be reflected more precisely. (ii) We introduce a new solution concept that enables bargainers to update preferences during a bargaining. (iii) We identify a set of fuzzy rules by psychological experiments, which are used to update bargainers' preferences in every bargaining round according to their regret degree, initial risk attitudes, and patience. This can also help to reflect the individualities of different bargainers.

The rest of the paper is organized as follows. Section 2 recaps the solution concept of simultaneous concession solution and fuzzy logic. Section 3 develops a new bargaining model. Section 4 discusses fuzzy rules for updating preference during a bargaining. Section 5 illustrates our model. Section 6 does our experimental analysis. Section 7 discusses related work. Finally, Section 8 concludes the paper with future work.

2   Preliminaries

This section will recap some basic concepts of simultaneous concession solution [10] and fuzzy set theory [12,13].

In Zhang's model [10], he built a logic-based bargaining model to specify arbitrary n-person bargaining games, which is formally defined as follows:

Definition 1. *An n-agent bargaining game is a tuple of $((X_1, \succcurlyeq_1), \cdots, (X_n, \succcurlyeq_n))$, which can be abbreviated as $((X_i, \succcurlyeq_i)_{i \in N})$, where $(X_i, \succcurlyeq_i)$ is the demand set of agent i in $\mathcal{L}$ and $\succcurlyeq$ is a total pre-order on X, which satisfies totality, reflexivity and transitivity, and $\mathcal{L}$ is a propositional language consisting of a finite set of propositional variables and standard propositional connectives $\{\neg, \vee, \wedge, \rightarrow\}$.*

Definition 2. *A bargaining solution f is a function that assigns to a bargaining game a possible outcome. For any $G = ((X_i, \succcurlyeq_i)_{i \in N}, f(G) = (f_1(G), \cdots, f_n(G))$, where $f_i(G) \subseteq X_i$. $f_i(G)$ denotes the i-th component of $f(G)$ and $A(G) = \bigcup_{i \in N} f_i(G)$ denotes the agreement of the game.*

For a bargainer, some bargaining demands have the same preference rank, then an equivalence relation $\approx$ is needed, which can be defined from ordering $\succcurlyeq$. Let $\{X_i^1, \cdots, X_i^{L_i}\}$ be the partition of X_i induced by equivalence relation $\approx$, and L_i is called the height of the hierarchy. For convenience, $X_i^{>k}$ is short for $\bigcup_{l>k} X_i^l$.

Definition 3. *A bargaining solution F to bargaining game* $G = (X_i, \succcurlyeq_i)_{i \in N}$ *is the simultaneous concession solution (SCS) if*

$$F(G) = \begin{cases} (X_1^{>\mu}, \ldots, X_n^{>\mu}) & \text{if } \mu < L, \\ (\varnothing, \ldots, \varnothing) & \text{otherwise,} \end{cases} \tag{1}$$

where $\mu = \min\{k \mid \cup_{i=1}^n X_i^{>k} \text{ is consistent}\}$ *and* $L = \min\{L_i \mid i \in N\}$. *We call* L *the height of* G *and* μ *is the minimal rounds of concessions of the game.*

From the definition of such a solution, we can see that in every round if the remaining demand set is inconsistent (i.e., demand A and demand $\neg A$), then every bargainer has to make a minimal concession at the cost of giving up the least preferred demands in their own demand preference hierarchy. The agreement comes out only when the remaining demands are consistent. Now we will give an example of a bargaining game solved by simultaneous concession solutions.

Example 4. A couple is bargaining for which fruit should buy. The husband prefers to buy apple, pear and peach, hate orange and durian, while the wife like orange, pear and grape, hate peach and durian. The demand hierarchies of the couple can be illustrated by Table 1. According to simultaneous concession solution, we get $F_{husband}(G) = \{apple, \neg durian\}$, $F_{wife}(G) = \{orange, \neg grape\}$. And $A(G) = F_1(G) \cup F_2(G) = \{apple, \neg durian, orange, grape\}$.

Table 1. An example about buying fruit

Husband'demands	Wife'demands	Agreement
apple	orange	
¬ durian	grape	✓
peach	¬ peach	✗
¬ orange	pear	✗
pear	¬ durian	✗

However, there are two disadvantages in such a solution concept: 1) the difficulty of estimating the degree of risk attitudes; and 2) the inability to capture the change of bargainers attitudes, which is very important for a bargainer to adjust his strategy to get a high outcome. So, in Section 3, we will revise SCS and introduce a dynamic solution concept that can adjust an agent's strategy by an FL-based system of his decision process. Thus, it is necessary to list some basic concepts of fuzzy set theory [12,13] as follows:

Definition 5. *A fuzzy set, denoted as A, on domain U is characterized by a membership function* $\mu_A : U \mapsto [0, 1]$, *and* $\forall u \in U$, $\mu_A(u)$ *is called the membership degree of u in fuzzy set A.*

The following definition is about the implication of the Mamdani method [13].

Definition 6. *Let* A_i *be a Boolean combination of fuzzy sets* $A_{i,1}, \cdots, A_{i,m}$, *where* $A_{i,j}$ *is a fuzzy set defined on* $U_{i,j}$ $(i = 1, \cdots, n; j = 1, \cdots, m)$, *and* B_i *be fuzzy set on* U'

$(i = 1, \cdots, n)$. *Then when the input is* $\mu_{A_{i,1}}(u_{i,1}), \cdots, \mu_{A_{i,m}}(u_{i,m})$, *the output of such fuzzy rule* $A_i \to B_i$ *is fuzzy set* B'_i, *which is defined as:* $\forall u' \in U'$,

$$\mu_i(u') = min\{f(\mu_{A_{i,1}}(u_{i,1}), \cdots, \mu_{A_{i,m}}(u_{i,m})), \mu_{B_i}(u')\}, \tag{2}$$

where f *is obtained through replacing* $A_{i,j}$ *in* A_i *by* $\mu_{i,j}(u_{i,j})$ *and replacing "and", "or", "not" in* A_i *by "min", "max", "$1 - \mu$", respectively. And the output of all rules* $A_1 \to B_1, \cdots, A_n \to B_n$, *is fuzzy set* M, *which is defined as:* $\forall u' \in U'$,

$$\mu_M(u') = max\{\mu_1(u'), \cdots, \mu_n(u')\}. \tag{3}$$

By Definition 6, the result what we get is still a fuzzy set. To defuzzify the fuzzy set, we need the following centroid method [13]:

Definition 7. *The centroid point* u_{cen} *of fuzzy set* M *given by formula (3) is:*

$$u_{cen} = \frac{\sum\limits_{j=1}^{n} u_j \mu_M(u_j)}{\sum\limits_{j=1}^{n} \mu_M(u_j)}. \tag{4}$$

Actually, u_{cen} in above is the centroid of the area that is covered by the curve of membership function μ_M and the horizontal ordinate.

3   The Bargaining Model

This section introduces a new bargaining model and the concept of dynamic solution to this kind of bargaining games.

To overcome the difficulty of assessing bargainers' risk attitudes, we distinguish a person's original demand preference from his dynamic one. The original one just reflects on his own favorites in his mind (before they start to bargain) without considering whether or not an agreement can be reached. After participating in a certain bargaining game, they give another preference order, called their dynamic preferences, which consider not only their own taste but also other game information, such as possible conflicting demands and opponents' risk attitudes. For example, when a couple (one is risk seeking and the other is risk averse) are bargaining about purchasing fruits (apple or orange), the risk seeking one likes apple more than orange and the risk averse one likes orange more than apple. However, when they are going to bargain through simultaneous concession, they need think about the other side's risk attitude and possible conflicting demands. After putting their thinking into account, probably the risk seeker will still prefer apple to orange but the risk averse one changes to prefer apple to orange because the risk averse one worries about no fruits to eat if the bargaining is broken down.

Moreover, in our new model, dynamic preference can be changed during a bargaining because the bargainers will get more information during the game, while the original preference will not change because it does not depend on the other bargainers' information. Thus, we define a parameter, called change degree (CD), to capture the degree of bargainers' attitude of changing their preference. It is calculated by our fuzzy logic system, which inputs are bargainers' risk attitude, patience descent degree and regret degree (we will detail it in the next section). Then, every bargainer will take a proper action to change preference accordingly. Formally, we have:

Definition 8. *An n-player bargaining game is a tuple of* $(N, \{X_i, \succcurlyeq_{i,o}, \succcurlyeq_{i,d}^{(0)}$ $, \mathcal{A}_i\}_{i \in N}, FLS)$, *where*

- *N is the set of players;*
- X_i *is the demand set of bargainer i;*
- $\succcurlyeq_{i,o}$ *is bargainer i's original demand preference order;*
- $\succcurlyeq_{i,d}^{(0)}$ *is bargainer i's initial dynamic demand preference ordering and* $\succcurlyeq_{i,d}^{(n)}$ *denotes bargainer i's dynamic demand preference ordering after n-th round of a bargaining game;*
- $\mathcal{A}_i$ *is bargainer i's action function; and*
- *FLS is a fuzzy logic system for calculating the preference change degree.*

In this paper, we choose a political example to illustrate our model and use our solution concept to solve the problem. There are two reasons for us to choose such a bargaining example. Firstly, in real life, some bargaining processes can be modelled well by existing axiomatic methods such as game theory and price theory. However, these existing methods cannot deal with multi-issue bargaining problems appropriately, especially for the bargaining situation in which numerical utilities are hard to elicit and quantitative analysis are subdued, such as political negotiation [14,10]. Secondly, in a political negotiation, each party is in favour of policies that safeguard their own group's profit on one hand and try their best to win votes or get agreement with other parties even though at the price of policies they espoused on the other. For example, a party has to latch onto environmental issues to win votes even though it prefers to open the factories to get more profit. So, in such example, the difference between original preference and dynamic preference is obvious and our model does well in such bargaining situations.

Now we have a look at a political example. Suppose two political parties are bargaining over some policies that will be written into the new planning. Party 1 supports homosexual marriage (HM), raising taxes (RT), building high-speed railways (BHR), creating job opportunities (CJO), increasing education investment (IEI) and lengthening paid annual vacation (LPAV); but opposes rescuing major bank (RMB), fighting with hostile country (FHC) and land reclamation (LR). Party 2 supports RMB, BHR, CJO and IEI; but opposes HM, RT, LPAV, FHC and LR. Thus, their demand sets are:

$$X_1 = \{HM, RT, BHR, CJO, IEI, LPAV, \neg RMB, \neg FHC, \neg LR\} ,$$

$$X_2 = \{\neg HM, \neg RT, BHR, CJO, IEI, \neg LPAV, RMB, \neg FHC, \neg LR\} .$$

As shown in Table 2, two parties have original preferences over their own policies, which just reflect their own voters' favorites rather than the other side's situation. However, when going to the bargaining, they will worry about their conflicting demands and thus adjust the preferences to form dynamic ones, hoping to reach an agreement

Table 2. Original and initial dynamic preferences of Parties 1 and 2

Rank	party 1		party 2	
	original	dynamic	original	dynamic
1	RT	RT	RMB	¬FHC
2	CJO	LPAV	¬HM	RMB
3	LPAV	¬RMB	¬FHC	¬HM
4	¬RMB	CJO	¬RT	CJO
5	¬LR	¬LR	CJO	¬RT
6	IEI	IEI	¬LR	¬LR
7	BHR	HM	¬LRAV	IEI
8	¬ FHC	BHR	IEI	¬LRAV
9	HM	¬ FHC	BHR	BHR

easilier meanwhile keeping their demands as many as possible. Specifically, conflicting demands refer to the demands from different bargainers that will lead to contradiction. In this example, party 1 demands RT but party 2 demands $\neg RT$, which is a contradiction. Intuitively, when a bargainer puts all the conflicting demands at the top levels of his initial dynamic preference hierarchy, with the simultaneous concession solution method [10], he may get most of his conflicting demands if his opponent is risk averse, but he might get the bargaining broken if his opponent is risk-seeking as well. On the contrary, he can show his lowest risk attitude when he puts all his conflicting demands in the bottom places of initial dynamic preference hierarchy. So, we can assess a bargainer's attitude towards risk by comparing his original preference with dynamic one. Formally, we have:

Definition 9. $\forall x \in X_i$, *let* $\rho_i(x)$ *and* $P_i(x)$ *denote the level of* x *in the original preference hierarchy and in initial dynamic preference hierarchy, respectively. Specifically,* $\rho_i(x) = 1$ *means bargainer i prefers x the most in the original preference and* $\rho_i(x) = L_i$ *means bargainer i prefers x the least in the original preference, where* $L_i = \max \{\rho_i(x) \mid x \in X_i\}$. *It is similar for* $P_i(x)$. *Then the initial risk degree is given by:*

$$RD_i = \sum_{c_i \in CDS_i} (\rho_i(c_i) - P_i(c_i)), \tag{5}$$

where CDS_i *is the conflicting demand set of bargainer i.*

For the political example, from Table 2, by formula (5), party 1's initial risk degree is

$$
\begin{aligned}
RD_1 &= (\rho_1(RT) - P_1(RT)) + (\rho_1(LPAV) - P_1(LPAV)) \\
&\quad + (\rho_1(\neg RMB) - P_1(\neg RMB)) + (\rho_1(HM) - P_1(HM)) \\
&= (1 - 1) + (3 - 2) + (4 - 3) + (9 - 7) \\
&= 4.
\end{aligned}
$$

Similarly, we can obtain $RD_2 = -4$, which means that he downgrades more inconsistent demands rather than upgrades them in his dynamic preference hierarchy. For example, party 2 downgrades the conflicting demand RMB from the top level to the second level and downgrades $\neg LPAV$ from the seventh level to the eighth level. So, party 2 is risk-averse. On the contrary, party 1 is risky.

In different games, the domain of RD might be different. Thus, to handle the parameters' level (*high*, *medium* or *low*) in a unified way, we need change parameter RD's domain from $[\underline{RD_i}, \overline{RD_i}\,]$ into $[-1, 1]$ by multiplying $\frac{1}{|\overline{RD_i}|}$ (when $RD_1 \geqslant 0$) and $\frac{1}{|\underline{RD_i}|}$ (when $RD_1 < 0$), and we denote it as $r_{i,3}$ (which means the third linguistic parameter of agent i). $\underline{RD_i}$ indicates the value of RD when bargainer i puts all his conflicting demands at the top levels of his initial dynamic preference hierarchies, and similarly, $\overline{RD_i}$ indicates the value of RD when bargainer i puts all his conflicting demands in the bottom places of initial dynamic preference hierarchy. That is,

$$
\begin{aligned}
\overline{RD_i} &= (\rho_i(c_{i,1}) - 1) + (\rho_i(c_{i,2}) - 2) + \cdots + (\rho_i(c_{i,N_i}) - N_i) \\
&= \sum_{m=1}^{N_i} (\rho_i(c_{i,m}) - m) \\
&= \sum_{c_i \in CDS_i} \rho_i(c_i) - \frac{(1 + N_i)N_i}{2},
\end{aligned} \tag{6}
$$

$$
\begin{aligned}
\underline{RD_i} &= (\rho_i(c_{i,1}) - L_i) + (\rho_i(c_{i,2}) - (L_i - 1)) + \cdots + (\rho_i(c_{i,N_i}) - (L_i - N_i + 1)) \\
&= \sum_{m=1}^{N_i} (\rho_i(c_{i,m}) - (L_i - m + 1)) \\
&= \sum_{c_i \in CDS_i} \rho_i(c_i) - N_i L_i + \frac{(N_i - 1)N_i}{2},
\end{aligned} \tag{7}
$$

where N_i is the number of bargainer i's conflicting demands and L_i is the number of bargainer i's preference hierarchies. For the example, by formula (6), we have:

$$
\begin{aligned}
\overline{RD_1} &= (\rho_1(RT) - 1) + (\rho_1(LPAV) - 2) + (\rho_1(\neg RMB) - 3) + (\rho_1(HM) - 4) \\
&= (1 - 1) + (3 - 2) + (4 - 3) + (9 - 4) \\
&= 7.
\end{aligned}
$$

Similarly, by formula (7), we can obtain $\underline{RD_1} = -13$. Then risk degree RD_1 ranges in $[-13, 7]$, and RD_2 ranges in $[-16, 4]$. After changing into $[-1, 1]$, we have $r_{1,3} = 4/7 = 0.57$ and $r_{2,3} = -4/16 = -0.25$.

To solve the second problem of how to change a bargainer's risk attitude, we need to deal with the whole bargaining process round by round. In this way, the bargainers can get more information for choosing a preference to increase the chance to reach an agreement. Suppose that bargainers can be aware of the concession he has given up in the last round, but not the others' because the ordering of dynamic demand preference is bargainers' private information in real life.[2]

The bargaining proceeds as follows. The bargainers all try to get an agreement by sacrificing their less preferred demands. Thus, the concession will begin at the bottom, but after the new concession, the bargainers can update his preference in the light of new information, by being more conservative to guarantee consistent demands to be reserved.

However, someone may say the most risk-seeking bargainers will still be risky and thus upgrades the inconsistent demands in a dynamic preference hierarchy. Then, the

---

[2] Of course, we can also take the others' concession information into account, but it is beyond the scope of this paper, and so we leave it as future work.

bargaining will become a Hawk-Dove game [15], in which the risk averse ones get less preferred demands than the risk-seeking ones. However, it is easy to lead to a breakdown consequence when there are more than one risky bargainer who insists on demands inconsistent with each other because the conflicting demands cannot be given up. So, to increase the chance to reach an agreement, we restrict the change action just towards the conservative direction. Formally, we have:

Definition 10. *The action function can be defined as follow:*

$$\mathcal{A}_i : N_{CD} \times X_i \rightarrow AC_i. \tag{8}$$

where N_{CD} is the set of all CD's values (which are denoted as cd), X_i is the demand set of bargainer i, $AC_i = \{$no change, declining the conflicting demands one level, declining the conflicting demands two levels, $\cdots$, declining the conflicting demands L_i levels$\}$.

In the political example, we suppose that each party's action functions are as follows:

$$\mathcal{A}_i(cd, x) = \begin{cases} \text{declining the conflicting demand two levels} \\ \quad \text{if } cd \geqslant 0.6, \text{ and } \exists \, x', x'' \notin CDS_i, x >_{i,d}^{(n)} x', x'' \\ \text{declining the conflicting demand one level} \\ \quad \text{if } 0.6 > cd \geqslant 0.2 \text{ and } \exists \, x' \notin CDS_i, \ x >_{i,d}^{(n)} x', \text{ or} \\ \quad cd \geqslant 0.6, \text{ and } !\exists \, x' \notin CDS_i, \ x >_{i,d}^{(n)} x' \\ \text{no change} \quad \text{otherwise} \end{cases} \tag{9}$$

where n means the n-th round of the bargaining game.

Suppose there is nothing worse than breakdown of bargaining. Then, one of our main purposes is to make all bargainers to reach an agreement. As we have discussed above, upgrading inconsistent demands will increase the chance of breakdown. So, upgrading inconsistent demands is not an option in the AC. After every bargaining round, if bargainers do not agree, the dynamic preferences will be updated to new ones by bargainers' action functions. Formally, we have:

Definition 11. *For each bargainer i, the $(\lambda + 1)$-th round demand $X_i^{\lambda+1}$ and preference $\succcurlyeq_{i,d}^{(\lambda+1)}$ is given by:*

$$(X_i^{\lambda+1}, \succcurlyeq_{i,d}^{(\lambda+1)}) = \mathcal{U}((X_i^{\lambda}, \succcurlyeq_{i,d}^{(\lambda)}), a_i), \lambda \geqslant 1, \tag{10}$$

That is, after the λ-th round, the demand set with dynamic demand preference $(X_i^{\lambda}, \succcurlyeq_{i,d}^{(\lambda)})$ of bargainer i will be updated into a new one, denoted as $(X_i^{\lambda+1}, \succcurlyeq_{i,d}^{(\lambda)})$, by a certain action through action functions, which input (i.e., CD) is determined by our fuzzy logic system. And after the updating, the demands of bargainers will be reduced because of the concession. So, $X_i^{\lambda} \subseteq X_i^{\omega}$ if $\lambda \geqslant \omega$.

Now, by Definitions 8 and 11, we can define our dynamic simultaneous concession solution (DSCS) as follows:

Definition 12. *A solution $\overline{F}$ to bargaining game $G = (N, \{X_i, \succcurlyeq_{i,o}, \succcurlyeq_{i,d}^{(0)}, \mathcal{A}_i\}_{i \in N}, FLS)$ is the dynamic simultaneous concession solution if*

$$\overline{F}(G) = \begin{cases} (X_1^v, \cdots, X_n^v) \text{ if } \forall i \in N, X_i^v \neq \varnothing, v < L, \\ (\varnothing, \ldots, \varnothing) \text{ otherwise,} \end{cases} \tag{11}$$

where $v = \min\{k \mid \cup_{i=1}^n X_i^k$ is consistent$\}$ (X_i^k is the set of demands of bargainer i after k rounds of the bargaining game). We call v the minimal rounds of concessions of the game.

4 Fuzzy Logic-Based System

From the previous section, we know that the change of preference order depends on the change degree. This section will discuss how to use fuzzy rules to calculate the degree.

4.1 The Fuzzy Parameters

According to our psychological experiment, which is based on [16,17,18], the change degree of preference mainly depends on the three human cognitive factors: (i) *Regret degree (r_1)*. It is represented by the percentage of the consistent demands, which have been removed at the expense of bargaining with each other because all the bargainers want the consistency demands. (ii) *Patience descent degree (r_2)*. It is measured by the percentage of the completed bargaining rounds because the closer to the deadline, the more likely the bargainers will change their current preference. (iii) *Initial risk degree (r_3)*. According to our psychological study, a risk-seeking bargainer will insist on his original preference, expecting the possible better outcome in the future. However, a risk averse one will become even more conservative.

In real life, these parameters are described in some linguistic terms, which is an effective way to describe people's perception [12]. There are several types of membership functions to model linguistic terms, such as trapezoidal, triangle, Gaussian and sigmoid [19]. In this paper, we employ the trapezoidal one as follows:

$$f(x, a, b, c, d) = \begin{cases} 0 & \text{if } x \leqslant a, \\ \frac{x-a}{b-a} & \text{if } a \leqslant x \leqslant b, \\ 1 & \text{if } b \leqslant x \leqslant c, \\ \frac{d-x}{d-c} & \text{if } c \leqslant x \leqslant d, \\ 0 & \text{if } x \geqslant d. \end{cases} \tag{12}$$

The reason for our choice of formula (12) is as follows. Its parameters (i.e., a, b, c, and d) can reflect well that different people could set the membership function of a same linguistic term differently. For example, when $a = b = c < d$, it reflects a decreasing tendency; when $a < b = c = d$, it reflects an increasing tendency; when $a < b = c < d$, it reflects a tendency that is increasing between a and b, decreasing between c and d; and when $a < b < c < d$, it reflects a tendency that is increasing between a and b, reaching the maximum level between b and c, and decreasing between c and d [20].

The meanings of linguistic terms of these parameters are as follows. The *low* regret degree indicates that a bargainer does not care much about the concession in the previous round and just regrets a little for the demands given up in the previous round. The *medium* regret degree means that a bargainer cares about the demands he has given up and regrets having insisted on the preference in the previous round. And the *high* regret degree means that a bargainer regrets very much for insisting on the preference in the previous round and more likely changes it because it leads a lot of consistent demands lost. Similarly, we can understand the linguistic terms of other two parameters.

For convenience, we represent the formula (12) as (a, b, c, d) for short. Thus, in this paper, linguistic terms of regret degrees can be expressed as $Low_{r_1} = (-0.36, -0.04, 0.1, 0.3)$, $Medium_{r_1} = (0.15, 0.4, 0.6, 0.85)$, and $High_{r_1} = (0.7, 0.9, 1.04, 1.36)$. Similarly, we can have $Low_{r_2} = (-0.36, -0.04, 0.2, 0.4)$, $Medium_{r_2} = (0.15, 0.3, 0.7, 0.85)$, and $High_{r_2} = (0.6, 0.8, 1.04, 1.36)$; $Low_{r_3} = (-1.72, -1.08, -0.7, -0.4)$, $Medium_{r_3} = (-0.7, -0.2, 0.2, 0.7)$, and $High_{r_3} = (0.4, 0.7, 1.08, 1.72)$; $Low_{CD} = (-0.36, -0.04, 0.2, 0.4)$, $Medium_{CD} = (0.2, 0.4, 0.6, 0.8)$, and $High_{CD} = (0.6, 0.8, 1.04, 1.36)$.

4.2   The Fuzzy Rules

We calculate a CD (change degree) from a bargainer's *Regret Degree* (r_1), *Patience descent degree* (r_2), and *Initial Risk Degree* (r_3) according to the fuzzy rules as shown in Table 3. Here we just explain fuzzy Rule 1 (and then others can be understood similarly): if a bargainer does not lose too many consistent demands, which makes him not regret, and his initial risk degree is not low (i.e., he is not a conservative person), then his change degree is low. So, no matter how much percentage of rounds have been completed, he is confident enough to insist on his preference.

Table 3. Fuzzy rules

If r_1 is *Low* and r_3 is not *Low* then *CD* is *Low*.
If r_1 is *High* and r_3 is not *High* then *CD* is *High*.
If r_1 is not *Low* and r_2 is *High* and r_3 is not *High* then *CD* is *High*.
If r_1 is not *High* and r_2 is *Low* and r_3 is not *Low* then *CD* is *Low*.
If r_1 is not *High* and r_2 is not *High* and r_3 is *High* then *CD* is *Low*.
If r_1 is not *Low* and r_2 is not *Low* and r_3 is *Low* then *CD* is *High*.
If r_1 is *Low* and r_2 is *Low* and r_3 is *Low* then *CD* is *Low*.
If r_1 is *Low* and r_2 is not *Low* and r_3 is *Low* then *CD* is *Medium*.
If r_1 is *Medium* and r_2 is *Low* and r_3 is *Low* then *CD* is *Medium*.
If r_1 is *Medium* and r_2 is *Medium* and r_3 is *Medium* then *CD* is *Medium*.
If r_1 is *Medium* and r_2 is *High* and r_3 is *High* then *CD* is *Medium*.
If r_1 is *High* and r_2 is *Low* and r_3 is *High* then *CD* is *Medium*.
If r_1 is *High* and r_2 is not *Low* and r_3 is *High* then *CD* is *High*.

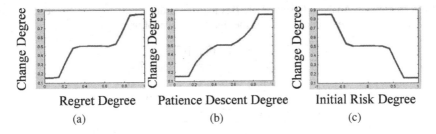

Fig. 1. CD's change with the r_1, r_2 and r_3, respectively

With the help of Matlab, we can see the relation between inputs and output in the Surface Viewer of Matlab (as shown in Fig. 1).[3] Fig. 1(a) shows that when losing a few consistent demands, the bargainer will still insist on his preference, but with the regret degree increasing, his change degree also increases. Fig. 1(b) indicates that when more bargaining rounds are completed, the bargainer's change degree increases as well. Fig. 1(c) implies that the bargainer's change degree is high (which means he will choose a more conservative preference in the next round) when he is risk averse, while being low (which means he probably not change his current preference) when he is a risk seeking person. Actually, this is in accord with our intuition and psychological study presented in the following subsection.

4.3 Psychological Study Validation

To verify these fuzzy rules, according to psychological methodology, we conducted a survey study and recruited 40 human subjects. Empirically, 30 is the minimal sample size required to conduct such a statistical analysis in our study, while more than 50 is pointless [20]. So, it is fine for us to choose 40 (18 females and 22 males). They ranged in age from 19 to 40, and varied in careers and educational levels. All the subjects volunteered to participate in our survey and completed our questionnaires.

Our questionnaire consists of four parts: (i) *Risk Orientation Questionnaire* [16]. The questionnaire uses 12 items to assess individuals' risk propensity and cautiousness. (ii) *Regret Scale* [17]. It consists of 5 items and is designed to assess how individuals deal with decision situations after the decision have been made, specifically the extent to which they experience regret. (iii) *Delay-discounting rate* [18]. It is used to assess individuals' patience level by offering subjects a series of choices between immediate rewards and larger but delayed rewards. (iv) *Maximisation Scale Short* [22]. It uses 6 items to assess individuals' tendency to optimise decisions. And people with more tendencies to optimise their decision would less likely change their original decisions in our bargaining game scenario.

We tested the effect of risk degree, regret degree and patience on how individuals approach their decision by conducting multiple regression analysis. The analysis results are reported in Table 4. Regret degree is significantly relevant to the tendency to change

---

[3] The calculations are actually based on Mamdani method [21], which is one of the standard fuzzy reasoning methods.

Table 4. Regression Analysis Results. Here β is the standardized regression coefficient; $S.E.$ is the standard error of the estimate; and p is the significant level of the t- test.

	β	S.E.	t value	p
Intercept	-12.38	6.59	-1.88	0.07
Regret degree	0.36	0.17	2.12	0.04
Impatience	1.18	2.16	0.55	0.59
Risk degree	-0.17	0.10	1.66	0.10

their decisions (i.e., β =0.36 and p=0.04). Those who experience more regret after the decision has been made are more likely to change their decisions. Risk degree is marginally significantly related to the change degree (i.e., β =-0.17 and p=0.10). Those, who prefer a higher level of risk, tend to insist on their original decisions. Patience level is also positively relevant to the change degree (i.e., β =1.18 and p=0.59).

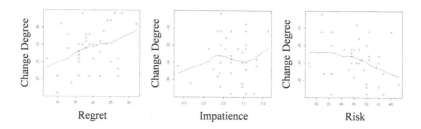

Fig. 2. CD's change with three parameters in experiments

As shown in Fig. 2, according to experiment results, we draw three scatter plots for CD's change with the regret, patience descent, and risk degrees, respectively. The curve was superimposed on each scatter plot using the scatter smoother function $lowess()$ of MASS package in the R system for statistical analysis.

5   Illustration

Now we continue our political example to show how our model solves it according to fuzzy rules in Table 2. During the bargaining, the changes of preference and parameters are shown in Tables 5 and 6, respectively.

More specifically, after the first round, their dynamic preferences are shown in the left first table in the first row (denoted as Round 1). By the parameters' definitions, we can obtain $r_{1,1} = 0.2, r_{1,2} = 0.11, r_{1,3} = 0.57, r_{2,1} = 0.2, r_{2,2} = 0.11$, and $r_{2,3} = -0.25$, respectively. Thus, by fuzzy rules in Table 2, based on Mamdani method (see Definition 5), we can obtain $CD_1 = 0.153$ and $CD_2 = 0.153$ in this round. Then, according to their action function (9), their preferences are updated into new ones shown in the next table in the first row (denoted as Round 1*). Because both choose "no change" in the first round, the two tables Round 1 and Round 1* in the first row are the same. Similarly, after the second round, their preferences are shown in Round 2. However, in

Table 5. Dynamic bargaining proceeding

Round 1

Rank	party 1	party 2
1	RT	¬FHC
2	LPAV	RMB
3	¬RMB	HM
4	CJO	CJO
5	¬LR	¬RT
6	IEI	¬LR
7	HM	IEI
8	BHR	¬LPAV

Round 1*

Rank	party 1	party 2
1	RT	¬FHC
2	LPAV	RMB
3	¬RMB	HM
4	CJO	CJO
5	¬LR	¬RT
6	IEI	¬LR
7	HM	IEI
8	BHR	¬LPAV

Round 2

Rank	party 1	party 2
1	RT	¬FHC
2	LPAV	RMB
3	¬RMB	HM
4	CJO	CJO
5	¬LR	¬RT
6	IEI	¬LR
7	HM	IEI

Round 2*

Rank	party 1	party 2
1	CJO	¬FHC
2	RT	CJO
3	LPAV	RMB
4	¬RMB	HM
5	¬LR	¬LR
6	IEI	¬RT
7	HM	IEI

Round 3

Rank	party 1	party 2
1	CJO	¬FHC
2	RT	CJO
3	LPAV	RMB
4	¬RMB	HM
5	¬LR	¬LR
6	IEI	¬RT

Round 3*

Rank	party 1	party 2
1	CJO	¬FHC
2	¬LR	CJO
3	RT	¬LR
4	LPAV	RMB
5	¬RMB	HM
6	IEI	¬RT

Round 4

Rank	party 1	party 2
1	CJO	¬FHC
2	¬LR	CJO
3	RT	¬LR
4	LPAV	RMB
5	¬RMB	HM

Round 4*

Rank	party 1	party 2
1	CJO	¬FHC
2	¬LR	CJO
3	RT	¬LR
4	LPAV	RMB
5	¬RMB	HM

Round 5

Rank	party 1	party 2
1	CJO	¬FHC
2	¬LR	CJO
3	RT	¬LR
4	LPAV	RMB

Table 6. Parameters

parameters	Round 1	Round 2	Round 3	Round 4
$r_{1,1}$	0.2	0.4	0.4	0.6
$r_{1,2}$	0.11	0.22	0.33	0.44
$r_{1,3}$	0.57	0.57	0.57	0.57
CD_1	0.153	0.269	0.31	0.31
$r_{2,1}$	0.2	0.2	0.4	0.4
$r_{2,2}$	0.11	0.22	0.33	0.44
$r_{2,3}$	-0.25	-0.25	-0.25	-0.25
CD_2	0.153	0.25	0.433	0.5

this round, both parties choose "declining the conflicting demand one level" to update their preferences. Thus, *RT, LPAV,* and *RMB* of party 1 and *¬RT, HM,* and *RMB* of party 2 are declined. The game ends after the 5th round because both of the parties have nothing in contradictory.

From Table 5, we can see that by our dynamic simultaneous concession method (see Definiton 12), the outcome of the game is: $\overline{F}_1(G) = \{CJO, ¬LR, RT, LPAV\}$ and $\overline{F}_2(G) = \{¬FHC, CJO, ¬LR, RMB\}$. So, their agreement is:

$$\overline{F}_1(G) \cup \overline{F}_2(G) = \{CJO, ¬LR, RT, LPAV, ¬FHC, RMB\}. \tag{13}$$

However, by the method of Zhang [10] (bargainers also do simultaneous concession but their preferences never change during the bargaining), the outcome is: $F_1(G) = \{RT, LPAV\}$ and $F_2(G) = \{¬FHC, RMB\}$, and so the agreement of two parties is:

$$F_1(G) \cup F_2(G) = \{RT, LPAV, ¬FHC, RMB\}. \tag{14}$$

By comparing (13) with (14), we can see that ours is more reasonable. This can be attributed to that our model can reflect bargainers' risk attitudes well and reach an agreement containing more demands. For example, in Zhang's model, the bargainers have to give up the demands *CJO* and *¬LR* (which are demands consistent with both bargainers) as cost of their bargaining risk attitudes.

6   Experimental Analysis

This section empirically analyses how well our model (i.e., DSCS) works against the one of [10] (i.e., SCS). On Matlab platform, we conduct two experiments to see how both models work when the number of conflicting demands and bargainers change, respectively. In both experiments, we do 1000 times under the setting that every bargainer's action function is formula (9) and the fuzzy rules are those in Table 3.

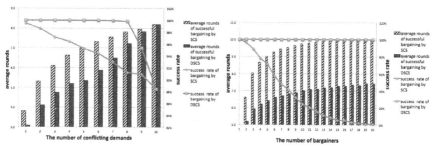

(a) Success rate and average rounds to reach agreements over the number of conflicting demands by SCS and DSCS, respectively

(b) Success rate and average rounds to reach agreements over the number of bargainers by SCS and DSCS, respectively

Fig. 3. The change of success rate and average rounds to reach agreements

In the fist experiment, we randomly generate 10 demands in different preference levels for two bargainers and randomly label N (changing from 1 to 10) of them as their conflicting ones. The bargaining is carried out in two models. From Fig. 3(a), we can see that the success rate of our bargaining model is much higher than that of Zhang, especially when the conflicting demands are increasing; and in our model the average rounds to reach agreements are lower.

In the second experiment, we randomly generate 10 demands in different preference levels for M bargainers (changing from 1 to 20) and randomly select 4 of them as the conflicting ones of all bargainers. Bargaining will proceed in both models. From Fig. 3(b), we can see our model can keep a high success rate of bargaining even when the number of bargainers increases and it also keeps lower rounds to reach agreements, while the success rate will decrease obviously with Zhang's model.

So, our model is able to not only reflect bargainers' individualities (i.e., risk, regret, patience), but also keep a high success rate of bargaining and high efficiency. This means that our model has a greater potential in practical applications.

7   Related Work

Like Zhang [10], Vo and Li [23] also build an axiomatic bargaining model, in which the bargaining situation is described in propositional logic language and the preference over outcomes is ordinal. Further, they handle the belief revision problem during a bargaining, which purpose is similar to ours. However, unlike our work in this paper, their model does not reflect the bargainers' risk attitudes and patience, which are very important factors for bargaining in real life.

In [24], a fuzzy logic based model is also introduced for a buyer to decide to accept or reject a seller's offer according to the proposed price, the belief about the seller's deadline, the remaining time, the demand relevancies, and so on. However, this model does not show how the risk attitude changes the bargainers' preferences, while our solution does via a fuzzy reasoning system.

In the bilateral negotiation model of [25], fuzzy logic is used for offering evaluation. Moreover, they distinguish three attitudes of bargainers in three concession strategies: greedy, anxious and calm. However, unlike our model, their model does not deal with risk attitudes of the bargainers. And their preferences on the demands are ranked by using real numbers. However, it is hard to articulate the preference in exact numbers [4]. Since our work uses ordinal preference, the problem does not exist in our model.

Fuzzy logic approaches are also used to solve other problems in bargaining games, for example: (i) to predict the negotiation strategy of the opponent [26]; and (ii) to calculate, in bargaining, the need for a project according to received revenues, future business opportunities, and level of competition [27]. However, none of them uses fuzzy logic to update the preference during a bargaining, while we do in this paper.

8 Conclusion

This paper develops a fuzzy logic based bargaining. It distinguishes the original demand preference from the dynamic one. By comparing the two preferences, we can estimate bargainers' initial risk attitudes. Moreover, fuzzy rules are used to calculate how the bargainers should update their preferences during a bargaining according to their regret, patience and risk factors. Many could be done in the future. For example, our action functions could be improved further to reflect that some people will become riskier when feeling the bargaining may be broken down.

Acknowledgements. The authors appreciate the anonymous referees for their insightful comments, which have been used to improve the paper. The authors also thank Yinping Yang for her valuable comments on the early version of this paper. Moreover, this paper is partially supported by National Natural Science Foundation of China (No. 61173019), Bairen Plan of Sun Yat-sen University, and Major Projects of the Ministry of Education (No. 10JZD0006) China. Finally, Kwang Mong Sim gratefully acknowledges financial support from the School of Computing at the University of Kent for supporting the visit of Xudong Luo to carry out this research from 15 October 2012 through 14 December 2012.

References

1. Nash, J.F.: The bargaining problem. Econometrica 18(2), 155–162 (1950)
2. Ni, Q., Zarakovitis, C.C.: Nash bargaining game theoretic scheduling for joint channel and power allocation in cognitive radio systems. IEEE Journal on Selected Areas in Communications 30(1), 70–81 (2012)
3. Zadeh, L.A.: The concept of a linguistic variable and its application to approximate reasoning – I. Information Sciences 8(3), 199–249 (1975)
4. Luo, X., Zhang, C., Jennings, N.R.: A hybrid model for sharing information between fuzzy, uncertain and default reasoning models in multi-agent systems. International Journal of Uncertainty, Fuzziness and Knowledge-Based Systems 10(4), 401–450 (2002)
5. Domshlak, C., Hüllermeier, E., Kaci, S., Prade, H.: Preferences in AI: An overview. Artificial Intelligence 175(7-8), 1037–1052 (2011)
6. Shubik, M.: Game theory in the social sciences: Concepts and solutions. MIT Press, Cambridge (2006)
7. Zhang, D., Zhang, Y.: An ordinal bargaining solution with fixed-point property. Journal of Artificial Intelligence Research 33(1), 433–464 (2008)
8. García-Gallego, A., Georgantzís, N., Jaramillo-Gutiérrez, A.: Gender differences in ultimatum games: Despite rather than due to risk attitudes. Journal of Economic Behavior & Organization 83(1), 42–49 (2012)
9. Lippman, S.A., McCardle, K.F.: Embedded Nash bargaining: Risk aversion and impatience. Decision Analysis 9(1), 31–40 (2012)
10. Zhang, D.: A logic-based axiomatic model of bargaining. Artificial Intelligence 174(16-17), 1307–1322 (2010)
11. Zadeh, L.A.: Fuzzy logic and approximate reasoning. Synthese 30(3), 407–428 (1975)
12. Zadeh, L.A.: Fuzzy sets. Information and Control 8(3), 338–353 (1965)
13. Nanda, S., Das, N.R.: Fuzzy Mathematical Concepts. Alpha Science Intl. Ltd. (2010)
14. Iklé, F.C., Leites, N.: Political negotiation as a process of modifying utilities. Journal of Conflict Resolution 6(1), 19–28 (1962)
15. Smith, J.M., Price, G.R.: The logic of animal conflict. Nature 246(2), 15–18 (1973)
16. Rohrmann, B.: Risk attitude scales: Concepts and questionnaires. Project report, University of Melbourne (November 2004)
17. Schwartz, B., Ward, A., Monterosso, J., Lyubomirsky, S., White, K., Lehman, D.R.: Maximizing versus satisficing: Happiness is a matter of choice. Journal of Personality and Social Psychology 83(5), 1178–1197 (2002)
18. Kirby, K.N., Maraković, N.N.: Delay-discounting probabilistic rewards: Rates decrease as amounts increase. Psychonomic Bulletin and Review 3(1), 100–104 (1996)
19. Piegat, A.: Fuzzy Modeling and Control. STUDFUZZ. Physica-Verlag HD (2010)
20. Luo, X., Jennings, N.R., Shadbolt, N.: Acquiring user tradeoff strategies and preferences for negotiating agents: A default-then-adjust method. International Journal of Man-Machine Studies 64(4), 304–321 (2006)
21. Mamdani, E.H., Assilian, S.: An experiment in linguistic synthesis with a fuzzy logic controller. International Journal of Man-Machine Studies 7(1), 1–13 (1975)
22. Nenkov, G.Y., Morrin, M., Schwartz, B., Ward, A., Hulland, J.: A short form of the maximization scale: Factor structure, reliability and validity studies. Judgment and Decision Making 3(5), 371–388 (2008)
23. Bao, V.N.Q., Li, M.: From axiomatic to strategic models of bargaining with logical beliefs and goals. In: Proceedings of the 11th International Conference on Autonomous Agents and Multiagent Systems, vol. 1, pp. 525–532 (June 2012)

24. Kolomvatsos, K., Anagnostopoulos, C., Hadjiefthymiades, S.: A fuzzy logic system for bargaining in information markets. ACM Transactions on Intelligent Systems and Technology 3(2), 32 (2012)
25. Zuo, B., Sun, Y.: Fuzzy logic to support bilateral agent negotiation in e-commerce. In: 2009 International Conference on Artificial Intelligence and Computational Intelligence, vol. 4, pp. 179–183 (November 2009)
26. Lee, W.M., Hsu, C.C.: An intelligent negotiation strategy prediction system. In: 2010 International Conference on Machine Learning and Cybernetics, pp. 3225–3229. IEEE (2010)
27. Yan, M.R.: A fuzzy logic enhanced bargaining model for business pricing decision support in joint venture projects. Journal of Business Economics and Management 12(2), 234–247 (2011)

Domain Ontology Enrichment Based on the Semantic Component of LMF-Standardized Dictionaries

Feten Baccar Ben Amar[1,2], Bilel Gargouri[2], and Abdelmajid Ben Hamadou[3]

[1] Preparatory Institute to Engineering Studies, Sfax University, Tunisia
[2] MIRACL Laboratory, FSEGS, Sfax University, Tunisia
feten.baccarb@mes.rnu.tn bilel,.gargouri@fsegs.rnu.tn
[3] MIRACL Laboratory, ISIMS, Sfax University, Tunisia
abdelmajid.benhamadou@isimsf.rnu.tn

Abstract. This paper proposes an approach for automatic domain ontology enrichment based on the semantic component of Lexical Markup Framework (LMF, ISO 24613)-standardized dictionaries. The originality of this research work lies in the use of simple and regularly-structured textual fields disseminated in the definitions and examples of lexical entries, thus facilitating the extraction of concepts and relations candidates. It also lies in exploiting the semantic component of the LMF dictionary in order to guide the extraction of further knowledge from the adopted source. Along with addressing the quality issue which is of a major importance in ontology engineering, the proposed approach has been implemented in a rule-based system, relying on lexico-syntactic patterns for ontology items extraction and applied to a case study in the field of astronomy. The experiment has been carried out on the Arabic language, whose choice is explained both by the great deficiency of research work on the Arabic ontology development and the availability within our research team of an LMF-standardized Arabic dictionary.

1 Introduction

The recent development of and increasing interest in semantic resources including lexica, dictionaries, thesauri and terminologies, has suggested that these resources can form an interesting basis for developing domain ontologies. It is in this context that we have proposed an approach for generating core domain ontologies from LMF (ISO 24613)-standardized dictionaries, taking advantage of its finely-structured content rich in lexical and conceptual knowledge [6]. In this paper, we are interested in the enrichment of the resulting core by exploiting the semantics conveyed by textual fields disseminated in the definitions and examples of lexical entries.

Ontology enrichment is the task of extending an existing ontology with additional concepts and semantic relations and placing them at the correct position in the ontology [23]. It is generally an incremental process which relies on text mining techniques to identify the potentially relevant knowledge from data encoded in unstructured textual forms provided by raw texts [9], Machine-Readable Dictionaries (MRDs) [20, 22] and web documents [24].

M. Wang (Ed.): KSEM 2013, LNAI 8041, pp. 404–419, 2013.

Among different text mining techniques, the pattern-based approaches, pioneered by [18], have inspired a great deal of research work and are increasingly attracting the attention of the scientific community [1]. However, the effectiveness of the proposed systems depends on the appropriate preparation (selection) of the text collections as well as the patterns that are dependent on particular domain and purpose of the application. Besides, they are considered as the most time-consuming activities requiring considerable effort and significant human (expert) involvement in the whole process [2, 23].

The main goal of this paper is to propose a domain ontology enrichment approach using an LMF-standardized dictionary, taking advantage of its textual content regularities and of the restricted vocabulary used, thus facilitating the automatic knowledge extraction as well as the lexico-syntactic patterns creation. With respect to the use of lexico-syntactic patterns and their projection over linguistically analyzed textual content emanating from dictionaries [3, 21] are concerned, the proposed approach is relatively classical. However, its originality lies in the fact that the semantic component of lexical entries relative to the new detected concepts can contribute to guide the extraction of further knowledge from the adopted source.

Moreover, the suggested process includes a validation stage so as to preserve the quality of the produced ontology throughout the enrichment stage. Furthermore, our experiments are conducted using the LMF-standardized dictionary of the Arabic language [8] as a case study evaluating the approach.

The remainder of this paper is structured as follows. Section 2 outlines the domain ontology generation approach using LMF-standardized dictionaries. Then, Section 3 describes the proposed approach for domain ontology enrichment based on the semantic component of LMF-dictionaries. Section 4 gives details of the system implementation. In Section 5, we present a case study relating to the astronomy domain investigating the LMF-standardized dictionary of Arabic language. As for Section 6, it is devoted to present related work of domain ontology enrichment based on MDRs' knowledge. Finally, Section 7 concludes the paper with opening perspectives for future work.

2 Generating Domain Ontologies from LMF-Standardized Dictionaries

This section briefly describes a general overview of the proposed approach for generating domain ontologies from LMF-standardized dictionaries with a special focus on how LMF structure [13] can help in systematically generating a core of the targeted domain ontology as well as guiding its enrichment process.

2.1 General Overview

The main contribution of the current research work is to propose a novel approach for the domain ontology generation starting from LMF (ISO-24613)-standardized dictionaries [19]. Firstly, it consists in building an ontology core. Secondly, the constructed core will be further enriched with additional knowledge included in the text available in the dictionary itself. Indeed, based on the LMF meta-model subtlety and power, lexical knowledge is expressed either directly, through the fields (e.g. *Context, SubjectField, SenseRelation*) related to the lexical entries or indirectly, through the text included in the semantic knowledge (e.g. definitions and examples).

Before proceeding to the core building, we begin with the creation of a dictionary fragment (henceforth domain dictionary), by extracting the relevant part of the whole dictionary. It firstly gathers the lexical entries of the related senses to the domain of interest as well as their semantically related words by accessing to the explicit information corresponding to the usage domain in the dictionary. Then, all along the building process, this domain dictionary is extended by importing the textual resource of new lexical entries pertaining to the concepts revealed by the text analysis (§ subsection 2.3).

2.2 LMF Structure-Based Ontology Acquisition: The Core Building

The ontological core within our context stands for all possible sets of basic objects in a specific domain that could be directly derived from the systematic organization of linguistic objects in an LMF-standardized dictionary [6]. According to our investigation on LMF structure, we managed to define a set of identification rules allowing for the elicitation of ontological entities. For instance, since a concept corresponds to a meaning of a word, we can directly deduce the concepts of the domain ontology from the particular instances (e.g., *Context, SenseExample*) attached to the *Sense* class.

Another example of rules can be illustrated by the fact that a semantic relationship (e.g., *synonymy, hypernymy, meronymy*) between the senses of two or several lexical entries by means of the *SenseRelation* class gives birth to an ontological relation linking the corresponding concepts. Figure 1 shows a portion of LMF meta-model including the classes we are interested in. We refer the reader to [13] for an in-depth study of the LMF standard.

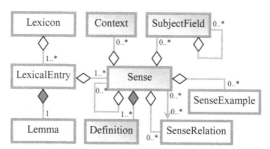

Fig. 1. Portion of LMF meta-model (described in UML)

Based on a variety of experiments carried out on the handled dictionary and having done a first evaluation in order to highlight the importance of the core construction stage, we claim that, in almost all cases, the number of concepts would represent around 85% of the total number of concepts. The rest cannot be identified in the core construction stage for two reasons. On the one hand, though tagged with another domain, many concepts are relevant to the targeted domain. On the other hand, certain entries of the dictionary have senses without usage domain indication. In addition, about 30% of taxonomic relationships can be easily identified with the help of the dictionary structure [5]. What is worth emphasizing is that the overlooked concepts as well as relations will be deduced from the text analysis in the enrichment stage.

2.3   How LMF-Semantic Component Can Be Exploited in the Ontology Enrichment

Although the whole lexical data in LMF-standardized dictionary cannot be directly accessed, its textual fields contain a profusion of semantic knowledge included in a more or less regularly-structured text (especially in the case of text definition), thus facilitating the Natural Language Processing (NLP). Based on the idea that the dictionary definition is a linguistic representation conveying all the necessary information to define a concept in a certain domain [25], we note that all new concepts candidates revealed by the text analysis (§ section 4.2) are relevant to the considered domain. Therefore, the textual fields of lexical entries pertaining to these concepts can be interesting in order to ensure the coverage of the targeted domain.

Accordingly, having provided an initial dictionary fragment for a specific domain, the idea is to extend this dictionary with additional textual fields and rather to limit their importation to only the ones yielding entities relevant to the considered domain. Because this task is very tedious and time-consuming if done manually, we define the Degree of Relevance (DR) as the needed information assigned to the concept for deciding in an automatic way the salient textual fields.

According to our investigation on the dictionary representation of the various units in the definitions, and as it is shown in Figure 2, we may distinguish three classes of degrees of relevance. Such classes are determined with the help of the semantic component knowledge in the dictionary (classes represented in Figure 1 by the gray color).

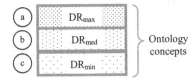

Fig. 2. Degrees of Relevance (DRs) classification

The first class of DR given in the top-part of Figure 2-a groups the concepts of the highest (maximum) DR (DR_{max}) together. They are the core concepts directly deduced from the « Context » class of the meaning attached explicitly to the « SubjectField » class in the dictionary which is the same as the considered ontology domain.

The second class of DR (Figure 2-b) gathers concepts qualified of DR medium (DR_{med}). These concepts are those whose:

- Label exists in « Context » class but also is attached to a « SubjectField » class different from the considered ontology domain. For example:

*Let us consider the domain of « Astronomy » and the concept label « equator »
which is assigned with the « geography » domain in the dictionary. This concept is more or less relevant to the « Astronomy » domain;*

- Label exists in « Lemma » class and neither « SubjectField » class nor « Context » classes are attached to a « Sense » class and, at least, there is a concept
  revealed by its text analysis that already exists in the ontology and assigned to
  DR_{max}. It is not the case of an incomplete dictionary; it is rather the case of a
  concept that belongs to numerous domains, including the considered one.

In the case of DR_{med}, the domain dictionary should be extended with the analyzed
text.

As far as the concepts of the third class of DR (Figure2-c) are concerned, they
have the minimum DR (DR_{min}) and are those whose:

- Label exists in « Lemma » class and neither « SubjectField » class nor « Context » class are attached to a « Sense » class, and all concepts revealed by its
  text analysis do not exist in the ontology. It can be the case of domain-
  independent concepts such as *Time* and *Instrument*.
- No matching can be found across explicit knowledge of LMF structure.

Therefore, only the current concept of DR_{min} is kept, which means that the analyzed text pertaining to domain-independent concepts should be overlooked. In the
present work, a concept and a relation are formally defined as follow:

Definition 1. *A **concept**, denoted by **C**, is defined as a couple, **C** = (N, DR), where N
is the label of C and DR denotes its degree of relevance whose value is equal either to
"DR_{max}" if C is a core concept, or to "DR_{med}" if both C and text associated with it
are concerned by the enrichment process, or to "DR_{min}" if only C is interesting to
enrich the ontology.*

Definition 2. *A **relation**, denoted by **R**, is defined by a triplet, **R**= (N, CD, CR), where
N is the name of the relation, CD is the domain of **R** and CR is the range of **R**, (where
CD and CR are two concepts).*

3   The Proposed Approach for Domain Ontology Enrichment

3.1   Overview

We propose an approach, whose overall process is depicted in Figure 3, for automatically enriching domain ontologies from LMF-standardized dictionaries. It consists of
an incremental process that iterates through four stages:

- Textual Fields (TF) conceptual annotation;
- Triple candidates' identification;
- Duplication check and validation;
- New TF importation.

Furthermore, owing to the linguistic nature and limited size of the handled textual data, the proposed approach relies on a set of lexico-syntactic patterns as well as NLP tools for text analysis. These patterns which are not restricted to any specific type of semantic relation can be extracted from the dictionary itself. What follows is the description of each stage.

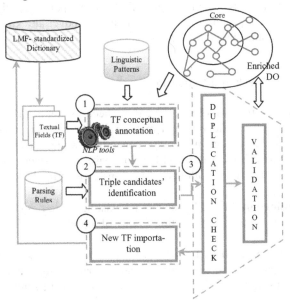

Fig. 3. Overview of the proposed process for domain ontology enrichment

3.2   TF Conceptual Annotation

There is an agreement in the NLP community that relational information is, at sentence level, typically conveyed by verbs. At the level of noun phrases, this role is dominantly played by prepositions. Regarding to concepts, they are generally denoted by noun phrases [28]. Consequently, the aim of this stage is to find out linguistic clues for ontology items in the entry text and annotate them with their corresponding tags (e.g., <concept> and <relation>) using a pattern-matching technique.

This annotation can be performed by means of a linguistic platform such as NooJ [26], GATE [11] and UIMA [14]. These environments provide a set of programming components along with various linguistic resources (e.g., dictionaries, lexicons, grammars and taxonomies) for natural text processing. They are also provided with formalisms in order to describe regular expressions and perform their annotation (e.g., based on the JAPE language for GATE and finite-state automata within NooJ).

3.3 Triple Candidates' Identification

The objective of this stage is to identify triple candidates like **(Rel, C1, C2)**, which systematically means that there is a binary relation, named *Rel*, between the concepts *C1* and *C2*. It makes use of a set of parsing rules which are applied to the annotated textual data of the previous stage in order to find concept pairs of each relationship. It is worth stating that the proposed solution is able to discover any kind of binary relationship; both taxonomical and non-taxonomical

3.4   Duplication Check and Validation

As its name indicates, the goal of this stage is to check for duplicated triple candidates for the purpose of their validation and may be further new TF importation. In the proposed process for core domain ontology generation, we have elaborated two stages for duplication checking and validation [6]. Likewise, these stages are needed in the enrichment stage. One of the following situations is identified for any learnt triple $t = (N, C1, C2)$:

- **Exact duplication**: t directly matches an existing triple $t' = (N', C1, C2)$, i.e. N and N' are the same thus reflecting a copy of a previously identified triple candidate;
- **Quasi-exact duplication**: t could be matched with an existing triple $t' = (N', C1, C2)$, i.e. N and N' are synonyms (The comparison for synonymy can be carried out by using the dictionary itself) thus reflecting the same relation between C1 and C2;
- **Implicit duplication**: t is a completely different candidate but whose knowledge can be inferred from the existing ontology elements. Formally, it is represented as follows, let $t' = (N', C1', C2')$ and $t'' = (N'', C1'', C2'')$ be two existing triples,

$$\left[\begin{array}{c} t = (N, C1, C2) \text{ is an implicit duplicated triple } \textbf{iff} \\ \dfrac{t' \qquad t''}{t' \wedge t'' \Rightarrow t} \end{array} \right.$$

In all duplication types, the duplicated candidates should be ignored. Therefore, the constructed domain ontology does not store unnecessary or useless entities. This quality criterion is also called *conciseness* [16].

Once duplication check is performed, a further validation stage is required to verify whether the resulting ontology remains coherent when the triple candidate is added to it. In fact, a triple construction may involve one of the following actions:

- The creation of only the relation if there are two concepts existing in the ontology whose labels are equal or synonym to those of C1 and C2 labels;
- The creation of the relation as well as its second argument if there is any concept in the given ontology whose label is equal or synonym to that of C2.

For validating non-duplicated triples, we are interested in the kinds of errors that can be automatically detected (i.e. without human expert involvement). Therefore, only inconsistency errors, particularly those of circularity and partition types [15], are addressed in the present work.

3.5   New TF Importation

This stage includes basically a module for assigning the appropriate DR to new concepts candidates. It is indicated above (§ Sub-section 2.3) that this task will be done automatically by exploiting the semantic component of the LMF-standardized dictionary. Let us remind that DR denotes the concept's degree of importance with the aim of guiding the extraction of further knowledge from the adopted source (i.e., LMF-standardized dictionary).

Therefore, Figure 4 shows the appropriate actions to extend TF according to the DR of a given concept.

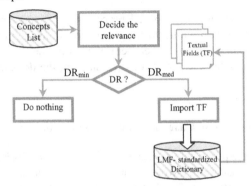

Fig. 4. Appropriate actions according to the Degree of Relevance (DR)

4   Implementation

The approach for domain ontology enrichment based on the semantic component of LMF-standardized dictionaries is implemented in a Java-based tool that enables users to automatically build domain ontologies formalized in OWL-DL, a sublanguage of OWL [12]. Indeed, an OWL-DL formalized ontology can be interpreted according to description logics, and DL-based reasoning software (e.g., RacerPro[1] or Pellet[2]) can be applied to check its consistency or draw inferences from it. To take advantage of this, we have decided to incorporate the Pellet reasoner into our system. It is an open-source Java-based OWL-DL reasoning tool [27]. Its consistency checking ensures that the produced ontology does not contain any contradictory facts. After the loading of the built OWL file, Pellet determines if the ontology is actually consistent by calling the isConsistent() method, whereby its boolean return would decide whether the addition operation could be performed in the resulting ontology.

Regarding NLP tools used for the purpose of text analysis and conceptual annotation, we have successfully implemented a set of linguistic patterns and an NLP chain [4] within the NooJ platform [26]. We have also managed to define and formalize necessary parsing rules allowing for triple candidates' identification with NooJ syntac-

---

[1] http://www.racer-systems.com
[2] http://pellet.owldl.com

tic grammars. Indeed, the use of NooJ is simpler and easier with a fairly quick seizure. Besides, its users are considered as a growing community that is, currently, developing expanded linguistic resources in several languages including Arabic.

Moreover, in order to visualize the created OWL domain ontology, we have utilized the Protégé ontology editor[3] with its jambalaya plug-in which allows displaying Arabic text.

5   A Case Study: The Astronomy Domain

In this section, we describe the proposed process for domain ontology enrichment through an illustrative case relating to the astronomy domain. In what follows, we firstly define the dictionary of the experiment as well as the domain of interest. Secondly, we report examples and results of each stage in the domain ontology enrichment approach.

5.1   Arabic Dictionary Use Case and Domain Definition

In order to illustrate the proposed process, a series of experiments has been conducted on different domains using an LMF-standardized Arabic dictionary. This choice is explained by two main motives. The first one is the great deficiency of work on Arabic ontology development and the second is the availability within our research team of an LMF- standardized Arabic dictionary. This dictionary is covering various domains, of which animals, plants, astronomy and sports are but a few. Besides, it contains about 40.000 lexical entries and about 30.000 relationships among which the most frequent are *synonymy*, *antonymy* and *hypernymy*.

In the remainder of this section, we report an illustrative case study pertaining to the astronomy field, whose choice is based on three main reasons. Firstly, Arabic scientists have soundly described the astronomy science « علم فلك » thanks to the richness in terms of concepts and relations of such domain. Secondly, it is quite popular in the ontology learning community. Thirdly, for this domain, the handled dictionary is quite rich and complete with lexical and semantic knowledge as well.

For the illustration, we consider a fragment of an LMF-standardized dictionary corresponding to the domain of interest, which deals with the study of celestial objects (such as stars, planets, comets, nebulae, star clusters and galaxies) and phenomena that originate outside the Earth's atmosphere (such as the cosmic background radiation). It is concerned with the evolution, physics, chemistry, meteorology, and motion of celestial objects, as well as the formation and development of the universe.

Starting from an LMF-standardized dictionary of the Arabic language comprising thousands of multi-domain lexical entries, we collect those corresponding to the domain of interest. Then, we build our domain dictionary comprising a hundred or so of lexical entries (around 350 senses).

---

[3] http://protege.stanford.edu/

As mentioned above, the extraction of all relevant senses is based on the explicit information associated with the domain-specific senses of lexical entries (i.e. *Subject-Field* class) to describe their meaning in a given domain. Subsequently, only the senses (or concepts) that are directly connected to the domain of the astronomy were considered (i.e. the concepts of mathematical, chemistry, physics, etc. domains were disregarded). It is to be noted here that the overlooked senses, even tagged with other domain, would be revealed at the core enrichment stage.

5.2   Domain Ontology Enrichment

As mentioned earlier, the implemented system for linguistic annotation relies on a set of patterns in order to automatically capture textual sequences resulting from the pattern-matching over the text and assign them with needed annotations, among them <Concept> and <Relation> tags. At present, we dispose of a lexico-syntactic pattern-base comprising 104 patterns for concepts and relations recognition expressed with NooJ syntactic grammars [4]. Additional rules can be defined manually in order to expand its coverage. Some examples of patterns are given in Table 1. We used the part-of-speech tag set provided by Tim Buckwalter and his colleagues [17] instead of the NooJ formalization so as to simplify the comprehension for non-experts in NooJ.

Table 1. Examples of patterns for the recognition of concepts and relations

	Pattern example	Examples
Concept	NOUN	Earth «الأَرْض»
	NOUN + NOUN	Solar_System « النِّظَام الشَّمْسِيّ »
Relation	VERB	Happen « تَجْرِي»
	VERB +PREP	Consists_of « يَتَكَوَّنُ مِنْ»
	ADJ+PREP	Surrounded_by «مُحَاطٌ بِـ»

As an illustration, let us consider the definition of "مِرِّيخ" (Mars) concept given in the top-part of Figure 5-a.

(a)	مِرِّيخ (mirīkh): كَوْكَبٌ سَيَّارٌ وَهُوَ الرَّابِعُ فِي النِّظَام الشَّمْسِيّ، أَصْغَرُ مِنَ الأَرْضِ وَيُعْرَفُ بِمَارس. Mars: a planet and it is the fourth in the solar system, smaller than Earth and is known as Mars
(b)	مِرِّيخ: كَوْكَبٌ سَيَّارٌ وَهُوَ الرَّابِعُ فِي النِّظَام الشَّمْسِيّ مِرِّيخ: أَصْغَرُ مِنَ الأَرْضِ وَيُعْرَفُ بِمَارس.

Fig. 5. The definition pertaining to " مِرِّيخ" concept

A preliminary step consists in the text splitting according to punctuation marks (full stop and comma) (Figure 5-b). Then, each sentence is linguistically annotated through the use of the existing NooJ dictionaries and the tokenization module based on morphological grammars.

For the conceptual annotation, we are interested in syntactic features of each token, for instance <N+i> indicates that the token is a noun and it is in the genitive case. Besides the *<Concept>* and *<Relation>* tags, we need other tags like *<refp>* corresponding to pronouns and/or conjunction words. Indeed, these tags will serve to find the relation argument(s) when encountering linguistic phenomena such as anaphora situation. For instance, "وَهُوَ" *wahouwa* ("and it") is assigned with *<refp>* tag and leads to refer to the concept domain of the previous relation (Figure 6).

[وَهُوَ] refp	[كَوْكَبٌ سَيَّارٌ] concept	[:] p	[مِرِّيخ] concept
	[النَّظَام الشَّمْسِيَّ] concept	[الرَّابِعُ فِي] relation	
[الأَرْضِ] concept	[أَصْغَرُ مِنْ] relation	[:] p	[مِرِّيخ] concept
[مَارِس] concept	[يُعْرَفُ بِ] relation	[وَ] refp	

Fig. 6. Conceptual annotation of the definition pertaining to the " مِرِّيخ " concept

At the end of this stage, the interesting information for triple elements' identification is provided in an output file described in XML, a more suitable format for automatic processing (Figure 7).

```
- <DEF>
  - <LU LEMMA="مِرِّيخ" CAT="CONCEPT">
      <CONCEPT>مِرِّيخ</CONCEPT>
    </LU>
    <P>:</P>
    <CONCEPT>كَوْكَبٌ سَيَّارٌ</CONCEPT>
    <REFP>وَهُوَ</REFP>
    <RELATION>الرَّابِعُ فِي</RELATION>
    <CONCEPT>النَّظَام الشَّمْسِيَّ</CONCEPT>
  - <LU LEMMA="مِرِّيخ" CAT="CONCEPT">
      <CONCEPT>مِرِّيخ</CONCEPT>
    </LU>
    <P>:</P>
    <RELATION>أَصْغَرُ مِنْ</RELATION>
    <CONCEPT>الأَرْضِ</CONCEPT>
    <REFP>وَ</REFP>
    <RELATION>يُعْرَفُ بِ</RELATION>
    <CONCEPT>مَارِس</CONCEPT>
</DEF>
```

Fig. 7. XML representation of the result of conceptual annotation of the definition pertaining to " مِرِّيخ " concept

Once the conceptual annotation is done, the parsing rules aiming at digging out triple elements candidates are applied. The following rules show the transformation when encountering the corresponding situation.

Rule 1:

[token]$_{concept}$ [:]$_p$ [token]$_{concept}$➜ < *is-a*, concept$_{left}$, concept$_{right}$ >

Rule 2:

[token]$_{concept}$ [:]$_p$ [token]$_{relation}$ [token]$_{concept}$➜ < *relation*, concept$_{left}$, concept$_{right}$ >

Rule 3:

[token]$_{refp}$ [token]$_{relation}$ [token]$_{concept}$➜ < *relation*, concept$_{leftprevious}$, concept$_{right}$ >

Figure 8 shows triple candidates extracted from the XML output described in Figure 8 after applying these parsing rules. Identified triple candidates are:

- (is-a, "مِرِّيخ (Mars)", "كَوْكَبٌ سَيَّارٌ (Planet)");
- ("الرَّابِعُ فِي (the_fourth_in)", "مِرِّيخ (Mars)", "النِّظَام الشَّمْسِيِّ (Solar_System)");
- ("أَصْغَرُ مِنْ (smaller_than)", "مِرِّيخ (Mars)", "الأَرْض (Earth)");
- ("يُعْرَفُ بِـ (is_known_as)", "مِرِّيخ (Mars)", "مَارس (Mars)").

```
<REL NAME="is-a" DOMAIN="مِرِّيخ" RANGE="كَوْكَبٌ سَيَّارٌ" />
<REL NAME="الرَّابِعُ فِي" DOMAIN="مِرِّيخ" RANGE="النِّظَام الشَّمْسِيِّ" />
<REL NAME="أَصْغَرُ مِنْ" DOMAIN="مِرِّيخ" RANGE="الأَرْض" />
<REL NAME="يُعْرَفُ بِـ" DOMAIN="مِرِّيخ" RANGE="مَارس" />
```

Fig. 8. XML representation of triple candidates' identification

Thanks to the implemented module of detecting duplication, the system will not classify the "مِرِّيخ" (Mars) concept under the "كَوْكَبٌ سَيَّارٌ" (Planet) concept because the corresponding triple is considered as a quasi-duplicated of an already existing triple (is-a, "مِرِّيخ", "كَوْكَبٌ سَيَّارٌ" (Planet)"). Therefore, the suggested system creates the new discovered concepts: "النِّظَام الشَّمْسِيِّ" (Solar_System), "الأَرْض" (Earth), and "مَارس" (Mars). It also creates the relations linking the "مِرِّيخ" (Mars) concept to new created concepts.

As mentioned earlier, the ontology is implemented with OWL standard. Therefore, Table 2 demonstrates the corresponding OWL encoding of generated ontological entities relating to the "مِرِّيخ" (Mars) example in the first iteration.

Table 2. An excerpt of OWL-generated file relating to the "مِرِّيخ" (Mars) example

```
<owl:Class rdf:ID="مِرِّيخ"/>
<owl:Class rdf:ID="كَوْكَبٌ سَيَّارٌ"/>
<owl:Class rdf:ID="النِّظَام_الشَّمْسِيِّ"/>
<owl:Class rdf:ID="مَارس"/>
<owl:Class rdf:ID="الأَرْض"/>
<owl:Class rdf:about="#مِرِّيخ">
    <rdfs:subClassOf rdf:resource="#كَوْكَبٌ سَيَّارٌ"/>
</owl:Class>
<owl:ObjectProperty rdf:ID="الرَّابِعُ_فِي">
    <rdfs:domain rdf:resource="#مِرِّيخ"/>
    <rdfs:range rdf:resource="#النِّظَام_الشَّمْسِيِّ"/>
</owl:ObjectProperty>
<owl:ObjectProperty rdf:ID="أَصْغَرُ_مِنْ">
    <rdfs:domain rdf:resource="#مِرِّيخ"/>
    <rdfs:range rdf:resource="#الأَرْض"/>
</owl:ObjectProperty>
<owl:ObjectProperty rdf:ID="يُعْرَفُ_بِـ">
    <rdfs:domain rdf:resource="#مِرِّيخ"/>
    <rdfs:range rdf:resource="#مَارس"/>
</owl:ObjectProperty>
```

In addition to creating and automatically validating these candidates, the system decides the DR of each new concept based on the semantic component of the handled LMF-standardized dictionary. Then, "النِّظَام الشَّمْسِيِّ" (Solar_System) and "الأَرْض" (Earth) concepts are assigned with DR_{med} while "مَارس" (Mars) is assigned with DR_{min} because the only one meaning related to "مَارس" (Mars) lemma describes it as the third month of the year and has no relation with the astronomy field. Therefore, the initial domain dictionary is extended by importing TF of "النّظَامِ الشَّمْسِيّ" (Solar_System) and "الأَرْض" (Earth).

Similarly, the system proceeds by analyzing the updated domain dictionary. As a result, a significant number of concepts and relations are defined after achieving several iterations. Figure 9 illustrates the ontology items related to the "مِرّيخ" (Mars) concept visualized by means of Protégé editor and its jambalaya plug-in which is successful in supporting visualization of Arabic characters.

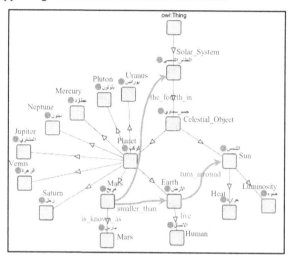

Fig. 9. The ontology items related to the "مِرّيخ" (Mars) concept

5.3   Results and Evaluation

As far as the evaluation of the obtained results is concerned, it can obviously be seen that besides its fully automated level, the proposed approach exhibits many other important benefits. Indeed, we can firstly point out that all the ontology concepts as well as relations holding between them are relevant to the considered domain. Secondly, the LMF-standardized dictionaries are undeniably widely-accepted and commonly-referenced resources; thereby they simplify the task of labeling concepts and relationships. Finally, there are no consistency errors since we managed to check the coherence of the generated ontology with a specialized tool.

Added to their satisfactory quality, we need to assess how good the automatically learned ontologies are to reflect a given domain. To achieve this goal, we established a comparison between the obtained ontology and a hand-crafted corresponding ontology, manually created by ontology engineers, to reach the conclusion that all acquired ontological items are relevant to the considered domain. Moreover, it is worthwhile to state that 100% of concepts are acquired while 12% of binary semantic relations are not. We are also aware that the obtained ontology contains some inconsistencies that cannot be automatically detected and should be manually corrected. Indeed, the research and exploration of data from a non-structured text are not error-free. We have found that the error situations are due to the failure of the first annotation process, which can be accredited to two factors. Actually, it can be due either to the non-recognition of the lexical item by NooJ or to the voyellation, which a source of syntactic error annotation that is sometimes partial. These errors are safe because they can be easily reviewed by experts. Other considered issues are related to linguistic phenomena like ellipsis. Finally, the main drawback of our approach is that it is currently restricted to binary relationships.

6   Related Work

Ontology enrichment is time-consuming, error prone and labor-intensive task when performed manually. Ontology learning can facilitate this process by using machine learning methods relying on text mining techniques to obtain potentially relevant knowledge from textual data. In this context, among different text mining techniques, the pattern-based approaches, pioneered by [18], have inspired the work of many and are getting more and more attention in the scientific community.

In the present work, we are interested in linguistic methods for knowledge extraction from Machine-Readable Dictionaries (MRDs). Early work on knowledge extraction from MRDs goes back to the 80s [10]. A lot of effort has been devoted to exploiting semantic knowldge MRDs in recent years [20, 21, 22]. The basic idea is to exploit the regularity of dictionary entries to initially find a suitable *hypernym* for the defined word (For more details, we refer the reader to [10]).

However, special interest has been given to taxonomic relationships (i.e., of the types *is-a* and *part-whole*) [24], and light has been particularly shed on the analysis of the head of the first noun phrase in the dictionary definition.

From another standpoint, the effectiveness of the proposed systems depends on the appropriate preparation of the text collections (selection) as well as the patterns which are both dependent on the particular domain and purpose of the application. Besides, they are considered as the most time-consuming activities requiring considerable effort and significant human (expert) involvement in the whole process [2, 23].

7   Conclusion and Future Work

This paper presents a domain ontology enrichment approach using LMF (ISO 24613)-standardized dictionaries. The originality of this approach lies in exploiting the semantic component of such valuable resources, rich in both lexical and conceptual knowledge, in order to guide the extraction of further knowledge from the adopted source. Ontology quality which is another challenging task of a major importance has also been addressed. For this reason, our concern was focused on error detecting during the enrichment process. Therefore, we have proposed an automatic and iterative process, whose ultimate goal is to reduce the need for time consuming data source preparation and to produce coherent and rich ontologies.

Furthermore, the proposed approach is proven to be reliable through the case study carried out on astronomy domain. In reality, more experiments have been conducted on different domains using an LMF-standardized Arabic dictionary, but without lack of generality, we have chosen the astronomy domain in the present study illustrations. In addition, both qualitative and quantitative evaluations have shown the richness of the LMF-standardized dictionary which we believe a promising and important source of productivity gains in ontology development.

The next challenges will be the focus on the *n-ary* relations extraction which has not received much attention in the knowledge acquisition community. We also plan to compare the results of our approach with a gold standard provided by our research team corresponding to the astronomy domain will be considered.

References

1. Auger, A., Barrière, C.: Pattern-based approaches to semantic relation extraction: A state-of-the-art. Terminology, Special Issue on Pattern-Based Approaches to Semantic Relations 14(1), 1–19 (2008), Auger A., Barriere, C. (eds.)
2. Aussenac-Gilles, N., Jacques, M.-P.: Designing and Evaluating Patterns for Relation Acquisition from Texts with CAMÉLÉON. Terminology, Special Issue on Pattern-Based approaches to Semantic Relations 14(1), 45–73 (2008), Auger A., Barriere C. (eds.)
3. Ayşe, Ş., Zeynep, O., Ilknur, P.: Extraction of Semantic Word Relations in Turkish from Dictionary Definitions. In: Workshop on Relational Models of Semantics, RELMS 2011, Portland, Oregon, USA, pp. 11–18 (2011)
4. Baccar Ben Amar, F., Gargouri, B., Ben Hamadou, A.: Annotation conceptuelle du contenu textuel des dictionnaires normalisés LMF. In: Terminologie & Ontologie: Théories et applications (TOTh 2013), Chambéry, France, June 6-7 (2013)
5. Baccar Ben Amar, F., Gargouri, B., Ben Hamadou, A.: LMF Dictionary-Based Approach for Domain Ontology Generation. In: Pazienza, M.-T., Stellato, A. (eds.) Semi-Automatic Ontology Development: Processes and Resources, ch. 5, pp. 106–130 (2012)
6. Baccar Ben Amar, F., Gargouri, B., Ben Hamadou, A.: Domain Ontology Generation Using LMF Standardized Dictionary Structure. In: 6th International Conference on Software and Data Technologies (ICSOFT), Seville, Spain, pp. 396–401 (2011)
7. Baccar Ben Amar, F., Gargouri, B., Ben Hamadou, A.: Towards Generation of Domain Ontology from LMF Standardized Dictionaries. In: 22nd International Conference on Software Engineering and Knowledge Engineering, USA, pp. 515–520 (2010)
8. Baccar Ben Amar, F., Khemakhem, A., Gargouri, B., Haddar, K., Ben Hamadou, A.: LMF standardized model for the editorial electronic dictionaries of Arabic. In: 5th International Workshop on Natural Language Processing and Cognitive Science (NLPCS), in conjunction with (ICEIS), Barcelona, Spain, pp. 64–73 (2008)
9. Buitelaar, P., Cimiano, P.: Ontology Learning and Population: Bridging the Gap between Text and Knowledge. Frontiers in Artificial Intelligence and Applications, vol. 167. IOS Press, Amsterdam (2008)
10. Cimiano, P.: Ontology Learning and Population from Text: Algorithms, Evaluation and Applications. Springer (2006)
11. Cunningham, H., Maynard, D., Bontcheva, K., Tablan, V.: Gate: An architecture for development of robust HLT applications. In: 40th Annual Meeting on Association for Computational Linguistics, pp. 168–175. ACL, Stroudsburg (2002)
12. Dean, M., Schreiber, G.: OWL web ontology language reference. W3C recommendation, W3C (2004)

13. Francopoulo, G., George, M.: Language Resource Management-Lexical Markup Framework (LMF). Technical report, ISO/TC37/SC4 (N330 Rev.16) (2008)
14. Ferrucci, D., Lally, A.: UIMA: an Architectural Approach to Unstructured Information Processing in the Corporate Research Environment. Natural Language Engineering 10, 327–348 (2004)
15. Gómez-Pérez, A.: Evaluation of Taxonomic Knowledge on Ontologies and Knowledge-Based Systems. In: North American Workshop on Knowledge Acquisition, Modeling, and Management, KAW (1999)
16. Gomez-Perez, A.: Ontology evaluation. In: Staab, S., Studer, R. (eds.) Handbook on Ontologies in Information Systems, 1st edn. International Handbooks on Information Systems, ch. 13, pp. 251–274. Springer, Berlin (2004)
17. Hajič, J., Smrž, O., Buckwalter, T., Jin, H.: Feature-based Tagger of Approximations of Functional Arabic Morphology. In: Treebanks and Linguistic Theories (TLT), Barcelona, Spain, pp. 53–64 (2005)
18. Hearst, M.A.: Automatic acquisition of hyponyms from large text corpora. In: 14th Conference on Computational Linguistics, Morristown, NJ, USA, pp. 539–545 (1992)
19. ISO 24613: Lexical Markup Framework (LMF) revision 16. ISO FDIS 24613:2008
20. Kurematsu, M., Iwade, T., Nakaya, N., Yamaguchi, T.: DODDLE II: A Domain Ontology Development Environment Using a MRD and Text Corpus. IEICE(E) E87-D(4), 908–916 (2004)
21. Malaisé, V., Zweigenbaum, P., Bachimont, B.: Detecting semantic relations between terms in definitions. In: 3rd International Workshop on Computational Terminology, CompuTerm, Geneva, Switzerland, pp. 55–62 (2004)
22. Nichols, E., Bond, F., Flickinger, D.: Robust ontology acquisition from machine readable dictionaries. In: International Joint Conference on Artificial Intelligence IJCAI 2005, Edinburgh, pp. 1111–1116 (2005)
23. Petasis, G., Karkaletsis, V., Paliouras, G., Krithara, A., Zavitsanos, E.: Ontology population and enrichment: State of the Art. In: Paliouras, G., Spyropoulos, C.D., Tsatsaronis, G. (eds.) Multimedia Information Extraction. LNCS (LNAI), vol. 6050, pp. 134–166. Springer, Heidelberg (2011)
24. Sanchez, D., Moreno, A.: Learning non-taxonomic relationships from web documents for domain ontology construction. Data & Knowledge Engineering 64(3), 600–623 (2008)
25. Seppälä, S.: Composition et formalisation conceptuelles de la définition terminographique, Mémoire de DEA en traitement informatique multilingue, Université de Genève, École de traduction et d'interprétation, Genève (2004)
26. Silberztein, M.: NooJ: a Linguistic Annotation System for Corpus Processing. In: HLT/EMNLP, pp. 1–11 (2005)
27. Sirin, E., Parsia, B., Grau, B.C., Kalyanpur, A., Katz, Y.: Pellet: A practical OWL DL reasoner. Journal of Web Semantics 5(2), 51–53 (2007)
28. Turenne, N.: Etat de l'art de la classification automatique pour l'acquisition de connaissances à partir de textes. Technical Report, INRA (2001)

Research on Relationships between Response Implementation Activities in Multiple Emergencies Based on Instantiated Knowledge Element[*]

Li Cui and Qiuyan Zhong

Faculty of Management & Economics, Dalian University of Technology,
Dalian, 116024, China
cli314@163.com, zhongqy@dlut.edu.cn

Abstract. Emergencies often accompanied by derivative and secondary disasters. In order to better response and dispose multiple disasters, give the attributes of knowledge elements assignment to obtain instantiated activity knowledge elements according to the characteristic of activity knowledge elements. Then analyze the relationship among each attribute to identify that of implementation in multiple emergencies. Induce this relationship and express it by a formalized mathematical logical symbol. Thus provide a method for emergency managers in emergency response and a basis for process integration of multiple emergencies.

Keywords: energency management, activity knowledge element, instantiated knowledge element, emergencies, relationship.

1 Introduction

"Emergency Response Law of the People's Republic of China" was implemented in November 1, 2007. Public emergencies were classified as natural disasters, accidents, public health incidents and social security incidents according to the process, properties and mechanism of Public emergencies [1]. In recent years, the emergency is a serious threat to the economic development and people's normal life, such as earthquake, explosion, terrorist attacks and so on. Therefore, how to effectively respond to emergencies and minimize the loss have become the problems that need to be solved urgently in emergency management field.

In the past few years, emergency management process continuously integrated with the process of other areas, such as computer technology, geographic information systems, combined knowledge management and used simulation technology, artificial intelligence [2], and so on. Using method and technology of knowledge management to solve the problems in emergency management has become a hot research aspect.

---

[*] This work is supported by National Natural Science Foundation of China (No.: 91024029) and Humanity and Social Science Youth foundation of Ministry of Education (No.:1YJC630023).

M. Wang (Ed.): KSEM 2013, LNAI 8041, pp. 420–430, 2013.

Michael Dinh et al. pointed out the importance to develop knowledge ability and improve organizational learning ability, and applied it to the every stage of disaster [3]. Josefa and Serrano applied knowledge model in environment emergency management, and put forward three stages of knowledge model: What is happening? What will happen? And what should be done? [4].The same study is also included in reference [5]. A knowledge management framework was constructed by Dongsong et al. to support emergency decision in order to reduce the effect of disasters [6]. Dautun et al. judged the deterioration degree of emergency evolution through trigger events and early warning signal so as to improve decision auxiliary process [7]. These studies all used existed knowledge and model in emergency management directly, neither from the activities in the emergency process nor the study literatures from the angle of knowledge to study the appropriate methods. On the other hand, the existed literatures more focus on the description and construction of the implementation process in a single incident [8] [9] [10]. However, it is known that emergency happening often accompanied by derivative or secondary disasters according to the characteristics of the emergency. Reference [11] is such an example. This means that in most cases, the actual emergency process is to respond to multiple disasters. This response is a complex task, not a simple single event linear summation. How to dispose multiple disasters effectively in emergency management has more and more realistic and important role. Therefore, this paper mainly studies the response disposal of multiple emergencies. For convenience, emergencies mentioned in this paper are all refer to multiple emergencies.

To sum up, this paper considers using knowledge element in implementation process of emergencies. Identify the relationship between activities in dealing with emergencies through analyzing attributes and attribute values of knowledge elements. Provide advices and methods for disposing emergencies from a new angle and play an important role in construction and integration of implementation process in the future.

2   Basic Knowledge

2.1   Knowledge Element

Knowledge element mentioned in this paper is proposed by professor Wang Yanzhang, who comes from Dalian University of Technology [12]. He uses an ordered triad

$$K_m = \left(N_m, A_m, R_m \right) \qquad \forall m \in M \tag{1}$$

for describing common knowledge, where K_m represents a common knowledge element; N_m is the concept and attribute names of corresponding things; A_m is attribute state set; R_m is mapping relationship set. And the things are cognitive, have $N_m \neq \phi$, $A_m \neq \phi, R_m \neq \phi$.

Let $a \in A_m (m \in M)$, then we have attribute knowledge element of the corresponding thing:

$$K_a = (p_a, d_a, f_a), \quad a \in A_m, \quad \forall m \in M \tag{2}$$

where p_a is measurable characteristic description, which represents attribute state of things can be described or not; If attribute state is measurable, then d_a represents measure dimensional set; If attribute state is measurable and state changing with time is distinguishable, then we have function $a_t = f_a(a_{t-1}, t)$, f_a may empty.

Let $r \in R_m (m \in M)$ is a mapping relation on $A_m \times A_m$, then we have relationship knowledge element:

$$K_r = (p_r, A_r^I, A_r^O, f_r), \quad r \in R_m, \quad \forall m \in M \tag{3}$$

where p_r is mapping attribute description; A_r^I is input attribute state set; A_r^O is output attribute state set; f_r is a specific mapping function, written as $A_r^O = f_r(A_r^I)$.

The above knowledge elements are common knowledge of any things, which means any things contain their respective attribute set and relationship set.

2.2   Activity Knowledge Element

Emergency response can be seen as a large complicated system. From the point of view of knowledge, emergency management can be divided into three systems: Emergency process system, objective things (disaster carrier) system and emergency management activity system. These three systems respectively correspond to three kinds of knowledge elements: emergency knowledge element, objective things knowledge element and emergency management activity knowledge element. The emergency management activity knowledge element also can be divided into decision-making activity knowledge element and implementation activity knowledge element. This paper mainly study relationship between activities in multiple emergencies from the perspective of implementation. Therefore activity knowledge elements mentioned in this paper all refer to the implementation activity knowledge elements of emergency management.

Definition 1. Activity knowledge element is also called emergency management activity knowledge element. It refers to the inseparable and constant minimum activity unit in the management field which can realize a certain target or achieve a certain purpose and can change the state of the objective things system during the emergency response process. We use an ordered triad

$$K_a = (N_a, A_a, R_a) \tag{4}$$

for describing activity knowledge element, where N_a represents the concept and attribute names of implementation activity; A_a is the attribute state set of activity knowledge element; R_a is the mapping relation set of attribute state changing and interaction.

Concept set, attribute state set and mapping relationship set are set up as follows:

(1) Names of concept and attribute

$$N_a = \{C_t, S, C_d, A\}$$

Where C_t represents the context of current activity; S is the current state of activity; C_d is the trigger condition of activity; A is the name of activity.

(2) Attribute state set

$$A_a = \{D, S_u, O_b, T, R_e, O_p, E\}$$

Where D represents the department of implementation activity; S_u is subject set; O_b is object set; T represents tools of activity, such as traffic tools; R_e are the resources involved in activity, including human resources and materials; O_p is operation of activity; E is used to evaluate activities.

(3) Mapping relation set

R_a represents the relationship of implementation activity attributes. In single event, it shows internal attribute relationship between activities. In multiple events, it shows attribute relationship of activities between events.

2.3   Instantiated Activity Knowledge Element

According to the degree of improvement of the activity knowledge element attribute value, activity knowledge element can be divided into common activity knowledge element, instantiated activity knowledge element and substantiated activity knowledge element. Give attributes of common activity knowledge element assignment to obtain instantiated activity knowledge element.

Substantiated knowledge element refers to that on the basis of instantiated knowledge element, consider specific factors such as time and place to form knowledge element. Level structure among the three is shown in figure 1:

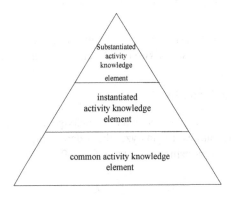

Fig. 1. Implementation activity knowledge element hierarchical chart

Common activity knowledge element is at bottom, which is most basic and abstract; Substantiated activity knowledge element is located in the top of the hierarchy structure, which is the most specific. Therefore, this paper studies instantiated activity knowledge element, which can reflect the characteristics of the implementation activities in the specific situation and the conclusion of which can be widely used in all kinds of emergency response activities.

3   Identification and Analysis of Relationship of Instantiated Activity Knowledge Element

3.1   Hypothesis

In order to better reflect the relationship of implementation activities in emergency management, and make it have universality, this paper gives hypothesis as follows:

(1) The concept of activity is clear.
Assume that emergencies have occurred. That's to say state or state value of concept set is known, not need to analyze.

(2) Only consider the attributes of disposal process.
In multiple attributes, only consider the attributes which have relation with disposal process, including subject, object of the implementation activity and used resources. Don't consider subject department and used traffic tools, etc. Assume that materials have arrived at the scene.

(3) Multiple emergencies occurre in the same period and the same area.
Mainly research the relationship of implementation activities in multiple emergencies which have influence on each other. Consider the relationship between source disaster and its secondary or derivative disaster.

(4) On the analysis of the relationships, if subjects or resources of disposing two emergencies have some same sets, assume subjects or resources are not enough.

3.2   Relation Recognition Model

According to the hypothesis, use activity knowledge element attributes including subject, object and resources to analyze the relationship between activities. The recognition process of relationship is shown in figure2:

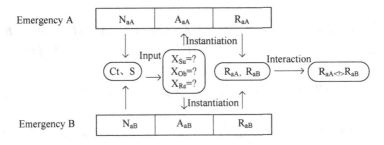

Fig. 2. Relation recognition model based on instantiated activity knowledge element

Have two emergencies A and B. Where Ct, S represent Context and State respectively; X = A or B. Firstly, from the concept set find context and state in both emergencies, then input them to the attributes of knowledge element as attribute values. Secondly, give subject, object and resources attributes assignment, instantiate the two knowledge elements. Then, two instantiated activity knowledge elements interact and form various sorts of relation. Finally output the relationship between two disposal activities.

3.3   Recognition and Analysis of the Relationship

According to the hypothesis and relation recognition model, attribute set of activity knowledge element is defined as:

$$K_a(X) = \{\text{subject of activity } (S_u), \text{ object of activity } (O_b), \text{ required resources of activity } (R_e)\},$$

relation set is R. Where, subject refers to the persons who engage in implementing emergency disposal activities; Object of activity is a problem to solve or a goal to complete in emergency activities; Resources refer to required resources when subjects dispose and complete object target in emergency activities.

Have two emergencies A and B occurred in the same area. A_{Su} and B_{Su} respectively is the subject for disposing emergency A and emergency B. A_{Ob} and B_{Ob} respectively is goal object of A and B achieved through response activity; A_{Re} and B_{Re} respectively are resources which are required by subject to complete object task in emergency A and B. In dealing with the process of implementation, according to subject, object and resources of different emergency activities whether have conflict or not, analyze and conclude the relationship between the activities as the following four categories:

(1) Non   conflict
In two emergencies, if the subjects use resources to complete some target object and there has not any conflicts, then the relationship between two activities is non conflict relationship. Include the following types and show by knowledge element:

IF

1) $A_{Su} \cap B_{Su} = A_{Su} = B_{Su} \neq \phi$, $A_{Ob} \cap B_{Ob} = A_{Ob} = B_{Ob} \neq \phi$, $A_{Re} \cap B_{Re} = A_{Re} = B_{Re} \neq \phi$.

2) $A_{Su} \cap B_{Su} = A_{Su} = B_{Su} \neq \phi$, $A_{Ob} \cap B_{Ob} = A_{Ob} = B_{Ob} \neq \phi$, $A_{Re} \cap B_{Re} = \phi$ or $A_{Re} \cap B_{Re} = \alpha$).

3) ($A_{Su} \cap B_{Su} = \phi$ or $A_{Su} \cap B_{Su} = \gamma$), $A_{Ob} \cap B_{Ob} = A_{Ob} = B_{Ob} \neq \phi$, $A_{Re} \cap B_{Re} = A_{Re} = B_{Re} \neq \phi$.

4) ($A_{Su} \cap B_{Su} = \phi$ or $A_{Su} \cap B_{Su} = \gamma$), $A_{Ob} \cap B_{Ob} = A_{Ob} = B_{Ob} \neq \phi$, ($A_{Re} \cap B_{Re} = \phi$ or $A_{Re} \cap B_{Re} = \alpha$).

5) $A_{Su} \cap B_{Su} = \phi$, ($A_{Ob} \cap B_{Ob} = \phi$ or $A_{Ob} \cap B_{Ob} = \beta$), $A_{Re} \cap B_{Re} = \phi$.

THEN $K_r = K_a$ (A) $\bigcirc K_a$ (B)

Where $\alpha \neq A_{Re}, B_{Re}, \phi$, $\beta \neq A_{Ob}, B_{Ob}, \phi$, $\gamma \neq A_{Su}, B_{Su}, \phi$.The symbol "$\bigcirc$" means non conflict.

For instance, emergency A represents fire incident caused by tank explosion; Emergency B represents water pollution incident as a result of disposal is not timely, crude oil into the sea. As spilled oil can cause more serious explosion one hand, on the other hand also can cause more serious water pollution, so cooling disposal activity of oil tanks should be executed in both emergencies. According to the style 1), activity subject is same, and perform the same goal using the same resources. So get instantiated activity knowledge elements:

K_a (A) = cool oil tank = (solders, oil tank, cooling agent)

K_a (B) = cool oil tank = (solders, oil tank, cooling agent)

There are not any conflicts between subjects, objects or resources in two emergencies. Thus the relationship between the two activities is non conflict relationship. Other types have similar explanation.

(2) Subject conflict

In two emergencies, the subjects of activities use resources to complete their corresponding target objects, conflict exist only when the number of subjects is not enough. Then the relationship between these two activities is subject conflict relationship. Types as follows:

IF

1) $A_{Su} \cap B_{Su} = A_{Su} = B_{Su} \neq \phi$,($A_{Ob} \cap B_{Ob} = \phi$ or $A_{Ob} \cap B_{Ob} = \beta$), $A_{Re} \cap B_{Re} = \phi$.

2) $A_{Su} \cap B_{Su} = \gamma$,($A_{Ob} \cap B_{Ob} = \phi$ or $A_{Ob} \cap B_{Ob} = \beta$), $A_{Re} \cap B_{Re} = \phi$.

THEN $K_r = K_a$ (A_{Su}) $\triangle K_a$ (B_{Su})

Where $\beta \neq A_{Ob}, B_{Ob}, \phi$, $\gamma \neq A_{Su}, B_{Su}, \phi$.The symbol "$\triangle$" means subject conflict.

For instance, there is an activity of "put out a fire" in fire emergency A, and an activity of "clean spilled oil" in water pollution emergency B. When the two activities occur at the same time, according to the type 1), instantiated activity knowledge elements are:

K_a (A) = put out a fire = (solders, fire, water cannon)

K_a (B) = clean spilled oil = (solders, crude oil, Oil absorbent boom)

From the knowledge elements above, the activity objects in two emergencies are all solders, but due to the shortage of subject number, there has conflict between the two activities. Thus the relationship between the two activities is subject conflict relationship. Type 2) is the same.

(3) Resource conflict

In two emergencies, the subjects of activities use resources to complete their corresponding target objects, conflict exist only when the number of resources is not enough. Then the relationship between these two activities is resource conflict relationship. Types as follows:

IF

1) $A_{Su} \cap B_{Su} = \phi$, ($A_{Ob} \cap B_{Ob} = \phi$ or $A_{Ob} \cap B_{Ob} = \beta$), $A_{Re} \cap B_{Re} = A_{Re} = B_{Re} \neq \phi$.

2) $A_{Su} \cap B_{Su} = \phi$, ($A_{Ob} \cap B_{Ob} = \phi$ or $A_{Ob} \cap B_{Ob} = \beta$), $A_{Re} \cap B_{Re} = \alpha$.

THEN $K_r = K_a$ (A_{Re}) ∇ K_a (B_{Re})

Where $\alpha \neq A_{Re}, B_{Re}, \phi$, $\beta \neq A_{Ob}, B_{Ob}, \phi$ .The symbol "∇" means resource conflict.

For instance, there is an activity of "isolate flowing fire" in fire emergency A. And an activity of "intercept crude oil" in water polluted emergency B. The two activities occur at the same time, according to the type 1), instantiated activity knowledge elements are:

K_a (A)=isolate flowing fire= (solders, flowing fire, sandbags)

K_a (B) = intercept crude oil = (volunteers, crude oil, sandbags)

From the knowledge elements above, the object of solders is flowing fire. The object of volunteers is crude oil. But they all need sandbags to implement. It means the required resources are same. But resources are not enough according to hypothesis 4. So there has conflict between the two activities. Thus the relationship between the two activities is resource conflict relationship. Type 2) is the same.

(4) Complete conflict

In two emergencies, the subjects of activities use resources to complete their corresponding target objects, conflict exist not only when the number of subjects is not enough, but also the resources are shortage. Then the relationship between these two activities is complete conflict relationship. Types as follows:

IF

1) $A_{Su} \cap B_{Su} = A_{Su} = B_{Su} \neq \phi$,($A_{Ob} \cap B_{Ob} = \phi$ or $A_{Ob} \cap B_{Ob} = \beta$), $A_{Re} \cap B_{Re} = A_{Re} = B_{Re} \neq \phi$.

2) $A_{Su} \cap B_{Su} = A_{Su} = B_{Su} \neq \phi$,($A_{Ob} \cap B_{Ob} = \phi$ or $A_{Ob} \cap B_{Ob} = \beta$), $A_{Re} \cap B_{Re} = \alpha$.

3) $A_{Su} \cap B_{Su} = \gamma$, ($A_{Ob} \cap B_{Ob} = \phi$ or $A_{Ob} \cap B_{Ob} = \beta$), $A_{Re} \cap B_{Re} = A_{Re} = B_{Re} \neq \phi$.

4) $A_{Su} \cap B_{Su} = \gamma$, ($A_{Ob} \cap B_{Ob} = \phi$ or $A_{Ob} \cap B_{Ob} = \beta$), $A_{Re} \cap B_{Re} = \alpha$.

THEN $K_r = K_a(A) \diamondsuit K_a(B)$

Where $\alpha \neq A_{Re}, B_{Re}, \phi$, $\beta \neq A_{Ob}, B_{Ob}, \phi$, $\gamma \neq A_{Su}, B_{Su}, \phi$ .The symbol "$\diamondsuit$" means complete conflict.

For instance, the same activities as example (3):"isolate flowing fire" and "intercept crude oil". When the two activities occur at the same time, according to the type a), instantiated activity knowledge elements are:

K_a (A)=isolate flowing fire= (solders, flowing fire, sandbags)

K_a (B) = intercept crude oil = (solders, crude oil, sandbags)

From the knowledge elements above, the activity subjects in two emergencies are all solders, and they use the same resources to implement different target. But due to the shortage of subjects and resources, there has conflict between the two activities. Thus the relationship between the two activities is complete conflict relationship. Other types are the same.

4   Method and Application of Relationship

4.1   Formalized Method on Creative Thinking

Creative thinking is a process that uses the characteristics of knowledge element. Whose attributes and attribute value as basis to identify relationship between activities, and expressing which through mathematical logical formalized description method. Thus effective combine the knowledge of knowledge management and mathematic field into emergency response process.

4.2   Application in Management

The application of identification method of relationship between activities in emergency management is to provide site disposal method for emergency managers, especially when the multiple emergencies occur. Through instantiating knowledge element to analyze the relationship between activities, we can figure out which factors do not enough in advance so that make effective prevention and response.

4.3   Application in Process Integration

From the characteristics of multiple emergencies, some implementation activities can be performed together to improve disposal efficiency when the emergencies occur at the same time and area. Where, some activities can be integrated into one activity so as to realize the integration of process. Through the analysis of the relationship between activities, some activities exist some conflicts in the process of implementation, but some not. So, according to some rules, integrate these activities to realize the process integration in emergency response.

5   Conclusions

This paper mainly studies the relationship of implementation activities in dealing with disasters when the secondary and derivative of emergencies occur. Give attributes assignment to instantiate the knowledge elements and then to recognize the relationship of implementation activities. This recognition of relationship is important to lead emergency response and process integration.

However, this recognition of relationship is on the basis of instantiation. Though this method would be general to a certain extent, it also has certain limitations. Occurrence and evolution of emergencies is a complicated process with context factors, such as climate, wind direction and so on also play an important role. So the next step in the research, context factors will be taken into consideration to substantiate knowledge element. Study how to fast and efficient response in the specific emergencies.

References

1. Wang, Y., Ye, X., Qiu, J., Wang, N.: Emergency Management Information System. Beijing Science Press (2010)
2. Yu, K., Wang, Q.Q., Rong, L.L.: Emergency Ontology Construction in Emergency Decision Support System. In: Service Operations and Logistics, and Informatics, Beijing, pp. 801–805 (2008)
3. Dinh, M., Tan, T., Bein, K., Hayman, J., Wong, Y.K., Dinh, D.: Emergency department knowledge management in the age of Web 2.0: Evaluation of a new concept. Emergency Medicine Australasia 23, 46–53 (2011)
4. Hernández, J.Z., Serrano, J.M.: Environmental emergency management supported by knowledge modeling techniques. AI Communications 14, 13–22 (2001)
5. Avouris, N.: Cooperating knowledge-based systems for environmental decision support. Knowledge-Based Systems 8(1), 39–54 (1995)
6. Zhang, D., Zhou, L., Nunamaker Jr., J.F.: A Knowledge Management Framework for the Support of Decision Making in Humanitarian Assistance/Disaster Relief. Knowledge and Information Systems 4, 370–385 (2002)
7. Dautun, C., Tixier, J., Chapelain, J., Fontaine, F., Dusserre, G.: Crisis management: Improvement of knowledge and development of a decision aid process. Institution of Chemical Engineers, 16–21 (2008)

8. Franke, J., Charoy, F.: Design of a Collaborative Disaster Response Process Management System. In: The 9th International Conference on the Design of Cooperative Systems, France, pp. 55–57 (2010)

9. Saidani, O., Nurcan, S.: Context-Awareness for Adequate Business Process Modelling. In: The Third International Conference on Research Challenges in Information Science, Morocco, pp. 177–186 (2009)

10. Rosemann, M., Recker, J.: Context-aware Process Design Exploring the Extrinsic Drivers for Process Flexibility. In: The 18th International Conference on Advanced Information Systems Engineering, Luxembourg, pp. 149–158 (2006)

11. Meng, K., Wang, B.: A enlightenment of "Jilin chemical explosion" for the preparation of emergency plan. Energy Research and Information 22(4), 198–203 (2006)

12. Wang, Y.: Knowledge and representation of model management. Journal of System Engineering 26(6), 850–856 (2011)

The Method of Emergency Rescue Program Generation in Civil Aviation Airport of China

Wang Hong[1,2], Zhou Wen-tao[1], and Wang Jing[1]

[1] College of Computer Science and Technology, Civil Aviation University of China,
Tianjin 300300, China
{hwang,j_wang}@cauc.edu.cn, w.t.zhou74@gmail.com
[2] College of Management, Tianjin University, Tianjin 300072, China

Abstract. Although, at present, a rather complete emergency rescue system of airport has been formed in civil aviation of China, most of the emergency rescue programs for disposing emergency are still formed manually by means of experiences of decision-makers. The paper presents an auto-generated method of emergency rescue program for airport based on contingency plans of airport and airport emergency business process management notation ontology (AE_BPMNO) which makes the programs generated automatically come true via the construction of SWRL semantic rules and semantic queries based on SQWRL and provides auxiliary emergency command and disposal of airport with methodological support.

Keywords: Civil Aviation Airport, Emergency Rescue Program, SWRL, semantic reasoning, SQWRL.

1   Introduction

Emergency management is an important part of civil aviation management, and the informatization in civil aviation of China is an important way for emergency management. In order to handle a variety of civil aviation emergencies, the enterprises and institutes in civil aviation all have been formulated relevant contingency plans and designed emergency rescue management system with different types of software and hardware platforms.

The kernel of emergency management is to make an effect strategic decision for emergency rescue. How to form an emergency rescue program with good contents and organizations at an express speed according to the relevant contingency plans possesses high referenced and decisional value is necessary for decision-makers to defuse crisis appropriately and creatively. While, the contents defined in the contingency plans are difficult to be carried out effectively since the existing contingency plans are exist in documents which embody information that cannot be understood and processed by computers that causes the contents cannot be reflected in existing emergency rescue management system when the emergencies happened. On the contrary, most of the emergency rescue programs are gained manually by means of decision-makers' experiences. In order to realize the programs generated automatically and then improve

M. Wang (Ed.): KSEM 2013, LNAI 8041, pp. 431–443, 2013.

the efficiency of emergency rescue for airport, the paper presents an auto-generated method of emergency rescue in airport based on contingency plans which has been realized in emergency rescue management system of civil aviation airport, and to some degree, which resolves situation that most of emergency rescue programs are generated rely on decision-makers' experiences in existing airport.

2   Content Analysis of Emergency Rescue Program in Civil Aviation Airport

Civil Aviation of China that has basically formulated an emergency response management system of China civil aviation which takes the files "National Public Emergencies Overall Contingency Plan" as general programme, "Civil Aviation Emergency Management Provisions of China" as guidance and "Civil Aviation Overall Plan", "Civil Aviation Special Plan", "Civil Aviation Unit Programme" as foundation, and the plans, like "Civil Aviation emergency management formulation of China", "Civil Airport operational and safety management regulations" that focus on emergency rescue of civil airport have been issued by civil aviation administration of china centred around the transport characteristics of civil aviation and the practical works of emergency management, which offers the enterprises and institutions to carry out effective emergency management with guarantee of system and mechanism.

The emergency rescue program is a kind of action plans which are used for different emergency happened around airport [1]. They are transparency and standardized emergency response procedures and precondition of which the rescue actions can process orderly according to advanced plans and effective implementation procedures, which are the basic guarantee for quick response and rescue effectively. According to contingency plans of civil airport, different emergencies correspond to different contingency plans, but the contents of which always include:

(1). The emergency rescue level activated, that is the incidental level and rescuing level of current emergency.

(2). The overall contingency plans activated, like: the plan "Aircraft Crashed Field contingency Plan" should be activated immediately when the aircraft crashed field happened.

(3). The list of rescue leaders, the members of headquarters, contacts and scene to be informed.

(4). The list of rescue departments, contacts and scene to be informed.

(5). The different rescue actions for different rescue departments and their corresponding plans.

Organizationally, the emergency rescue programs almost always should follow: 1) Unity of direction, namely all the rescue departments and personals should step up relief orderly and reasonably under the unified direction of control center; 2) Well arranged, namely the lower level should be subordinate to the higher level strictly; 3) Coordinate comprehensively, namely department to department, person to person and department to person should perform tasks together and in harmony; 4) Department division, namely each department should perform its own task strictly under the unified command and dispatch of commanders.

3   Airport Emergency Business Process Management Notation Ontology

AE_BPMNO[2] is an ontology in civil airport that combines with civil airport emergency domain ontology (CAEDO)[3] with Business Process Management Notation Ontology (BPMNO)[4], and generated from semantic Civil Airport Emergency Business Process Diagram (CAE_BPD) that realized by adopting semantic annotation, in which, CAEDO and BPMNO are used to describe the domain concepts and their relevant constraints, structures and their relevant constraints of rescue processes respectively. The basic ideas are as follows:

(1). According to civil aviation emergency rescue manuals and their relevant formulations, the beginning of process for semantic annotation is to construct CAE_BPD via Eclipse BPMN Modeler on a basis of business process management and annotation specification and then to annotate objects in the existed diagram by making use of the concepts in CAEDO.

(2). To define merging axioms of ontologies via description logic, in which, CAEDO is domain mode, while the BPMNO is regarded as structural model of emergency rescue processes.

(3). To realize semantic annotation through adopting the method presented by C. Di Francescomarino et.al[5-7], that is to implement the semantization of CAE_BPD. The detailed implementation has been discussed in [2], and the structural of AE_BPMNO is shown partially for the space's sake, which is as follows:

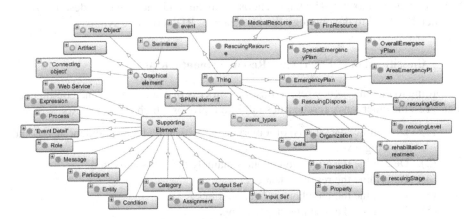

Fig. 1. The structure of AE_BPMNO

AE_BPMNO combines CAEDO with BPMNO, and the Fig. 1 shows the kernel concepts of AE_BPMNO, like: Graphical element Class, Supporting element Class and their subclasses used for describing structure of rescuing processes; RescuingDisposal Class used for describing emergency disposal; RescuingResource Class used for desecribing resources; rescuingAction Class used for rescuing actions and so on, and the object properties like: has_sequence_flow_target_ref used for describing rescuing order; has_message_flow_target_ref used for describing the message passing directions etc. To reason against AE_BPMNO based on OWL DL can generate part of emergency rescue program since the OWL has its own semantic rules.

4   Construction of Civil Aviation Airport Emergency Rescue Rules Based on SWRL

AE_BPMNO is an airport emergency business process domain ontology based on OWL-DL, which realize the semantization of CAE_BPD that can be understood and disposed by computer, and then lay a foundation of generation of emergency rescue program based on SWRL[8-10] and SQWRL(Semantic Query-enhanced Web Rule Language)[11]. Not only does it describe the concepts, data properties and their inter-relationship in the emergency management field of civil aviation, but describes the sequences of execution and message passing among rescuing departments. In order to get more reasonable and abundant results, the paper describes more extra comprehensive logical relations in emergency rescue on a basis of reasoning against OWL based on OWL.

SWRL that is a proposal for a Semantic Web rules language based on a combination of the OWL DL and OWL Lite adopt the subset of RuleML (Rule Markup Language) that only contains unary and binary predicate logics based on Horn clause. The syntax mainly includes Abstract syntax, Human readable syntax, XML concrete syntax and RDF concrete syntax etc. The paper chooses human readable syntax to describe rules, and then be added to the ontology file by means of RDF concrete syntax.

4.1   Classification of Emergency Rescue Semantic Rules

According to the emergency rescue business process of civil aviation and contingency plans[12], the emergency rescue rules fall into 6 groups which are as follows:

    (1). Determine the status of emergency, which includes incident level and rescuing level.
    (2). Determine the contingency plans, namely the plans activated.
    (3). Determine the rescuing leaders.
    (4). Determine the members of headquarter.
    (5). Determine the rescuing departments.
    (6). Determine the rescuing resources and their number.

The rules described in SWRL for understandable sake are as shown in the following table.

Table 1. Emergency rescuing rules of civil aviation airport

Rules Type	Description of Rules in SWRL	Instructions	Rules Number
RULE-1:Judgment of status	Emergency(?x)∧belongToEventType(?x,?y) ∧ damagedCondition(?x,?z)∧hasCasualty(?x,?m) ∧ swrlb:lessThan(?m,39) →activateRescuingLevel(?x,?n)∧ activateincidentLevel(?x,?p)	?x: name of emergency; ?y: type of emergency	18
RULE-2: Judgment of Plans Activated	Emergency(?x)∧belongToEventType(?x,?y) →activateRescuingPlan(?x,?p)	?p: plan	64
RULE-3: Judgment of Rescuing Leaders	Emergency(?x)∧belongToEventType(?x,?y) ∧ hasOfficeDuty(?l,?d) → toReportLeader(?y, ?n)	?d: Duty of Leaders; ?n: Name of Leaders	6
RULE-4: Judgment of Headerquarter's member	Emergency(?x)∧belongToEventType(?x,?y) ∧ hasOfficeDuty(?y,?d)∧ toReportLeader(?z,?y) → headquartersTitle(?y, ?t)	?t: Duty of Headerquarter's Member	6
RULE-5: Judgment of Rescuing Departments	Emergency(?x)∧belongToEventType(?x, ?y) ∧ hasOfficeDuty(?y, ?z) → toReportUnit(?y, ?u)	?u: Rescuing Departments	6
RULE-6:Judgment of Rescuing Resources	Emergency(?x)∧belongToEventType(?x, ?y) ∧ hasCasualty(?x,?cv)∧ swrlb:greaterThan(?cv,?z) → needRescuerNo(?u, ?n)	?c: Casualty; ?cv: Critical Value of ?c; ?n: Number of Rescue Workers	6

4.2   Implementation of Emergency Rescue Rules

The Human Readable syntax is used for describing the rules, and the rules are stored in the ontology file through RDF concrete syntax. To take one specific rule of RULE-1 as example, how to implement the emergency rescue rules are shown as following:

"The incident level will be set as '*Serious Accident*' if the aircraft is badly damaged or ditched at the place that it cannot be shipped out, or there are injuries and deaths whose number is less or equal than 39. The rescue level should be set as '*Emergency Dispatch*' if the type of emergency is '*Aircraft Crashed Field Emergency*'"--- 205 Contingency Plans of Civil Aviation Airport [13].

$$Emergency(E) \land belongToEventType(E,?x)$$
$$\land damagedCondition(E,badly\_damaged) \land hasCasualty(E,?y) \land$$
$$swrlb:greaterThan(?y,39)$$
$$\rightarrow activateRescuingLevel(E,Emergency\_Dispatch) \land$$
$$incidentLevel(E,Serious\_Aviation\_Accident)$$

In which, the class *'Emergency'* indicates emergency of civil aviation airport; the object properties *'belongToEventType'* , *'activateRescuingLevel'*, *'activateIncidentLevel'* indicate the types of some specific emergency, the corresponding rescue level and the corresponding incident level respectively. The data properties *'damagedCondition'* , *'hasCasualty'* indicate the extent of damage and the corresponding casualty respectively. The variables *x*, *y* indicate the type of emergency, casualty respectively.

The rule described in RDF concrete syntax is as follows (Part of the whole given for the space's sake).

```
<swrl:Variable
    rdf:about="http://www.owl-ontologies.com/domain.owl#
    y"/>
......
    <swrl:Variable rdf:about="urn:swrl#z"/>
      <swrl:Imp>
        <swrl:body>
              ......
                    <swrl:IndividualPropertyAtom>
                      <swrl:propertyPredicate
        rdf:resource="http://www.owl-ontologies.com/do
        main.owl#belongToEventType"/>
                          <swrl:argument1
        rdf:resource="http://www.owl-ontologies.com/do
        main.owl#x"/>
                          <swrl:argument2
        rdf:resource="http://www.owl-ontologies.com/do
        main.owl#Aircraft Crashed Field Emergency"/>
                    </swrl:IndividualPropertyAtom>
      ......
          </swrl:body>
      </swrl:Imp>
```

In which, the contents included in the label <swrl:Imp></swrl:Imp> indicates a rule, the label <body></body> and the label <head></head> form a rule. _head is true iif _body is true. <swrl:Variable/> defines variables; the elements like _Atomlist, _first are used for describing concrete contents of rule.

5   Fulfillment of Emergency Rescue Program Generation Method

According to the contingency plans, this section will illustrate the fulfillment process of program and its detailed implementation on a basis of AE_BPMNO and emergency rescue rules.

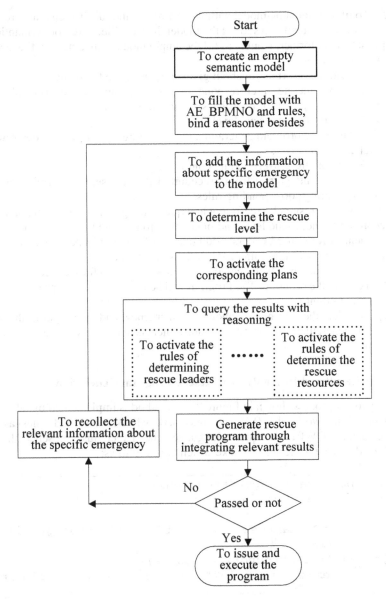

Fig. 2. Fulfillment process of emergency rescue program based on contingency plans

5.1   Fulfillment Process of Emergency Rescue Program

As shown in Figure 2, the process includes the following procedures:

(1). To create an empty semantic model.

The purpose is to offer an operable address to semantic data storing and semantic reasoning and query.

(2). To fill the semantic model with AE_BPMNO and rule, which bind a reasoner.

There are a couple of ways to fill the model, like via file, or remote semantic data. The paper select the former, namely to fill the empty model with ontology file, and then bind it with pellet reasoner.

(3).   To add the information about given emergency to the model.

In order to generate a corresponding rescue program, the facts about the emergency should be given for the rules selecting and reasoning.

(4).   To determine the rescue level

Before the rescue program generating, the first thing to be done is to determine the rescue level according given facts.

(5).   To activate the corresponding contingency plans.

To take the results gained from procedure 4 as premise, to determine the plans activated by executing corresponding rules.

(6).   To get the results by reasoning and querying against the semantic data.

According to the rescue level and plans, then to activate the corresponding rules, like 'Judgment of Rescuing Leaders', 'Judgment of Rescuing Departments' and so on, query the results through reasoning against the semantic data.

(7).   To integrate the results gained from procedure 6 to form expected program.

(8).   To audit the program. The Program will be issued only if the program is given a green light.

(9).   Recollect the information about the emergency, and then input to the system for generating a revising program which is more reasonable, then turn to procedure 3.

5.2   The Result and Its Analysis of Rescue Program Generation

According to the process shown in Figure 2, the method is implemented by taking Jena as semantic framework, the Pellet as reasoned, AE_BPMNO and rules as basic semantic data, SQWRL as query language, the paper just presents the core implemented codes for the space's sake which are as follows:

```
FileInputStream inputStream=null;
  OntModel model=null;
  try{
      inputStream=new FileInputStream("ontology file
        path");
  }catch(FileNotFoundException e){
      System.err.println("Given file is an invalid input
file");
  }
  Reasoner
reasoner=PelletReasonerFactory.theInstance().create();
  Model infModel=ModelFactory.createInfModel(reasoner,
ModelFactory.createDefaultModel());
  model=ModelFactory.createOntologyModel(OntModelSpec.
OWL_DL_MEM,infModel);
      model.read(inputStream,null,"RDF/XML");
  ......
```

In order to further analyze the method that can be applied to generate the corresponding rescue program according the specific emergency, the paper gives two examples for contrastive analysis, one is 'Aircraft Crashed Field Emergency' (Which is marked E1), and another is 'Ice and Snow Emergency' (Which is marked E2)

(1). To determine the rescue level

To begin with, for emergency E1, according to RULE-1, the rule of determining rescue level described in SWRL is shown as following:

$$Emergency(?x)^\wedge belongToEventType(?x,Aircraft\ Crashed\ Enfield$$
$$Emergency)^\wedge damagedCondition(?x,?y)^\wedge hasCasualty(?x,?z)^\wedge swrlb:lessThan(?z,39)$$
$$\rightarrow activateRescuingLevel(?x,Emergency\_Dispatch)$$
$$^\wedge activateIncidentLevel(?x,Serious\_Aviation\_Accident)$$

In which, assume that the casualty of E1 is 20, namely, the value of *hasCasualty* is 20. while, as for emergency E2, the rule is shown as follows:

$$Emergency(?x)\ ^\wedge belongToEventType(?x,Ice\ and\ Snow\ Emergency)$$
$$^\wedge hasCasualty(?x,?y)\ ^\wedge swrlb:lessThan(?y,20)$$
$$\rightarrow activateRescuingLevel(?x,Emergency\_Dispatch)$$
$$^\wedge activateIncidentLevel(?x,Major\ Public\ Emergency)$$

With respect to E2, assume that the casualty is 0, namely the value of *hasCasualty* is 0. According to the above the rules, we can get the incident level and rescue level of E1, E2 respectively through executing the follow SQWRL language with reasoning.

$$activateIncidentLevelActivated(?x,?y)\ ^\wedge activateRescuingLevelActivated(?x,\ ?z)$$
$$\rightarrow sqwrl:select(?x,\ ?y,\ ?z)$$

(2). To determine the rescuing plans activated

According to RULE-2, for E1, The corresponding SWRL rule is:

$$Emergency(?x)\ ^\wedge belongToEventType(?x,Aircraft\ Crashed\ Enfield\ Emergency)$$
$$\rightarrow activateRescuingPlan(?x,TBIA2005-2001-002 "Aircraft\ Crashed\ Enfield$$
$$Emergency\ Plan")$$

While for E2, the rule is as follows:

$$Emergency(?x)\ ^\wedge belongToEventType(?x,Ice\ and\ Snow\ Emergency)$$
$$\rightarrow activateRescuingPlan(?x,TBIA205-503-001 "Large\ Area\ Abnormal\ Flight\ of$$
$$Civil\ Airport\ Enfield\ Disposal\ Plan")$$

According to the above the rules, we can get the plan activated of E1, E2 respectively through executing the follow SQWRL language with reasoning.

$$activateRescuingPlan(?x,?y) \rightarrow sqwrl:select(?x,\ ?y)$$

(3). To determine the rescue leaders

According to RULE-3, for E1, The corresponding SWRL rule is:

$$Emergency(?x)\ ^\wedge belongToEventType(?x,\ Aircraft\ Crashed\ Enfield\ Emergency)$$
$$^\wedge hasOfficeDuty(?l,\ President\ of\ Airport)\ ^\wedge hasOfficeDuty(?l,\ Vice\text{-}president\ of\ Airport)$$
$$^\wedge hasOfficeDuty(?l,\ Shift\_Leader)\ ^\wedge hasOfficeDuty(?l,\ No.\ 2\ Shift\_Leader)$$
$$^\wedge hasOfficeDuty(?y,\ Staff\ of\ TAMCC)$$
$$\rightarrow toReportLeader(?y,?l)$$

While for E2, the rule is as follows:

$$Emergency(?x) \wedge belongToEventType(?x, Ice\ and\ Snow\ Emergency)$$
$$\wedge hasOfficeDuty(?l, Vice\text{-}president\ of\ Airport)$$
$$\wedge hasOfficeDuty(?l, Manager\ of\ Meteorological\ Department)$$
$$\wedge hasOfficeDuty(?l, Shift\_Leader) \wedge$$
$$hasOfficeDuty(?l, Manager\ of\ Airport\ Operation\ Department)$$
$$\wedge hasOfficeDuty(?y, Staff\ of\ TAMCC)$$
$$\rightarrow toReportLeader(?y, ?l)$$

According to the above the rules, we can get the rescue leaders to be informed of E1, E2 respectively through executing the follow SQWRL language with reasoning.

$$toReportLeader(?y, ?l) \wedge hasPhoneNumber(?l, ?tel) \rightarrow sqwrl:select(?l, ?tel)$$

(4). To determine the rescue departments
According to RULE-5, for E1, The corresponding SWRL rule is:

$$Organization(?o) \wedge Emergency(E)$$
$$\wedge belongToEventType(E, Aircraft\ Crashed\ Enfield\ Emergency)$$
$$\wedge hasOfficeDuty(?y, Staff\ of\ TAMCC)$$
$$\wedge hasOrganizationName(?o, Fire\ Brigade\ of\ Airport)$$
$$\wedge hasOrganizationName(?o, Medical\text{-}aid\ center\ of\ Airport)$$
$$\wedge hasOrganizationName(?o, Pubic\ Security\ Sub\text{-}bureau\ of\ Airport)$$
$$\wedge hasOrganizationName(?o, Aviation\ Safety\ Department)$$
$$\wedge hasOrganizationName(?o, Maintenance\ Security\ Department)$$
$$\wedge hasOrganizationName(?o, Transport\ Services\ Department)$$
$$\rightarrow toReportUnit(?y, ?o)$$

While for E2, the rule is as follows:

$$Organization(?o) \wedge Emergency(E)$$
$$\wedge belongToEventType(E, Air\ traffic\ Control\ Sub\text{-}bureau)$$
$$\wedge hasOfficeDuty(?y, AirField\ Department)$$
$$\wedge hasOrganizationName(?o, Transport\ Services\ Department)$$
$$\wedge hasOrganizationName(?o, Airline\&Agency\ Company)$$
$$\rightarrow toReportUnit(?y, ?o)$$

According to the above the rules, we can get the rescue departments to be informed of E1, E2 respectively through executing the follow SQWRL language with reasoning.

$$toReportUnit(?y, ?o) \wedge hasPhoneNumber(?o, ?tel) \wedge hasManager(?o, ?m)$$
$$\rightarrow sqwrl:select(?o, ?m)$$

(5). To determine the rescue actions
Different emergency has different rescue processes that are decided by process description in AE_BPMNO, so for E1 (Take the fire brigade of airport for example), we can get the rescue actions of fire brigade of airport through executing the follow SQWRL language with reasoning.

$$fireDeapartment(?fd) \wedge BPMN\_Ontology:task(?t)$$
$$\wedge BPMN\_Ontology:process(?p) \wedge ownTask(?fd, ?t)$$
$$\wedge BPMN\_Ontology:has\_process\_graphical\_elements(?p, ?t)$$
$$\wedge BPMN\_Ontology:has\_flow\_object\_name(?t, ?name)$$
$$\wedge hasReferencePlan(?task, ?plan)$$
$$\rightarrow sqwrl:select(?name, ?plan)$$

While for E2 (Take the airfield department for example), we can get the rescue actions of airfield department through executing the follow SQWRL language with reasoning.

$$airfieldDepartment(?ad) \wedge BPMN\_Ontology:task(?t)$$
$$\wedge BPMN\_Ontology:process(?p) \wedge ownTask(?ad, ?t)$$
$$\wedge BPMN\_Ontology:has\_process\_graphical\_elements(?p, ?t)$$
$$\wedge BPMN\_Ontology:has\_flow\_object\_name(?t, ?name)$$
$$\wedge hasReferencePlan(?task, ?plan)$$
$$\rightarrow sqwrl:select(?name, ?plan)$$

To integrate all of the above results, we can get the whole contents of rescue program for E1 and E2 respectively. The paper just illustrates the comparative results of two different emergencies for the space's sake.

Table 2. The comparative results of E1 and E2

Emergency Types	Incident Level	Rescue Level	Plans Activated	Leaders to be Informed	Departments to be Informed	Rescue Actions and their referenced plans
Aircraft Crashed Enfield Emergency (E1)	Serious_Aviation_Accident	Emergency_Dispatch	TBIA2005-2001-002"Aircraft Crashed Enfield Contingency Plan"	Yan xin, Xu Huanran et.al	Fire Brigade of Airport, Medical-aid center of Airport and so on.	To execute the mission according to the Fire commander-TBIA303-300-010 "Fire Fighting working procedure"
Ice and Snow Emergency (E2)	Major Public Emergency	Emergency_Dispatch	TBIA205-503-001"Large Area Abnormal Flight of Civil Airport Enfield Disposal Plan"	Xu Huanran, Gong Linyun et.al	Air traffic Control Sub-bureau, Airfield Department and so on.	To execute the work of snow removing according to the instructions-TBIA203-300-003"Snow Removing Plan of Aircraft Movement Area"

The table 2 shows that the method not only can generate rescue program instead of manual operation, but can get the corresponding rescue program according to the specific emergency.

To combine with manual operation by experiences from decision-makers, the method presented has the following advantages.

(1). Organizationally, the method presented is in accord with manual operation, which is convenient for reading and executing.

(2). In the content, the method presented based on contingency plans can find out the exact corresponding contents and data about rescue through reasoning and querying, which can avoid human errors gained from manual operation, furthermore, aid generate rescue program, besides, can improve the program's precision.

Besides, not only can the method presented generate the corresponding rescue program, but can adjust the program automatically to some degree. For example: During the rescue process, for the 'Aircraft Crashed Enfield Emergency', the incident level should be set as 'Special Aviation Accident' if the casualty is up to more than 39, like 40. According to the RULE-1, namely: *Emergency(?x)^belongToEventType(?x,Aircraft Crashed Enfield Emergency) ^ hasCasualty(?x, ?y) ^swrlb:greaterThan(?y, 39)→ incidentLevel(?x, Special Aviation Accident),* the incident level will be set as '*Special Aviation Accident*' automatically after executing the above rule, and then, the whole rescue program will be adjusted accordingly.

6   Conclusion

According to the civil emergency rescue system and plans, the paper gives the processes of creating emergency rules based on SWRL and generating rescue program on a basis of AE_BPMNO, and then shows how to gain the detailed contents of rescue program by means of querying against reasoning results via SQWRL query language, which proves that the method can realize generation and adjustment of rescue program automatically. The method has been applied to the emergency rescue management system of civil aviation airport, which has partly solved the problem that rescue program can only be gained through manual operation from decision-maker's experiences effectively. In order to realize more sophisticated reasoning, the describing language and construction of rules should be further studied.

Acknowledgements. This work was supported by the National Natural Science Foundation of China under Grant No. 61079007, The Scientific Research Foundation of Civil Aviation University of China under Grant No. 09QD04X.

References

[1] Yang, T., Zhang, J.: Airport Operation and Command. Chinese Aviation Express (2008)
[2] Zhou, W., Wang, H., Wang, J., et al.: Research on Constructing Semantic Model of Emergency Decision-making Plan for Civil Aviation. Application Research of Computers 30(1) (January 2013)
[3] Wang, H., Gao, S., Pan, Z., Xiao, Z.: Application and research of non-taxonomic relation extractionmethod based on NNV association rule. Application Research of Computers 29(10) (October 2012)
[4] Ghidini, C., Rospocher, M., Serafini, L.: A formalisation of BPMN in description logics. Technical Report TR 2008-06-004, FBK-irst (2008)
[5] Di Francescomarino, C., Tonella, P.: Supporting Ontology-Based Semantic Annotation of Business Processes with Automated Suggestions. International Journal of Information System Modeling and Design 1(2), 59–83 (2010)
[6] Di Francescomarino, C., Ghidini, C., Rospocher, M., Serafini, L., Tonella, P.: Semantically-aided business process modeling. In: Bernstein, A., Karger, D.R., Heath, T., Feigenbaum, L., Maynard, D., Motta, E., Thirunarayan, K. (eds.) ISWC 2009. LNCS, vol. 5823, pp. 114–129. Springer, Heidelberg (2009)

[7] Rospocher, M., Di Francescomarino, C., Ghidini, C., Serafini, L., Tonella, P.: Collaborative Specification of Semantically Annotated Business Processes. In: Business Process Management Workshops (BPM 2009), 3rd International Workshop on Collaborative Business Processes (CBP 2009) (2009)

[8] SWRL Submission, http://www.w3.org/Submission/SWRL/

[9] SWRL Built-in Specification, http://www.daml.org/rules/proposal/builtins.html

[10] SWRLTab Plugin, http://protege.cim3.net/cgi-bin/wiki.pl?SWRLTab

[11] O'Connor, M., Das, A.: SQWRL: A Query Language for OWL (2009)

[12] Civil Aviation Emergency Management Provisions of China, Civil Aviation Administration of China (2008)

[13] 205 Contingency Plans of Civil Aviation Airport, Tianjin Binhai International Airport, China (2009)

Detection of Article Qualities in the Chinese Wikipedia Based on C4.5 Decision Tree

Kui Xiao[1,2], Bing Li[1,*], Peng He[1], and Xi-hui Yang[1]

[1] State key laboratory of software engineering,
Wuhan University, Wuhan 430072, China
bingli@whu.edu.cn
[2] School of computer and software,
Wuhan Vocational College of Software and Engineering,
Wuhan 430205, China

Abstract. The number of articles in Wikipedia is growing rapidly. It is important for Wikipedia to provide users with high quality and reliable articles. However, the quality assessment metric provided by Wikipedia are inefficient, and other mainstream quality detection methods only focus on the qualities of the English Wikipedia articles, and usually analyze the text contents of articles, which is also a time-consuming process. In this paper, we propose a method for detecting the article qualities of the Chinese Wikipedia based on C4.5 decision tree. The problem of quality detection is transformed to classification problem of high-quality and low-quality articles. By using the fields from the tables in the Chinese Wikipedia database, we built the decision trees to distinguish high-quality articles from low-quality ones.

Keywords: Wikipedia, Article quality, Data ming, Decision tree, Application of supervised learning.

1 Introduction

Wikipedia, as the most popular free online-encyclopedia, contains more than 25,000,000 articles in over 280 languages. Contributors around the world are attracted by its free and open styles, collaborative editing and multi-languages supporting policies. Along with rapid growth of the number of articles, article quality is becoming increasingly important. In Wikipedia, some articles can provide professional, outstanding and thorough information, but other articles may provide very little meaningful contents. There is a wide difference in qualities between articles.

There is a system provided by Wikipedia, which is based on peer review methods. Articles in Wikipedia are divided into seven classes according to their qualities. They are FA-Class (featured articles), A-Class, GA-Class (good articles), B-Class, C-Class, Start-Class, Stub-Class. However, the promotion process is a little complicated and inefficient. For example, when an article needs to be

---

* Corresponding author.

M. Wang (Ed.): KSEM 2013, LNAI 8041, pp. 444–452, 2013.

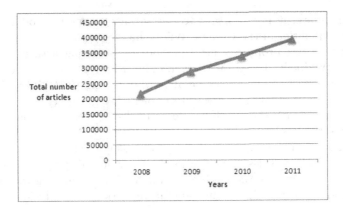

Fig. 1. Evolvement of the total number of articles in the Chinese Wikipedia

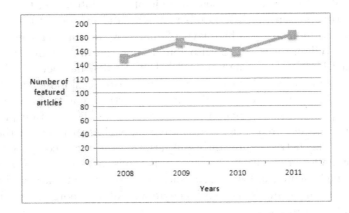

Fig. 2. Evolvement of the number of featured articles in the Chinese Wikipedia

promoted to FA-Class, it has to be nominated by a Wikipedia autoconfirmed user first. Then in the next two weeks, all the autoconfirmed users can vote for supporting or opposing its promotion. If the supporters win, the article can be marked with a featured article symbol.

Although the peer review methods can generate precise article qualities, the process is inefficient. Figure 1 and Figure 2 illustrate the evolvements of total number of articles and the number of featured articles in the Chinese Wikipedia from the end of 2008 to the end of 2011. The results suggest that the total number of articles almost doubled during the three years, but the number of featured articles increased by 20% in the same period.

There are a lot of article quality assessing approaches proposed by researchers. However, almost all of the approaches are applied on the English Wikipedia. Moreover, many of them assesse article qualities by analyzing the text of articles. As is well-known, there is huge difference between the patterns of English and

Chinese. Therefore, it is important to design an efficient method in order to detect article qualities in the Chinese Wikipedia.

In this paper, we propose a new method for detecting article qualities in the Chinese Wikipedia based on the C4.5 algorithm. There are two contributions in this paper: 1) finding out several attributes from the Chinese Wikipedia database, which are related to the criteria of high-quality articles; 2) building the decision trees with the attributes so as to distinguish high-quality articles from low-quality ones.

The remainder of the paper is organized as follows. Section 2 introduces the related work. Section 3 describes the method and learning process of training sets in detail. In section 4, the method is evaluated by comparing with the word count method. In section 5, we conclude future work and conclusions.

2   Related Work

Article quality assessment is a research topic of great interest in the Wikipedia fields. Lih [1] proposed that the number of edits and number of unique editors provides a good indicator of a high level of quality within the Wikipedia project. In addition, Wilkinson et al. [2] also found that the high-quality articles are distinguished by a marked increase in number of edits, number of editors, and intensity of cooperative behavior. Besides, Blumenstock [3] found that the word count method can also be employed to measure article quality. In his work, the problem of quality measurement was transformed to classification problem of featured and non-featured articles. In spite of its striking simplicity, the author showed that the metric significantly outperforms the more complex methods. Furthermore, a lot of quality measurement methods make use of the word count method as the benchmark. It is clearly that very few attributes are used to assess article qualities in the methods mentioned above. Adding more article attributes may yield better results.

On the other hand, a lot of researchers analyze the text of articles so as to measure the qualities. Zeng et al. [4] explored the hypothesis that revision information can be used to compute a measure of trustworthiness of revised documents, and then developed a revision history-based trust model for computing and tracking the trustworthiness of the articles in Wikipedia. The trust model is represented in a dynamic Bayesian network (DBN). Lim et al. [5] proposed three article quality measurement models that make use of the interaction data between articles and their contributors derived from the article edit history, including Basic model, PeerReview model, ProbReview model. The Basic model is designed based on the mutual dependency between article quality and their author authority. The PeerReview model introduces the review behavior into measuring article quality. Finally, The ProbReview models extend PeerReview with partial reviewership of contributors as they edit various portions of the articles. Lipka et al. [6] employed various trigram vector representations along with a classifier in order to identify featured articles. Lex et al. [7] suggested a simple fact-related quality measure, factual density, which measured the relative number of document facts and thus indicates a documents informativeness. Their experiments on a subset of the English Wikipedia revealed that featured/good articles can be separated from non-featured articles with a high confidence even if the articles are similar in length. However, analyzing text of articles can be a very time-consuming process. In this paper, we employed several common attributes extracted from the Chinese Wikipedia database to assess article qualities. This will be more efficient than text analyzing methods.

In addition, by making use of the one-class classification technologies, Anderka et al. [8] developed a quality flaw model and employ a dedicated machine learning approach to predict Wikipedias most important quality flaws. Li et al. [9] discussed several article quality assessing methods. The methods mainly include two kinds: the correlation-based analysis and cooperation modeling. Furthermore, the authors presented the open problems of automatic quality evaluation and the possible promotions of collective intelligence.

Most of the quality assessment methods mentioned above focus on the qualities of the English Wikipedia articles. In the current work, we focus on the qualities of the Chinese Wikipedia ones.

3   Approach

In this paper, we propose a new method for detecting article qualities in the Chinese Wikipedia based on C4.5 algorithm. The problem of quality detection is transformed to classification problem of high-quality and low-quality articles. Our method, in contrast to many other methods, does not analyze the text contents of article history versions, but use the fields from the tables in the Chinese Wikipedia database to build decision trees, which can be used to distinguish high-quality articles from low-quality ones.

3.1   C4.5 Algorithm

In this paper, we employed the C4.5 algorithm to classify high-quality and low-quality articles. This classification is a supervised learning task. In other words, each instance in training sets has a notion of class label. Both the C4.5 algorithm and the Naive Bayes algorithm are suited to this kind of tasks. As is well-known, Normal distribution is an important precondition to the Naive Bayes classifier. However, our previous experiments suggested that the values associated with some article attributes are not distributed according to a Normal distribution, such as the number of edits and the number of editors. For this reason, we selected the C4.5 algorithm to classify articles.

C4.5 is an algorithm used to generate a decision tree developed by Quinlan [10], which can be used for classification. C4.5 is an extension of ID3 algorithm. The decision tree can be learned from the training sets with the algorithm.

C4.5 algorithm selects the attribute with the highest normalized information gain to make the decision of classification every time. The normalized information gain is defined as:

$$GainRatio(\alpha) = \frac{Gain(\alpha)}{Entropy(\alpha)} \tag{1}$$

α is an attribute of instances. $Entropy(\alpha)$ is the information entropy of the attribute, which is defined as:

$$Entropy(\alpha) = \sum_{i=1}^{c} -p_i * \log(p_i) \tag{2}$$

where p_1, p_2, ..., p_c are properties of the attribute on c values. $Gain(\alpha)$ is the information gain of the attribute, which is calculated as:

$$Gain(\alpha) = Entropy(\alpha) - \sum_{v} \frac{|D_v|}{|D|} * Entropy(\alpha \ in \ D_v) \tag{3}$$

where v is the set of possible values, D denotes the entire dataset, D_v is the subset of the dataset for which attribute α has that value, and the notation $|D|$ denotes the size of a dataset (in the number of instances).

3.2 Attributes Selection

In order to distinguish high-quality articles and low-quality ones, we had to study the attributes of high-quality articles. Featured articles are considered to be the best articles in Wikipedia. Their criteria can be used as features of high-quality articles.

Among these requirements, only part of them can be quantified with fields from the tables in the Chinese Wikipedia database. So we made use of seven attributes to describe features of high-quality articles, including page length, number of external links, number of images, number of internal links, number of edits, number of editors. Note that, internal links should be divided into in-links and out-links. In addition, number of editors contains both registered users and anonymous users who are identified by IPs.

Page length came from table *page* (page_len); Number of external links came from table *externallinks*; Number of images came from table *imagelinks*; Number of in-links and out-links came from table *pagelinks*. Number of edits and number of editors came from a XML file *pages − meta − history* which should be parsed first and stored locally. Although we used the article edit history file, we did not analyze text contents of article history versions. So the efficiency of our method would not be reduced practically.

3.3 Training Sets

In this paper, we built decision trees with training sets which contains only two kinds of articles, they are featured articles (FAs) and Start-Class articles (SAs). We chose the sample dataset from the snapshot of the Chinese Wikipedia from December 2011. Up to 12/9/2011, there were 183 featured articles and about 10,000 Start-Class articles in the Chinese Wikipedia. All the 183 featured articles were selected as the sample articles. Besides, we also randomly selected 549 Start-Class articles, which were added to the sample dataset as well.

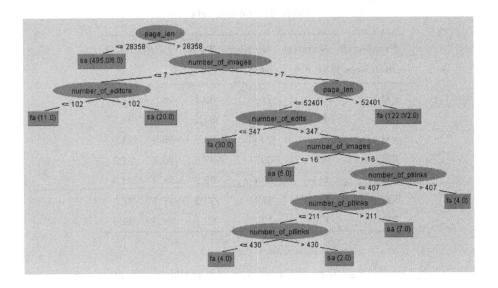

Fig. 3. The decision tree of 700 instances

We successively select 100, 200, 300, 400, 500, 600, 700 articles at random from the sample dataset in order to create seven training sets. An article denotes an instance, and each of the training sets was used to build a decision tree with the WEKA data mining suite [11]. It is noteworthy that we used the J48 classifier, which is WEKAs implementation of C4.5. With 10-fold cross-validation the decision tree of 700 instances is illustrated in Figure 3.

The experimental results are shown in Table 1. Here, FA represents featured articles, SA represents the Start-Class articles. TPR represents the True Positive Rate, which is the ratio between the number of true positives and the sum of the number of true positives and the number of false negatives. The TPR of featured articles is defined as:

$$TPR_f = \frac{T_f}{T_f + F_f} \tag{4}$$

Here, T_f denotes the number of true featured articles which were judged by the decision tree correctly. F_f denotes the number of false featured articles which were judged by the decision tree mistakenly.

In the same way, the TPR of the Start-Class articles is defined as:

$$TPR_s = \frac{T_s}{T_s + F_s} \tag{5}$$

Here, T_s denotes the number of true Start-Class articles which were judged by the decision tree correctly. F_s denotes the number of false Start-Class articles which were judged by the decision tree mistakenly.

Table 1. Training sets

Number of instances	Number of FAs	Number of SAs	TPR_f	TPR_s	ACC
100	29	71	93.1	98.6	97.0
200	44	156	84.1	96.2	93.5
300	70	230	84.3	96.5	93.7
400	96	304	87.5	96.4	94.3
500	115	385	92.2	98.2	96.8
600	147	453	93.2	98.7	97.3
700	175	525	89.7	98.7	96.4

ACC represents classification accuracy, which is the classification precision of the decision tree. It is defined as:

$$ACC = \frac{T}{T + F} \tag{6}$$

Here, T denotes the number of articles which were judged by the decision tree correctly. F denotes the number of articles which were judged by the decision tree mistakenly.

Results suggest that the TPR of featured articles were between 84% and 94%, and the TPR of Start-Class articles were between 96% and 99%. The average ACC is over 95.6%.

4   Evaluation

In order to evaluate our method, we made use of Blumenstocks word count method [3] as the benchmark, which is the most effective metric currently known. Note that, only the words in the title and the main body of articles were counted. The words in other parts of articles, such as the references, were not counted in our experiments.

Table 2. Comparision between the two methods

	TPR_f	TPR_s	ACC
decision tree	91.8	96.0	94.6
word count	81.6	91.0	87.9

149 articles were chosen from the sample dataset, including 49 featured articles and 100 Start-Class articles. In practice, we counted the Chinese characters of the articles. After that, the articles were sorted from most to least.

The 49th article had 7915 Chinese characters. In the first 49 articles, there were 40 featured articles and 9 Start-Class articles. The comparison results between our method and the word count method were listed in Table 2. In all the items, our method outperforms the word count method.

5   Conclusions

In this paper we presented a method for detecting article qualities in the Chinese Wikipedia based on the C4.5 algorithm. The problem of quality assessment was transformed to classification problem of high-quality and low-quality articles. Experiments revealed a high classification accuracy.

In future research we plan to continue to study the classification-based quality detection method. On the one hand, we can try to apply more attributes in decision tree. For example, high-quality articles usually own a number of corresponding articles written in other languages, so the number of languages may be used as an attribute. On the other hand, we only classify the featured articles and Start-Class articles in current work. Other five Classes can also be taken into account in the next step.

Acknowledgements. This work is supported by the National Nature Science Foundation of China under Grant No.61273216, the Key Technologies R & D Program of Wuhan under Grant No. 201210621214, the open foundation of JiangShu Provincial Key Laboratory of Electronic Business under Grant No. JSEB2012-02.

References

[1]   Lih, A.: Wikipedia as Participatory Journalism: Reliable Sources? Metrics for evaluating collaborative media as a news resource. In: Proc. of the 5th International Symposium on Online Journalism (2004)

[2]   Wilkinson, D.M., Huberman, B.A.: Cooperation and Quality in Wikipedia. In: Proc. of the International Symposium on Wikis, pp. 157–164 (2007)

[3]   Blumenstock, J.E.: Size Matters: Word Count as a Measure of Quality on Wikipedia. In: Proc. of the 17th International Conference on World Wide Web, pp. 1095–1096 (2008)

[4]   Zeng, H.L., Alhossaini, M.A., Ding, L., et al.: Computing Trust from Revision History. In: Proc. of the International Conference on Privacy, Security and Trust (2006)

[5]   Hu, M.Q., Lim, E.P., Sun, A.X., et al.: Measuring Article Quality in Wikipedia: Models and Evaluation. In: Proc. of the 16th ACM Conference on Information and Knowledge Management, pp. 243–252 (2007)

[6] Lipka, N., Stein, B.: Identifying Featured Articles in Wikipedia: Writing Style
     Matters. In: Proc. of the 19th International Conference on World Wide Web, pp.
     1147–1148 (2010)
[7] Lex, E., Voelske, M., Errecalde, M., et al.: Measuring The Quality of Web Content
     Using Factual Information. In: Proc. of the 2nd Joint WICOW/AIRWeb Workshop
     on Web Quality, pp. 7–10 (2012)
[8] Anderka, M., Stein, B., Lipka, N.: Towards Automatic Quality Assurance in
     Wikipedia. In: Proc. of the 20th International Conference on World Wide Web,
     pp. 5–6 (2011)
[9] Li, D.Y., Zhang, H.S., Wang, S.L., Wu, J.B.: Quality of Articles in Wikipedia.
     Geomatics and Information Science of Wuhan University 36(12), 1387–1391 (2011)
[10] Quinlan, J.R.: C4. 5: Programs for Machine Learning. Morgan Kaufmann (1993)
[11] Witten, I.H., Frank, E.: Data Mining: Practical Machine Learning Tools and Tech-
     niques with Java Implementations. Morgan Kaufmann (1999)

The Category Structure in Wikipedia: To Analyze and Know Its Quality Using K-Core Decomposition

Qishun Wang[*], Xiaohua Wang, and Zhiqun Chen

Institute of Cognitive and Intelligent Computing,
Hangzhou Dianzi University, Hangzhou 310018, China
qishun.wang@gmail.com, wxhhie@sohu.com,
25792523@qq.com

Abstract. Wikipedia is a famous and free encyclopedia. A network based on its category structure is built and then analyzed from various aspects, such as the connectivity distribution, evolution of the overall topology. As an innovative point of our paper, the model that is on the base of the k-core decomposition is used to analyze evolution of the overall topology and test the quality (that is, the error and attack tolerance) of the structure when nodes are removed. The model based on removal of edges is compared. Our results offer useful insights for the growth and the quality of the category structure, and the methods how to better organize the category structure.

Keywords: Wikipedia, complex network, overall topology, quality, k-core.

1 Introduction

Wikipedia has become the largest and most popular encyclopedia in the world [1]. It is applied to improve related fields' performance, such as information extraction, information retrieval, ontology building, natural language processing and so on [2]. As a result, firstly we want to study how the category structure developed from 2004 to 2012, using a method based on the k-core decomposition to measure the evolutionary trend of the structure's overall topology.

As everyone knows, when the nodes with a lot of edges in the network are removed gradually, the network would crash, in other words these nodes play an important role in the network. In this paper we want to find whether the nodes exist which play more important roles than the nodes with a lot of edges, in order to search the key nodes which maintain the structure stable. Furthermore we verify whether the strong edges play a more important role in maintaining the network's integrity than the weak ones, in order to understand the importance of the weak edges.

2 Category Structure

Wikipedia provides many useful structures, such as the structure of categories, pages and images. They are useful in the area of data mining, a lot of useful knowledge can

---

[*] Corresponding author.

M. Wang (Ed.): KSEM 2013, LNAI 8041, pp. 453–462, 2013.
© Springer-Verlag Berlin Heidelberg 2013

be mined and results can be used in related fields to improve the performance. In this paper we mainly analyze the structure of categories, in order to know more about the organization of knowledge in Wikipedia, the evolutionary trend of the structure from 2004 to 2012 and the quality of the structure under attacks and failures.

Wikipedia seeks to create a summary of all human knowledge in the form of an on-line encyclopedia, with each topic of knowledge covered encyclopedically in one article, so in order to represent information and knowledge hierarchy, the category structure is set up. When authors write new articles, they are encouraged to assign some useful categories to their articles, in order to organize and sort articles easily, both articles and categories can belong to more than one category [2]. For example, the article Teacher falls in the category Educators, and the category Educators also belongs to the following twenty-four: Schoolteachers, Academics, Lecturers, Deaf educators and so on. The categories also can assign themselves to other more general categories, for instance, Language teacher training belongs to Teacher training, which in turn belongs to Teaching. The structure is a graph in which multiple organization schemes coexist, not a simple tree-structured taxonomy.

We download a complete snapshot of Chinese Wikipedia dated August 2012 from the official website. It contains many meaningful structures, such as category structure and page structure, and the former is applied in our paper. Each record in this dataset has a property which tells us when it was created or modified, offering a chance to study how the network grows and evolves over time. To build the network which represents organization of this knowledge, we traverse all the links and filter the links only when its type is category. The structure is transformed and then used to build a complex network, which its node is a category and its edge is the relation between the categories. The resulting network consists of $N \approx 2.5*10^5$ nodes and $E \approx 5.3*10^6$ edges, the vast majority (99.6%) of these nodes belonging to a single connected cluster [giant component (GC)].

3 Topological Features

In this section, in order to know more about the relation between categories and how to organize categories efficient, some topological features are analyzed in detail, such as connectivity distribution, cluster distribution, the overall topology and evolution of the overall topology [3].

3.1 Connectivity Distribution

The connectivity distribution (also called the degree distribution) $P(k)$ which indicates the probability that a randomly selected node has k connections shows the essential statistical character of large-scale, complex networks. It represents the importance and connectivity of reality networks, that is, the influence of the node in the network, the node with large connectivity is more important.

The connectivity distribution and the cumulative connectivity distribution $P(K>k)$ which is the fraction of nodes with degree greater than or equal to k, as can be seen in Fig. 1. Some networks, for instance, the category structure and the page structure in

Wikipedia, the Internet and the ecological network are found to have a skewed degree distribution that approximately follows a power law with a fat tail: $P(k) \sim k^{-\alpha}$, where α is a constant, in the category structure α is equal to 2.987. If we look closely at Fig. 1, this network is inhomogeneous: in the nodes about 78% have less than five edges and about 94% have less than ten edges, but only 1% have a large number of edges (hubs), guaranteeing that the system is fully connected (more than 99% of the nodes are connected). So according to the fact we can make the conclusion that the category structure is a small-world and scale-free network.

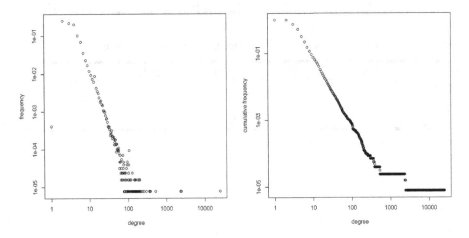

Fig. 1. Connectivity distribution of the category structure

3.2   Evolution of the Overall Topology

Here we propose a method based on the k-core decomposition to measure the evolutionary trend of the network's overall topology. The k-core of a network is the maximal subgraph in which every node has at least degree k, each one obtained by recursively removing all the nodes of degree smaller than k, until the degree of all remaining nodes is larger than or equal to k [4]. Normally the node which has a large coreness clearly has a more central position in the network's structure, with a large degree. But our experiment finds a phenomenon that the node has a large coreness even its degree is small, because it is the bridge between nodes which have large degree, the node is also called the bridge node.

Even Wikipedia was founded in 2001, in Chinese Wikipedia the earliest category was created in 2004, because the input page lacked of the Chinese input function until 2004. In order to know how the category structure grows and evolves with the development of Wikipedia year by year, the network is divided into nine networks according to the creation time of the edges, then these networks are analyzed in two ways: the basic attributes such as the topology, the number of nodes and edges, the

Table 1. The k-core of the category structure from 2004 to 2012

k-core	2004	2005	2006	2007	2008	2009	2010	2011	2012
1	109	3784	5618	8891	13614	21193	27044	32573	36039
2	0	3072	7963	11658	17979	22672	35460	44216	38731
3	0	6	44	248	2025	4040	11252	22914	40199
4	0	0	0	0	65	146	356	1537	12077
5	0	0	0	0	0	3	3	78	914
6	0	0	0	0	0	13	13	15	47
7	0	0	0	0	0	0	0	0	18

Table 2. The coreness change of the nodes in category structure

id	2008	2009	2010	2011	2012
34714	1	1	6	6	7
29226	1	1	6	6	7
42619	1	6	6	6	7

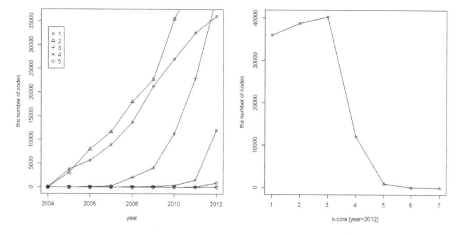

Fig. 2. Distribution of the number of the nodes in k-cores (k = 1, 2, 3, 4, 5) from 2004 to 2012 (left), and distribution of the number of the nodes in k-cores in 2012 (right)

average path length and the average clustering coefficient [5], and the evolutionary trend of the overall topology. In our paper the later is analyzed in detail. The k-core of the networks between 2004 and 2012 is calculated and the results are shown in Table 1. In order to study the evolutionary trend clearly, the distribution of the number of the nodes in k-cores from 2004 to 2012 is drawn in Fig. 2, in particular the data in 2012 is drawn individually.

As illustrated in Fig. 2, the growing trend of the network's overall topology is very obvious, for example, the number of nodes in each shell grows approximately linearly over time, the number of nodes decreases when the k-core increases. A node may become more important in the network's structure, in Table 2 we can find that the node's coreness becomes larger and larger during the period from 2004 to 2012. Observing the nodes whose coreness is smaller than four, their increasing speed is fastest. This phenomenon indicates that the category which added to the structure recently aims at expanding the related categories and making the category structure completely.

4   Quality of the Structure

4.1   Model

In order to represent information and knowledge hierarchy, the category structure was set up by Wikipedia. We aim at analyzing how much impact on the structure when some categories are deleted or some relations are removed. Considering the category structure in 2012, we present a descriptive model that is on the base of k-core decomposition to study the error and attack tolerance of the network. We also present another model to study the quality of the network when a percentage of the edges are removed.

In terms of the nodes, we start the node model with the full structure. Then at each time-step:

1. Each node's coreness and degree is calculated, and the nodes are sorted according to the coreness, if the nodes have same coreness, they are sorted by their degree.
2. About 0.5 percent of the nodes are deleted, starting with the large coreness and then moving toward the smaller ones, the edges related to the nodes are also removed.
3. To observe how the largest cluster decomposes when some nodes are removed from the cluster, the relative size of the largest cluster (S) is calculated. The average size of the isolated clusters (<s>) is calculated to observe how much impact on the isolated clusters when the largest cluster is split into many parts.

In terms of the edges, we also start the edge model with the full structure. Then at each time-step:

1. We measured the relative topological overlap of the neighborhood of two categories v_i and v_j, representing the proportion of their common knowledge representation $O_{ij} = n_{ij}/((k_i - 1) + (k_j - 1) - n_{ij})$, where n_{ij} is the number of common neighbors of v_i and v_j, and k_i (k_j) denotes the degree of node v_i (v_j). If v_i and v_j have no common knowledge representation, then $O_{ij} = 0$, the edge between i and j representing potential bridges between two different communities. If i and j are part of the same circle of knowledge representation, then $O_{ij} = 1$ [6]. The edges are sorted according to this property.
2. We choose two methods to remove about 0.5 percent of the edges, one is removing the large O_{ij} edges and then toward the smaller ones, the other is starting with the small O_{ij} edges and then moving toward the larger ones.

3. The following two arguments are calculated to observe the stability of the category structure to the edge removal: the relative size of the giant component $R_{GC}(f) = N_{GC}(f)/N_{GC}(f = 0)$ where N_{GC} is the size of the giant component, and the value $S = \sum_{s<smax} n_s s^2/N$ where n_s is the number of clusters containing s nodes [6].

4.2   Quality of the Nodes

It has been long known that a lot of networks show resilience to random node removal, but are fragile to the removal of the hub nodes. In order to prove this conclusion, many researchers analyze all kinds of networks from various aspects. The following networks are tested, for example, the scale-free networks [7], the evolving networks [8], the small-world networks [9], and the communication networks [10]. Many parameters are calculated in order to study the error and attack tolerance of complex networks, such as the evolution of cooperation [7], the small-worldness index [8], the connectivity [9], the global and local efficiency [10-11], the diameter and the size of the clusters [12].

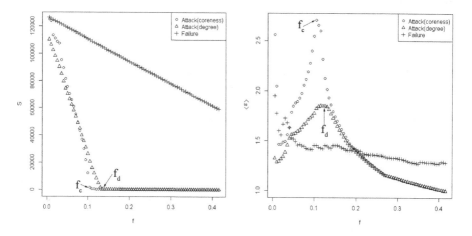

Fig. 3. Network fragmentation under coreness attacks, degree attacks and random failures. The control parameter f denotes the fraction of removed nodes, the parameter S is the relative size of the largest cluster and the parameter <s> is the average size of the isolated clusters ($f_c = 0.105$, $f_d = 0.130$)

Many authors had made the conclusion that most of the networks are fragile to the removal of the important nodes [7-12]. The important nodes may be the nodes which have a large degree, and if these nodes are removed from the network, it is a disaster to the network's structure. But in our paper the nodes which have a large coreness are defined as the important nodes, they play a more important role in the network than the nodes which have a large degree. The nodes are removed from largest to smallest by two different parameters which are the coreness and degree, or randomly removed. The results are compared and drawn in Fig. 3. The cluster size distribution is drawn in

Fig. 4 when f (f = 0.1, 0.25, 0.4) percent of the nodes are removed, this shows how the structure breaks down when the network is under attacks or failures.

In terms of the coreness attacks and degree attacks, the nodes which are important to the network are cut off from the main cluster firstly [12]. In order to understand the impact of attacks and failures on the network structure better, we next measure some significant parameters to investigate this fragmentation process, for example, the size of the largest cluster S and the average size of the isolated clusters <s> (that is, all the clusters except the largest one). As illustrated in Fig. 3, we find that for the category structure under coreness attacks, when f reaches the f_c = 0.105, S tends to zero which displays a threshold-like behavior, similar behavior is observed when we monitor <s>, finding that <s> increases rapidly until <s> ≈ 2.7 at f_c, and decreases rapidly to <s> ≈ 1. As f increases, the size of the fragments which come off the main cluster increases, leading to an increasing <s>. At f_c the structure collapses, the main cluster breaks into small pieces, leading to S ≈ 0 and appearing an obvious peak at <s>. As the nodes are continued to be removed (f = f_c), these isolated clusters break into smaller clusters, leading to a decreasing <s>. These behaviors delay until f_d = 0.13 when the network is under degree attacks, but doesn't occur under random failures, because the coreness attack has a destructive effect on the structure than the degree attack, and the random failure has little effect on the structure.

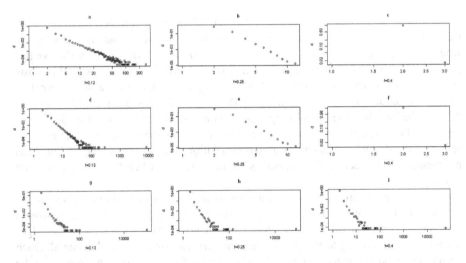

Fig. 4. Summary of the response of a network to coreness attacks, degree attacks and random failures (a-i). The control parameter f denotes the fraction of removed nodes, the cluster size distribution is drawn when f percent of the nodes are removed. (a-c) is coreness attacks, (d-f) is degree attacks, (g-i) is random failures.

We observe clearly how the cluster size distribution changes when some percent of the nodes are removed in Fig. 4. The network under coreness attacks and the network under degree attacks behave similarly. For small f, clusters of different sizes break down, but a large cluster still exists in the network. When f increases, in the case of attacks the largest cluster breaks apart, forming many isolated clusters, but in the case of random failures the largest cluster still exists even a lot of nodes are removed. We suppose $f_c < f_1 < f_d < f_2 < f_3$ ($f_c = 0.105$, $f_d = 0.13$, $f_1 = 0.12$, $f_2 = 0.25$, $f_3 = 0.4$), when f_1 ($f_c < f_1 < f_d$) percent of the nodes are removed, the network breaks into small fragments between sizes 1 and 200, the large cluster in coreness attacks disappears (a) but still exists in degree attacks (d). At even larger f_2 ($f_c < f_d < f_2$), the large cluster disappears both in coreness attacks (b) and degree attacks (e), the clusters are further fragmented into single nodes or clusters of size smaller than ten. Only the clusters whose size is smaller than three leave when f_3 percent of the nodes are removed (c, f). This is supported by the cluster size distribution in Fig. 4. The cluster's decomposition speed under coreness attack is faster than the speed under degree attack, but at the moment of random failures, even for an unrealistic high rate of $f_3 = 0.4$, the large cluster still exists (g-i).

According to above analysis and evidences, we can draw the conclusion that the network is fragile to the removal of the hub nodes and more fragile to the removal of the bridge nodes which connect the hub nodes in the network, the bridge nodes are more important than the hub nodes.

4.3   Quality of the Edges

Many authors had analyzed the error and attack tolerance of complex networks when the nodes are removed by some specific rules, but it is rare to analyze the situation when the edges are removed by some given regulations. In terms of the nodes we had come to the conclusion that the node with a large degree has a more central position in the network's structure. So we expect that the strong edges play a more important role in maintaining the network's integrity than the weak ones. Then in order to prove this conclusion, the edge model is applied and the results are drawn in Fig. 5. But our result of this experiment shows the opposite effect in the category structure: The removal of the weak edges causes a phase transition-like network collapse, but it has no measurable impact on the network's overall integrity when the strong edges are removed.

In order to prove the conclusion that network structure is related to edge's strength, we measure the relative size of the giant component ($R_{GC}(f)$) to explore the network's resistance to the removal of either strong or weak edges. When f percent of the edges are removed, according to $R_{GC}(f)$ we can know the fraction of nodes that can all reach each other through connected paths. As illustrated in Fig. 5, we find that removing the smallest O_{ij} edges and then toward the larger ones leads to the network's sudden disintegration at $f_{low} = 0.87$. On the contrary, removing first the largest O_{ij} edges will result in the network's gradual shrinking but will not precipitously break it apart, and it has little impact on the network's overall integrity. Monitoring $S = \sum_{s<smax} n_s s^2 / N$ to forecast the position at which the network disintegrates, the critical point in R_{GC} is same with the peak in S, furthermore this phenomenon appears only when we start with the smallest O_{ij} edges.

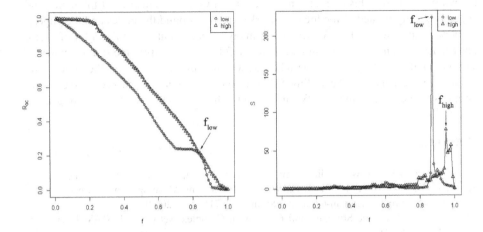

Fig. 5. The stability of the category structure to edge removal. The control parameter f denotes the fraction of removed edges, R_{GC} denotes the relative size of the giant component, S equals with the value $\sum_{s<smax} n_s s^2 / N$ ($f_{low} = 0.87$, $f_{high} = 0.95$).

According to above analysis, we can draw the conclusion that the weak edges appear to be crucial for maintaining the network's structural integrity, but strong edges play an important role in maintaining local communities. Further experimental study also shows that the weak edges play a significant role in the link prediction problem [13-14].

5   Discussion and Conclusions

Although the study of complex networks and Wikipedia has a long history, only a few researchers analyze and study Wikipedia under the guidance of the complex network theory. In this paper, taking advantage of the dataset offered by the official website of Wikipedia, we mainly analyze the connectivity distribution, the evolutionary trend of the category structure, and the quality of the structure under attacks and failures

The category structure in Wikipedia is analyzed in many respects. The connectivity distribution of the network shows that it is a small-world and scale-free network. The method based on the k-core decomposition is used to measure the overall topology of the network. We can clearly understand how the category structure grows and evolves from 2004 to 2012.

The quality of the nodes and edges in the network is analyzed in detail, that is, the resilience to the removal of the nodes or edges. In terms of the nodes, this network shows resilience to random node removal, but is fragile to the removal of the hub nodes and more fragile to the removal of the bridge nodes in the network. In terms of the edges, our results show that the removal of strong edges results in the network's gradual shrinking and has little impact on the network's overall integrity, but the removal of the weak edges results in a sudden and phase transition-driven collapse of the whole network. This conclusion reveals the laws that the quality of the nodes and edges is closely related to the local network structure around the nodes and the edges.

In our paper, we analyze Wikipedia in many respects, but it still has a lot of useful knowledge to mine. What we should do in the future is to analyze the future growth of Wikipedia after 2013 and find the methods which could resist and fix the attacks and failures to the network. According to the research results we can give some useful opinions to the development of Wikipedia in the future. So this work is meaningful.

References

1. The Introduction to Wikipedia, http://en.wikipedia.org/wiki/Wikipedia
2. Medelyan, O., Milne, D., Legg, C., Witten, I.H.: Mining Meaning from Wikipedia. International Journal of Human-Computer Studies 67, 716–754 (2009)
3. Newman, M.E.: The Structure and Function of Complex Networks. J. SIAM Review 45, 167–256 (2003)
4. Alvarez-Hamelin, J.I., Dall'Asta, L., Barrat, A., Vespignani, A.: K-core Decomposition: A Tool for the Visualization of Large Scale Networks. arXiv preprint cs/0504107 (2005)
5. Allamanis, M., Scellato, S., Mascolo, C.: Evolution of a Location-based Online Social Network: Analysis and Models. In: Proceedings of ACM Internet Measurement Conference, IMC 2012 (2012)
6. Onnela, J.P., Saramäki, J., Hyvönen, J., Szabó, G., Lazer, D., Kaski, K., Barabási, A.L.: Structure and Tie Strengths in Mobile Communication Networks. Proceedings of the National Academy of Sciences 18, 7332–7336 (2007)
7. Perc, M.: Evolution of Cooperation on Scale-free Networks Subject to Error and Attack. New Journal of Physics 3, 033027 (2009)
8. Sun, S., Liu, Z., Chen, Z., Yuan, Z.: Error and Attack Tolerance of Evolving Networks with Local Preferential Attachment. Physica A: Statistical Mechanics and Its Applications 373, 851–860 (2007)
9. Jalili, M.: Error and Attack Tolerance of Small-worldness in Complex Networks. Journal of Informetrics 3, 422–430 (2011)
10. Crucitti, P., Latora, V., Marchiori, M., Rapisarda, A.: Error and Attack Tolerance of Complex Networks. Physica A: Statistical Mechanics and Its Applications 1, 388–394 (2004)
11. Crucitti, P., Latora, V., Marchiori, M., Rapisarda, A.: Efficiency of Scale-free Networks: Error and Attack Tolerance. Physica A: Statistical Mechanics and Its Applications 320, 622–642 (2003)
12. Albert, R., Jeong, H., Barabási, A.L.: Error and Attack Tolerance of Complex Networks. Nature 6794, 378–382 (2000)
13. Lü, L., Zhou, T.: Role of Weak Ties in Link Prediction of Complex Networks. In: CIKM 2009/CNIKM 2009, pp. 55–58. ACM Press, New York (2009)
14. Lü, L., Zhou, T.: Link Prediction in Weighted Networks: The Role of Weak Ties. Europhys. Lett. 89, 18001 (2010)

Learning to Map Chinese Sentences to Logical Forms

Zhihua Liao[1] and Zili Zhang[2,3]

[1] MFLETC, Foreign Studies College, Hunan Normal University, CS 410081, China
[2] Faculty of Computer and Information Science, Southwest University, CQ 400715, China
[3] School of Information Technology, Deakin University, VIC 3217, Australia
liao.zhihua61@gmail.com, zzhang@deakin.edu.au

Abstract. This paper addresses the problem of learning to map Chinese sentences to logical forms. The training data consist of Chinese natural language sentences paired with logical representations of their meaning. Although many approaches have been developed for learning to map from some western natural languages to two different meaning representations, there is no such approached for Chinese language. To this end, a Chinese dataset with 880 (Chinese sentence, logical form) pairs was developed. Then, the unification-based learning (UBL) approach which induces a probabilistic Combinatory Categorial Grammar (CCG) with higher-order unification is applied to the task of learning. Experimental results show high accuracy on benchmark datasets in Chinese language with two different meaning representations.

1 Introduction

An important purpose in natural language processing is to learn a mapping from natural language sentences to formal representations of their meaning. Recently, a large amount of research has addressed this problem by learning semantic parsers given sentences paired with logical meaning representations[2,11,12,13,14,15,16,17,18,19]. Furthermore, some approaches have been designed for multiple languages with a wide variety of logical representations of linguistic meaning such as English, Spanish, Turkish as well as Japanese[20,21]. For example, consider the following sentences paired with the logical form representing their meaning in the training example:

1. *English*
   sentence: what is the capital of utah
   logical form: (capital: c utah:s)
2. *Spanish*
   sentence: que es la capital de utah
   logical form: (capital:c utah:s)
3. *Turkish*
   sentence: utah in baskenti nedir
   logical form: (answer (capital (loc_2 (stateid utah:e))))
4. *Japanese*
   sentence: yuutaa no chuto wa nan desu ka
   logical form: (answer (capital (loc_2 (stateid utah:e))))

M. Wang (Ed.): KSEM 2013, LNAI 8041, pp. 463–472, 2013.

The first two examples consist of English and Spanish sentences paired with lambda-calculus meaning representations, respectively. The latter ones consist separately of Turkish and Japanese sentences corresponding to a variable-free logical expression.

For Chinese language, there are many characteristics different from other languages such as English, Spanish or Turkish. One distinctive characteristic is that Chinese sentences are composed of strings of characters without word boundaries that are marked by spaces like English. Therefore, word segmentation must be used in order to identify words in Chinese language processing. Secondly, meaning understanding of the Chinese sentence is usually much complex. Mostly, a sentence contains more than one meaning. For a sentence, people may interpret it from different perspectives, or with different backgrounds and experiences. Sometimes, just slightly changing the sentence structure can make much difference of meaning. Hence, the reason for generalizing to Chinese language is obvious.

Our work focuses on developing a Chinese dataset (Geo880) which consists of Chinese sentences paired with logical representations of their meaning. The original database domain: Geo880 includes a set of 880 English queries to a database of United States geography. For the dataset, we first manually translate them into Chinese sentences. Then, they are parsed to based-word sentences by Chinese parser ICT-CLAS2011 [1]. Afterward, we complete the phonetic format-*pinyin*. Finally, we build the experimental dataset. We test the benchmark data set by employing UBL method which uses probabilistic CCG Grammar from logical form with higher-order unification. The experimental results demonstrate high accuracy in Chinese language.

2 Related Work

There has been a significant amount of work on the problem of learning to map sentences to meaning representations. Researchers have developed some approaches using models and algorithms from statistical machine translation, inductive logic programming, probabilistic push-down automata, to CCG grammar induction techniques and inducing probabilistic CCG Grammar with higher-order unification.

Zelle and Mooney develop one of the earliest examples of learning system for *NLIDBs*, which describes a learning algorithm called *CHILL*[11]. Thompson and Mooney design a system that learns a lexicon for *CHILL* that perform almost as well as the original system[12]. Wong and Mooney develop the WASP system that uses statistical machine translation techniques to learn synchronous context free grammars containing both words and logic[15]. Later, they extend to a variant of WASP (λ-WASP) that has been designed for the lambda-calculus representations[16]. Kate and Mooney develop the *KRISP* system that is a discriminative approach where meaning representation structures are constructed from the natural language strings hierarchically and is built on the top of SVM struct with string kernels[13,14]. Lu et al. present a generative model that builds a single hybrid tree of words, syntax and meaning representation[22,23]. Zettlemoyer and Collins present CCG grammar induction techniques where lexical items are proposed according to a set of hand-engineering lexical templates. These algorithms are all language independent but representation specific[17,18,19]. More recently, Kwiatkowski et al. describe a more general method that induces a probabilistic CCG grammar with higher-order unification that represents the meaning of individual words and defines how these meanings can be combined to analyze complete sentences[20,21]. This method can be generalized to multiple languages and a variable-free logical expression. But these languages need to be word-based languages. That is, based-character Chinese language is still not explored. At a result, the techniques they employed are not applicable directly to learning Chinese language.

3   An Overview of the UBL Approach

The unification-based learning (UBL) approach is a more general method that induces a probabilistic CCG grammar that represents the meanings of individual words and defines how these meanings can be combined to analyze complete sentences [20]. This method employs higher-order unification to define a hypothesis space containing all grammars consistent with the training data, and uses an online learning algorithm that efficiently searches this space while simultaneously estimating the parameters of a log-linear parsing model. This section provides an introduction to this approach of using lambda calculus and higher-order unification to construct meaning representation. Besides, it also reviews the probabilistic CCG grammar and learning algorithm.

3.1   Lambda Calculus and Higher-Order Unification

Sentence meanings being represented as logical expressions, logical expressions can be constructed from the meaning of individual words by using the operations defined in the lambda calculus[4]. For lambda-calculus expression, the meaning of words and phrases can contain constants, quantifiers, logical connectors, and lambda abstractions. Its advantages lie in the generality which the meaning of individual words and phrases can be arbitrary lambda expressions while the final meaning for a sentence can be different forms.

The higher-order unification involves finding a substitution for the free variables in a pair of lambda-calculus expressions[3,6]. When applied, it can make the expressions equal. In the grammar induction process the restricted version of higher-order unification is tractable. This limited form can allow us to define the way to split a given expression into subparts that can be recombined with CCG parsing operations.

3.2   Probabilistic CCG

Combinatory Categorial Grammar (CCG) is a convenient linguistic formalism that tightly couples syntax and semantics[9,10]. For demonstration a CCG grammar that includes a lexicon Λ with entries is shown as follows:

alasijia:-NP : alaska:s
alasijiazhou:-NP : alaska:s
alasijia:-NP : alaska:n
alasijiazhou:-NP : alaska:n
zhijiage:-NP : chicago:c

zhijiageshi:-NP : chicago:c
zhijiage:-NP : chicago:n
zhijiageshi:-NP : chicago:n

It can model a wide range of language phenomena. CCG combines categories by using a set of combinatory rules. The probabilistic CCG (PCCG) is the result of generalizing CCG. The motivation of extending CCG to PCCG lies in that it can deal with ambiguity by ranking alternative parses for a sentence in order of probability. In general, given a CCG lexicon Λ, there will be many possible parses y for each sentence x. Suppose the parameters θ and lexicon Λ, the most likely logical form z can be determined [5] as: $f(x) = arg \max p(z|x; \theta, \Lambda) = arg \max \sum_y p(y, z|x; \theta, \Lambda)$, where $p(y, z|x; \theta, \Lambda)$
$= \frac{e^{\theta \cdot \phi(x,y,z)}}{\sum_{(y',z')} e^{\theta \cdot \phi(x,y',z')}}$.

3.3 Splitting Lexical Items

Before listing a complete learning algorithm, we first review the procedure of splitting lexical items by using higher-order unification. This splitting process is used to expand the lexicon during learning algorithm. Suppose lexical entry $w_{0:n} \vdash A$, with word sequence $w_{0:n} = \langle w_0, \cdots, w_n \rangle$, and CCG category A, the set of splits can be defined as: $S_L(w_{0:n} \vdash A) = \{(w_{0:i} \vdash B, w_{i+1:n} \vdash C)|0 \leq i < n \wedge (B, C) \in S_C(A)\}$ where it enumerates all ways of splitting the words sequence $w_{0:n}$ and aligning the subsequences with categories in $S_C(A)$.

3.4 Learning Algorithm

Although the splitting procedure can break apart overly specific lexical items into smaller ones that may generalize better to unseen data, the space of possible lexical items is too large to explicitly enumerate. So we need a trade-off between the splitting process and the space of lexical items. Instead, the PCCG meets such requirements. Given a learned lexicon, the parameters of a PCCG need to be learned to guide the splitting process and to select the best parse.

The unification-based learning algorithm (UBL) steps through the data incrementally and performs two steps for each training example[20]. First, new lexical items are induced for training instances by splitting and merging nodes in the best correct parse, given the current parameters. Next, the parameters of the PCCG are updated by making a stochastic gradient update on the marginal likelihood, given the updated lexicon. The UBL learning algorithm is listed as following:

Initialization :
  – Set $\Lambda = \{x_i \vdash S : z_i\}$ for all $i = 1 \ldots n$.
  – Set $\Lambda = \Lambda \cup \Lambda_{NP}$
  – Initialize θ using coocurrence statistics.
Algorithm :
  For $t = 1 \ldots n, i = 1 \ldots n$
  Step 1: Update Lexicon
      – Let $y^* = arg \max_y p(y|x_i, z_i; \theta, \Lambda)$

- Set $\Lambda = \Lambda \cup NEW - LEX(y_*)$

Step 2: Update Parameters
- Let $\gamma = \frac{\alpha_0}{1+c\cdot k}$ where $k = i + t \cdot n$.
- Let $\Delta = E_{p(y|x_i,z_i;\theta,\Lambda)}[\phi(x_i,y,z_i)] - E_{p(y,z|x_i;\theta,\Lambda)}[\phi(x_i,y,z)]$
- Set $\theta = \theta + \gamma\Delta$

Output : Lexicon Λ and parameters θ.

4 Data Domain - Chinese Geo880 Corpora

In order to evaluate the UBL algorithm on the Chinese Geo880 domain, we need to build this Chinese corpus that contains Chinese language queries paired with logical representation of each query's meaning. The original Geo880 is a set of 880 queries to a database of U.S. Geography which annotated with Prolog style semantics. Later, Kwiatkowski et al. manually convert them to equivalent statements in the lambda calculus and free variable meaning representations[20]. The full Geo880 dataset contains 880 (English sentence, logical form) pairs, and is split into a development set of 600 pairs and a test set of 280 pairs. The Geo250 dataset is a subset of Geo880 that contains 250 sentences and is divided into 10-fold cross validation for evaluation. For a direct comparison among different languages, we follow the same folds as Zettlemoyer & Collins[18] and Kwiatkowski et al.[20] implemented.

For Geo880 dataset with examples illustrated in Figure 1, we first collect all 880 English sentences and translate them to Chinese sentences. Next, we use Chinese parser ICTCLAS2011[1] to parse these sentences[1]. That is, the Chinese character string is segmented into based-word ones. Then, we manually check the word segmentations and correct wrong segmentations. Having done these steps, we implement the phonetic process - *pinyin* format[2]. Figure 2 illustrates the workflow about transforming the English sentence to Chinese *pinyin* string. At the same time, we collect all country, state, capital, city, county, river, valley and mountain names appeared in these sentences and build a list of (English, Chinese) word pairs in order to construct a seed lexical entry and lexical items. Finally, we match each Chinese *pinyin* sentence paired with its corresponding logical representation where the English sentence can be translated into this Chinese *pinyin* sentence as in Figure 3. Likewise, the variable-free logical form paired with its Chinese *pinyin* string can be also constructed. After accomplished these procedures, we have constructed both training and test examples as showed in Figure 4 and 5, respectively.

(lambda $0 e(loc:t (argmax $1 (and (place:t $1)(loc:t $1 montana:s))(elevation:i $1)) $0))
en: where is the highest point in montana

Fig. 1. Lambda-calculus form and its English sentence in the Geo880 dataset

---

[1] http://ictclas.org/
[2] http://py.kdd.cc/

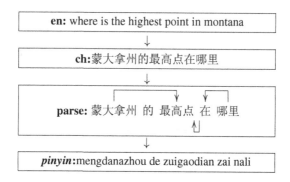

Fig. 2. Transforming the English sentence to Chinese *pinyin* string.

(lambda $0 e(loc:t (argmax $1 (and (place:t $1)(loc:t $1 montana:s))(elevation:i $1)) $0))
pinyin: mengdanazhou de zuigaodian zai nali

Fig. 3. Lambda-calculus form and its Chinese *pinyin* string in the Chinese Geo880 dataset

mianji zui da de zhou de zuigaodian shi nali
(argmax $0 (and (place:t $0) (loc:t $0 (argmax $1 (state:t $1) (area:i $1))))) (elevation:i $0))

youtazhou de shoufu shi nali
(capital:c utah:s)

meiguo neige zhou zui xiao
(argmin $0 (and (state:t $0) (loc:t $0 usa:co)) (size:i $0))

luodedaozhou jingnei de zhuyao chengshi you neixie
(lambda $0 e (and (major:t $0) (city:t $0) (loc:t $0 rhode_island:s)))

meiguo de zuigaodian shi nali
(argmax $0 (and (place:t $0) (loc:t $0 usa:co)) (elevation:i $0))

neihuadazhou de zuigaodian you duoshao mi
(argmax $0 (and (place:t $0) (loc:t $0 nevada:s)) (elevation:i $0))

misulihe you duo chang
(len:i missouri_river:r)

neixie zhou you mingwei aositing de chengshi
(lambda $0 e (and (state:t $0) (exists $1 (and (city:t $1) (named:t $1 austin:n) (loc:t $1 $0)))))

Fig. 4. Example with lambda-calculus meaning representations in the Chinese Geo880 dataset

mianji zui da de zhou de zuigaodian shi nali
(answer (highest (place (loc_2 (largest_one (area_1 (state all:e)))))))

youtazhou de shoufu shi nali
(answer (capital (loc_2 (stateid utah:e))))

meiguo neige zhou zui xiao
(answer (smallest (state (loc_2 (countryid usa:e)))))

luodedaozhou jingnei de zhuyao chengshi you neixie
(answer (major (city (loc_2 (stateid rhode_island:e)))))

meiguo de zuigaodian shi nali
(answer (highest (place (loc_2 (countryid usa:e)))))

neihuadazhou de zuigaodian you duoshao mi
(answer (highest (place (loc_2 (stateid nevada:e)))))

misulihe you duo chang
(answer (len (riverid missouri:e)))

neixie zhou you mingwei aositing de chengshi
(answer (state (loc_1 (city (cityid austin:e _:e)))))

Fig. 5. Example with variable-free meaning representations in the Chinese Geo880 dataset

Next, we begin to include a set of lexical features.We initialize the weight of the lexical features using the coocurrance statistics estimated with the Giza++ implementation of IBM model 1 [8]. The translation scores for (Chinese word, constant) pairs that cooccur in examples of the Chinese training data can be computed. The weights of the seed lexical entry in Λ_{NP} are set to 10 which are equivalent to the highest possible coocurrance score. Besides, we initialize the weights of the semantic features to zero. Following Zettlemoyer & Collins and Kwiatkowski et al., we use the same learning rate $\alpha_0 = 1.0$ and cooling rate $c = 10^{-5}$ in all training examples and run the algorithm for $T = 100$ iterations. Thereby, such values are selected with cross validation on the Geo880 development set.

5   Experiment

Evaluation. We report the results of Chinese Geo880 corpus for two different representations which use the standard measures of recall, precision and F1. Recall is the percentage of test sentences assigned correct logical forms, precision is the percentage of logical forms returned that are correct, and *F1* is the harmonic mean of precision and recall. Besides, we also report the results with a two-pass parsing strategy (UBL-s) to investigate the trade-off between precision and recall. That is, if the parser fails to return an analysis for a test sentence due to novel words or usage, the sentence will be reparsed and the parser will be allowed to skip such words with a fixed cost. Accordingly, the skipping words may potentially increase recall.

Results. Table 1, 2, 3 and 4 show the results for both Chinese Geo250 and Geo880 datasets with two different logical expressions. Table 1 and 2 compare the performance of Chinese Geo250 with ones of English, Spanish, Japanese and Turkish. For lambda-calculus expression, Japanese achieves the highest recall while Chinese achieves the best precision and *F1*. For variable-free logical form, Japanese obtains the highest recall again whereas English instead of Chinese achieves the best precision and *F1*. From Table 3, it indicates that the performances on our Chinese Geo880 slightly outperforms the ones on English with lambda-calculus expression. Table 4 shows that English Geo880 corpus achieves the better results compared with our Chinese Geo880 corpus with variable-free meaning representations. From these four tables, it can be seen that Chinese is easier to be learned the lambda-calculus expressions than variable-free ones. More important, is that for lambda-calculus logical form Chinese is also easier to be learned than other languages such as English, Spanish and Turkish as well as Japanese.

Table 1. Performance across languages on Geo250 with lambda-calculus meaning representations

Language	UBL (%)			UBL-s (%)		
	Rec.	Pre.	F1	Rec.	Pre.	F1
English	78.0	93.2	84.7	81.8	83.5	82.6
Spanish	75.9	93.4	83.6	81.4	83.4	82.4
Turkish	67.4	93.4	78.1	71.8	77.8	74.6
Japanese	78.9	90.9	84.4	83.0	83.2	83.1
Chinese	75.8	98.5	85.6	81.9	86.6	84.1

Table 2. Performance across languages on Geo250 with variable-free meaning representations

Language	UBL (%)			UBL-s (%)		
	Rec.	Pre.	F1	Rec.	Pre.	F1
English	78.1	88.2	82.7	80.4	80.8	80.6
Spanish	76.8	86.8	81.4	79.7	80.6	80.1
Turkish	70.4	89.4	78.6	74.2	75.6	74.9
Japanese	78.5	85.5	81.8	80.5	80.6	80.5
Chinese	75.7	87.0	80.8	78.6	79.3	78.9

Table 3. Performance between English and Chinese on Geo880 with lambda-calculus meaning representations

	UBL (%)			UBL-s (%)		
Language	Rec.	Pre.	F1	Rec.	Pre.	F1
English	85.0	94.1	89.3	87.9	88.5	88.2
Chinese	82.6	98.8	90.0	88.1	90.8	89.4

Table 4. Performance between English and Chinese on Geo880 with variable-free meaning representations

	UBL (%)			UBL-s (%)		
Language	Rec.	Pre.	F1	Rec.	Pre.	F1
English	81.4	89.4	85.2	84.3	85.2	84.7
Chinese	78.2	88.2	82.9	82.0	84.0	83.0

6   Conclusion and Future Work

In this paper, we addressed the problem of learning to map Chinese sentences to logical forms. We built a corresponding Chinese Geo880 corpus as Zettlemoyer & Collins and Kwiatkowski et al. provided Geo880 dataset. At the same time, we employed the UBL algorithm to learn the training examples and to test the testing set. In the experiments, we showed that the performance on Chinese language with two different meaning representations can work well as other languages such as English, Spanish and Turkish as well as Japanese and achieve the high accuracy.

For the future work, we will focus on extending the UBL approach to many other eastern languages or applying it to a range of larger Chinese dataset. Larger Chinese dataset should improve the recall performance, and allow us to develop a more comprehensive set of rules and to create a robust intelligent system. Besides, we are also interested in designing similar grammar induction techniques for multiple language context-dependent understanding problems, such as Spanish, Turkish, Chinese and Japanese.

References

1. Zhang, H.-P., Yu, H.-K., Xiong, D.-Y., Liu, Q.: HHMM-based Chinese Lexical Analyzer ICTCLAS. In: Proceedings of the Second SIGHAN Workshop Affiliated with 41th ACL, Sapporo Japan, pp. 184–187 (2003)
2. Bos, J., Clark, S., Steedman, M., Curran, J.R., Hockenmaier, J.: Wide-coverage Semantic Representations from a CCG Parser. In: Proceedings of the International Conference of the Association for Computational Linguistics (2004)
3. Buszkowski, W., Penn, G.: Categorial Grammars Determined from Linguistic Data by Unification. Studia Logica 49, 431–454 (1990)

4. Carpenter, B.: Type-logical Semantics. The MIT Press (1997)
5. Clark, S., Curran, J.R.: Wide-coverage Efficient Statistical Parsing with CCG and Log-linear Models. Computational Linguistics 33(4), 493–552 (2007)
6. Huet, G.: A Unification Algorithm for Typed λ-calculus. Theoretical Computer Science 1, 27–57 (1975)
7. Miller, S., Stallard, D., Bobrow, R.J., Schwartz, R.L.: A Fully Statistical Approach to Natural Language Interfaces. In: Proceedings of the Association for Computational Linguistics (1996)
8. Och, F.J., Ney, H.: A Systematic Comparison of Various Statistical Alignment Models. Computational Linguistics 29(1), 19–51 (2003)
9. Steedman, M.: Surface Structure and Interpretation. The MIT Press (1996)
10. Steedman, M.: The Synactic Process. The MIT Press (2000)
11. Zelle, J.M., Mooney, R.J.: Learning to Parse Database Queries Using Inductive Logic Programming. In: Proceedings of the National Conference on Artificial Intelligence (1996)
12. Thompson, C.A., Mooney, R.J.: Acquireing Word-meaning Mappings for Natural Language Interfaces. Artificial Intelligence Research (2002)
13. Kate, R.J., Wong, Y.W., Mooney, R.J.: Learning to Transform Natural to Formal Languages. In: Proceedings of the National Conference on Artifical Intelligence (2005)
14. Kate, R.J., Mooney, R.J.: Using String-kernels for Learning Semantic Parsers. In: Proceedings of the 44th Annual Meeting of the Association for Computational Linguistics (2006)
15. Wong, Y.W., Mooney, R.J.: Learning for Semantic Parsing with Statistical Machine Translation. In: Proceedings of the Human Language Technology Conference of the NAACL (2006)
16. Wong, Y.W., Mooney, R.J.: Learning Synchronous Grammars for Semantic Parsing with Lambda Calculus. In: Proceedings of the Association for Computational Linguistics (2007)
17. Zettlemoyer, L.S., Collins, M.: Learning to Map Sentences to Logical Form: Structured Classification with Probabilistic Categorial Grammars. In: Proceedings of the Conference on Uncertainty in Artificial Intelligence (2005)
18. Zettlemoyer, L.S., Collins, M.: Online Learning of Relaxed CCG Grammars for Parsing to Logical Form. In: Proceedings of the Joint Conference on Empirical Methods in Natural Language Processing and Computational Natuaral Language Learning (2007)
19. Zettlemoyer, L.S., Collins, M.: Learning Context-dependent Mapping from Sentences to Logical Form. In: Proceedings of the Joint Conference on of the Association for Computational Linguistics and International Joint Conference on Natural Language Processing (2009)
20. Tom, K., Luke, Z., Sharon, G., Mark, S.: Inducing Probabilistic CCG Grammars from Logical Form with Higher-order Unification. In: Proceedings of the Conference on Empirical Methods in Natural Language Processing (EMNLP), Cambridge, MA (2010)
21. Tom, K., Luke, Z., Sharon, G., Mark, S.: Lexical Generalization in CCG Grammar Induction for Semantic Parsing. In: Proceedings of the Conference on Empirical Methods in Natural Language Processing (EMNLP), Edinburgh, UK (2011)
22. Lu, W., Hwee, T.N., Wee, S.L., Luke, Z.: A Generative Model for Parsing Natural Language to Meaning Representations. In: Proceedings of the Conference on Empirical Methods in Natural Language Processing (EMNLP), Honolulu, Hawaii, USA (2008)
23. Lu, W., Hwee, T.N.: A Probabilistic Forest-to-String Model for Language Generation from Typed Lambda Calculus Expressions. In: Proceedings of the Conference on Empirical Methods in Natural Language Processing (EMNLP), Edinburgh, Scotland, UK (2011)

Representation and Verification of Attribute Knowledge

Chunxia Zhang[1], Zhendong Niu[2], Chongyang Shi[2], Mengdi Tan[1],
Hongping Fu[2], and Sheng Xu[1]

[1] School of Software, Beijing Institute of Technology, Beijing, China
[2] School of Computer Science and Technology, Beijing Institute of Technology, Beijing, China
{cxzhang,zniu,cy_shi,mdtan,fhongping,xusheng}@bit.edu.cn

Abstract. With the increasing growth and popularization of the Internet, knowledge extraction from the web is an important issue in the fields of web mining, ontology engineering and intelligent information processing. The availability of real big corpora and the development of technologies of internet network and machine learning make it feasible to acquire massive knowledge from the web. In addition, many web-based encyclopedias such as Wikipedia and Baidu Baike include much structured knowledge. However, knowledge qualities including the incorrectness, inconsistency, and incompleteness become a serious obstacle for the wide practical applications of those extracted and structured knowledge. In this paper, we build a taxonomy of relations between attributes of concepts, and propose a taxonomy of attribute relations driven approach to evaluating the knowledge about attribute values of attributes of entities. We also address an application of our approach to building and verifying attribute knowledge of entities in different domains.

Keywords: Taxonomy of attribute relations, attribute values, knowledge verification, ontology verification.

1 Introduction

With the increasing growth and popularization of the Internet, the problem of how to extract knowledge from the World Wide Web is an important issue in the fields of web mining, ontology engineering, intelligent information processing, and question answering. The availability of real big corpora and the development of technologies of internet network and machine learning make it feasible to acquire massive knowledge from the web. Moreover, many web-based encyclopedias such as Wikipedia and Baidu Baike comprise a great deal of structured knowledge.

However, knowledge qualities including the inconsistency and incompleteness become a serious obstacle for the wide practical applications, such as information retrieval, of those extracted and structured knowledge. Hence, this paper focuses on how to verify the inconsistency and incompleteness of knowledge about attribute values of attributes of entities. Verification about attribute values of entities is also an indispensable part of ontology evaluation.

To the verification of knowledge bases, the present verification techniques consist of structural testing approaches and functional testing approaches. The former kind of

M. Wang (Ed.): KSEM 2013, LNAI 8041, pp. 473–482, 2013.

approaches mainly includes decision tables based methods, decision trees based methods, logical methods, and graph-oriented methods. The latter kind of approaches comprises machine learning methods, relational methods, and refinement methods [1-3]. The detected primary anomalies of knowledge bases contain inconsistency, redundancy, deficiency or incompleteness.

As to the ontology verification, present approaches can be classified into the following methods: (1) rule-based methods. They use rules to detect errors in ontologies [4]. (2) Metric-based methods. Lozana-Tello et al. [5] built a hierarchical framework OntoMetric which provides a qualitative analysis to the quality and suitability of ontologies. (3) A linguistics-based method. This method is to calculate qualitative measures about occurrence features of components of ontologies in corpora to evaluate ontologies [6]. (4) Application or task based methods. They are to evaluate ontologies according to the capability of solving practical problems which ontoloiges are used to handle [7, 8]. (5) A gold standard methods. It is to compare a built ontology with a standard ontology. The contents of ontology verification include concepts, attributes, structures, contexts, and applications [8].

Most verifying approaches of knowledge-based systems are strongly depend on knowledge representation languages of knowledge bases [1], hence it is difficult for those approaches to evaluate large scale heterogeneous knowledge. In this paper, we build a taxonomy of relations between attributes of concepts, and propose a taxonomy of attribute relations driven approach to evaluating the knowledge about attribute values of entities in different domains.

The main contributions of this paper are given as follows. (1) On the aspect of ontology construction, we built a taxonomy or classification framework of attribute relations. (2) On the aspect of knowledge and ontology verification, we propose a method based on the taxonomy of attribute relations to evaluate the inconsistency and incompleteness of knowledge about attribute values of entities. Our approach has nothing to do with the representation languages and sources of attribute knowledge, and is independent of domains and concepts to which entities belong.

The rest of the paper is organized as follows. Section 2 introduces the taxonomy of attribute relations. The verification method of attribute values and its application are presented in section 3. Section 4 concludes the paper and gives future works.

2   Taxonomy of Relations between Attributes

Ontologies mainly consist of concepts, instances, attributes, and relationships. A concept can be defined by its extension or intension in ontologies. The extension of a concept is the set of all instances of this concept. The intension of a concept is the set of properties which all instances of this concept possess. Attributes and attribute values of a concept specify the intension of this concept. This section is intended to build a taxonomy of relations between attributes of concepts. We use the first-order predicate calculus as the representation language of attribute knowledge.

In this paper, we use c for concepts, x for instances, a for attributes, and V for attribute value sets. An attribute knowledge representation framework consists of

(1) Two predicates Instanceof(x, c) and Valueof(x, a, V) to denote that x is an instance of c, and V is an attribute value set of a of x. The predicate Belongto(v, V) to describe that v is an element (i.e., an attribute   value) of the set V.

(2) Predicates Equal(v_1,v_2), Equivalent(v_1,v_2), Include(v_1,v_2) and Imply(v_1,v_2) to represent that (a) v_1 and v_2 have the same strings; (b) v_1 and v_2 have different strings but bear the same meanings; (c) v_1 includes v_2; (d) v_1 implies v_2. These predicates represent four types of relations between attribute values: equal relations, equivalent relations, inclusion relations and implication relations.

For instance, Instanceof(China, Country) and Valueof(China, Establishment Time, {A.D.1949}) means that China is an instance of "Country", and the value of the attribute "Establishment Time" of China is "A.D.1949". An attribute value set V may contain multiple elements or attribute values.

As illustrations, (a) for predicates Valueof(Nokia 1280, Shape,{straight plate}) and Valueof(Nokia 1280, Appearance,{straight plate}), we have Equal(straight plate, straight plate). (b) To predicates Valueof(Dairy of A Madman, Author, {Luxun}) and Valueof(Dairy of A Madman, Author, {Shuren Zhou}), Equivalent(Luxun, Shuren Zhou) is true. (c) For Valueof(Antirrhinum jajus, Distribution Area, {France, Portugal, Turkey, Morocco, Lyon}), we have Include(France, Lyon). (d) To Valueof(Luxun, Age, {Fifty-five years old}) and Valueof(Luxun, Age, {More than fifty years old}), Imply(Fifty-five years old, More than fifty years old) holds.

Based on relations between attribute values, relations between attributes can be divided into equivalence relations, inheritance relations, inclusion relations, implication relations, and antonymous relations, as shown in Fig.1.

Definition 1. For a concept c, its two attributes a_1 and a_2, if the formula (1) holds, and f is a one-to-one mapping from V_1 to V_2, then we say the relation between a_1 and a_2 is an equivalent relation, written as AttEquivalent(c, a_1, a_2).

$$AttEquivalent(c,a_1,a_2) \Leftrightarrow \forall x \forall V_1 \forall V_2 \forall v_1 \forall v_2 (Instanceof(x,c)$$
$$\wedge Valueof(x,a_1,V_1) \wedge Valueof(x,a_2,V_2) \wedge Belongto(v_1,V_1) \qquad (1)$$
$$\wedge Belongto(v_2,V_2) \rightarrow (f(v_1)=v_2) \wedge (Equal(v_1,v_2) \vee Equivalent(v_1,v_2)))$$

For instance, the attribute "*Shape*" of the concept "Phone" is equivalent to "*Appearance*" of this concept, that is, AttEquivalent(Phone, Shape, Appearance).

Definition 2. For a concept c, its two attributes a_1 and a_2, if the formula (2) holds, then we say that a_2 inherits a_1, written as AttInherit(c, a_2, a_1), a_1 is called a super-attribute, a_2 is called a sub-attribute.

$$AttInherit(c,a_1,a_2) \Leftrightarrow \forall x \forall V_1 \forall V_2 \forall v (Instanceof(x,c)$$
$$\wedge Valueof(x,a_1,V_1) \wedge Valueof(x,a_2,V_2) \wedge Belongto(v,V_2) \qquad (2)$$
$$\rightarrow Belongto(v,V_1) \vee \exists w(Belongto(w,V_1) \wedge Equivalent(w,v)))$$

For example, the attribute "*The Largest City*" of "Country" inherits "*City*", that is, AttInherit(Country, The Largest City, City), because if v is an attribute value of "The Largest City" of a country, then v is an attribute value of "City" of this country.

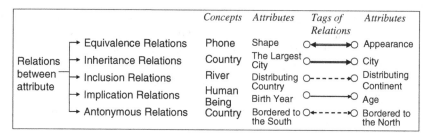

Fig. 1. Five types of relations between attributes and their examples

Definition 3. For a concept c, its two attributes a_1 and a_2, if the formula (3) holds, then we say that a_1 includes a_2, written as AttInclude(c, a_1, a_2), a_1 is called a whole-attribute, a_2 is called a part-attribute.

$$AttInclude(c, a_1, a_2) \Leftrightarrow \forall x \forall V_1 \forall V_2 \forall v (Instanceof(x, c)$$
$$\wedge Valueof(x, a_1, V_1) \wedge Valueof(x, a_2, V_2) \wedge Belongto(v, V_2)$$
$$\rightarrow \exists w(Belongto(w, V_1) \wedge Include(w, v)))$$
$$\wedge \forall s(Belongto(s, V_1) \rightarrow \exists t(Belong(t, V_2) \wedge Include(s, t)))$$

(3)

For instance, the attribute a_1, i.e., "*Distributing Continent*" of the concept c, i.e., "River" includes its attribute a_2, i.e., "*Distributing Country*", that is, AttInclude(c, a_1, a_2). As an illustration, the value of a_1 of the river "Changjiang" is "Asia", the value of a_2 of this river is "China", then Asia includes China.

Definition 4. For a concept c, and its two attributes a_1 and a_2, if the formula (4) is true, then we say that a_1 implies a_2, written as AttImply(c, a_1, a_2), and a_1 is called an antecedent attribute, and a_2 is called a consequent attribute.

$$AttImply(c, a_1, a_2) \Leftrightarrow \forall x \forall V_1 \forall V_2 \forall v (Instanceof(x, c)$$
$$\wedge Valueof(x, a_1, V_1) \wedge Valueof(x, a_2, V_2) \wedge Belongto(v, V_2)$$
$$\rightarrow \exists w(Belongto(w, V_1) \wedge Imply(w, v)))$$

(4)

For example, the attribute "*Birth Year*" of the concept "Human Being" implies its attribute "*Age*", that is, AttImply(Human Being, Birth Year, Age), because the age of any person can be calculated according to his or her birth year.

Definition 5. For a concept c, its two attributes a_1 and a_2, if the formula (5) holds, then we say that a_1 is antonymous to a_2, written as AttAntonymous(c, a_1, a_2).

$$AttAntonymous(c, a_1, a_2) \Leftrightarrow \forall x_1 \forall x_2 \forall V (Instanceof(x_1, c)$$
$$\wedge Instanceof(x_2, c) \wedge Valueof(x_1, a_1, V)$$
$$\wedge Belongto(x_2, V) \rightarrow Valueof(x_2, a_2, \{x_1\}))$$

(5)

For instance, the attribute "*Bordered to the north*" of the concept "Country" is antonymous to its attribute "*Bordered to the south*", that is, AttAntonymous(Country, Bordered to the north, Bordered to the south).

3   Verification of Attribute Values

3.1   Knowledge Verification about Single Attribute Relations

This subsection presents how to validate the consistency and completeness of attribute values whose attributes are involved in a single kind of attribute relations. Table 1 illustrates types of errors about attribute values and the verification approaches.

Table 1. Different types of errors and their verification approaches

	Types of relations		Types of Errors	Verification Methods
Inconsistency errors	Equivalence relation	(1)	Error about the number of attribute values	Theorem 1
		(2)	Error of mapping between attribute value sets of equivalent attributes	Definition 1
	Inheritance relation	(3)	Error about the number of attribute values	Theorem 2
	Inclusion relation	(4)	Error about the number of attribute values	Theorem 3
		(5)	Error of mapping between attribute value sets of attributes with inclusion relations	Theorem 3
	Implication relation	(6)	Error of implication between attribute values of antecedent attributes and consequent attributes	Definition 4
	Antonymous relation	(7)	Error of attribute values of attributes with antonymous relations	Theorem 5
Incompleteness errors	Equivalence relation	(8)	Deficiency of attribute values	Theorem 1
	Inheritance relation	(9)	Deficiency of attribute values of super-attributes	Definition 2
	Inclusion relation	(10)	Deficiency of attribute values of whole-attributes and part-attributes	Definition 3
	Implication relation	(11)	Deficiency of attribute values of antecedent attributes	Definition 4
	Antonymous relation	(12)	Deficiency of attribute values of attributes with antonymous relations	Theorem 5

Theorem 1. For a concept c, its two attributes a_1 and a_2, if AttEquivalent(c, a_1, a_2) is true, then the formula (6) holds. Here, EqualNumber(V_1, V_2) means that the number of elements of V_1 is equal to that of V_2.

$$\forall x \forall V_1 \forall V_2 (AttEquivalent(c, a_1, a_2) \wedge Instanceof(x, c) \tag{6}$$
$$\wedge Valueof(x, a_1, V_1) \wedge Valueof(x, a_2, V_2) \rightarrow EqualNumber(V_1, V_2))$$

Proof. According to definition 1, there is a one-to-one mapping f from V_1 to V_2. Hence, we can obtain that EqualNumber(V_1, V_2) is true.

To two equivalent attributes a_1 and a_2, theorem 1 and definition 1 can ensure that the Error (1), (2), and (8) in table 1 don't happen. (a) Error (1), i.e., *"Error about the number of attribute values"*, of the inconsistency error occurs when EqualNumber(V_1, V_2) does not hold. (b) Error (2), i.e., *"Error of mapping between attribute value sets of equivalent attributes"*, of the inconsistency error happens while there does not exist a one-to-one mapping from V_1 to V_2. (c) Error (8), i.e., "*Deficiency of attribute values*", of the incompleteness error encounters when $\mathsf{EqualNumber}(V_1, V_2)$ is not true, that is, there are some attribute values missing in V_1 or V_2.

Theorem 2. For a concept c, its two attributes a_1 and a_2, if $\mathsf{AttInherit}(c, a_1, a_2)$ hold, then the formula (7) is true. Here, $\mathsf{NotGreaterNumber}(V_1, V_2)$ shows that the number of elements of V_1 is not greater than that of V_2.

$$\forall x \forall V_1 \forall V_2 (AttInherit(c, a_1, a_2) \wedge Instanceof(x, c)$$
$$\wedge Valueof(x, a_1, V_1) \wedge Valueof(x, a_2, V_2) \rightarrow NotGreaterNumber(V_1, V_2)) \tag{7}$$

Proof. Based on definition 2, we can infer that for any v in V_1, there is at least an element in V_2 which is equal or equivalent to v. Therefore, we can conclude that the number of elements of V_1 is same as or smaller than that of V_2.

For a_1 and a_2 with the inheritance relation, if their attribute values satisfy theorem 2, then the Error (3) and (9) in table 1 can be avoided. (a) Error (3), i.e., "*Error about the number of attribute values*", is made when $\mathsf{NotGreaterNumber}(V_1, V_2)$ is not true. (b) Error (9), i.e., "*Deficiency of attribute values of super-attributes*", encounters when there exist an element v in V_1, for any w in V_2, w is not equal or equivalent to v.

Theorem 3. For a concept c, its two attributes a_1 and a_2, if $\mathsf{AttInclude}(c, a_1, a_2)$ is true, then the formula (8) holds, and there exists a one-to-many mapping of including relations between values from V_1 to V_2.

$$\forall x \forall V_1 \forall V_2 (AttInclude(c, a_1, a_2) \wedge Instanceof(x, c) \wedge Valueof(x, a_1, V_1)$$
$$\wedge Valueof(x, a_2, V_2) \rightarrow NotGreaterNumber(V_1, V_2)) \tag{8}$$

Proof. Based on definition 3, for any element v in V_2, there is at least one element in V_1 which includes v; for any element in w in V_1, there is at least one element in V_2 which is included by w. In addition, since attribute values in V_1 do not have the same parts, these values include different elements in V_2. Hence, there is a one-to-many mapping from V_1 to V_2. Hence, $\mathsf{NotGreaterNumber}(V_1, V_2)$ is true.

To attributes a_1 and a_2, a_1 includes a_2, their values can avert the Error (4), (5) and (10) in table 1, if they meet theorem 3 and definition 3. (a) Error (4), i.e., "*Error about the number of attribute values*", happens when $\mathsf{NotGreaterNumber}$ (V_1, V_2) does not hold. (b) Error (5), i.e., "*Error of mapping between attribute value sets of attributes with inclusion relations*", encounters while there does not exist a one-to-many mapping from V_1 to V_2. (c) Error (10), i.e., "*Deficiency of attribute values of whole-attributes and part-attributes*", occurs when there is an attribute value of a_2 which is not included by any attribute value of a_1, and vice versa.

Theorem 4. For a concept c, its two attributes a_1 and a_2, if $\mathsf{AttImply}(c, a_1, a_2)$ is true, then the formula (9) holds, where $\mathsf{NonNull}(V)$ means that V is a non-null set.

$$\forall x \forall V_1 \forall V_2 (AttImply(c, a_1, a_2) \wedge Instanceof(x, c) \wedge Valueof(x, a_1, V_1)$$
$$\wedge Valueof(x, a_2, V_2) \wedge NonNull(V_2) \rightarrow NonNull(V_1)) \tag{9}$$

Proof. For any v in V_2, there exists at least an element in V_1 which implies v, based on definition 4. Therefore, if V_2 is a non-null set, then V_1 is also a non-null set.

For attributes a_1 and a_2, a_1 implies a_2, if their attribute values satisfy theorem 4 and definition 4, then the Error (6) and (11) in table 1 can be avoided. (a) Error (6), i.e., *"Error of implication between attribute values of antecedent attributes and consequent attributes"*, encounters when the values of a_2 are wrong. (b) Error (11), i.e., *"Deficiency of attribute values of antecedent attributes"*, occurs when there exists an attribute value of a_2 which is not implied by any attribute value of a_1, or when Non-Null$(V_2) \wedge \neg$NonNull(V_1) holds.

Theorem 5. For a concept c, its two attributes a_1 and a_2, if AttAntonymous(c, a_1, a_2) is true, then the formula (9) holds.

$$\forall x_1 \forall x_2 (AttAntonymous(c, a_1, a_2) \wedge Instanceof(x_1, c)$$
$$\wedge Instanceof(x_2, c) \wedge Valueof(x_1, a_1, \{x_2\}) \rightarrow Valueof(x_2, a_2, \{x_1\})) \tag{10}$$

Proof. According to the condition, we obtain that Valueof$(x_1, a_1, \{x_2\})$ holds. Hence, if Valueof(x_1, a_1, V) is true, then x_2 belongs to V. Further, we can infer that Valueof$(x_2, a_2, \{x_1\})$ holds based on definition 5.

To two attributes with the antonymous relation, their attribute values can refrain from the Error (7) and (12) in table 1, if they meet theorem 5. (a) Error (7), i.e., *"Error of attribute values of attributes with antonymous relations"*, happens when Valueof$(x_1, a_1, \{x_2\}) \wedge \neg$Valueof$(x_2, a_2, \{x_1\})$ holds. (b) Error (12), i.e., *"Deficiency of attribute values of attributes with antonymous relations"*, occurs while there exist an attribute value v of a_1 of x_1, the attribute value set of a_2 of v does not include x_1.

3.2   Knowledge Verification about Hybrid Attribute Relations

This subsection discusses how to verify attribute values whose attributes are involved in multiple kinds of relations between attributes of concepts.

Theorem 6. For a concept c, and its any four attributes a_1, a_2, a_3, and a_4, we have formulae (11) and (12) which are two properties about equivalent and inheritance relations between attributes. Analogously, we can obtain similar properties about equivalent and inclusion relations, properties about equivalent and implication relations, and properties about equivalent and antonymous relations.

$$\forall a_1 \forall a_2 \forall a_3 (AttEquivalent(c, a_1, a_2) \wedge AttInherit(c, a_1, a_3) \rightarrow AttInherit(c, a_2, a_3)) \tag{11}$$

$$\forall a_1 \forall a_3 \forall a_4 (AttEquivalent(c, a_3, a_4) \wedge AttInherit(c, a_1, a_3) \rightarrow AttInherit(c, a_1, a_4)) \tag{12}$$

Proof. For an instance x of c, let V_1, V_2, V_3 are attribute value sets of a_1, a_2, and a_3 of x, respectively. For any element v in V_2, since AttEquivalent(c, a_1, a_2) is true, there exists an element u in V_1 which is equal or equivalent to v. Further, AttInherit(c, a_1, a_3) holds, so there is at least an element w in V_3 which is equal or equivalent to u. Hence, for any v in V_2, there is at least the element w in V_3 which is equal or equivalent to v. Based on definition 2, we can infer that AttInherit(c, a_2, a_3) holds. The property of formula (12) and other properties can be proved in analogous way.

To multiple attributes with equivalence and inheritance relations, the incompleteness error "*Deficiency of inheritance relation*" occurs when AttInherit(c, a_2, a_3) or AttInherit(c, a_1, a_4) does not exist in the knowledge base. This error can be avoided by theorem 6. Other inconsistent and incompleteness errors are same as Error (3) and (9) in table 1. Attributes with equivalent and inclusion relations, or with equivalent and implication relations, or with equivalent and antonymous relations have the analogous errors and the corresponding checking methods.

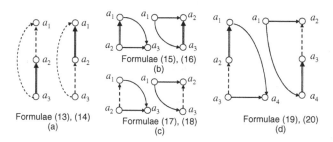

Fig. 2. Properties about different kinds of relations between attributes

Theorem 7. For a concept c, and its any three attributes a_1, a_2 and a_3, we have the formulae (13) and (14), which are two properties about inheritance relations and inclusion relations, as shown in Fig.2(a).

$$\forall a_1 \forall a_2 \forall a_3 (AttInclude(c,a_1,a_2) \wedge AttInherit(c,a_3,a_2) \rightarrow AttInclude(c,a_1,a_3)) \quad (13)$$

$$\forall a_1 \forall a_2 \forall a_3 (AttInherit(c,a_2,a_1) \wedge AttInclude(c,a_2,a_3) \rightarrow AttInclude(c,a_1,a_3)) \quad (14)$$

Proof. For an instance x of c, let V_1, V_2, V_3 are attribute value sets of a_1, a_2, and a_3 of x, respectively. For any w in V_3, since AttInherit(c, a_3, a_2) holds, there is at least an element v in V_2 which is equal or equivalent to w. Further, AttInclude(c, a_1, a_2) holds, hence there exists at least an element u in V_1 which includes v. Therefore, for any w in V_3, there is at least u in V_1 which includes w. According to definition 3, we can infer that AttInclude(c, a_1, a_3) is true. Formula (14) can be proved in similar way.

Theorem 8. For a concept c, and its any three attributes a_1, a_2 and a_3, we have formulae (15) and (16), which are two properties about inheritance and implication relations (shown in Fig.2(b)). Moreover, we deduce the formulae (17) and (18) which are two properties about inclusion and implication relations (shown in Fig.2(c)).

$$\forall a_1 \forall a_2 \forall a_3 (AttInherit(c,a_2,a_1) \wedge AttImply(c,a_2,a_3) \rightarrow AttImply(c,a_1,a_3)) \quad (15)$$

$$\forall a_1 \forall a_2 \forall a_3 (AttImply(c,a_1,a_2) \wedge AttInherit(c,a_3,a_2) \rightarrow AttImply(c,a_1,a_3)) \quad (16)$$

$$\forall a_1 \forall a_2 \forall a_3 (AttInclude(c,a_1,a_2) \wedge AttImply(c,a_2,a_3) \rightarrow AttImply(c,a_1,a_3)) \quad (17)$$

$$\forall a_1 \forall a_2 \forall a_3 (AttImply(c,a_1,a_2) \wedge AttInclude(c,a_2,a_3) \rightarrow AttImply(c,a_1,a_3)) \quad (18)$$

Proof. For any w in V_3, because AttImply(c, a_2, a_3) holds, there is at least an element v in V_2 which implies w. Further, AttInherit(c, a_2, a_1) is true, so there is at least an element u in V_1 which is equal or equivalent to v. Hence, for any w in V_3, there is at least the element u in V_1 which implies w. Based on definition 4, we can deduce that AttImply(c, a_1, a_3) holds. Formulae (16), (17) and (18) can be proved in similar way.

For multiple attributes with inheritance and inclusion relations, the incompleteness error *"Deficiency of inclusion relation"* happens while AttInclude(c, a_1, a_3) could not be found in the knowledge base. This error can be averted by theorem 7. Other inconsistent and incompleteness errors are same as Error (4), (5) and (10) in table 1. Attributes with inheritance and implication relations, or with inclusion and implication relations, have the analogous errors and the corresponding resolving methods. To attributes with inheritance, inclusion, and implication relations, we can deduce the property of theorem 9 based on theorem 8.

Theorem 9. For a concept c, and its any four attributes a_1, a_2, a_3 and a_4, we have the formulae (19) and (20), as shown in Fig.3(d).

$$\forall a_1 \forall a_2 \forall a_3 (AttInherit(c,a_2,a_1) \wedge AttInclude(c,a_2,a_3)$$
$$\wedge AttImply(c,a_3,a_4) \rightarrow AttImply(c,a_1,a_4)) \quad (19)$$

$$\forall a_1 \forall a_2 \forall a_3 (AttImply(c,a_1,a_2) \wedge AttInclude(c,a_2,a_3)$$
$$\wedge AttInherit(c,a_4,a_3) \rightarrow AttImply(c,a_1,a_4)) \quad (20)$$

To multiple attributes with inheritance, inclusion, and implication relations, the incompleteness error *"Deficiency of implication relation"* is made when AttImply(c, a_1, a_4) does not exist in the knowledge base. This error can be avoided by theorem 9. Other errors are same as Error (6) and (11) in table 1.

3.3   Building and Verification of Attribute Knowledge in Different Domains

We selected entities which are instances of eight concepts in different domains to detect errors of attribute values of those entities. Those concepts are *Country, City, Movie, Chinese Revolutionist, Web Site, Navy Warship, Archaeological Site*, and *Museum*. Attributes of the former four concepts and attribute values of entities of those concepts were downloaded from Wikipedia in Chinese, while those of the latter four concepts in Chinese were previously built by our team. We chose average about 100 entities and average about 40 attributes of the former four concepts, and average about 80 entities and average about 30 attributes of the latter four concepts. Our approach was used to evaluate attribute values of those attributes of those entities.

Below we will give some examples of attribute relations and errors found by our verification approach. (1) An example of equivalent relations is given as follows. To the concept c, i.e., "City", we found that there are five attributes equivalent to the attribute a, i.e., "Established Time of City(建城时间)". And we have AttEquivalent(c, a, Established

Time (建置时间)), AttEquivalent(*c*, *a*, Time of Set Up(设置时间)), AttEquivalent(*c*, *a*, Founded Time(成立时间)), AttEquivalent(*c*, *a*, The First Year of Establishment (设置始年)), and AttEquivalent(*c*, *a*, Establishing City(建城)).

(2) One inconsistent error about implication relations is given as follows. Attributes "Population" and "Area" of "City" imply the attribute "Population Density". To the city "Anqing", two predicates Valueof(Anqing, Population, {6,186,500}) and Valueof(Anqing, Area, {15,398 square kilometers}) are true. Thus, the value of "Population Density" of "Anqing" should be "401.8/km^2", not "344.9/km^2" in Wikipedia. This error is the Error (6) in table 1 which was found by definition 4.

(3) We will illustrate an incompleteness error about antonymous relations. In Wikeipedia, the predicate Valueof(Sun Yat-sen, Wife, {Dayuexun}) hold in the webpage of "Sun Yat-sen". Since the attribute "Wife" is antonymous to the attribute "Husband", the predicate Valueof(Dayuexun, Husband, {Sun Yat-sen}) is also true. However, this piece of knowledge missed in the webpage of "Dayuexun", and this error is the Error (12) and was checked by theorem 5.

4   Conclusion

Representation and verification of attribute knowledge is one of fundamental issues in the fields of knowledge and ontology verification. This paper has two points of focus. The first is that we build a classification framework or taxonomy of relations between attributes of concepts. The second is that we propose the taxonomy of attribute relations driven approach to verify attribute values of entities. An application in different domains shows the effectiveness of our approach. In the future, we will evaluate attribute values of multiple entities which are instances of different concepts.

Acknowledgments. This work was supported by the National Natural Science Foundation of China (61272361).

References

1. Tsai, W.T., Vishnuvajjala, R., Zhang, D.: Verification and Validation of Knowledge-Based Systems. IEEE Transactions on Knowledge and Data Engineering 11, 202–212 (1999)
2. Owoc, M.L., Ochmanska, M., Gladysz, T.: On Principles of Knowledge Validation. In: 5th European Symposium on Validation and Verification of Knowledge Based Systems, pp. 25–35 (1999)
3. Leemans, P., Treur, J., Willems, M.: A Semantical Perspective on Verifica-tion of Knowledge. Data and Knowledge Engineering 40, 33–70 (2002)
4. Arpinar, I.B., Giriloganathan, K., Aleman-Meza, B.: Ontology Quality by Detection of Conflicts in Metadata. In: 4th Workshop on Evaluation of Ontologies on the Web (2006)
5. Lozano-Tello, A., Gomez-Perez, A.: ONTOMETRIC: A Method Choose the Appropriate Ontology. Journal of Database Management 15, 1–18 (2004)
6. Gangemi, A., Catenacci, C., Ciaramita, M., Lehmann, J.: A Theoretical Framework for Ontology Evaluation and Validation. In: 2nd Italian Semantic Web Workshop on Semantic Web Applications and Perspectives (2005)
7. Song, D.: Research on Some Problems about Ontology Evaluation. Research on Library Science 35, 6–9 (2011) (in Chinese)
8. Porzel, R., Malaka, R.: A Task-based Approach for Ontology Evaluation. In: ECAI Workshop on Ontology Learning and Population (2004)

Knowledge and Ontology Engineering for Smart Services

Roberto Carlos dos Santos Pacheco[1], José Francisco Salm Junior[2],
Viviane Schneider[3], and Karine Koller[4]

[1] Federal University of Santa Catarina, Brazil
[2] University of Santa Catarina State, Brazil
[3] Program of Post-Graduate Knowledge Engineering and
Manager at Federal University of Santa Catarina, Brazil
[4] University of Santa Catarina State, Brazil
Modeling Subproject by FINEP / FAPESC, Brazil
Development in Institute Stela
Address: Rua Prof. Ayrton Roberto de Oliveira, 32, 7° andar - Itacorubi
Florianópolis - Santa Catarina - Brasil - CEP: 88034-050
{pacheco,viviane.sch,salm,karine}@stela.org.br

Abstract. This work points out different aspects of a methodology that is concerned with how hypermedia content can be shared in real time through ubiquitous network and in systems context-awareness, mediated by knowledge engineering. The software components that are shared within these environments all share non-functional requirements of high reliability, availability and security. Addition to techniques of knowledge engineering and ontology engineering, these non-functional requirements need to be addressed also by the methodology. Furthermore, that type of service application also should address other modeling aspects that are related to inter-organizational aspects like culture, power and relationship. These methodological aspects are described also in this work. The benefits of applying this approach is the added value for those who need to model not just the ubiquitous software application but also the knowledge engineering aspects that allow the Hypermedia content in to be share in an ubiquitous network.

1   Introduction

In 2006 during the Seminar called "The Grand Challenges for Computing Research in Brazil: 2006-2016"[1], researchers identified five challenges for the next years, in the Brazil's computing. This seminar was held in São Paulo and it was organized by the Brazilian Computer Society. At that time, the researcher discussed the following challenges:

---

[1] "*Grand Challenges in Computer Science Research in Brazil – 2006 – 201. Workshop Report – May 8-9, 2006*". Available in <
http://www.sbc.org.br/index.php?option=com_jdownloads&Itemid=1
95&task=finish&cid=12&catid=50>.

M. Wang (Ed.): KSEM 2013, LNAI 8041, pp. 483–492, 2013.

- information management in large volumes of distributed multimedia data;
- computational modeling for three kinds of systems: (i) complex and artificial; (ii) natural and socio-cultural; and (iii) human-nature interaction;
- impacts in computer science due to the transition from silicon to new technologies;
- universal and participatory access to knowledge for Brazilian citizens; and
- quality in technological development: available, accurate, secure, scalable, persistent and ubiquitous systems.

Besides pointing out these challenges, the researchers highlighted the importance of studies in ubiquitous information systems, digital convergence, and other developments in Information Technology and Communication (ITC). In this paper, the gathered results from our research brought together the methodological and conceptual components for a framework that assists activities for knowledge modeling related to ubiquitous system, whose need to be persistent, reliable, secure, and highly available. As research scenarios we have chosen use-cases for helping some standards for welfare.

In this context, in last two years, twelve research labs have cooperated in this STN environment, in a project financed by the Brazilian innovation agency FINEP[2], in cooperation with the state foundation in Santa Catarina (FAPESC[3]). This project is called "Information Technology and Communication Services for Large-Scale Multiplatform: Ubiquitous Systems in Service Quality of Life". It intends to address some of the main computer researches challenges in Brazil: ubiquitous systems, knowledge modeling and complex applied systems. In this context, the main goal of our work is to present a Framework for conduce development of a knowledge system, with characteristics cited above. The challenge for this research is to have a methodology that is wide-ranging enough to apply in different system scenarios, and didactic to add value to the developers. In the next section is introducing the next step of methodological framework.

2   State of Art

In order to develop such framework, it was required a combination of methods and tools from the following disciplines: (i) General System Theory; (ii) Knowledge Engineering; (iii) Artificial Intelligence; and (iv) Ontology Engineering.

This project started off, with some premises that are:

(i) System analysis and modeling cannot be based only on software engineering techniques. Software engineering technics can be combined from engineering practices in order to produce added results, in particular, the cultural and political aspects;

---

[2] FINEP - *Financiadora de Estudos e Projetos* (Financier of Studies and Projects). The FINEP finances our researches.
[3] FAPESC – *Fundação de amparo à pesquisa do Estado de Santa Catarina* (*Foundation for support research of Santa Catarina State*). FAPESC has a responsibility for to manage the resources.

(ii) Understanding the project structure, as a system requires a systematic approach to ontological and that can characterize the project itself as a system and relate it to others elements at different levels (super system and subsystem);

(iv) In particular, considering the project's focus on structuring advanced technologies, it is important that the analysis and modeling of systems represent also knowledge – commands and inferences that can be used for troubleshooting problems what needed knowledge intensive for itself solution.

In this perspective, the focus of state-of-art was to find techniques to make formal knowledge representation for it to be using in the smart service, which is connected with a ubiquitous system and hypermedia contents, distributed in the Web. Basically, the outcomes from state-of-art are the requirements from literature for create knowledge modeling (we used ontology for this) and to represent the environment that it will operate. Two academic studies supported this research: Doctorate work [1] and Master work [2], by postgraduate program – Knowledge Management and Engineering.

Next section presenting the outcomes from state-of-art which was elements key for compose the framework.

3   Smart Ubiquitous Service Framework (SUSF)

As illustrated Figure 1, we discovered four approaches for build the framework: **(i)** Approach for **systems**, from General System Theory – in this topic we look for guidelines in order to organize the system, by macro vision. Also, was developed one research [2], which treats about way to represent every knowledge system in a general format [3]; **(ii)** Approach for **environment** (context), from Knowledge Engineering – this topic are based also in studies [2] that explain about context modeling, which articulates a Systemic Vision [3]. For understand states more specifies from context, we used CommonKADS Methodology [4] and KAMMET II [5], in order to represent some context keys elements; **(iii)** approach to **formalize knowledge**, from Ontology Engineering - in this stage our aim was to represent computational knowledge. To do it we use some main references, which explain about Ontology Engineering, particularly NeOn Methodology [2, 6, 7]; **(iv)** approach for **Smart Web Service**, from Artificial Intelligence – in order to build the framework, we needed to make prototyping of one technological artifacts. This is required for validate these methodology. We also find requirements to complete it, in the literature [9, 10, 11, 12, 13, 14, 15, 16, 17].

The last part of Framework is the evolution of knowledge model [18]. The characteristics by Framework, exposed in four phases was based in another Framework that was explained a study case, in a similar research [19]. As illustrate Figure 1, there are domains of development.

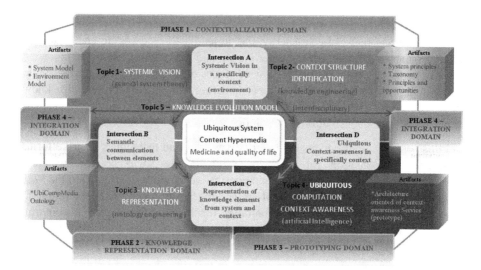

Fig. 1. Smart Ubiquitous Web Service Framework – Macrostructure

In the central block of Figure 1 is described the mains topics of research, and the topic "Medicine and quality of life" represent where we expect impact. The first topic, from context domain (phase 1) explain the macro vision about knowledge system. This macro vision can show the impact from system in the context, can find new opportunities, and also consider the culture of people will use the web smart service. These are important aspects, because on ubiquitous system the people and respective quality and invisible interaction with "machines" are the main focus of development. In Figure 2 we explain the connection about elements illustrated in Figure 1.

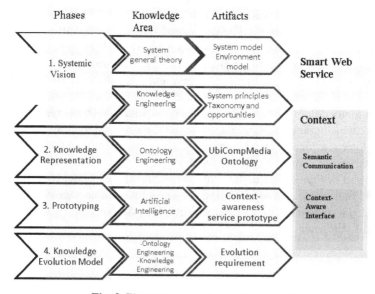

Fig. 2. Elements connection by Framework

In Phase 1 - Contextualization Domain – the main goal is understanding application context. To effect it, is necessary modeling some knowledge system aspects, which usually are not considered, as the influence of social and cultural environment. The first step is formulating system questions main that will to direct whole development of knowledge system. These questions are based in 5W1H– Who, What, Where, Why, Which, When and How [19].

These questions are applied for to set narratives that will compose knowledge model. After this task, this information was organized in a systemic model. The systemic vision is categorized in four parts: Components, Environment, Structure and Mechanism [3]. In order to facilitate this task we made four questions:

1. Components: Which is the collection of system parts? It means, which are the macro elements, and their leaders who make the system work?

2. Environment: Which is the collection of items that DO NOT belongs to the system, but currently suffers some action by one or more than one system components?

3. Structure: Which is the collection of links (bonds) between the components of the system, or between these and their environment?

4. Mechanism: Which elements or combination of elements, that make the system behave the way it works? You will need to describe the basic functioning of system. After these tasks, the next step is understood what kind of application can emerge by this environment. In this point, we used CommonKADS Methodology (we used Worksheet OM-1, problems and opportunities and Worksheet OM-2 – Variant Aspects) and KAMET II [5]. Also, CommonKADS Methodology [4] gave us some tools for identify the feasibility of application (we use Worksheet OM-5 – Checklist for Feasibility Decision Document), as illustrate Figure 3.

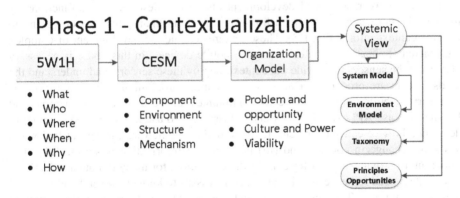

Fig. 3. General Vision of Phase 1 of Framework

Thus, in this Phase we can know if the system will be viable or not. If it will not viable, the work can stop. In case the knowledge system is viable (second the criterions from Worksheet OM-5) the Phase 2 can be started. In Phase 2 of framework – knowledge representation – we used a computational ontology. In order to effect it we use basically NeOn Methodology[4]. As illustrated Figure 4, this phase has four main activities. The first activity is about the ontology scope specification, as layout and it layers.

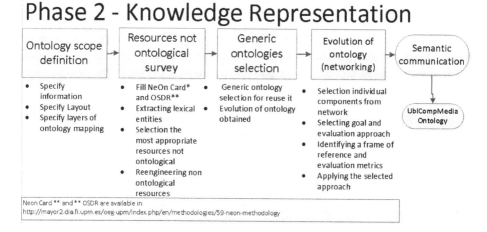

Fig. 4. General Vision of Phase 2 of Framework

Phase 2 can be started in parallel to Phase 3, because some specification by application can interfere on knowledge model (Ontology), and vice versa. Thus, in Phase 3 is developed application that will use knowledge model (UbiCompMedia Ontology). This knowledge model will make the communication between Internet, device, people and environment.

In Phase 3 we recommend development also a middleware, whose functionality for asynchronous messaging, publish / subscribe event, management, storage and to manage context information [25]. These features enable the development of complex multimedia applications, context-sensitive mobile devices. In this Phase is necessary also to set the sensors to provide the context, deploy these sensors, and understand the kinds of data are produced. If not exist application programming interface (API) is necessary to build some interface for communication between devices, people and Internet. In case API exist, is necessary to learn how to use it. Others tasks are: (i) to determine how to consult the sensor, (ii) how the Middleware can be notified when has some change in context, and (iii) how context information is storing [25]. It is important to provide support for context dissemination, for many remote applications. The context need to be treated, it means, is necessary to know how send and received message from it. The ontology (knowledge model) will promote the semantic communication on web smart service, as illustrate Figure 5.

---

[4] In order to set tasks from NeOn Methodology, we used especially these references: [20, 21]. In reference [22] we used specifically pages: 149-151. In reference [23] we used specifically pages: 193, 196-199, 199-202, 202-203. In reference [24] we used specifically pages: 122-124.

Phase 3 - Prototyping

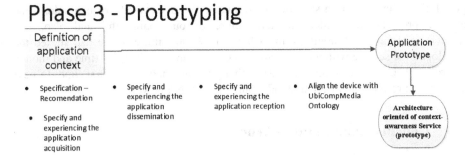

Fig. 5. General Vision of Phase 3 of Framework

The protocol used in smart web service, should interact with mobile phone also, because this device is, currently, the most important artifact in Digital Convergence. The idea is that objects like smartphones, tablets, televisions, computers, key chains, RFID tags, pens and other common objects, keep permanently connected to Internet, for sending and receiving information and incorporate our habits with advanced technologies, in order to facilitate our day. Through connections with 3G and 4G signal quality, data sharing using concepts of cloud computing and automatic data synchronization via network, between every kind of objects. The advantage from these protocols is it is independence of physical network, for data transmitted.

In last Phase 4, we explain about integration and evolution of knowledge system and the service which it provided, as illustrate Figure 6. In order to make it, is necessary to test then with a group research, whose needed will present the way for evolve the smart service and knowledge system.

Phase 4 – Integration and Evolution

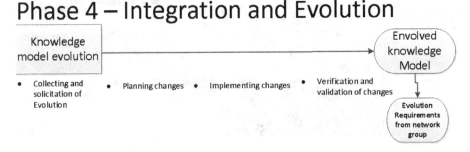

Fig. 6. General Vision of Phase 4 of Framework

This methodological framework was been tested in Knowledge System projects. The studies to effect the framework happened in doctoral work [1], master work [2], in Subproject Modeling. Part of framework was applied in E-Government Project and in others two projects, in order to identify the context and demand by client. The next step by research is to development the next phases in a telemedicine application, for validate whole Framework (proof of concept). The Phase 1 was already developed. The outcomes show how a detail plan about context can help us understand several possibilities that exist in a specific environment. We discovered, for example, some another projects we needed to start before of our main aim, and also, this methodology gave us guidelines about laws and local culture. If we had started the developed the service without these guidelines, probably we had lost time and money in project, because we figure more work than we imagined, in the first time. In next section we present conclusion and main outcomes by Framework.

4   Conclusion and Future Work

This work is part of a research whose shape was created to help researchers and developer of computer science, in order to develop smart applications for web services, a challenge by next years.

In this context, our challenge is development a Methodological Framework which can explain how Hypermedia can be manipulated and disseminated between devices & Web & System Context-Awareness & People, in more natural way. Basically, the goal is make possible development Ubiquitous and Smart Web Service, considering local culture and environment. How illustrate Figure 7, our Framework should guide developer to set a Hypermedia Architecture.

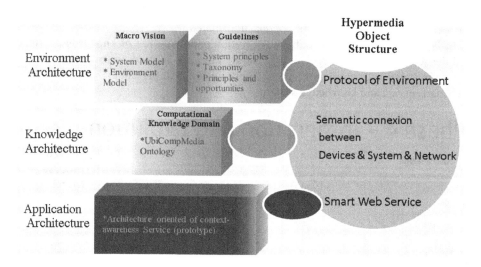

Fig. 7. General Vision outcomes expected by Framework product

This architecture consists in a structure of Meta Data from Hypermedia content, and include also knowledge environment, which this content will interact.

At moment, the outcomes are: the phase 1 applied on telemedicine gave us a new possibilities for development new projects that need to exist for support the first aim. In this application we obtain guidelines for to make some political changes (law changes too). The phase 1 also gave us one plan for next years, for evolution of knowledge system and environment where it will work. This plan show us that

Brazil's infrastructures needs for to have one smart web service for telemedicine, like the one we was projected. It is an interesting deal for the country; because the government can plan some advanced technologies based part on this plan. Particularly in Brazil, we believe the results of this project will help to face some of the national computer research challenges pointed out at the São Paulo seminar in 2006.

Acknowledgements. This research was funded by FINEP and FAPESC, Brazil, and implemented by Post Graduate Program from Federal University of Santa Catarina State - Knowledge Engineering and Management (in Portuguese *Programa de Pós-Graduação Engenharia e Gestão do Conhecimento*), and support by Institute Stela. Our acknowledgements to each these entities.

References

1. Salm Jr., J.F.: Ontology Design Pattern for adding references from the New Public Service in Open Government Platforms. Thesis (Doctor degree in Engineering and Knowledge Management) - Knowledge Engineering and Management Post-Graduation Program, Federal University of Santa Catarina, Florianopolis, Brazil (2012) (in Portuguese)
2. Schneider, V.: Method for modeling strategic context for knowledge-based systems. Dissertation of Master's Degree in Engineering and Knowledge Management. Federal University of Santa Catarina (2013) (in Portuguese)
3. Bunge, M.: Mechanism and explanation. Philosophy of the Social Sciences 27(4), 410–465 (1997); Bunge, M.: How Does It Work? The search for explanatory mechanisms. Philosophy of the Social Sciences 34(2), 182–210 (2004); Bunge, M.: Systemism: the alternative to individualism and holism. Jornal of Socio-Economics 29(6), 147–157 (2006)
4. Schreiber, G., et al.: Knowledge Engineering and Management: the CommonKADS Methodology. MIT Press, Cambridge (2002)
5. Cairó, O., Guardati, S.: The KAMET II methodology: Knowledge acquisition, knowledge modeling and knowledge generation. Expert Systems with Applications 39, 8108–8114 (2012)
6. Villazón-Terrazas, B., Gomez-Pérez, A.: Charper 6 – Reusing and Re-engineering Non-ontological Resources for Building Ontologies. In: Suárez-Figueroa, et al. (org.) Ontology Engineering in a Networked World, New York (2012)
7. Sabou, M., Fernandez, M.: Chapter 9 - Ontology (Network) Evaluation. In: Suárez-Figueroa, et al. (Org.) Ontology Engineering in a Networked World, New York (2012)
8. Levashova, T., et al.: Product design network self-contextualization: enterprise knowledge-based approach and agent-based technological framework. In: International Conference on Industrial Applications of Holonic and Multi-Agent Systems, Linz. Proceedings..., vol. 4, pp. 61–71. Springer, Berlin (2008/2009)
9. Brézillon, P.: Context in artificial intelligence: I. A survey of the literature. Computers and Artificial Intelligence 18(4), 321–340 (1999a) ISSN: 02320274; Brézillon, P.: Context in artificial intelligence: II. Key elements of contexts. Computers and Artificial Intelligence 18(5), 425–446 (1999b) ISSN: 02320274; Brézillon, P.: Context in problem solving: a survey. The Knowledge Engineering Review 14(1), 1–34 (1999c)
10. Zhdanova, A.V., et al.: Context acquisition, representation and employment in mobile service platforms. In: Workshop on Capturing Context and Context Aware Systems and Platforms, Mykonos. Proceedings... [s.n.], [S.l.] (2006)

11. Morse, D.R., Armstrong, S., Dey, A.K.: The What, Who, Where, When and How of Context-Awareness (2000), `http://www-static.cc.gatech.edu/fce/context toolkit/pubs/CHI2000-workshop.pdf` (access date: October 10, 2012)
12. Bazire, M., Brézillon, P.: Understanding Context Before Using It. In: International and Interdisciplinary Academic Conference, Buenos Aires. Proceedings..., vol. 5, pp. 29–40. Springer, Paris (2005)
13. Madkour, M., Maach, A., Elghanami, D.: Context-aware middleware for services retrieval and adaptation. International Review on Computers and Software 7(1), 166–176 (2012)
14. Gwi-Hwan, J.I., Ryum-Duck, O.H.: Wetland ontology modeling to apply the context-aware technology for application service environment. Journal of Computer Science 8(3), 342–347 (2012)
15. Dey, A.K., Abowd, G.: The Context Toolkit: aiding the development of context-aware applications. In: Workshop on Software Engineering for Wearable and Pervasive Computing, Limerick. Proceedings...University of Washington, Seatle (2000)
16. Pessoa, R.M., et al.: Aplicação de um *Middleware* Sensível ao Contexto em um Sistema de Telemonitoramento de Pacientes Cardíacos. In: Congresso Da Sociedade Brasileira De Computação. Anais..., vol. 26, pp. 32–46. SEMISH, Campo Grande (2006)
17. Maass, W., Janzen, S.: Pattern-Based Approach for Designing with Diagrammatic and Propositional Conceptual Models. In: International Conference on Service-Oriented Perspectives in Design Science Research (DESRIST), Milwaukee. Proceeding..., vol. 6, pp. 192–206. Springer, Berlin (2011)
18. Palma, et al.: Charter 11 – Ontology Evolution. In: Suárez-Figueroa, et al. (org.) Ontology Engineering in a Networked World, New York (2012)
19. Kim, J., Son, J., Baik, D.: CA 5W1H onto: ontological context-aware model based on 5W1H. International Journal of Distributed Sensor Networks (2012)
20. Alexakos, C., et al.: A Multilayer Ontology Scheme for Integrated Searching in Distributed Hypermedia. In: Sirmakessis, S. (ed.) Studies in Computational Intelligence: Adaptive and Personalized Semantic Web, vol. 14, Springer, [S.l.] (2006)
21. Hardman, L., Bulterman, D.C.A.: Document Model Issues for Hypermedia. In: Grosky, W.I., Jain, R., Mehrotra, R. (eds.) The Handbook of Multimedia Information Management. Prentice Hall, [S.l.] (1997)
22. Suárez-Figueroa, M.C., et al.: Ontology Engineering in a Networked World. Springer, [S.l.] (2012) ISBN 978-3-642-24794-1
23. Sabou, M., Fernandez, M.: Chapter 9 - Ontology (Network) Evaluation. In: Suárez-Figueroa, et al. (org.) Ontology Engineering in a Networked World. Springer, New York (2012)
24. Villazón-Terrazas, B., Gomez-Pérez, A.: Charper 6 – Reusing and Re-engineering Non-ontological Resources for Building Ontologies. In: Suárez-Figueroa, et al. (org.) Ontology Engineering in a Networked World. Springer, New York (2012)
25. Madkour, M., Maach, A., Elghanami, D.: Context-aware middleware for services retrieval and adaptation. International Review on Computers and Software 7(1), 166–176 (2012)

Learning to Classify Short Text with Topic Model and External Knowledge

Ying Zhu, Li Li, and Le Luo

Faculty of Computer and Information Science
Southwest University, Chongqing 400715, China
{lily}@swu.edu.cn

Abstract. Many methods have been developed to utilize topic analysis models to deal with the noises and sparseness of the text. However, the use of a topic model solely sometimes unable to achieve the expected high performance, it is very necessary to improve the current topic model to cope with the characteristic of texts and specific requirements. In this paper, we focus on two tasks. One is to make use of different external corpus to identify topics from texts for better categorization. The other is to add the weight of a few features in texts to get some other topics from those of topic model. We further evaluate the performance of the two tasks with baseline results. The experiments show that our proposed method can achieve a higher accuracy in text classification. The approach can find truly representative words which may contribute to wide acceptance of topic models in micro-blog analysis.

Keywords: text classification, LDA model, topic analysis, twitter.

1 Introduction

Researchers have made extensive efforts to develop and leverage topic models these years. Topic models have been applied to increase the accuracy of classification [1] and also to develop applications in social networks [2][3].

Generally, stop-words which selected manually are removed from texts before doing classification tasks. When dealing with short and extensive texts, many leaning methods like support vector machines (SVMs) and Maximum Entropy(MaxEnt) suffered from limited information in texts[4]. Recent researches show that finding hidden topics from external corpus is a success way to enrich the text semantic. But Choosing an appropriate external knowledge is a very difficult task as external corpus affects the result a lot. How to utilize the result of topic model to get a satisfied classification result is also an inevitable problem.

Today, people increasingly prefer to publish short messages (tweets) about their everyday activities on Twitter. Tweets are limited to 140 characters. Additionally, the number of tweets does not equally distributed for every topic because people pose tweets by following their own inclinations. So extracting topics broadly and accurately is a non-trivial problem.

In this paper, we mainly focus on two tasks. Firstly, we investigate the benefits of using muti-source external corpus in the classification and how to leverage topics to classification can get a better outcome. Secondly, we proposed a method

M. Wang (Ed.): KSEM 2013, LNAI 8041, pp. 493–503, 2013.

to increase the weight of some feature words in tweets so that more hidden information can be identified by the topic model.

The rest of this paper is organized as follows. We firstly discuss texts classification and topic extraction. In Section 3 we introduce the background of topic model. We systematically validate the method over two datasets in Section 4. Finally, section 5 is about the review of relevant works.

2   Problem Description

The limited information in the text cause difficulties in managing and analyzing texts based on the bag-of-words representation only. And increasing the weight of features in short texts can achieves a comprehensively outcome. In this study, we use Latent Dirichlet Allocation (LDA) model to extract topics.

2.1   Extract Hidden Topics from External Knowledge

In [4], the authors use single corpus to extracted hidden topics. But in this paper, we try to use two or more corpuses to infer hidden topics and analyze the influence of different external datasets to classification.

The underlying idea of our framework is that for each classification task, we collect one or more large-scale external data collection called universal dataset, and analyze topics of the universal data with LDA model. Based on the results conducted by LDA model, we then proceed with topic inference on both the test and training data also with LDA model. To estimate the effect of external datase bring to classification, we build a classifier on both a (small) set of labeled training data and a rich set of hidden topics discovered from data collection mentioned above..

2.2   Integration of Topics and Texts for Classification

We expand the texts by appending some words to text based on the semantic relatedness between words [4][5]. Here is how we integrate topics into texts.

After doing topic inference with LDA model, we can get θ_m that is a topic distribution for document m. The value in θ_m is the probability of topics. But the classifier we use is MaxEnt which requires discrete feature attributes. So the value in θ_m need to be dispersed as names of topics. The name of a topic appears once or several times depending on the probability of that topic. Here is an example of integrating the topic distribution into its bag-of-word vector.

- $W_m = \{$rushdie islamic law beauchaine jaeger supports...propaganda $\}$
- $\theta_m = \{\vartheta_{m,1} = 0.1075, ..., \vartheta_{m,5} = 0.0047, ..., \vartheta_{m,11} = 0.0140, ..., \vartheta_{m,1} = 0.1262\}$
- Applying different discretization intervals according to different number of topics
- $V_m \bigcup W_m =$ combination of topics and texts

W_m is the original text. The value in θ_m is discreted according to the discretization intervals which is affected by the number of topics. As the number of topics increase, the interval becomes shorter. The result shows that the performance of classification improves by applying different intervals.

2.3   Aggregation of Similar Tweets

Standard LDA may not work well with tweets because the information tweets contain is too limited. To overcome this difficulty, we proposed to put similar tweets of a user together.

Formally, we assume that there are T topics in Twitter, each represented by a word distribution. For a tweet posed by user U, we can judge which topic the tweet belong to according the word distribution. So we calculate the similarity S between the tweets of a user. Subsequently, we combine tweets while $S \subset m$. m is the interval to control the result. Here we use cosine similarity because it is relatively simple and fast to run. Let t_{u_i} denotes the vector of tweet i posed by user u, and $i \subseteq [1, n]$. The dimensionality of t_{u_i} and t_{u_j} depends on the number of different words in tweets i and j. And the value of t_{u_i} depends on the number of occurrences of every word in tweet i. The similarity of two tweets can be represented as follow.

$$S(\overrightarrow{t_{u\_i}}, \overrightarrow{t_{u\_j}}) = \frac{\overrightarrow{t_{u\_i}} \cdot \overrightarrow{t_{u\_j}}}{|\overrightarrow{t_{u\_i}}| \cdot |\overrightarrow{t_{u\_j}}|} \tag{1}$$

From formula above, we can get that the S values higher when the two tweets t_{u_i} and t_{u_j} have more same words. Considering that the same words of tweets t_{u_i} and t_{u_j} is sometimes noises or the less important words, the outcomes must be poorly. So in the experiment, we modify the weight according to the importance of word. In this paper, we consider that the entity words is more important than other words.

Given the content of Twitter messages we increase the weight of the entity word so that the important information can be identified easily. In the paper, we simply double its original occurrence. Here we utilize OpenCalais[1] to identify entities from tweets. OpenCalais allows for detection and identification of 39 different types of entities such as persons, events, products or music groups.

3   Topic Model

Models such as pLSA, LDA and their variants have been more successful applied in document & topic modeling [6], categorization [7], opinion mining [8] and many more. However, it has been pointed out that LDA is more complete than pLSA in such a way that if follows a full generation process for document collection [9].

3.1   Latent Dirichlet Allocation

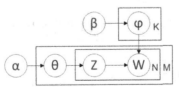

Fig. 1. LDA: a generative graphical model

---

[1] OpenCalais, http://www.opencalais.com/

LDA is a generative graphical model. It can be used to model and discover underlying topic structures of any kind of discrete in which text is a typical example. LDA was developed based on an assumption of document generation process depicted in both Figure 1 and Table 1.

In LDA, a document $w_m = \{w_m, n\}_{n=1}^{N_m}$ is generated by first picking a distribution over topics v_m from a Dirichlet distribution ($Dir(\alpha)$), which determines topic assignment for words in that document. Then the topic assignment for each word placeholder [m, n] is performed by sampling a particular topic $z_{m,n}$ from multinomial distribution $Mult(\theta_m)$. And finally, a particular word $w_{m,n}$,n is generated for the word placeholder [m, n] by sampling from multinomial distribution $Mult(\varphi_{z_{m,n}})$.

Table 1. Parameters and variables for LDA model

M	the number of documents
K	the number of topics
V	vocabulary size
α, β	Dirichlet pareameters
θ_m	topic distribution for document m
$\Theta = \theta_m$	a $M \times K$ matrix
φ_k	word distribution for topic k
$\Phi = \varphi_k$	a $K \times V$ matrix
N_m	the length of document m
$z_{m,n}$	topic index of nth word in document m
$\omega_{m,n}$	a particular word for word placeholder [m, n]

From the generative graphical model depicted in Fig. 1, we can write the joint distribution of all known and hidden variables given the Dirichlet parameters as follows.

$$p(w_m, z_m, \theta_m, \Phi | \alpha, \beta) = p(\Phi | \beta) \prod_{n=1}^{M_m} p(w_{m,n} | \varphi_{z_{m,n}}) p(z_{m,n} | \theta_m) p(\theta_m | \alpha) \quad (2)$$

And the likelihood of a document w_m is obtained by integrating over θ_m, Φ and summing over z_m as follows.

$$p(w_m | \alpha, \beta) = \int \int p(\theta_m, \alpha) p(\Phi, \beta) \prod_{m=1}^{M_m} p(z_{m,n} | \theta_m, \Phi) d\Phi d\theta_m \quad (3)$$

Finally, the likelihood of the whole data collection W $= \{w_m, n\}_{n=1}^{N_m}$ is product of the likelihoods of all documents:

$$p(W | \alpha, \alpha) = \prod_{m=1}^{M} p(w_m | \alpha, \beta) \quad (4)$$

3.2   LDA Estimation

The methods such as Variational Methods and Gibbs Sampling can be used to estimate parameters for LDA approximately. Gibbs Sampling is a special case of Markov-chain Monte Carlo (MCMC) and often yields relatively simple algorithms for approximate inference in high-dimensional models such as LDA [9]. So in this study, we prefer to use Gibbs sampling to estimate the parameters. After finishing Gibbs Sampling, two matrices Θ and Φ are got. Θ is the topic distribution for documents. Φ is the word distribution for topics.

4   Experiments

In order to evaluate the proposed model in the paper, we conduct a series of experiments. For the classification task, we choose MaxEnt as machine learning method because MaxEnt is robust and has been applied successfully to a wide range of NLP (NonLinear Programming) tasks. Besides, MaxEnt is very fast in both training and inference compared to SVMs.

4.1   Data Preparation

Table 2. Statistics of test and train data

20Newsgroups	**Raw data**: 6.33MB test data |docs|=7532
	Final data: 4.83MB test data |docs|=7532
	Raw data: 10.20MB train data |docs|=11311
	Final data: 7.32MB train data |docs|=11311
Wikipedia	**Raw data**: 4.3GB |docs|=486,391
	Final data: 356MB |docs|=91,754 |words|=30,501,211 and |vocabulary|=60,944
News articles	**Raw data**: 2.4GB |docs|=15,351
	Final data: 250MB data |docs|=12,352 |words|=5,213,476 and |vocabulary|=27,657
Twitter data	**Raw data**: |users|=6033 |docs|=479088
	Raw data: |users|=3931 |docs|=374491 |words|=101777 and |vocabulary|=11174

- **20Newsgroups** We choose the 20Newsgroups[2] as texts to be classified. As showed in Table 2, after removing non-words and links, we got 4.83M test data and 7.32M train data.
- **Wikipedia** Since Wikipedia is comprehensive and conveniently available, we crawled relevant pages from Wikipadia using JWikiDocs[3]. Each crawling transaction is limited to 4 depth of hyperlink. We remove duplicate documents, HTML tags, links, and rare and stop-words from the raw data.
- **News articles** Inspired by the idea that linking Twitter posts with related news to contextualize Twitter activities [10] and considering that 20 Newsgroups dataset is about news, we choose it as the external corpus. To extract the main content of news articles we use Rss facilitates BoilerPipe [11]. By removing non-words and noisy text, we finally obtained 340MB data.

---

[2] 20Newsgroups, http://people.csail.mit.edu/jrennie/20Newsgroups/
[3] JWikiDocs, http://jwebpro.sourceforge.net

- **Twitter** The Twitter collection we use contains 479,088 tweets posed by 6,033 users from January to February in 2010 (two months). The number of Twitter messages posted per user follows a power-law distribution. As the methods we use is combining similar tweets of a user, we remove these who contributed less than 20 tweets in total and tweets that contain less than 3 words. Table 2 shows some details of this dataset.

4.2 Effect of Topic Analysis Model

To evaluate the effect of the topic model, we extracted topics from two types texts of 20Newsgroups using LDA model. One is raw texts without removing stop-words, denoted as "DeTxt". The other is texts with removing stop-words, denoted as "DeTxt". We estimated LDA models for 20NewGroups with different numbers of topics (from 20 to 100, 150 and 200). Fig. 2 shows most likely words of some sample topics of raw texts. The number of the topic is 50.

T0: space nasa earth launch moon orbit shuttle mission solar henry satellite ...
T1: 0d ah air 2di a86 75u 0t 7u pl 6um b8f b8 lq b9r 6ei 6t 3t mk bh 34u...
T6: jason ticket revolver cop kratz safety gang shooting freeman trigger...
T8: image jpeg file color format images bit files gif quality convert program...
T9: god jesus christ bible christian christians sin lord hell church heaven...
T12: printer print hp font fonts laser postscript paper printing use printers ...
T20: dos windows os run pc network microsoft software unix ibm version...
T27: medical disease health cancer patients doctor treatment medicine pain...
T44: drive scsi hard disk drives ide controller bus data floppy pc card...
T48: said went did came told saw time started left took know says got does...

Fig. 2. Most likely words of some topics of raw texts from 20Newsgroups

The Fig. 2 indicates that the LDA model is a very successful topic model. These most likely words of each topic are very related, such as "T0", "T8", "T20" etc. LDA model makes the texts more topic-oriented by extracted topics from the texts. More important is that the words in "T1" and "T48" are usually considered as stop-words which should be removed from texts.

Classification With Topics and Texts. For evaluate how to leverage topics can get satisfied results, we did many classification tasks to "RaTxt" and "De-Txt" by using topics only and combining the topics and texts. First, different number of topics (from 20 to 100, 150 and 200) are extracted from the two types of texts separately. Then MaxEnt classifiers are built on the training data according to different numbers of topics. Besides, a baseline classifier is built by directly using LDA model. The change of classification accuracy according to the number of topics is depicted in Fig. 3 and Fig. 4.

From the Fig. 3, we can see that the accuracy of "RaTxt" and "DeTxt" around 40 topics is much more higher than that of the baseline. Although the accuracy of "RaTxt" is much lower than the "DeTxt", but when the number of topic is more than 40, the "RaTxt" and "DeTxt" get a similar accuracy. The findings indicate that utilizing topic analysis models to classify texts without removing stop-words can still achieve a satisfactory result.

As showed in Fig. 4, the accuracy of "RaTxt" even with 20 topics is higher than the baseline in Figure 3. Around 70 and 80 topics, the classification accuracy of the "RaTxt" is little less than the "DeTxt". And when the number of

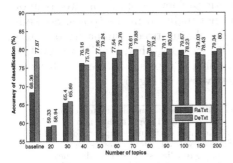

Fig. 3. classification with topics

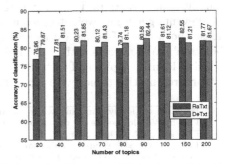

Fig. 4. classification with topics and texts

topics reaches 100 and 150, respectively, the "RaTxt" achieves highest results. Compared to Fig. 3, classification with topics and words can produce a better result compared to classification only with topics. And this method is also useful to the texts without removing stop-words.

4.3 Effect of External Knowledge

To analyze whether utilizing multi-original external corpus can benefit categorization, we choose the external dataset from two sites: Wikipedia and BBC or CNN news articles. We first use Wikipedia and news articles as external knowledge separately and then combine the two corpus together to enrich semantics. The results are showed in Table 3.

Table 3. Different accuracy according to different external dataset

External Dataset	Accuracy(%)	External Data	Accuracy(%)
Wikipedia	75.66	News+Texts	81.37
Wikipedia+Texts	81.23	News+Wikipedia	83.01
News	77.02	News+Wikipedian+texts	82.83

From Table 3, we can see that when using single coupus as external knowledge, the accuracy is even lower than the classified texts directly. This may be caused by two reasons: 1) the external dataset does not cover the topics that the texts have; 2) the external dataset contains too much irrelevant information. When we combine the Wikipedia or news articles with texts, the accuracy is close to that by utilizing topics from texts to classify. The result becomes satisfactory as we integrate Wikipedia and news articles and texts together.

4.4 Extract Topics from Twitter

As the tweets are too short, using tweets directly to extract topics may lead to two problems: inaccurate and incomplete. The tweets user published are determined by the user's interests, so they are not divided equally into every topics.

Usually, some events in Twitter get attention of few people. The words about these events get a low weight or even are not included in the topics. So we increase the weight of entity words when calculating tweets similarity of a user, then combine the tweets with high similarity score together. Finally, we use LDA model to identify topics in Twitter. The number of topics is 70. Here we list some most representative topics with most likely words in Fig. 5.

T0: black red hardy shirt blue wearing wear dress bag shoes pink size tattoo color...
T2: android blog google car phone iphone toyota recall ipad apple mobile app...
T14: speople haiti help chat joined million relief donate earthquake support food aid...
T22: ytet ydy sfile vampire diaries episode charice pyramid vampirediaries iyaz...
T25: days mandy savvy tour shows youtube song getonellen savvyandmandy miley...
T26: gproject group manager design salary website sonic local director senior...
T48: workout weight muscle fat bicep build good chest leg gym burning exercises...
T60: palin sarah fox serving blog news chicken soup ground lounge cheese tea black...
T66: today thank pisces jesus member disney life libra leo friends capricorn dear...
T67: happy birthday baby day wish gaga lady amazing oprah valentines valentine...

Fig. 5. most representative topics with most likely words from Twitter

It is intuitive to see that the words in every topics listed in Fig. 5 are belong to same class. We also extract topics from Twitter directly as baseline.

As many studies have proved that most of event-oriented topics can be linked to news articles [12][13], we use BBC news articles and Google Trends to evaluate the results. BBC website and Google Trends[4] both have a record about events happened over a period. The BBC news articles we use are published from 1st Jan. to 28th Feb. , 2012. Table 4 is the 16 hot event topics selected from Google Trends and BBC.

Table 4. Selected hot event topics from Google Trends and BBC

Toyoto recall	Vampire Diaries	Charice pyramid
Mandy&Savry song	Michael Jackson	Olympic Canada
Sarah Palin Fox News	Lady gaga Oprah	Haiti earthquake
Justin Bieber	Lady gaga Grammys	Valentines Apple ipod
Nigeria Adua leader	Google Android	World Warcraft Game

We use 3 methods to produce topics with LDA: using LDA directly (baseline), using LDA and cosine similarity (cos+LDA), and the method (cos+entity+LDA) proposed in this paper. The number of topics is 70, and each topic selects 20 most likely words. After comparison , we find the topics produced by our methods have advantages below. Table 5 shows the comparison data.

- Our method can get more entities than the baseline. We calculate how many entities the topics have with Opencalais. The result is introduced in Table 5. The Table 5 shows that more entities can extracted from tweets when the weight of entity words increased in calculate the similarity of tweets. So the topic have a high coverage when more entities extracted.

---

[4] Google Trends, http://www.google.com/trends/

- The topics extracted by our method have a high coverage on hot event topics. As the topics showed in Fig. 5, "toyoto recall" in T2, "charice pyramid" and "vampire diaries" in T22 can all be linked to events according to BBC and Google Trends. But they can not be found in topics of baseline because these event may be not attractive as Haiti earthquake. So our methods can get 87.5% event coverage as showed in Table 5.
- Our methods can get more meaningful and distinct topics as showed in Table 5. If two topics that have less than 10 same words, they are distinct topics. And meaningful topics is these have more than 15 words that can be recognized by people. Topics produced by LDA directly have more topics about user daily life. But topics extracted by our method not only have daily life topics, but also have special topics such as T66 which contains constellation (libra, leo, capricorn) correspond to January and February.

Table 5. Comparing topics extracted through different methods

Method	Entity Number	Event Coverage	Meaningful & Distinct Topic
LDA	67	62.5%	49
cos+LDA	32	50%	41
cos+entity+LDA	81	87.5%	53

5   Related Work

Our work is related to topic identification techniques such as LDA, and Latent Semantic Analysis (LSA) and pLSA. Recently, these topic models are widely used to text classification and Twitter Analysis [6][14].

Integrating external knowledge to text mining has recently attracted more attention. Text categorization performance is improved by augmenting the bag-of-words representation with new features from ODP and Wikipedia [4]. And Hu et al. [15] cluster short texts (i.e., Google snippets) by first extracting the important phrases and expanding the feature space by adding semantically close words or phrases from WordNet and Wikipedia. But for some texts with its special features, it may not suitable to use the universal external collections. The difference between these studies and our work comes that the external knowledge we use is from different sources and theirs comes only from one source. And we also estimate the effect of topic models on text categorization by using various methods to exploit topic models.

As micro-blogging became highly popular, researchers started to develop topic models to investigate the character of Twitter [16][14]. Works [3][13] focus on leveraging external database as feature to enrich the semantic of tweets. However the feature selection is manually, and topics in Twitter are changing over time, so are features. Identifying events in Twitter may be another trend. Popescu et al. [2] try to extract events in Twitter by using NLP techniques. But events only depict one aspect of Twitter, some topics about user daily life will be missed. In this paper, we will introduce the method that can extract topics more broadly. In addition, the features used in the paper are obtained automatically.

6   Conclusions

In this paper, we first introduce and analyze strategies that utilize LDA model to classify texts and extract topics from Twitters. The topics extracted from the short texts indicate that the LDA model can precisely make the texts topic-oriented. Two methods are presented to utilize topics to classify texts and estimate the influence of muti-original external datasets to classification. Experimental results show that by combining several different external corpus together can improve the performance of classification. Adjusting the weight of features captured from Twitter is applied in our experiments on Twitter. Compare to identifing topics with LDA directly, the topics produced by our method is more broadly and accurately.

In the future, we will consider how to combine the different external dataset more efficiently and appropriately. How to extract topics from large scale datasets efficiently is an important problem to be solved. Comparison with other learning models based on large scale dataset will be investigated further.

Acknowledgment. This work is supported by Natural Science Foundations of CQ(No.CSTC2012JJB 40012) and National Natural Science Foundations of China(No.61170192). L. Li is the corresponding author for the paper.

References

1. Sebastiani, F.: Machine learning in automated text categorization. ACM Computing Surveys 34(1), 1–47 (2002)
2. Popescu, A., Pennacchiotti, M., Paranjpe, D.: Extracting events and event descriptions from twitter. In: Proc. WWW, pp. 105–106 (2011)
3. Sriram, B., Fuhry, D., Demir, E., Ferhatosmanoglu, H., Demirbas, M.: Short text classification in twitter to improve information filtering. In: Proc. SIGIR, pp. 841–842 (2010)
4. Phan, X.H., Nguyen, C.T., Le, D.T., Nguyen, L.M., Horiguchi, S., Ha, Q.T.: A hidden topic-based framework toward building applications with short web documents. IEEE Trans. Knowl. Data Eng. 23(7), 961–976 (2011)
5. Strube, M., Ponzetto, S.P.: Wikirelate! computing semantic relatedness using wikipedia. In: Proc. the 21st National Conference on Artificial Intelligence, pp. 1419–1424 (2006)
6. Gabrilovich, E., Markovitch, S.: Overcoming the brittleness bottleneck using wikipedia: enhancing text categorization with encyclopedic knowledge. In: Proc. 21st National Conference on Artificia Intelligence, vol. 2, pp. 1301–1306 (2006)
7. Blei, D., Ng, A., Jordan, M.: Latent dirichlet allocation. The Journal of Machine Learning Research 3, 993–1022 (2003)
8. Xue, L., Xiong, Y.: Stock trend prediction by classifying aggregative web topic opinion. In: Proc. PAKDD, Gold Coast, Australia, pp. 173–184 (2013)
9. Heinrich, G.: Parameter estimation for text analysis. Technical report, Tech. Rep (2005)
10. Abel, F., Gao, Q., Houben, G.-J., Tao, K.: Semantic enrichment of twitter posts for user profile construction on the social web. In: Antoniou, G., Grobelnik, M., Simperl, E., Parsia, B., Plexousakis, D., De Leenheer, P., Pan, J. (eds.) ESWC 2011, Part II. LNCS, vol. 6644, pp. 375–389. Springer, Heidelberg (2011)

11. Kohlschütter, C., Fankhauser, C., Nejdl, P.: Boilerplate detection using shallow text features. In: Proc. 3rd ACM Web Search and Data Mining, New York, pp. 441–450 (2010)
12. Zhao, W.X., Jiang, J., Weng, J., He, J., Lim, E.-P., Yan, H., Li, X.: Comparing twitter and traditional media using topic models. In: Clough, P., Foley, C., Gurrin, C., Jones, G.J.F., Kraaij, W., Lee, H., Mudoch, V. (eds.) ECIR 2011. LNCS, vol. 6611, pp. 338–349. Springer, Heidelberg (2011)
13. Ramage, D., Dumais, S., Liebling, D.: Characterizing microblogs with topic models. In: Proc. ICWSM (2010)
14. Kwak, H., Lee, C., Park, H., Moon, S.B.: What is twitter, a social network or a news media? In: Proc. WWW, pp. 591–600 (2010)
15. Hu, X., Sun, N., Zhang, C., Chua, T.S.: Exploiting internal and external semantics for the clustering of short texts using world knowledge. In: Proc. 18th ACM Conference on Information and Knowledge Management, pp. 919–928 (2009)
16. Hopcroft, J.E., Lou, T., Tang, J.: Who will follow you back? reciprocal relationship prediction. In: Proc. CIKM, pp. 1137–1146 (2011)

An Efficient Hierarchical Graph Clustering Algorithm Based on Shared Neighbors and Links

Zhang Huijuan, Sun Shixuan, and Cai Yichen

School of Software Engineering, Tongji University, Shanghai, China
mszhj@tongji.edu.cn, {shixuansun89,cyc074280}@gmail.com

Abstract. Community structure is an important property of networks. A number of recent studies have focused on community detection algorithms. In this paper, we propose an efficient hierarchical graph clustering algorithm based on shared neighbors and links between clusters to detect communities. The basic idea is that vertices in the same cluster should have more shared neighbors than that in different clusters. We test our method by computer generated graphs and compare it with MCL algorithm. The performance of our algorithm is quite well.

Keywords: Community Structure, Hierarchical Graph Clustering, Shared Neighbors.

1 Introduction

Nowadays, many complex systems can be represented as networks which vertices are linked together by edges [1]. In these networks, connections between some vertices are dense while the others are loose. This feature of networks is called community structure or clustering [2]. These groups of vertices whose connections are dense are called communities, clusters or modules and this phenomenon exists in a lot of networks.

Identifying these communities can help us a lot to understand and optimize networks. For example, in our social networks, person in the same community are most likely to have the same interest and background. In the Internet, websites linking with each other densely have the same topic than others. So, detecting communities in networks is becoming a hot topic of research.

The aim of community detection in networks is to identify the clusters by only using the information encoded in the graph topology [3]. And researchers have presented some algorithms to do it. The traditional methods are that apply traditional hierarchical clustering algorithms to the graph [4]. These methods can be divided into two different kinds: top-down or bottom-up.

However, recently compared with traditional methods, some new heuristic algorithms are proposed. These methods are based on some properties of networks to propose some rules to detect communities. And these rules can be accepted. The widely used Girvan-Newman algorithm is based on the theory that edge betweenness between clusters should be more than that in the same cluster [2]. Stijn Marinus van

M. Wang (Ed.): KSEM 2013, LNAI 8041, pp. 504–512, 2013.

Dongen develops MCL algorithm based on Markov Chains [5]. Gary William Flake et al provides an algorithm based on minimum cut trees [6]. Besides, MFC algorithm [7], Wu-Huberman algorithm [8] and improved Girvan-Newman algorithm [9] are all computed based on different heuristic theories. At the same time, some researchers present several methods to evaluate these graph clustering algorithms [10, 11, 12].

But in above algorithms, it is existed some problems. The first one is the efficiency of algorithms. For example, the Girvan-Newman algorithm takes about $O(m^2n)$ time where m is the number of edges and n is the number of vertices in the network. For massive graphs, it is too slow. The second one is the correctness of algorithms. Most of algorithms run well when sizes of communities are nearly equal. However, when sizes of communities are quite different, some algorithms such as MCL algorithm failed to detect the clusters.

This paper proposes a new hierarchical graph clustering algorithm that based on fraction of shared neighbors to detect communities. It's called Improved Shared Neighbors Graph Clustering algorithm (ISNGC). The main idea is that vertices sharing more neighbors would have much more probability in the same cluster. For example, in our society, two friends of the same person have much more possibility know each other than two ones chosen at random from the population. This algorithm is quite fast and can detect communities whose sizes are quite different between each other. In section 2, we present our algorithm in details. In section 3, we do several experiments to evaluate the algorithm and compare it with MCL algorithm. Finally, in section 4, we give a summary of our work and possible future development of community detection.

2   ISNGC Algorithm

In this section, we present ISNGC Algorithm in details. This algorithm is based on the general knowledge that vertices shared more neighbors have much more probability in the same cluster. The process of ISNGC Algorithm is a process of bottom-up hierarchical clustering based on fraction of shared neighbors and links between clusters. Firstly, we give some basic definitions.

2.1   Basic Definitions

Let $G = \{V, E\}$ be an undirected and unweighted graph where V is the set of vertices and E is the set of edges. $|V|$ is the number of vertices and $|E|$ is the number of edges. $C(G) = \{c_1, \dots, c_k\}$ is a partition of vertices in graph G. c_i is the set of vertices in cluster i. The subgraph of cluster i is represented as $G(i) = \{c_i, E(c_i)\}$ where $E(c_i) = \{\{v, u\} \in E \mid v, u \in c_i\}$.

We define neighbors of vertex v in Eq. 1. It's the set of vertices directly linking to vertex u.

$$\Gamma(v) = \{u \mid \{v, u\} \in E\} \tag{1}$$

The set of shared neighbors of vertex v and u are the vertices that both directly connect to vertex v and u. It's defined in Eq. 2.

$$SN(v,u) = \{\Gamma(v) \cap \Gamma(u)\} \tag{2}$$

All neighbors of vertex v and u are the union of vertices that connect to vertex v or u. It is defined in Eq. 3.

$$AN(v,u) = \{\Gamma(v) \cup \Gamma(u)\} \tag{3}$$

In order to measure closeness between vertex v or u, we define $R(v,u)$ in Eq. 4. It is based on fraction of shared neighbors between the two vertices.

$$R(v,u) = \frac{SN(v,u)}{AN(v,u)} \tag{4}$$

Besides, during the hierarchical clustering process, we should merge the sub-clusters based on the closeness between different clusters. The links between cluster i and j is defined in Eq. 5. It's the number of edges that between cluster i and j.

$$Links(i,j) = \left| \{ \{v,u\} \in E \mid v \in c_i, u \in c_j \} \right| \tag{5}$$

The closeness between cluster i and j are defined in Eq. 6.

$$RC(i,j) = \frac{Links(i,j)}{|c_i| \times |c_j|} \tag{6}$$

2.2   Process of ISNGC Algorithm

The process of ISNGC Algorithm is a process of bottom-up hierarchical clustering. It has only one parameter which is the target number k of clusters. It is simply stated as follows:

1. Apply K-Nearest Neighbors Graph algorithm to graph in order to get small sub-clusters. In KNNG algorithm, we use $R(v,u)$ defined above to measure the distance between vertices.
2. Then calculate the closeness $RC(i,j)$ between sub-clusters which we get from step 1.
3. Merge the clusters which have the maximum value of $RC(i,j)$.
4. Repeat from step 2 until there are only k clusters left in the graph.

2.3   Work Principles of ISNGC Algorithm

At the beginning, there are n vertices in graph G. Because each vertex is a single cluster, there are n different clusters. It's a general knowledge that if vertices are in the same cluster, they tend to have more shared neighbors. So, we execute step 1 of ISNGC algorithm to connect single vertex to get small clusters.

However, we should notice that we cannot use Eq. 4 as the measurement of hierarchical clustering process, because this value has close relationship with the size of clusters. It is proved as follows. We suppose there is a graph G which contains n vertices. And cluster i contains x vertices. The probability that vertices in the same cluster link with each other is P_in. And the probability that vertex link with vertices in the other cluster is P_out. P_in is greater than P_out. Vertices u and v are all in cluster i. $R(v, u)$ can be estimated by Eq. 7. Because $SN(v, u)$ is nearly $x \times P\_in$ and $AN(v, u)$ is nearly $(n - x) \times P_{out} + x \times P\_in$. It can be easily proved that $R(v, u)$ is positive correlated with x.

$$f(x) = \frac{x \times P\_in}{(n - x) \times P\_out + x \times P\_in} \tag{7}$$

So if we use Eq. 4 as measurement of hierarchical clustering process, it tends to merge vertices in large clusters first. So, we just use Eq. 4 to execute KNNG algorithm in order to get small sub clusters. In Eq. 6, we not only consider links between clusters but also consider the affect of sizes of clusters. But, Eq. 6 can not measure closeness between clusters correctly if the size of clusters is too small. As a result, we firstly execute step 1 to get sub clusters then do the hierarchical clustering.

Furthermore, through step 1, vertices are combined into small clusters, this operation can cut the number of iteration times when doing hierarchical clustering. It can make algorithm more efficient. So, by combine step 1 and step 2, we can make the algorithm fast and detect communities of different sizes.

2.4   Computational Complexity Analysis

Firstly, we assume there are n vertices and l edges in graph G and the target number of clusters is k. The value of k is much smaller than n. In step 1, computing m nearest neighbors for n vertices will take about $O(n^2 log n)$ at the worst case. And there are improved KNNG algorithms to make it more efficient [13].

After step 1, at the worst case, it would leave n/m clusters. We assume at each iteration sizes of clusters are nearly equal. So the computational complexity of hierarchical clustering is shown in Eq. 8. It's about $O(n^3)$.

Combine with step 1, the computational complexity of ISNGC algorithm is $O(n^3)$. Its performance is quite well.

$$Complexity = \sum_{i=0}^{n/m-k} \binom{n/m-k-i}{2} \left(\frac{n}{n/m-k-i}\right)^2 \tag{8}$$

3   Experiments

In this section, we present a number of tests generated by computer to evaluate the performance of ISNGC algorithm. In each case, we find our algorithm can detect communities correctly in these graphs.

Firstly, we introduce two widely-used testing graph models. One is the Relaxed Caveman Graph. It is an early attempt in social sciences to capture the clustering properties of social networks. And it is produced by linking together a ring of small complete graphs by moving small number of the edges in each small graph to point another graph [14]. It can be represented as $Graph(l, k, p)$. The graph generated contains $l * k$ vertices. l is the number of clusters and k is the number of vertices in each cluster. p is the possibility that edges moving to point to other clusters.

The other one is Random Partition Graph [3]. It is a graph of communities with different sizes predefined. Nodes in the same cluster are connected with probability P_in and nodes in different clusters are connected with probability P_out. It is represented as $Graph(Sizes, P\_in, P\_out)$ where $Sizes$ is a list of size of different clusters.

Sizes of communities generate by Relaxed Caveman Graph are equal. So we use graphs generated Random Partition Graph as supplement. It can test whether the algorithms can detect communities correctly when sizes of clusters are quite different.

3.1   Experiment 1

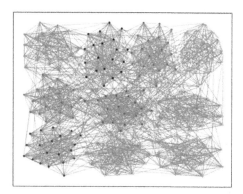 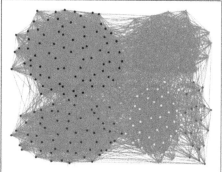

Fig. 1. The left picture is the result clustering of $Graph(10, 20, 0.2)$ and the right one is the result clustering of $Graph(\{20,40,50,60,80\}, 0.6, 0.05)$

In this experiment, we generate two test cases. The first one is generated by Relaxed Caveman Graph and its parameters are $Graph(10, 20, 0.2)$. The second one is generated by Random Partition Graph. It is $Graph(\{20,40,50,60,80\}, 0.6, 0.05)$. The result clustering got by ISNGC algorithm is shown in Fig. 1. As we can see from it, ISNGC can detect communities in these graphs correctly.

3.2   Experiment 2

In this experiment, in order to compare ISNGC algorithm with MCL algorithm we do a list of tests. Firstly, graphs generated by Relaxed Caveman Graph are used as test cases. These graphs are listed in Table 1.

Table 1. Graphs generated by Relaxed Caveman Graph

No.	l	k	p
1	10	20	0.05
2	10	20	0.10
3	10	20	0.15
4	10	20	0.20
5	10	20	0.25
6	10	20	0.30
7	10	20	0.35
8	10	20	0.40

We both apply ISNGC algorithm and MCL algorithm to these graphs. For comparison, the fraction of vertices classified correctly is used as evaluation standards. We show the testing results in Fig. 2. As it's shown, both ISNGC algorithm and MCL algorithm can detect these communities correctly when sizes of clusters are nearly equal.

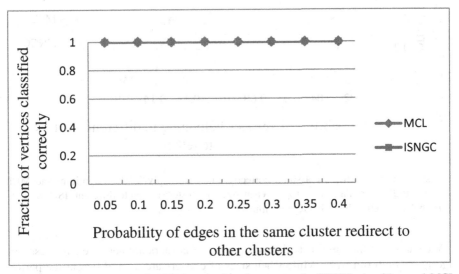

Fig. 2. Fraction of vertices classified correctly when we apply ISNGC algorithm and MCL algorithm to Relaxed Caveman Graph. When sizes of clusters are nearly equal, both the two algorithms can detect clusters correctly.

Then we use graphs generated by Random Partition Graph to test the performance of these two algorithms when sizes of clusters are quite different. These graphs used as test cases are listed in Table 2. And the testing results are shown in Fig. 3.

Table 2. Graphs generated by Random Partition Graph

No.	Sizes	P_in	P_out
1	{20, 40, 50, 60, 80}	0.6	0.02
2	{20, 40, 50, 60, 80}	0.6	0.04
3	{20, 40, 50, 60, 80}	0.6	0.06
4	{20, 40, 50, 60, 80}	0.6	0.08
5	{20, 40, 50, 60, 80}	0.6	0.10
6	{20, 40, 50, 60, 80}	0.6	0.12
7	{20, 40, 50, 60, 80}	0.6	0.14
8	{20, 40, 50, 60, 80}	0.6	0.16

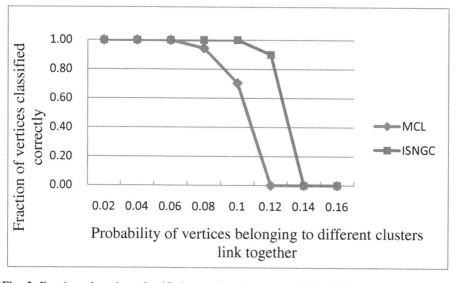

Fig. 3. Fraction of vertices classified correctly when we apply ISNGC algorithm and MCL algorithm to Random Partition Graph. When sizes of clusters are quite different, ISNGC algorithms behaves better than ISNGC algorithms.

We can see from these tests that when sizes of clusters are nearly equal these two algorithms both run well. While when sizes of clusters are quite different the performance of ISNGC algorithm is better than MCL algorithm. ISNGC algorithm detects the small clusters in the graphs successfully while MCL algorithm fails.

3.3   Experiment 3

Besides these above two experiments, we also test the performance of ISNGC algorithm when it's applied to large networks. We use Relaxed Caveman Graph to generate a test case that contains 5000 vertices. It is shown in the left picture of Fig. 4. And then we use ISNGC algorithm to process it. The result clustering is quite well and is shown in the right picture of Fig. 4.

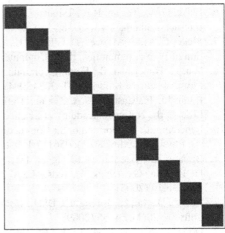

Fig. 4. The adjacency matrix of the 5000 vertices graph that contains 10 clusters. The picture is 5000×5000 pixel. If vertex u and v are connected, then pixel at position (u, v) is black. Otherwise, it's white. In the left picture, the vertices are randomly ordered and community structure is not obvious. In the right picture, the vertices are reordered based on results got by ISNGC algorithm. The community structure is quite clear. And the ten clusters are all detected correctly.

4   Conclusions

In this article, we introduce an efficient hierarchical graph clustering algorithm to detect community structure. Unlike previous works, this algorithm is mainly based on shared neighbors and links. We have tested our algorithm through a number of graphs generated by Relaxed Caveman Graph and Random Partition Graph. The performance of our algorithm is quite well.

But, a number of extensions or improvements of ISNGC algorithms may be possible. Firstly, it can be extended to be suitable for weighted and directed graph. We can add weight to the formulas. And by improving measurement methods, it can detect community structures which are not so clear in networks.

This methodology can be widely used in social network analysis, computer network analysis, bioinformatics analysis, traffic optimization analysis, etc. We hope our ideas and methods can help solve some existing problems in these areas.

References

1. Strogatz, S.H.: Exploring Complex Networks. Nature 410(6825), 268–276 (2001)
2. Girvan, M., Newman, M.E.J.: Community Structure in Social and Biological Networks. PNAS 99(12), 7821–7826 (2002)
3. Fortunato, S.: Community Detection in Graphs. Physics Report 486, 75–274 (2010)
4. Schaeffer, S.E.: Graph Clustering. Computer Science Review 1(1), 27–64 (2007)
5. van Dongen, S.M.: Graph Clustering by Flow Simulation. Ph.D. Thesis (2000)
6. Flake, G.W., Tarjan, R.E., Tsioutsiouliklis, K.: Graph Clustering and Minimum Cut Trees. Internet Mathematics 1(4), 385–408 (2004)
7. Flake, G.W., Lawrence, S., Giles, C.L., Coetzee, F.M.: Self-Organization and identification of Web communities. IEEE Computer 35(3), 66–71 (2002)
8. Wu, F., Huberman, B.: Finding communities in linear time: a physics approach. European Physical Journal B 38(2), 331–338 (2004)
9. Radicchi, F., Castellano, C., et al.: Defining and identifying communities in networks. Proc. of the National Academy of Science 101(9), 2658–2663 (2004)
10. Lancichinitti, A., Fortunato, S.: Community Detection Algorithms: A Comparative Analysis. Physical Review 80(5), 056117:1–056117:11 (2009)
11. Brandes, U., Gaertler, M., Wagner, D.: Experiments on Graph Clustering Algorithms. In: Di Battista, G., Zwick, U. (eds.) ESA 2003. LNCS, vol. 2832, pp. 568–579. Springer, Heidelberg (2003)
12. Bo, Y., Dayou, L., Jiming, L., Di, J.: Complex Network Clustering Algorithms. Journal of Software 20(1), 54–66 (2009)
13. Dong, W., Charikar, M., Li, K.: Efficient K-Nearest Neighbor Graph Construction for Generic Similarity Measures. In: Proceedings of the 20th International Conference on World Wide Web, pp. 577–586 (2001)
14. Watts, D.J.: Small Worlds. Princeton University Press, Princeton (1999)

A Correlation-Based Approach to Consumer Oriented Evaluation of Product Recommendation

Jing-Zhong Jin[1,*], Yoshiteru Nakamori[1], and Andrzej P. Wierzbicki[2]

[1] School of Knowledge Science, Japan Advanced Institute of Science and Technology, 1-1 Asahidai, Nomi City, Ishikawa, 923-1292 Japan
{jinjingzhong,nakamori2012}@gmail.com
[2] National Institute of Telecommunications, Szachowa 1, 04-894 Warsaw, Poland
A.Wierzbicki@itl.waw.pl

Abstract. The paper deals with the multi-attribute aggregation problem in product recommendation, which focuses on consumers' preferences and personal tastes. Specifically, an ontological structure is used for aggregating the attributes selected by consumers as their preferences; in particular this paper also takes the correlation effects between the selected attributes into account. Consequently, this paper presents two kinds of aggregation models based on ontological structure and correlation effect for comparison, and to recommend a ranking list to consumers, four ranking methods are also examined. To make an evaluation of the aggregation models and the ranking methods, a recommendation system was developed and a comparison test was conducted.

Keywords: Kansei engineering, ontological engineering, target-oriented fuzzy method, multi-attribute.

1   Introduction

In today's increasingly competitive market, consumers usually have to face to a huge number of products with different designs but having the same use [1]. Therefore, the design and sense of a product is increasingly important in marketing, since consumers put more attention to their sensibilities or the feelings of the products, not only to the quality of the products. Because consumers accept a tacit understanding that the quality of product is appropriate (consumers cannot know the quality before they bought the product), the design and sense of the products would be the visible factors for consumers to select their prefer product, so the design is now becoming as important as the quality of a product [2]. Thus it is necessary to further develop a consumer-oriented approach in recommending products; such approach focuses on the satisfactions of the consumers' requirements. This issue has received increasing attention in the research of consumer-oriented design and Kansei Engineering since the 1970s. Kansei Engineering, which was invented by *Nagamachi* at Hiroshima University in the 1970s, and defined as "translating technology of a consumer's feeling and image for a product

---

[*] Corresponding author.

M. Wang (Ed.): KSEM 2013, LNAI 8041, pp. 513–525, 2013.
© Springer-Verlag Berlin Heidelberg 2013

into design elements"[3], has been proved as an efficient and successful approach in many fields[4, 5], such as automotive, home electronics, office machines, cosmetics, food and drink, packaging, building products, and other sectors [6].The word Kansei expresses the subjective feelings of a product by people immanent phenomenological perception using all senses, viewing, hearing, touching, smelling and other ways [5]. In Kansei related researches, the most common method of the collection of data is the semantic differential (SD) method [9], which uses a set of adjectives and asks evaluators to express their feeling to an object with those words [5, 7, 8].

When consumers express their requirements for purchasing products, they usually have a multi-attribute expression; this means that an aggregation method of those attributes becomes an important part in matching consumer's requirements. We can suppose that the requirement of the consumer is a whole entity, and we can use the Kansei words selected by consumers to describe the entity, so it would become an ontological issue [1]. The consumers' requirements would be then interpreted as consumers' ontological profile. In previous research of multi-attribute aggregation problems in recommendation issues, the internal relationship between the attributes has not been concerned; this kind of relationship should be taken into account, since the relationship would make effects to the recommendation results.

In this paper, we will focus on evaluation of Japanese traditional crafts, and personalization of consumers' requirements by an ontological profile described by consumers; some aggregation methods of the multi-attribute aspects and some ranking methods will be also discussed, including the comparison of them.

2 Fuzzy Target Oriented Decision Making

2.1 Measurement of the Attributes of the Products

We will describe in this paper a consumer oriented evaluation problem with Kansei data and Context data for traditional crafts (for detail, see [10, 11, 12]). Let us denote by O the collection of specific crafts (products), a finite set, and let us denote its cardinality by $|O| = N$. There are two kinds of data in this paper, Kansei data and Context data. Kansei data is used to describe consumer's feelings regarding to traditional crafts, they are defined by pairs of adjective characters, for example, {Cute, Bitter}; Context data is used for describing the application situations, they are defined by pairs of phrases, For example, {For young, For senior}. Let:

1. $\{F_1, F_2, ..., F_K\}$ be the set of selected Kansei attributes.
2. $\{C_1, C_2, ..., C_M\}$ be the set of selected Context attributes.

---

[1] Ontology is understood here as a word borrowed from philosophy to describe matching words and concepts in computer science; more precisely, it is sometimes called ontological engineering. On the other hand, Kansei engineering has also connections to another part of philosophy called phenomenology, meaning perceiving the world as a whole, by all senses.

Attribute F_k consists of a pair of Kansei words, F_k^l and F_k^r mean the left side Kansei word and right side Kansei word of attribute F_k. We denote $F = \{ F_k^l, F_k^r \mid k = 1, \dots, K\}$ as the set of Kansei attributes; Attribute C_m consists of a pair of Context words, C_m^l and C_m^r mean the left side Context word and right side Context word of attribute C_m. We denote $C = \{ C_m^l, C_m^r \mid k = 1, \dots, K\}$ as the set of Context attributes, we can also denote X as the set of all attributes, where $F \subset X, C \subset X$.

Then we use semantic differential method to make a questionnaire to collect both Kansei data and Context data. We use *M-point* method to make the questionnaire, specifically, to the left side word of the paired Kansei attribute we attach *point 1*, and the other side of the paired Kansei attribute we attach *point M*, in this case, $M = 7$. Let us denote the *M-point* scale by $V = \{v_1, v_2, \dots, v_M\}$.

The questionnaire is given to the evaluators to express their emotional assessments, we denote E as the set of evaluators, where $|E| = P$. Specifically, for a certain object $o_i \in O$, evaluator e_j gives his/her marks on every Kansei attribute and Context attribute, we denote $x_{jk}(o_i)$ as the mark of evaluator e_j to attribute k of o_i, attribute k can be Kansei attributes or Context attributes. Here $x_{jk}(o_i) \in V$. Then we can denote $f_{ik}(v_h)$ as the distribution of the evaluators, which is given by

$$f_{ik}(v_h) = \left| \{e_j \in E : x_{jk}(o_i) = v_h\} \right| \tag{1}$$

2.2   Specification of Consumer's Personal Preference

When the database is settled, we can use it to describe how an object meets an average consumer's preference. Let us denote R as the set of individual consumers' requirements, here $R \subset X$. Suppose we concentrate on a single selected attribute $F_k \in R$. Suppose the consumer target of F_k can be expressed as $v_h` \in V$. To calculate how $F_k$ of $o_i$ meets the target $v_h`$, a fuzzy target based personalization method was proposed in Nakamori [13], with this approach, the fitness value of a single attribute will be calculated, for detail, see [13]. The fitness value means how the object meets the consumer's tastes on a certain attribute. Here we will use this idea and make some modifications.

For each attribute (generally, we use a Kansei attribute F_k as an example), a different evaluator may have different feels on the same object. This means that some evaluators may mark the attribute F_k in the F_k^l area, but others may mark it in the F_k^r area. Therefore, when we calculate the fitness value of an attribute for a certain object, we should separate the paired attribute into two single parts, which we call left part (F_k^l area) and right part (F_k^r area), and calculate the fitness value of each part by the fuzzy target method. We denote SC_{ik}^l as the significance coefficient of F_k when it is in the left part; relatively, SC_{ik}^r means the significance coefficient when F_k is in right part. Formula is as follow:

$$SC_{ik}^l = \frac{\sum_h v_h \cdot f_{ik}(v_h)}{|E|} \qquad h = 1, \dots, (M+1)/2 \tag{2}$$

$$SC_{ik}^r = \frac{\sum_h v_h \cdot f_{ik}(v_h)}{|E|} \qquad h = (M+1)/2, \dots, M \tag{3}$$

These two coefficients interpret the significance of the two sides, in other words, they can measure the distribution of the evaluation data obtained from evaluators. Then we can multifly these two coefficients to the fitness value. We denote the revised fitness value as $g_{k^l}(o_i, v`)$ or $g_{k^r}(o_i, v`)$. It depends on the selection of customer: if consumer's target is set in the left part of the paired attribute, we can use $g_{k^l}(o_i, v`)$ as the fitness value; correspondingly, if the target is set in the right part of the paired attribute, we can use $g_{k^r}(o_i, v`)$ as the fitness value.

3   Aggregation Models Based on Ontological Structure

According to the attributes of the products, the consumer' preference profile may contain several attributes of the products. In other words, we face to a multi-attribute requirement, thus the aggregation of these attributes would be very important in obtaining a scalar measure of consumer's preference. As we have discussed above, there are two kinds of attributes to describe a product: Kansei attributes and Context attributes. Kansei attributes are usually expressed by the adjective words; they are usually used to describe the sensibilities of a consumer about an object. These sensibilities usually have a vague nature for a human to describe. Correspondingly, Context attributes usually express the product's purpose of use or characteristic of users. They usually include some short phrases, and they are different from Kansei attributes: a Context attribute usually has an explicit meaning for a consumer. In particular, we can use an ontological structure to describe them: we will split consumer's requirement into several sub-requirement entities according to the selected Context attributes; and we use Kansei attributes to describe these sub-requirement entities, then we could integrate these sub-requirement entities as consumer's personal requirement. In the following section we will propose two aggregation models.

The overview of this recommendation system is shown in Fig. 1. There are 4 parts in this system: Interface, Specification module, Aggregation module, Database. The consumer can select attributes and set their levels and importance coefficients in the interface to describe his/her requirement; the specification module can measure the satisfactions of the selected attributes; the aggregation module can aggregate the selected attributes and make a recommendation list to consumer.

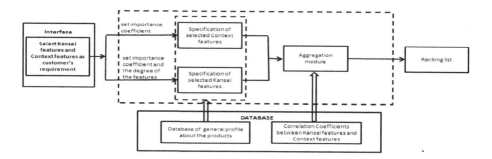

Fig. 1. Overview of the recommendation system

3.1   Consumer Target Based Aggregation Model

Given $R = R_c \cup R_k$ as the requirements set of consumer, where R_c means the Context requirements, R_k means the Kansei requirements: $R_c = \{rc_1, \ldots, rc_x\}$, $R_k = \{rk_1, \ldots, rk_y\}$. We have calculated how the object meets the selected attributes separately, and then we will use a method to aggregate them to see how the object meets the consumer's preference. Here we use an ontological structure to describe consumer's preference. Specifically, we concentrate on the Context attributes, and use the selected Kansei attribute set R_k to describe a selected Context attribute rc_x. This is some kind of enlarged Context attribute or we can say it is a enlarged concept, we denote it as $erc_x \in eR_c$, where $erc_x = \{rc_x, R_k\}$, and eR_c is the set of enlarged Context attributes.

When we use Kansei attributes to describe the Context attribute, we should know how important a Kansei attribute to a Context attribute is, We denote $f(rc_x, rk_y)$ as the importance coefficient between rc_x and rk_y, and we can use the correlation coefficient of attributes average values to describe a possible strength of their relationships. Generally, we define CC_{xy} as the correlation coefficient between Kansei attribute y and Context attribute x, and we assume that the Kansei attribute is more efficient in describing the Context attribute when the correlation effect is significant; and it would be not suitable to describe the Context attribute when the correlation effect is lower than a special value. Here we assume that the special value is 0.2. Consequently, we can map the correlation coefficient to the importance coefficient. Specifically, if their correlation coefficient is bigger than *0.2*, we assume that there is linear relationship between the importance coefficient and the correlation coefficient, the linear relationship is described by the segment *AB* in Fig. 2. The composite line *CDB* indicates a reasonable upper limit to the importance coefficients set by a consumer; hence we define it as the upper recommended transformation line for an adjustment range, defined to meet consumer's personal requirements, which consumer can adjust the importance coefficient in a certain degree. This range is defined in a sense objectively, by taking into account the correlation coefficients obtained from the database. The lower limit of the range is expressed by the composite line *CAB*: specifically, if the correlation coefficient is bigger than *0.2*, we assume that line segment *AB* defines a reasonable lower limit for importance coefficients set by the consumer. The vertical lines in the shadowed area in Fig. 2 are indicated the range of possible importance coefficients that can be selected by the consumer; as reasonable starting points, we can suggest to the consumer the middle values of these ranges, denoted on Fig. 2 by a broken line.

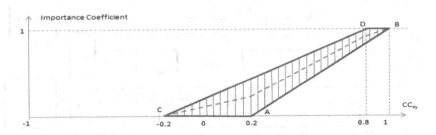

Fig. 2. Conversion from correlation coefficient to importance coefficient

Thus, consumers can set their preferred importance of a Kansei attribute to a certain Context attribute, which we denoted as $s(rc_x, rk_y)$. We define $f_l(rc_x, rk_y)$ and $f_u(rc_x, rk_y)$ as the recommended lower transformation function and upper transformation function (see Fig. 2), and we define $f_r(rc_x, rk_y)$ as the middle value function which we recommend to consumer as starting values, in detail as follows:

$$f_l\left(rc_x, rk_y\right) = \begin{cases} 0 & -1 \leq x < 0.2 \\ 1.25x - 0.25 & 0.2 \leq x \leq 1 \end{cases} \tag{4}$$

$$f_u\left(rc_x, rk_y\right) = \begin{cases} 0 & -1 \leq x < -0.2 \\ x + 0.2 & -0.2 \leq x < 0.8 \\ 1 & 0.8 \leq x \leq 1 \end{cases} \tag{5}$$

$$f_r\left(rc_x, rk_y\right) = \begin{cases} 0.5x + 0.1 & -0.2 \leq x < 0.2 \\ 1.125x - 0.025 & 0.2 \leq x < 0.8 \\ 0.625x + 0 & 0.8 \leq x \leq 1 \end{cases} \tag{6}$$

According to $s(rc_x, rk_y)$ and formula (4)-(6), we can define $f(rc_x, rk_y)$ as in (7). The equation can make the adjusted importance coefficient in the shadow part of Fig. 2.

$$f(rc_x, rk_y) = \begin{cases} f_r(rc_x, rk_y) + s(rcx, rky) \cdot \left(f_r(rc_x, rk_y) - f_l(rc_x, rk_y)\right) & -1 \leq s(rcx, rky) \leq 0 \\ f_r(rc_x, rk_y) + s(rcx, rky) \cdot \left(f_u(rc_x, rk_y) - f_r(rc_x, rk_y)\right) & 0 < s(rcx, rky) \leq 1 \end{cases} \tag{7}$$

For the enlarged context attribute erc_x, the fitness of the object o_i can be calculated by

$$g(o_i, erc_x) = g_{rc_x}\left(o_i, v_{rc_x}\right) + \sum_{rk_y \in R_k} f\left(rc_x, rk_y\right) \cdot g_{rk_y}\left(o_i, v_{rk_y}\right) \tag{8}$$

Here v_{rc_x} and v_{rk_y} are the targets of selected Context attribute and Kansei attribute.

As it is shown in Fig. 3, each selected Context attribute can be treated as a sub-requirement entity, and for each sub entity, we use all selected Kansei attributes to describe it according to the importance coefficients and correlation coefficients. For example, see Fig. 4, if there are 8 Kansei attributes (k_1, k_2, ⋯, k_8) for describing the sub-requirement entity, and the consumer selects several Kansei attributes (e.g. k_1, k_4 and k_8 were selected.) to describe his/her requirements, then he/she can adjust the importance coefficients of the selected Kansei attributes within the adjustment range to meet his tastes.

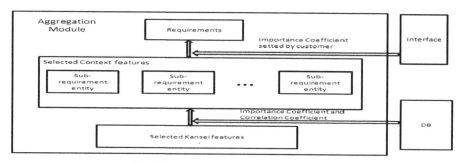

Fig. 3. Multi-attribute aggregation process

Then we should aggregate all $erc_x \in eR_c$, to see how the object meets consumer's preference. For each selected Context attribute, the consumer can also set the importance which describe that how important of the Context attribute to his/her preference. We denote it by $f(erc_x)$.

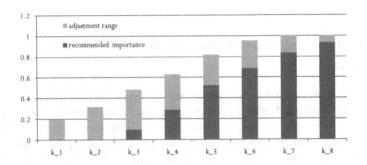

Fig. 4. Adjustment range of the selected Kansei attributes

If we assume that all selected Context attributes have a fuzzy logical relation *"OR"*, it means all Context attributes have no difference of importance, and then the aggregation method is as follow:

$$g(o_i) = max_{erc_x \in eR_c} f(erc_x) \cdot g(o_i, erc_x) \qquad (9)$$

If we assume the fuzzy logical relation *"AND"* rules all Context attributes, it means consumer wants every selected Context attribute to meet his/her requirements, then the aggregation method is:

$$g(o_i) = min_{erc_x \in eR_c} f(erc_x) \cdot g(o_i, erc_x) \qquad (10)$$

For the selected Context words, there might be some very small fitness value to an object, which will make the final score indiscernible, so we will introduce two other ranking methods to solve this problem. These two methods follow the *compensable criteria* and *essential criteria*. The first one means that a large value of criteria will compensate a small one, and it is some kind of weighted average approach; the second one means that all criteria should have reasonably large values, and it is some kind of *reference point method* which was proposed by Wierzbicki [14]. First, we should compute the statistical mean to see the average fitness of all objects for a given selected Context attribute:

$$g_{av}(erc_x) = \sum_{o_i \in O} f(erc_x) \cdot g(o_i, erc_x)/|O| \qquad (11)$$

where $g(o_i, erc_x)$ is computed as in *(8)* and $|O|$ is the number of the objects. Then we can define the *compensable criteria* and the *essential criteria*:

$$g_{com}(o_i) = \sum_{erc_x \in eR_c} (f(erc_x) \cdot g(o_i, erc_x) - g_{av}(erc_x)) \qquad (12)$$

$$g_{ess}(o_i) = min_{erc_x \in eR_C}(f(erc_x) \cdot g(o_i, erc_x) - g_{av}(erc_x)) + \varepsilon \cdot g_{com}(o_i) \quad (13)$$

where $g(o_i, erc_x)$ is also computed as in (8) and the coefficient $\varepsilon > 0$ in (13) indicates a compromise between interpreting the relations between the selected Context attributes as a fuzzy logical "AND" operation and interpreting them as compensable criteria

3.2   An Indirect Aggregation Model Using the Idea of a Prototype System PrOnto

The prototype system *PrOnto* was developed in the *Requested Research Project of Poland* in the National Institute of Telecommunications, entitled *"Teleinformatic Services and Networks of Next Generation – Technical, Applied and Market Aspects "*, The system *PrOnto* is based on radically personalized ontological profiles of users, and takes into account the interaction with different users (see [14]). If we use the idea of *PrOnto system* that a certain user-defined *concept* usually has a set of *keywords* to describe it, and a user can adjust the importance coefficients of these keywords, we can assume that for a certain Context attribute, treated as a *concept*, there is also a set of Kansei attributes, treated as a set of *keywords*, that can describe it commonly (we can call it a description set). And when we want to measure the satisfaction of a Context attribute, we can measure it by the related Kansei attributes, and personalize it by consumer's wishes. Specifically, we can make a set of Kansei attributes, which have higher correlations to that Context attribute, and instead of specifying the Context attribute by fuzzy method, we can use the fitness values of the set of Kansei attributes to describe the Context attribute indirectly, there is at least an advantage that we can distinguish different Context words in detail, because instead of using evaluation data only, there are many other Kansei attributes that can show their differences. If we want to take a typical selected Kansei attribute into account for the sub-requirement entity (a selected Context attribute), we can reset the correlation coefficient between this pair of Kansei attribute and Context attribute. For example, see Fig.5, the selected Kansei attributes were k_2, k_4 and k_5, we can just reset the correlation coefficients in a rational range to meet consumer's special tastes.

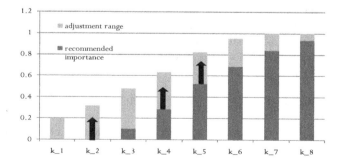

Fig. 5. Interior of a sub-requirement entity

The algorithm is similar to the method we have mentioned above, the difference is that how to calculate the fitness of the enlarged context attribute erc_x. We just use the Kansei attributes to describe the enlarged context attribute in this model, and we will not take the fitness of selected Context attribute erc_x into account, the reason is that all Kansei attributes have special relationships with a certain context attribute, we can just use the Kansei attributes which have significant relations with that context attribute to distinguish context attributes. We denote K_{rc_x} as the Kansei attributes set, which have significant relations with rc_x, and as mentioned above, R_k is the selected Kansei attributes set. Then $g(o_i, erc_x)$ is given by

$$g(o_i, erc_x) = \sum_{k_y \in K_{rc_x}, k_y \in R_k} f_r(rc_x, k_y) \cdot g_{rk_y}\left(o_i, v_{k_y}\right) + \sum_{rk_y \in R_k} f(rc_x, rk_y) \cdot g_{rk_y}\left(o_i, v_{rk_y}\right) \quad (14)$$

The first part on the right side of the equation is the satisfaction of the selected Context attributes calculated by the description set; the second part on the right side of the equation is the satisfaction of the additional selected Kansei attributes, which are not included in the description set.

4   Evaluation and Comparison of Different Aggregation Models

For evaluation and comparison, a group of samples called *"Kutani ware"* were used to test the aggregation models and ranking methods. These samples are groups of cups made by traditional methods, and for each sample, we use 6 pairs of Context words and 20 pairs of Kansei words to describe it. To evaluate them, 60 volunteers participated in. They were given some kind of questionnaire with 26 pairs of attributes (6 pairs of Context words and 20 pairs of Kansei words) on it. For each attribute, there are 7 degrees to select for describing how you feel about the samples.

We have also programmed a recommendation system for personalized consumers' requirements. It includes all aggregation models and ranking methods, which we have discussed above. According to the database and the personal requirements selected by consumers, this program can recommend a list of the cups' number for consumers, which can help them to select their favorite cups (see Fig. 6).

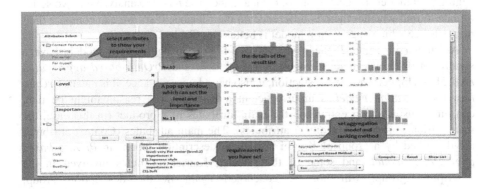

Fig. 6. The interface of the recommendation system

To make a comparison of different aggregation models and ranking methods, we will make an evaluation of the recommendation system to test the different aggregation models and ranking methods. In this evaluation, we had 25 volunteers for 49 times' test. Let them select some of Context attributes and Kansei attributes and set the level and importance of them as their requirements, then according to their selections, they should pick up the most preferred sample. We can use the recommendation system with these descriptions to test which models and which ranking methods are better.

There are 30 coffee cups in this test. With the recommendation system and consumer's special description of the requirement, we could get a recommendation list of the coffee cups for each time's test. Compare to this list with consumer's favorite coffee cup, we can know the position of the favorite coffee cup in this list. For example, if the number of the consumer's favorite cup is 23, and this cup is at the 3th position of the list, then we mark it as 3 (see Table 1). According to the marks, we can know the satisfaction of the recommendation results roughly, for example, if one of the marks in Table 1 is 5, the satisfaction would be *(30-5)/30 = 0.83*.

Table 1. Comparison results of different aggregation models and ranking methods

No.	Consumer target based model				PrOnto based model			
	ESS	COM	AND	OR	ESS	COM	AND	OR
1	10	9	10	9	13	13	13	14
2	10	19	25	15	3	4	19	3
3	3	2	2	4	4	4	4	6
			...					
49	6	6	16	7	5	5	18	5

As shown in Table 1, for each time of evaluation, there are two different aggregation models and 4 different ranking methods. At first, we compared the satisfactions of different aggregation models. Specifically, we have got the average marks of different aggregation models for each items of data, and then use *t-Test* to analyze the two series of data to find whether there is difference on their means, if so, we can compare their means to see which aggregation model is better (see Table 2).

As shown in Table 2, the mean values of the two aggregation models have a significant difference under the possibility of *90%*, and *PrOnto based aggregation model* has a higher average satisfaction, it also has a lower variance, this means *PrOnto based aggregation model* is more efficient and stable. The reason might be that, consumers usually concentrate on the main attributes of a product (in this case, the main attributes should be Context attributes). The main attributes (Context attribute) are described by a group of Kansei attributes (description group) in *PrOnto based aggregation model*, and some of the selected Kansei attributes involved in the description group, will not be concerned again in the description group; correspondingly, the main attributes are calculated by target-based fuzzy method in

Consumer target based aggregation model, and the selected Kansei attributes are calculated separately. Therefore, a duplicate calculation would happen when selected Kansei attributes have higher correlations to the main attributes. That would make the main attributes more easily to be affected by selected Kansei attributes in *Consumer target based aggregation model*. The evaluation data of all attributes expresses the general understandings/feelings of the object, so there might be some deviations when we use them to calculate the satisfactions of consumers. In *PrOnto based aggregation model*, the satisfaction of the Context attribute is calculated by description set (a set of Kansei attribute with higher correlation to that Context attribute) instead of using the evaluation data of the Context attribute, it means we used several pieces of evaluation data to measure the satisfaction of the selected Context attribute, therefore, the deviations might be averaged. That may be the reason why *PrOnto based aggregation model* is more stable.

Table 2. Comparison of different aggregation models

Aggregation models	Average Satisfaction	variance	t-Test
Consumer target based aggregation model	0.72	18.53	0.097
PrOnto based aggregation model	0.77	12.49	

As shown in Table 3, the two ranking methods with *essential criteria (ESS)* and *compensable criteria (COM)* have significant difference from the ranking method based on logical operation "AND" under the possibility of 95%. That means we can compare the average Satisfaction of the ranking method *ESS* and the ranking method based on logical operation "AND" (including the ranking method *COM* and the ranking method based on logical operation "AND") to see which one is better.

Table 3. Comparison of different ranking methods

	ESS	COM	AND	OR	Average Satisfaction	variance
ESS		0.5383	0.0059	0.3072	0.8051	13.8355
COM	0.5383		0.0481	0.7083	0.7867	25.1769
AND	0.0059	0.0481		0.1102	0.7153	30.9358
OR	0.3072	0.7083	0.1102		0.7738	27.1458

We can see that the ranking methods of *ESS*, *COM* and the ranking method based on logical operator "OR" have higher average satisfactions than the method based on logical operator "AND", but from the analysis above, we can only say that *ESS* and *COM* is better than the ranking method based on logical operator "AND". According to the variance value, we found that ranking method *ESS* has a lower variance; that means the ranking method *ESS* is more stable than others, so we can say ranking method *ESS* is better than other methods with using *PrOnto based aggregation model.*

5   Conclusion

We have discussed and tested two correlation-based multi-attribute aggregation models and four ranking methods for the recommendation problem of Japanese traditional crafts in this paper; we have taken consumer personalized preferences about the attributes of the objects into account. We have also compared different aggregation models and ranking methods with using the recommendation system we have developed. According to the analysis of the comparison, we found that *PrOnto based aggregation method* is more efficient and stable, and under using this aggregation model, the ranking method of *ESS* results in a better performance. We indicated also some problems, such as the matching problem between general understanding and personal understanding: the data we used in this paper is based on some kind of general understanding evaluated by a set of volunteers and recorded in the system database, so when consumers set their requirements, they might express some deviations between the database and the needs which consumers really want.

References

1. Jiao, J., Zhang, Y., Helander, M.: A Kansei mining system for affective design. Expert Systems with Applications 30(4), 658–673 (2006)
2. Petiot, J.-F., Yannou, B.: Measuring consumer perceptions for a better comprehension, specification and assessment of product semantics. International Journal of Industrial Ergonomics 33(6), 507–525 (2004)
3. Nagamachi, M.: Kansei engineering: A new ergonomic consumer-oriented technology for product development. International Journal of Industrial Ergonomics 15(1), 3–11 (1995)
4. Nagamachi, M.: Kansei engineering as powerful consumer-oriented technology for product development. International Journal of Industrial Ergonomics 33(3), 289–294 (2002)
5. Schütte, S.: Engineering emotional values in product design—Kansei engineering in development. Ph.D. dissertation, Inst. Technol., Linköpings Univ., Linköpings, Sweden (2005)
6. Childs, T., De Pennington, A., Rait, J., Robins, T., Jones, K., Workman, C., Warren, S., Colwill, J.: Affective design (Kansei engineering) in Japan. Faraday Packaging Partnership. Univ. Leeds, Leeds (2001)
7. Jindo, T., Hirasago, K.: Application studies to car interior of Kansei engineering. International Journal of Industrial Ergonomic 19(2), 105–114 (1997)
8. Nakada, K.: Kansei engineering research on the design of construction machinery. International Journal of Industrial Ergonomics 19(2), 129–146 (1997)
9. Osgood, C., Suci, G., Tannenbaum, P.: The Measurement of Meaning. University of Illinois Press, Urbana (1957)
10. Huynh, V.N., Yan, H.B., Nakamori, Y.: A target-based decision making approach to consumer-oriented evaluation model for Japanese traditional crafts. IEEE Trans. Engineering Management 57(4), 575–588 (2010)

11. Yan, H.B., Huynh, V.N., Nakamori, Y.: A probability-based approach to consumer oriented evaluation of traditional craft items using kansei data. In: Huynh, V.N., et al. (eds.) Interval/Probabilistic Uncertainty and Non-Classical Logics, pp. 326–340. Springer, Heidelberg (2008)
12. Huynh, V.-N., Nakamori, Y., Yan, H.: A Comparative Study of Target-Based Evaluation of Traditional Craft Patterns Using Kansei Data. In: Bi, Y., Williams, M.-A. (eds.) KSEM 2010. LNCS, vol. 6291, pp. 160–173. Springer, Heidelberg (2010)
13. Nakamori, Y.: Kansei Information Transfer Technology. In: Tang, Y., Huynh, V.-N., Lawry, J. (eds.) IUKM 2011. LNCS, vol. 7027, pp. 209–218. Springer, Heidelberg (2011)
14. Chudzian, C., Granat, J., Klimasara, E., Sobieszek, J., Wierzbicki, A.P.: Personalized Knowledge Mining in Large Text Sets. Journal of Telecommunications and Information Technology 3, 123–130 (2011)

The Failure Prediction of Cluster Systems Based on System Logs

Jungang Xu and Hui Li

University of Chinese Academy of Sciences, Beijing, China
xujg@ucas.ac.cn, lihui211@mails.ucas.ac.cn

Abstract. The failure prediction of cluster systems is an effective approach to improve the reliability of the cluster systems, which is becoming a new research hotspot of high performance computing, especially with the growth of cluster systems and applications both in scale and complexity. A classification sequential rule model is proposed to predict cluster system failures. The system logs of BlueGene/L, Red Storm, and Spirit are used as experimental datasets to predict cluster system failures. The results show that sequential rule approach outperforms SVM and HSMM in terms of precision and F-measure in 5hr prediction window, and in 1hr or 12hr prediction window, sequential rules, SVM and HSMM have their own strengths and weaknesses respectively.

Keywords: Cluster systems, Failure prediction, Sequential rule, SVM, HSMM.

1 Introduction

With the extension of applications of cluster systems, the failure of cluster systems is becoming a serious problem. More and more attentions are focused on cluster system failure prediction based on cluster system log analysis with data mining and machine learning technologies, which is an effective approach to improve the reliability of cluster systems.

At present, the main research areas of the cluster system failure prediction include: (1) the cluster system proactive failure analysis which aims to build a proactive prediction and control system based on time-series prediction method, rule-based prediction method and Bayesian network prediction method for a cluster system to alleviate the burden of system administrator in a complex cluster system environment [1][2][3]; (2) cluster system online failure prediction, which is different from traditional methods that enhance cluster system reliability, and is based on the current running state of the cluster system and uses the current system status and historical experience to predict cluster system failures. The basic methods and procedures for a variety of online application failure prediction methods [6] include SVM-based methods [4], rule-based methods, time-series methods [5] and decision tree-based methods [2]; (3) cluster system failure prediction model, in reference [8], an experiment on a 350 node cluster system for failure prediction is performed and the experimental results show that the hidden Markov models and Bayesian methods have relatively better accuracy and

M. Wang (Ed.): KSEM 2013, LNAI 8041, pp. 526–537, 2013.

recall rate. In reference [9], three simple failure prediction methods are presented by Liang based on the spatial and temporal distribution of cluster system and achieve good experiment results in the BlueGene/L system logs. Fulp et al [10] proposed a failure prediction method based on a new spectrum function support vector machine, which can predict hard disk failure with an accuracy of 73% two days in advance; (4) meta-learning failure analysis, Gujrati et al [11] proposed a three-stage failure prediction method that can automatically handle Remote Access Service (RAS) logs and discover the failure mode, and a meta-learning method is discussed to adaptively select the basic methods to improve the accuracy of failure prediction. A dynamic learning cluster system log failure prediction engine is proposed in reference [12] by Gu et al, which can perform dynamic training, dynamic testing and dynamic prediction.

Although the failure prediction of cluster systems has been studied for many years, it is still an outstanding issue. The low prediction accuracy is the main problem of the failure prediction of cluster systems. Most of the existing researches focus only on the specific methods that can capture and find failures. In fact, the system failure of cluster systems is extremely complex, therefore, it is unrealistic to capture all the system failures only with a single method. For example, many rule-based failure prediction approaches are only concerned about the relationship between the warning messages and the failure messages [11] [13]. In these works, experiment results show that there are only a small number of failure events occurred after warning events. Therefore, these methods cannot provide an effective failure prediction.

In our work, we focus on the failure prediction of cluster systems based on classification sequential rules. Classification sequential rules are obtained by analyzing the training logs among the prediction window before the FATAL-window. Then we predict whether system failure will happen in prediction window based on the classification sequential rules obtained in the period of observation window.

The rest of this paper is organized as follows. The next section briefly introduces the basic concepts and methods of the cluster system failure analysis, classification sequential rules and prediction methods. Subsequently, an experiment is designed for failure prediction of cluster systems based on classification sequential rules in section 3. In section 4, we apply classification sequence rules, SVM and HSMM to predict system failure with the system logs of BlueGene/L, Red Storm and Spirit. Finally, in section 5, we give the conclusion and future work.

2   Failure Analysis and Sequential Rules

2.1   Failure Prediction Definition

The basic idea of our failure prediction method of cluster systems is shown in Fig. 1. The aim of the failure prediction of a cluster system is to judge whether there will be failures in prediction window based on the events happened in the observation window. In our work, we attempt to predict whether the fatal incident will occur in the prediction window. We consider the prediction window as FATAL if fatal events are predicted to occur during the period, and consider the prediction window as NONFATAL if no fatal event happened in prediction window.

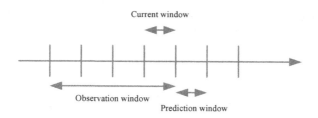

Fig. 1. The basic idea of the failure prediction method

2.2   Sequential Rules

Definitions. Definition 1: classification sequence database, let m be the number of classification clusters, $C = \{c_i | 1 \le i \le m\}$ is the item set that is contained in classification cluster respectively, c_i is the event item. The total sequence database is $D = D_1 + D_2 + \cdots + D_m$, D_i is a sub-sequence database where the sequences belong to cluster c_i.

Definition 2: classification sequential rules, $\alpha \Rightarrow c_i [\text{sup}, conf]$ is supposed to be a classification sequential rule, if α is a max frequent sequence, c_i is a classification cluster that the sequential rules belong to, $\text{sup}(\alpha, D_i)$ is the support of the classification sequential rules, $conf(\alpha) = \text{sup}(\alpha, D_i)/\text{sup}(\alpha, D)$ is the confidence of the classification sequential rules. A sequence rule is considered as a strong sequence rule when its confidence is greater than the minim confidence threshold.

Definition 3: sequential match, let α be a sequence, if there exists a sequence $\beta = (\beta_1 \alpha \cdots \beta_i \cdots \alpha \cdots \alpha \beta_k), 1 \le k, 1 \le i \le k$, where β_i is a sub-sequence of β. α is matched by β if α is not any sub-sequence β_i, and k is the times that α is matched by β.

Classification Sequential Rule. Firstly, sequential rule-based classification approach can produce a sequential rule database R with prefixspan sequence mining algorithm. And then, we match the unclassified sequence S with each classification rule in R. In our work, we choose the Average Confidence of Matched Sequence Rule (ACMSR) to classify a given sequence into a given class (FATAL or NONFATAL). We define $ACMSR_c^s$ as the ACMSR of the given sequence s in class c as shown in formula (1).

$$ACMSR_c^s = \left(\sum_{r \in R_c} r.cnt * r.conf \right) / |R_c| \tag{1}$$

In formula (1), R_c is one rule set that matches the given sequence S in sub-sequence rule database of class c. r is a classification rule in R_c. $r.cnt$ is the number of r matched sequence S. $r.conf$ is the confidence of rule r. A sequence s belongs to class c_i when c_i satisfies formula (2) as follows.

$$ACMSR_{c_i}^s = \underset{k}{MAX}\left(ACMSR_{c_k}^s\right) \tag{2}$$

3   Failure Prediction

3.1   Log Dataset

In this study, the Bluegene/L event logs are collected from BlueGene/L at Lawrence Livermore National Laboratory (LLNL) over a period of 215 days since 06/03/2005, which stood at number 1 in the top500 supercomputer list in 2004.Then Red Storm event logs are collected from Red Storm at Sandia National Laboratory (SNL) over a period of 104 days since 03/19/2006. And the Spirit event logs are collected from Spirit cluster system at Sandia National Laboratory (SNL) over a period of 558 days since 01/01/2005. An overview of the three cluster system logs is presented in Table 1.

Table 1. Log dataset

System	Start Date	Days	Size(GB)	Compressed (GB)	Messages	Alerts	Categories
BlueGene/L	2005-06-03	215	1.207	0.118	4747963	348460	41
Red Storm	2006-03-19	104	29.990	1.215	219096168	1665774	12
Spirit	2005-01-01	558	30.289	1.678	272298969	172816564	8

3.2   Log Sequence

In our work, each log in all the three systems is formatted into a five-tuple log item like (logtime, date, nodeid, severity degree, keywords). Logtime stands for the time of the log occurred, date stands for the date stamp of the log recorded and nodeid stands for the number of the node that events happened. We get logid by combining each nodeid, severity degree and keywords. And one log sequence is generated by sorting the logs according to the time. Log sequence information of the three systems is shown in Table 2.

Table 2. Log sequence information

System	BlueGene/L	Red Storm	Spirit	Liberty
Node number	64	107	486	223
Key words	28	6	6	5
Sequence	1856	749	3402	1338

3.3   Sequential Rule Based Prediction

Fig. 2 shows the process of the sequential rule based prediction. Firstly, prefixs-pan-based sequence mining algorithm is used to generate classification sequential rules from the training sequence database. Then, rule-based classifier classifies the sequence into NONFATAL or FATAL cluster.

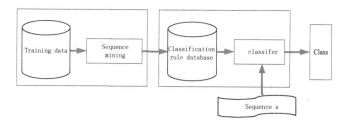

Fig. 2. Sequential rule based prediction

3.4   SVM-Based Prediction

In our work, Support Vector Machine (SVM) is the second classifier we used, which is a general linear classifier. The primary advantage of SVM is that the actual prediction can be fast after the training phase is completed. A drawback of SVM is that training cost can be expensive if there are a large number of training samples. In this study, we chose LIBSVM [14] with the Radial Basis Function (RBF) kernel and the three-fold CV as showed in reference [15].

3.5   HSMM-Based Prediction

The HSMM-based predictor is the third classifier we used in this study. We divide the training data into two parts: NON-FATAL training data and FATAL training data. For NON-FATAL training data, we established a NON-FATAL Hidden Semi-Markov Model (NONFATAL HSMM). Furthermore, a FATAL Hidden Semi-Markov Model is generated from FATAL training data. In our work, we chose HSMM described in reference [16] to predict cluster system failure.

4   Experiments

We use classification sequential rule to predict failure of cluster systems including Blue-Gene/L, Red Storm and Spirit, and contrast it with SVM and HSMM at the same time. The experiment results in time windows of 1hr /5hr/12hr are shown in Fig. 3 to Fig. 11.

In Fig. 3, in time window with duration of 1 hour on BlueGene/L, the prediction precision rate of sequential rule based predictor is 5.28%, better than that of SVM-based predictor and HSMM-based predictor with4.84% and 4.19% respectively. And the recalls of sequential rule based predictor, SVM-based predictor and

HSMM-based predictor are 1.67%, 1.85% and 1.53%. As to the benchmark of F-measure, SVM-based predictor defeats sequential rule based predictor and HSMM-based predictor with F-measure 2.68%.

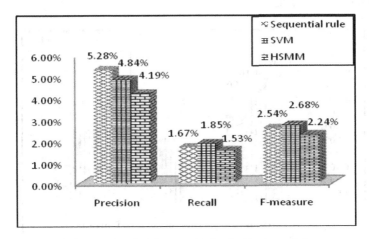

Fig. 3. BlueGene/L failure prediction in 1hr time window

In Fig. 4, in time window with duration of 1 hour on Red Storm, the prediction precision rate of HSMM-based predictor is 4.39%, better than that of sequential rule based predictor and SVM-based predictor with 4.16% and 4.38% respectively. And the recalls of sequential rule based predictor, SVM-based predictor and HSMM-based predictor are 0.82%, 2.06% and 0.80%. As to the benchmark of F-measure, SVM-based predictor defeats sequential rule based predictor and HSMM-based predictor with F-measure 2.80%.

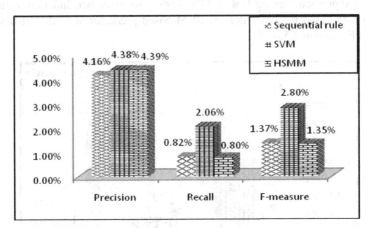

Fig. 4. Red Storm failure prediction in 1hr time window

In Fig. 5, in time window with duration of 1 hour on Spirit system, the prediction precision rate of sequential rule based predictor is 4.73%, better than that of HSMM-based predictor and SVM-based predictor with4.06% and 4.16% respectively.

And the recalls of sequential rule based predictor, SVM-based predictor and HSMM-based predictor are 0.78%, 1.94%, 0.82%. As to the benchmark of F-measure, SVM-based predictor defeats sequential rule based predictor and HSMM-based predictor with F-measure 2.65%.

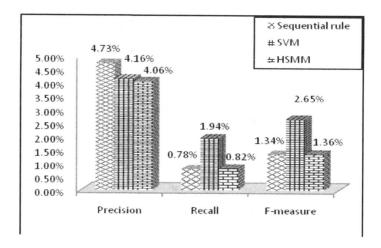

Fig. 5. Spirit failure prediction in 1hr time window

In Fig. 6, in time window with duration of 5 hours on BlueGene/L system, the prediction precision rate of rule-based predictor is 34.78%, better than that of HSMM-based predictor and SVM-based predictor with 22.68% and 24.37% respectively. And the recalls of the sequential rule based predictor, SVM-based predictor and HSMM-based predictor are 20.31%, 2.35%, 1.96%. As to the benchmark of F-measure, sequential rule based predictor defeats SVM-based predictor and HSMM-based predictor with F-measure 25.64%.

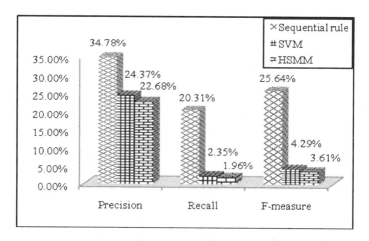

Fig. 6. BlueGene/L failure prediction in 5hr time window

In Fig. 7, in time window with duration of 5 hours on Red Storm, the prediction precision rate of sequential rule based predictor is 20.13%, better than that of HSMM-based predictor and SVM-based predictor with 17.92% and 18.19% respectively. And the recall of sequential rule based predictor, SVM-based predictor and HSMM-based predictor are 18.15%, 2.26% and 2.02%. As to the benchmark of F-measure, sequential rule based predictor defeats SVM-based predictor and HSMM-based predictor with F-measure 19.09%.

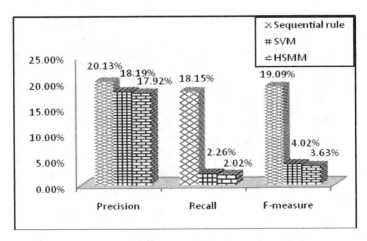

Fig. 7. Red Storm failure prediction in 5hr time window

In Fig. 8, in a time window with duration of 5 hours on Spirit, the prediction precision rate of sequential rule based predictor is 24.51%, better than that of HSMM-based predictor and SVM-based predictor with 18.20% and 18.57% respectively. And the recall of the sequential rule based predictor, SVM-based predictor and HSMM-based predictor are 19.01%, 2.14%, 1.84%. As to the benchmark of F-measure, sequential rule based predictor defeat SVM-based predictor and HSMM-based predictor with F-measure 21.41%.

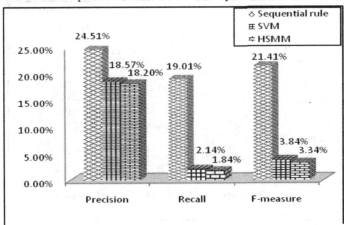

Fig. 8. Spirit failure prediction in 5hr time window

In Fig. 9, in a time window with duration of 12 hours on BlueGene/L, the prediction precision rate of sequential rule based predictor is 51.24%, better than that of HSMM-based predictor and SVM-based predictor with 48.02% and 47.51% respectively. And the recalls of sequential rule based predictor, SVM-based predictor and HSMM-based predictor are 58.72%, 66.52%, 65.01%. As to the benchmark of F-measure, SVM-based predictor defeats sequential rule based predictor and HSMM-based predictor with F-measure 55.43%.

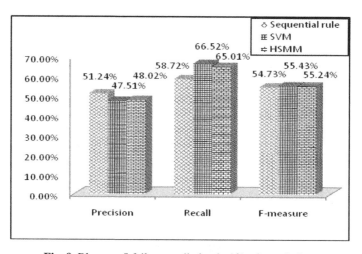

Fig. 9. Bluegene/l failure prediction in 12hr time window

In Fig. 10, in time window with duration of 12 hours on Red Storm, the prediction precision rate of sequential rule based predictor is 41.52%, better than that of HSMM-based predictor and SVM-based predictor with 40.21% and 38.82% respectively. And the recalls of sequential rule based predictor, SVM-based predictor and

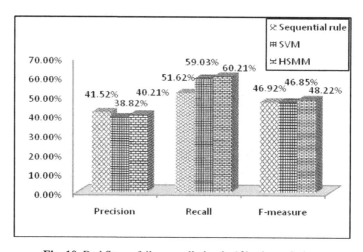

Fig. 10. Red Storm failure prediction in 12hr time window

HSMM-based predictor are 51.62%, 59.03% and 60.21%. As to the benchmark of F-measure, HSMM-based predictor defeats sequential rule based predictor and SVM-based with F-measure 48.22%.

In Fig. 11, in time window with duration of 12 hours on Spirit, the prediction precision rate of sequential rule based predictor is 39.73%, better than that of HSMM-based predictor and SVM-based predictor with 37.23% and 35.79% respectively. And the recalls of the sequential rule based predictor, SVM-based predictor and HSMM-based predictor are 48.23%, 68.32%, 70.10%. As to the benchmark of F-measure, HSMM-based predictor defeats sequential rule based predictor and SVM-based predictor with F-measure 48.63%.

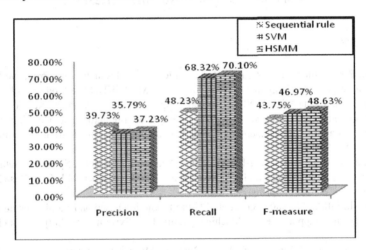

Fig. 11. Spirit failure prediction in 12hr time window

In a word, in time window with duration of 1 hour, the sequential rule based predictor is slightly better than HSMM-based predictor and worse than SVM-based predictor. And the same situation also appears in the Red Storm and Spirit cluster system. But in time window with duration of 5 hours, sequential rule based predictor is much better than HSMM-based predictor and SVM-based predictor for all three cluster systems. In addition, HSMM-based predictor is slightly better than sequential rule based predictor and SVM-based predictor in time window with duration of 12 hours.

5   Conclusion and Future Work

In this paper, we apply classification sequential rule based prediction approach to predict failure of BlueGene/L, Red Storm and Spirit systems, and meanwhile compare with SVM-based prediction approach and HSMM-based prediction approach. The precision, recall and F-measure are selected as the benchmarks to evaluate these three failure prediction approach. The experiment results show that classification sequential rule based approach defeats SVM-based and HSMM-based failure prediction approach in time window with duration of 5 hours, and in 1hr and 12hr prediction time window, sequential rules, SVM and HSMM have their own strengths and weakness respectively.

We just studied offline failure prediction approach in this paper. However, the online failure prediction is more significant for failure prediction of cluster system. In the future, we will apply the approaches used above into online failure prediction. In addition, dynamic meta-learning failure prediction approach is another research point for failure prediction of cluster system.

Acknowledgements. This work was partially supported by the National Key Technology R&D Program of China under Grant No. 2012BAH23B03.

References

1. Sahoo, R.K., Oliner, A.J., Rish, I., et al.: Critical Event Prediction for Proactive Management in Large scale Computer Clusters. In: 9th ACM SIGKDD Conference on Knowledge Discovery and Data Mining (SIGKDD 2003), pp. 426–435. ACM Press, New York (2003)
2. Guan, Q., Zhang, Z., Fu, S.: Ensemble of Bayesian Predictors and Decision Trees for Proactive Failure Management in Cloud Computing Systems. Journal of Communications 7(1), 52–61 (2012)
3. Pecchia, A., Cinque, M.: Log-Based Failure Analysis of Complex Systems: Methodology and Relevant Applications. In: Innovative Technologies for Dependable OTS-Based Critical Systems, pp. 203–215. Springer, Milan (2013)
4. Fronza, I., Sillitti, A., Succi, G., et al.: Failure Prediction based on Log Files using Random Indexing and Support Vector Machines. Journal of Systems and Software 86(1), 2–11 (2012)
5. Fu, X., Ren, R., Zhan, J., et al.: LogMaster: Mining Event Correlations in Logs of Large-scale Cluster Systems. In: 2012 IEEE 31st Symposium on Reliable Distributed Systems (SRDS 2012), pp. 71–80. IEEE Press, New York (2012)
6. Salfner, F., Lenk, M., Malek, M.: A Survey of Online Failure Prediction Methods. Journal of ACM Computing Surveys 43(3), 22–29 (2010)
7. Wenjian, W., Changqian, M., Weizhen, L.: Online Prediction Model based on Support Vector Machine. Neurocomputing 71(4-6), 550–558 (2008)
8. Zhenghua, X., Xiaoshe, D., et al.: A Survey on Failure Prediction of Large-scale Server Clusters. In: 8th ACIS International Conference on Software Engineering, Artificial Intelligence, Networking, and Parallel/Distributed Computing (SNPD 2007), pp. 733–738. IEEE Press, New York (2007)
9. Yinglung, L., Yanyong, Z., Jette, M.: BlueGene/L Failure Analysis and Prediction Models. In: Annual IEEE/IFIP International Conference on Dependable Systems and Networks (DSN 2006), pp. 425–434. IEEE Press, New York (2007)
10. Fulp, E.W., Fink, G.A., Haack, J.N.: Predicting Computer System Failures Using Support Vector Machines. In: Workshop on the Analysis of System Logs (WASL 2008), p. 5. ACM Press, New York (2008)
11. Gujrati, P., Yawei, L., Zhiling, L., et al.: A Meta-Learning Failure Predictor for Blue Gene/L Systems. In: International Conference on Parallel Processing (ICPP 2007), pp. 40–47. IEEE Press, New York (2007)

12. Jiexing, G., Ziming, Z., Zhiling, L.: Dynamic Meta-Learning for Failure Prediction in Large-Scale Systems: A Case Study. In: 2008 International Conference on Parallel Processing, pp. 157–164. IEEE Press, New York (2008)
13. Joshi, M., Agarwal, R., Kumar, V.: Mining Needle in a Haystack: Classifying Rare Classes via Two-phase Rule Induction. In: 2001 ACM SIGMOD International Conference on Management of Data, pp. 91–102. ACM Press, New York (2001)
14. Chih-Chuang, C., Chih-Jen, L.: LIBSVM: a Library for Support Vector Machines (2001), Software available at http://www.csie.ntu.edu.tw/cjlin/libsvm
15. Yinglung, L., Yanyong, Z., Hui, X.: Failure Prediction in IBM BlueGene/L Event Logs. In: IEEE Conference on Data Mining (ICDM 2007), pp. 583–588. IEEE Press, New York (2007)
16. Salfener, F., Malek, M.: Using Hidden Semi-markov Models for Effective Online Failure Prediction. In: 26th IEEE International Symposium on Reliable Distributed Systems, pp. 161–174. IEEE Press, New York (2007)

Analysis on Current Research of Supernetwork through Knowledge Mapping Method[*]

Wu Ke, Xi Yunjiang[**], Liao Xiao, and Li Weichan

School of Business Administration
South China University of Technology
Guangzhou 510641, P.R. China
yjxi@scut.edu.cn

Abstract. This paper mainly analyzes the current research on supernetwork in China and overseas. Methods of knowledge mapping, including bibliometric analysis, cluster analysis, strategic diagram analysis, co-word network analysis, etc, are used in this paper to demonstrate the knowledge structure of supernetwork and the research tendency via SPSS19.0, Ucinet 6 and NetDraw software. The findings show that researches on supernetwork focus on four knowledge clusters: (1) network equilibrium and optimization based on variational inequality, (2) supernetwork model and its application based on hypergraph, (3) knowledge management based on supernetwork, and (4) other applications of supernetwork scattered in different fields. Furthermore, the research maturity degree of each knowledge cluster and their relationships are also analyzed by utilizing the strategic diagrams and co-word analysis. And on basis of above analysis, the concept of supernetwork is redefined, which seems more accurate and clear to discride the main knowledge clusters of supernetwork research.

Keywords: Supernetwork, Knowledge mapping, co-word analysis, strategic diagrams.

1   Introduction

The research of complex network has attracted a great deal of attention over the past decade and obtained series of remarkable achievements, in which network model has been proved to be a valid tool to describe many kinds of systems. However, some special complex network systems, which may be composed of one or more network

---

[*] Supported by the Major Program Fund of National Science Foundation of China (Project No. 71090403/71090400), the National Natural Science Foundation of China (70801028,71003034), Youth Project in Humanities and Social Sciences of Education Ministry of China (08JC630027), and Fundamental Research Funds for the Central Universities, SCUT (x2gsD2118140)

[**] Corresponding author.

M. Wang (Ed.): KSEM 2013, LNAI 8041, pp. 538–550, 2013.

systems and with different types of interactions or connections in or between these networks, could not be described by using the general complex network model. Therefore, the concept of "supernetwork" was proposed to describe this particular complex network structure. The word of "supernetwork" was introduced firstly by Y.Sheffi in 1985, and was used on urban traffic planning [1]. Nagurney defined supernetworks as "the networks above and beyond existing networks", and applied it to solve the decision-making problems with a variety of related networks [2]. Besides, some researchers called the networks which can be described with hypergraph as hypernetwork [3].

Supernetwork can describe the structure of systems composed of multi-type or multi-tier networks, including the multi-type interactions in or between these networks. As a valid new tool for the study of complex network systems, supernetwork has got more and more concerned from researchers. Up to June 2012, there are nearly 200 relevant papers. Professor Wang Zhongtuo, has also pointed out that supernetwork model would be of significant value for the research of complex network systems [4,5]. In recent years, the relevant studies of supernetwork present the trend of explosive growth. This indicates that supernetwork would come to be one of the next research hot spots. Therefore, it is necessary to track and analyze the current researches of supernetwork. Furthermore, although many different types of supernetwork have been used by researchers, there is so far no accurate and clear concept for it which can describe the main knowledge clusters of current research.

According to the above mentioned problems, the paper aims to indicate the knowledge structure, the focus and tendency of supernetwork research by using knowledge mapping methods. Firstly, a brief overview of supernetwork research in China and overseas is presented through the bibliometric analysis. Then further deep analysis is conducted by using the integrated methods of knowledge mapping, including: (1) Using hierarchical cluster analysis method to divide the disciplinary structure and discover knowledge clusters in the field of supernetwork, (2) Using strategic diagrams to analyze the developmental tendency of the knowledge clusters, and (3) Discovering the hot spots of supernetwork research through the combination of K-Core Analysis and the Multidimensional Scaling Analysis in the Social Network Analysis(SNA) method. In the end the concept of supernetwork is also discussed and redefined based on the above analysis.

2   Data Sources

The data used in this paper was collected from Science Citation Index (SCI) and China National Knowledge Infrastructure (CNKI) database up to June 2012 through the object terms of "supernetwork" or "hypernetwork" in SCI database for the English papers and the object term of "supernetwork" in CNKI database for the Chinese papers. After removing the papers with little correlations, there are 148 papers in total, with 66 effective Chinese papers and 82 effective English papers respectively.

3   Research Methods and Software Tools

3.1   Co-word Analysis

Co-word analysis, one of the content analysis technique, is an important method of bibliometrics [6]. The main theory of co-word analysis is to count the co-occurrence frequency of a pair of words in one article which reflects the strength of relationships between words, and then cluster words into groups and display in network maps based on the co-word matrixes, by using the factor analysis method, the clustering method and some other multivariate statistical analysis techniques,, so as to reflect the knowledge structure and hot spots of the disciplines represented by these words.

3.2   Strategic Diagram Analysis

Strategic diagram analysis is to describe the internal relationship and interaction between different study areas in the form of coordinate graphs [7]. The strategic diagram often uses X-axis to represent the centrality, Y-axis to represent the density. The original point is the average of centrality and density of all points. Centrality is used to measure the degree of interaction among clusters and density is to measure the strength of internal relationship. The meaning of every quadrant in strategic diagram is shown in Figure 1.

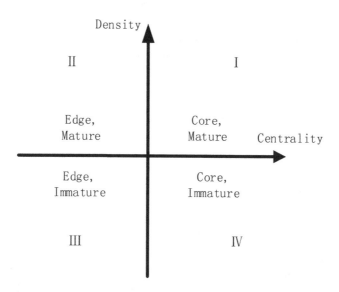

Fig. 1. Strategic diagram

3.3   Social Network Analysis

Social network analysis views the analysis objects as the set of individual actors within the network and their connections. In other words, a social network consists of nodes (representing individual actors) and edges (representing the connections between the individuals) [8]. In this research, two main analysis methods of social network analysis, the K-core Analysis and the Multidimensional Scaling analysis (MDS) are used to discover core subjects in supernetwork research and to show the knowledge structure of supernetwork field respectively.

3.4   Software Tool

Main software tools used in this paper include SPSS 19.0, Ucinet 6 and NetDraw. SPSS is used to do the hierarchical cluster analysis. The multidimensional scaling and k-core analysis are also conducted through Ucinet 6 and NetDraw which are used to draw knowledge mappings of supernetwork research.

4   Main Research Procedures

(1) The literature statistics and annual distribution analysis. The annual essay publication was calculated to discover the development tendency of supernetwork study.

(2) Keyword preprocessing. In order to standardize the keywords, this research used the English keywords both from the Chinese and English papers. Papers without keywords have been added some according to experts' opinion. Based on this, the keyword preprocessing was conducted. Firstly, to merge similar keywords, for example, "social network" was replaced by "social networks", "supernetwork" and "super-network" were replaced by "supernetwork", "variation inequality formulation", "variational inequality theory" and "variational inequality" were replaced by "variational inequalities", "optimization method" was replaced by "optimization", etc. Secondly, to split the composite keywords, for example, "knowledge supernetwork" was split into "knowledge network" and "supernetwork". Because "supernetwork" and "hypernetwork" both appear in English papers, they are regarded as different keywords in this paper in order to find their difference and connection.

(3) Statistics of word frequency and generation of the corresponding word-paper matrix and co-word matrix. Word frequency was calculated via Excel. According to experts' experience and opinion, the words whose frequency was not less than 3 were chosen to be high-frequency words. Consequently, there were 27 high-frequency words in total. Then, a 27*148 word-paper matrix and a 27*27 co-word matrix were generated as the material for next steps.

(4) Cluster analysis, strategic diagrams analysis and co-word network analysis were conducted based on the above word-paper matrix and co-word matrix. In the hierarchical cluster analysis, the cluster method adopts the "Between Groups Linkage" method, the "Binary" measurement standard and "Ochiai" coefficient to generate the tree diagram. The Strategic diagram analysis procedure is to calculate the centrality and density of every cluster according to the cluster analysis results, and then, draw the strategic diagrams via Excel. The Co-word network analysis procedure is: input co-word matrix into Ucinet 6, draw co-word network graphs via NetDraw, and perform

the k-core analysis and the MDS. Finally, change the shape of nodes according to the keywords in every cluster generated by cluster analysis. In other words, nodes of one cluster must differ from another.

5   Analysis Results and Discussion

5.1   Bibliometric Analysis

The annual distribution of papers related to supernetwork in China and overseas, as shown in Figure 2:

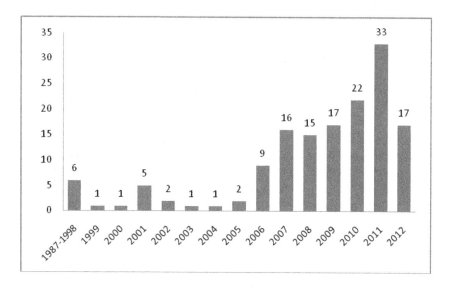

Fig. 2. Papers annual distribution diagram

(The data in figure 2 was collected from CNKI and SCI database, which don't include the full amount of journals. So, some papers may be left out.)

Supernetwork was proposed in 1985. The first paper related to supernetwork appeared in 1987 and only 6 papers from 1987 to 1998 which studied through hypergraph. From 1999 to 2005, supernetwork had been developing slowly. From 2006 to 2010, the relevant papers presented an increasing trend. From 2011(the data up to June 2012), it has shown an explosive growth. In consideration of the lagging of paper publication and record, there are still some papers not recorded or published during 2011 and 2012. Therefore, we can conclude that since 2011, the relevant study of supernetwork has attracted a great deal of attention, and most likely become one of the next valuable research focus.

5.2 Word Frequency Statistics and the Constructing of Word-Paper Matrix and Co-word Matrix

Word frequency statistics were calculated via Excel. The high-frequency words are shown in Table 1. Word-paper matrix and co-word matrix is omitted.

Table 1. High-frequency keywords

Seq.	Keyword	Freq.	Seq.	keyword	Freq.
1	Supernetwork	73	15	knowledge management	5
2	variational inequalities	36	16	complex network	4
3	Hypernetwork	25	17	e-commerce	4
4	Optimization	13	18	express network	4
5	supply chain	13	19	phylogenetic network	4
6	Equilibrium	10	20	closed-loop supply chain	3
7	network equilibrium	10	21	consensus network	3
8	Hypergraph	8	22	DNA computing	3
9	supply chain network	8	23	knowledge service	3
10	transportation network	7	24	logistics	3
11	knowledge network	6	25	reticulate evolution	3
12	social networks	6	26	supertrees	3
13	weighted network	6	27	traffic network	3
14	directed hypergraph theory	5			

High-frequency keywords can simply reflect the hotspot in supernetwork research. As is shown in Table 1, the major hotspots are: variational inequalities, optimization, supply chain, equilibrium, hypergraph, etc.

5.3 Cluster Analysis

The result of Cluster analysis is shown in Figure 3:

System cluster analysis reveals that the research of supernetwork includes 4 knowledge clusters (named by the author):

(1) Network equilibrium and optimization based on variational inequalities

Related keywords in this cluster are equilibrium, closed-loop supply chain, network equilibrium, traffic network, supply chain network, transportation network, optimization, express network, variational inequalities, supply chain, e-commerce, supernetwork, etc. We can see that this knowledge group mainly focuses on the

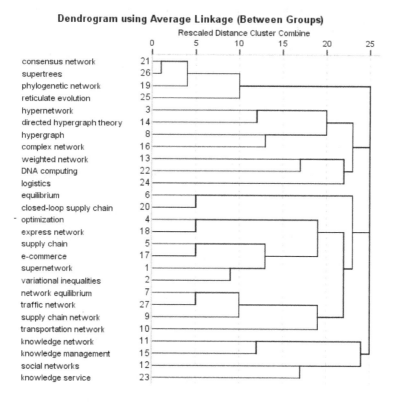

Fig. 3. Keyword cluster tree

equilibrium research of supply chain network [9-14], transportation network [15-17], and optimization of express network [18,19]. Supply chain network includes closed-loop supply chain and e-commerce, etc. "supernetwork" is used to describe this kind of network[4]. Keyword "supernetwork" was classified to this knowledge cluster which indicates that it connects tightly to this knowledge cluster.

(2) Supernetwork model and its application based on hypergraph.

Related keywords used in this knowledge cluster are hypernetwork, directed hypergraph theory, hypergraph, complex network, weighted network, DNA computing and logistics. Combining the word frequency data in Table 1, we can see that this knowledge cluster mainly focuses on the basic theory of hypergraph [20,21], the description and construction of supernetwork model[22], and the application of hypergraph theory in other areas[23-25]. "hypernetwork" is used to describe this kind of network [26]. From Figure 3, we can also see that keyword "hypernetwork" is classified to this knowledge group which indicates that the research related to "hypernetwork" mainly focuses on supernetwork which is based on hypergraph.

(3) Knowledge management research based on supernetwork

Related keywords in this knowledge cluster are knowledge network, knowledge management, social networks and knowledge service. This kind of supernetwork is generally named "knowledge supernetwork". In this cluster, supernetwork model is mainly used to describe the structural relationship of organizational knowledge systems, construct supernetwork model, and solve the problems in knowledge management. The current researches on knowledge supernetwork mainly include the basic nature of knowledge supernetwork and the connections between networks [27-29], the application of knowledge supernetwork to the knowledge transmission [30], sharing [31], and transferring [32] in knowledge management.

(4) Interdisciplinary and cross field research.

Related keywords used are consensus network, supertrees, phylogenetic network and reticulate evolution. Researches on this knowledge cluster are dispersed and mainly focus on the application in cross field via supernetwork model.

5.4   Strategic Diagram Analysis

The result is shown in Figure 4:

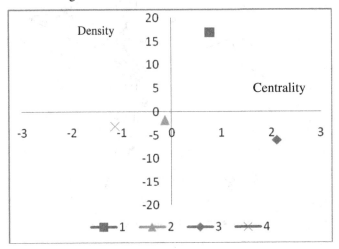

Fig. 4. The result of strategic diagram analysis

The meaning of centrality and density for each knowledge cluster:

(1) Cluster 1: network equilibrium and optimization based on variational inequalities. It locates in the first quadrant with high centrality, high density and tight inner structure. This indicates that research on Cluster 1 is relatively mature and is the core-field in supernetwork study.

(2) Cluster 2: supernetwork model and application based on hypergraph. It locates in the third quadrant and is close to the original point which indicates that the inner structure in Cluster 2 is a bit dispersed and has some relationship with other knowledge clusters. It implies that research on this cluster has space for further development. And in fact, it is still in the phase of model building and description.

(3) Cluster 3: knowledge management research based on supernetwork. It locates in the forth quadrant and has low density and loose inner structure. But its centrality is higher than Cluster 1, so it has more connections than other knowledge clusters. This indicates that Cluster 3 is also one of the core-fields, but not much matured.

(4) Cluster 4: interdisciplinary and cross field research. It locates in the third quadrant and has little connections with other knowledge clusters. This means that it is still in preliminary phase.

5.5   Co-word Network Analysis

The result of Co-word network analysis is shown in Figure 5:

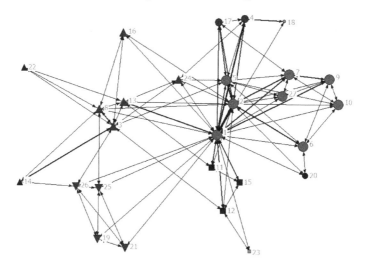

Fig. 5. Co-word network analysis

In this figure, red nodes represent core subjects, blue nodes represent second core subjects, black nodes represent the subject between core subjects and edge subjects, and grey nodes represent edge subjects. The size of tie line is relevant to the tightness of relationship between nodes. And the shape of the nodes represents the four knowledge clusters: circle nodes for Cluster 1, up triangle for Cluster 2, square for Cluster 3, and down triangle for Cluster 4.

We can find the core subjects and edge subjects by k-core analysis, as shown in Figure 5. Core subjects keywords are node 1 (supernetwork), node 2 (variational inequalities), node 5 (supply chain), node 6 (equilibrium), node 7 (network equilibrium), node 9 (supply chain network), node 10 (transportation network), node 18 (express network), node 20 (closed-loop supply chain) and node 27 (traffic network). On the basis of cluster analysis result, we can see that all the core subjects belong to Cluster 1, and the lines between node 2 (variational inequalities) and node 1, node 5, node 6, node 7 are thicker which indicates that node 2 (variational inequalities) has higher co-occurrence frequency with other nodes, and their relationship is tight. Node 1 (supernetwork) has lots of tie lines with others which indicates that node 1 has more connections with other clusters. Therefore, we can get the conclusion that Cluster 1 is the core knowledge cluster, the same conclusion with strategic diagram analysis. That means the research on network equilibrium and optimization based on variational inequalities is the core-field research of supernetwork.

MDS analysis presents the knowledge structure in the research field of supernetwork. The connections between knowledge clusters can also be revealed by combining the methods of MDS analysis and hierarchical cluster analysis. Firstly, Cluster 3 and 4 are both on the application of supernetwork in other fields. They both connect with Cluster 1 by the keyword "supernetwork". Secondly, Cluster 1 and 2 are exactly two different types of supernetworks, that is, the researches on supernetwork model are quite different to that on hypergraph. However, as seen in Figure 5, there are some cross researches between the two clusters. For example, node 1 (hypernetwork) connects with node 8 (hypergraph) and node 13 (weighted network), node 3 (hypernetwork) connects with node 5 (supply chain). Thus we can conclude that there is no absolute border between these two fields: supernetwork and hypergraph.

6   Discussion and Redefinition for the Concept of Supernetwork

From the above analysis, the supernetwork research includes four knowledge clusters. The concept of supernetwork is relatively clear in the first three knowledge clusters, while not in the forth. Therefore, it is reasonable to discuss and redefine the concept of supernetwork based on the first three knowledge clusters, and a more accurate and clearer definition is aimed to be supposed for it.

Through analyzing the types of nodes and edges contained in supernetwork in the first three knowledge clusters, the definition of supernetwork can be given as follows:

Definition 1: Network with different types of nodes or different types of edges (i.e. heterogeneous nodes or edges), or hyperedge, is called supernetwork.

In Cluster 1, supernetwork is mainly composed of multilayer networks. There are a variety of links between different layers, for example, financial flow and information flow may generate 2 different types of edges in the supernetwork. In Cluster 2, hyperedge exists in supernetwork. In Cluster 3, there are both heterogeneous nodes and heterogeneous edges in supernetwork. Therefore Definition 1 synthesizes all the concepts of supernetwork in the three knowledge clusters and seems more accurate and clearer.

Furthermore, as we know, bipartite network, as a special type of complex network, also contains two types of nodes, but only has one type of edges. To distinguish the similar research on complex network, Definition 1 can be further modified as followed:

Definition 2: Network with different types of edges, or hyperedge, is called supernetwork.

Definition 2 is more accurate which can not only summarize the concepts in the major knowledge clusters, but also differ from the similar research in complex network while definition 1 is relatively intuitionistic and easier to understand. Therefore, both of them are significant for researchers.

7   Conclusions

In this paper, the current research of supernetwork has been analyzed by using the integrated methods of knowledge mapping, which including bibliometric analysis, word frequency statistics, hierarchical cluster analysis, strategic diagram analysis, co-word network analysis, etc. Word frequency analysis shows that the current researches focus on variational inequalities, optimization, supply chain, equilibrium, hypergraph and so on. Cluster analysis reveals that the research fields include four knowledge clusters. Strategic diagrams analysis shows the current research and the research maturity degree of each knowledge clusters, and the developmental tendency of supernetwork from macroscopic view: Cluster 1 is the core field and matured, and connects tightly to "supernetwork". Cluster 2 is in the edge of supernetwork research and has the space for further development, and we can also figure out that most "hypernetwork" researches are based on hypergraph theory. Cluster 3 is also the core area, but not matured, while Cluster 4 is the most marginal and most immature. Finally, social network analysis and k-core analysis were also used to reveal the knowledge structure of supernetwork from microscopic view: K-core analysis reveals that the core subject keywords in supernetwork research are "supernetwork", "variational inequalities", "supply chain" etc. These keywords all belong to Cluster 1. The combination of the MDS analysis and cluster analysis results shows the connections among the knowledge clusters: Cluster 3 and Cluster 4 are application research of supernetwork in other fields while Cluster 1 and Cluster 2 are focus on two different types of supernetwork, and the researches on them have not exact borders. Based on the above analysis, the concept of supernetwork is discussed and redefined, and two definitions are presented, which is summarized from the main knowledge clusters of its research and can distinguish the similar concepts in complex network research.

As the research is based on the papers that have been published and recorded, some papers which are hard to be retrieved may be omitted, and further promotions may be conducted in the future.

Acknowledgment. The work described in this paper are substantially supported by the Major Program Fund of National Science Foundation of China (Project No. 71090403/ 71090400),the National Natural Science Foundation of China (70801028,71003034), Youth Project in Humanities and Social Sciences of Education Ministry of China (08JC630027), and Fundamental Research Funds for the Central Universities, SCUT (x2gsD2118140).

References

[1] Sheffi, Y.: Urban transportation networks: Equilibrium analysis with mathematical programming methods. Printice-Hall, NJ (1985)
[2] Nagurney, A., Dong, J.: Supernetworks: Decision-Making for the Information Age. Edward Elgar Publishing, Cheltenham (2002)

[3] Estrada, E., Rodrigues, V.R.: Subgraph centrality and clustering in complex hyper-networks. Physical A 364, 581–594 (2006)

[4] Wang, Z., Wang, Z.: Elementary Study of Supernetworks. Chinese Journal of Management 5(1), 1–8 (2008)

[5] Wang, Z.: Reflection on supernetwork. Journal of University of Shanghai for Science and Technology 33(3), 229–237 (2011)

[6] Qiu, J., et al.: An Empirical Analysis of the Library Science Research in China during 1999-2008 (Part I). Journal of Library Science in China 05, 72–79 (2009)

[7] Qiu, J., et al.: An Empirical Analysis of the Library Science Research in China during 1999-2008 (Part II). Journal of Library Science in China 06, 79-87+118 (2009)

[8] Liang, X.: Review of Mapping Knowledge Domains. Library Journal 06, 58–62 (2009)

[9] Yang, G.: The Optimization of the Closed-loop Supply Chain Supernetwork Recycled by the Retailers. Systems Engineering 27(6), 42–47 (2009)

[10] Wang, Z., Feng, Z.: Multi-objective Optimization of Supply Chain Supernetwork with Electronic Commerce. In: Jiang, L. (ed.) ICCE2011. AISC, vol. 112, pp. 45–53. Springer, Heidelberg (2011)

[11] Deng, Q., Dang, Y.: Supernetwork Model of Return Supply Chain and Electronic Commerce. In: 4th International Conference on Wireless Communications, Networking and Mobile Computing, WiCOM 2008, pp. 1–4. IEEE (2008)

[12] Nagurney, A., Liu, Z., Cojocaru, M.-G., Daniele, P.: Dynamic electric power supply chains and transportation networks: An evolutionary variational inequality formulation. Transportation Research Part E 43(5), 624–646 (2007)

[13] Wang, Z.P., Zhang, F.M., Wang, Z.T.: Research of Return Supply Chain Supernetwork Model Based on Variational Inequalities. In: Proccedings of IEEE International Conference on Automation and Loglistics, Jinan, China, pp. 25–30 (2007)

[14] Cruz, J.M., Nagurney, A., Wakolbinger, T.: Financial Engineering of the Integration of Global Supply Chain Networks and Social Networks with Risk Management. Naval Research Logistics 53(7), 674–696 (2006)

[15] Liu, Z., Nagurney, A.: Multiperiod competitive supply chain networks with inventorying and a transportation network equilibrium reformulation. Optimization and Engineering 13(3), 471–503 (2012)

[16] Yamada, T., Imai, K., Nakamura, T., Taniguchi, E.: A supply chain-transport supernetwork equilibrium model with the behaviour of freight carriers. Transportation Research Part E: Logistics and Transportation Review 47(6), 887–907 (2011)

[17] Xu, M., Gao, Z.: Supply chain network equilibrium model and its equivalent supernetwork based model: Complementarity formulations and comparisons. In: IEEE International Conference on Service Operations and Logistics, and Informatics, IEEE/SOLI 2008, vol. 2, pp. 2002–2007. IEEE (2008)

[18] Huang, J., Dang, Y.: Optimization method of multicriteria express supernetwork based on time threshold. Systems Engineering-Theory & Practice 30(12), 2129–2136 (2010)

[19] Huang, J., Dang, Y.: Express Supernetwork Model and the Cost-Based Optimization Method. Journal of Systems & Management 19(6), 689–695 (2010)

[20] Ruji, H.: Directed K-hypertree Method for Hypernetwork Analysis. Journal of Electrinics 9(3), 244–255 (1987)

[21] Ruji, H.: Development and Application of the Directed Hypergraph Theory. Geological Science and Technology Management (3) (1995)

[22] Volpentesta, A.P.: Hypernetworks in a directed hypergraph. European Journal of Operational Research 188, 390–405 (2008)

[23] Zhang, B.-T., Kim, J.-K.: DNA hypernetworks for information storage and retrieval. In: Mao, C., Yokomori, T. (eds.) DNA12. LNCS, vol. 4287, pp. 298–307. Springer, Heidelberg (2006)

[24] Maeshiro, T., Maeshiro, M., Shimohara, K., Nakayama, S.-i.: Hypernetwork model to represent similarity details applied to musical instrument performance. In: Jacko, J.A. (ed.) HCI International 2009, Part I. LNCS, vol. 5610, pp. 866–873. Springer, Heidelberg (2009)

[25] Ha, J.-W., Kim, B.-H., Lee, B., Zhang, B.-T.: Layered hypernetwork models for cross-modal associative text and image keyword generation in multimodal information retrieval. In: Zhang, B.-T., Orgun, M.A. (eds.) PRICAI 2010. LNCS, vol. 6230, pp. 76–87. Springer, Heidelberg (2010)

[26] BERGEC.Graphs and Hypergraphs. Elsevier, NewYork (1973)

[27] Xi, Y.-J., Dang, Y.-Z.: The Method to Analyze the Robustness of Knowledge Network based on the Weighted Supernetwork Model and Its Application. Systems Engineering-Theory & Practice (4), 134–140 (2007)

[28] Xi, Y.-J., Dang, Y.-Z., Liao, K.-J.: Knowledge supernetwork model and its application in organizational knowledge systems. Journal of Management Sciences in China 12(3), 12–21 (2009)

[29] Yang, Y.: Researches on Knowledge Supernetwork in Organizational Knowledge Management. Doctoral Dissertation, DaLian Universityof Technology (2009)

[30] Yu, Y., Dang, Y., Wu, J., et al.: Supernetwork-based-Analysis of Knowledge Diffusion Trend. Journal of the China Society for Scientific and Techical Information 29(2), 356–361 (2010)

[31] Pan, X., Wang, Y., Yang, Y.: The optimization step-Integrated knowledge sharing service in knowledge supernetwork environment. Science Research Management 32(18005), 87–93 (2011)

[32] Xu, S., Zou, H.: Analysis of Dynamics of Knowledge Transfer Based on Supernetwork. Journal of Intelligence 3007, 94–98 (2011)

Dealing with Trust, Distrust and Ignorance

Jinfeng Yuan, Li Li, and Feng Tan

Faculty of Computer and Information Science
Southwest University, Chongqing 400715, China
{jinfeng8,lily,tanfeng}@swu.edu.cn

Abstract. With the rapidly growing amount of information available to applications and users on the web, the question of whom and what to trust has become an increasingly important challenge, and effective trust models already play an important role in many intelligent web applications. In this paper, we present six propagation schemes for inferring both trust and distrust. (1) Our schemes are based on a trust score space and preserve trust provenance by simultaneously representing partial trust, partial distrust, partial ignorance and partial inconsistency, and treating them as differen and related concepts. (2) Trust information is obtained through a trusted third party. (3) Experiments based on three datasets give some interesting insight into the performance of propagation schemes. It is shown how prediction error of propagation schemes changes as more and more edges are removed.

Keywords: trust network, distrust, propagation models, error rate.

1 Introduction

With the rapidly growing amount of information available to applications and users on the web, the question of whom and what to trust has become an increasingly important challenge. A trust web allows a user to develop an opinion about another unknown user through creating links in a web of trust network. Effective trust models already play an important role in many intelligent web applications. For example, Xue Li et al. [1] proposed a seven-layer radio frequency identification (RFID) trust framework on the basis of the analysis of traditional trust framework in Supply-Chain Management (SCM).

Many models have been developed to deal with trust problem. Howerer, there still have some unsolved problems. Firstly, the existing many models usually deal with trust in a binary way, like two-valued boolean values, which is either completely true or completely false[2][3]. But in practice, it is often accepted that things can be either partly trust or partly distrust in some cases. This is reflected in our everyday language when we say for example "this source is rather trustworthy" or "I trust this source very much". Secondly, trust networks are facing two important challenges. On one hand, it is likely that many users do not know each other in large networks, so there is an abundance of ignorance. On the other hand, different users might provide different and even contradictory information owing to the lack of a central authority, so inconsistency may occur.

M. Wang (Ed.): KSEM 2013, LNAI 8041, pp. 551–560, 2013.
© Springer-Verlag Berlin Heidelberg 2013

In this paper, we take the above situations into account and propose six propagation schemes for inferring both trust and distrust. (1) Our schemes are based on a trust score space discussed in Section 2, which preserves trust provenance by simultaneously representing partial trust, partial distrust, partial ignorance and partial inconsistency, and treating them as differently. (2) Trust information is obtained through a trusted third party — In cryptography, a trusted third party (TTP) is an entity which facilitates interactions between two parties who both trust the third party. In a web of trust, propagation operators are used to handle the problem of establishing trust information in an unknown user by inquiring through TTPs. The simplest case, atomic propagation, is depicted in Fig. 1: if the trust score of user a in user b is p, and the trust score of b in user c is q, what information can be derived about the trust score of a in user c that is unknown to him? (3) Experiments based on three datasets give some interesting insight into the performance of propagation schemes. It is shown how prediction error of propagation schemes changes as more and more edges are removed.

The rest of the paper is organized as follows: Section 2 is the problem formulation. Section 3 presents our propagation schemes. Studies on the performance of the propagation schemes is evaluated in Section 4. In Section 5, some related work is discussed. Section 6 concludes the paper with future researches.

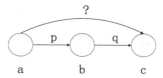

Fig. 1. Example of trusted third party

2   Problem Definition

We assume there are n users, User a may express some level of trust or distrust for user b. The estimate trust/disturst of n users can be viewed as a real-valued matrix. We group the entries into two matrices, one for trust and the other for distrust. The collection of users' trusts can be represented as a $N \times N$ matrix M_T, each element t_{ij} is the trust value that one user holds for each other user, t_{ij} is in the range $[0,1]$ where 0 is full absence of trust and 1 is full presence of trust. Similarly, the collection of users' distrusts can be represented as a $N \times N$ matrix M_D, each element d_{ij} is the distrust value that one user holds for the other user, d_{ij} is in the range $[0,1]$ where 0 is full absence of distrust and 1 is full presence of distrust. Pairs such as (t, d) are used in the following sections in which t corresponds to a trust degree, d to a distrust degree.

We introduce a new structure, called trust score space $\mathcal{BL}^{\square}$ in [4]. The model can be used to represent the trust a user may have other users in a given domain.

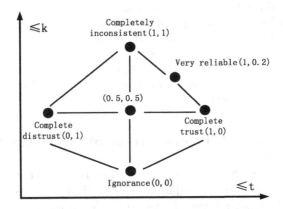

Fig. 2. Trust score space $\mathcal{BL}^\square$, a bilattice-based trust model that enables us to compare trust scores according to the available trust-distrust content ($\leq t$) and to evaluate the uncertainty that is involved ($\leq k$)

Definition 1. *(Trust Score Space) The trust score space*

$$\mathcal{BL}^\square = \left([0,1]^2 \leq t, \leq k \right) \tag{1}$$

consists of the set $[0,1]^2$ *of trust scores and two orderings defined by*

$$(t_1, d_1) \leq t\,(t_2, d_2) \quad \text{iff} \quad t_1 \leq t_2 \quad \text{and} \quad d_1 \geq d_2 \tag{2}$$

$$(t_1, d_1) \leq k\,(t_2, d_2) \quad \text{iff} \quad t_1 \leq t_2 \quad \text{and} \quad d_1 \leq d_2 \tag{3}$$

for all (t_1, d_1) *and* (t_2, d_2) *in* $[0,1]^2$. *In the trust score* (t_1, d_1), t_1 *is called the trust degree, while* d_1 *is the distrust degree.*

Fig. 2 shows $\mathcal{BL}^\square$, along with some examples of trust scores. The lattice $\left([0,1]^2 \leq t \right)$ orders the trust scores going from complete distrust $(0,1)$ to complete trust $(1,0)$. The lattice $\left([0,1]^2 \leq k \right)$ evaluates the amount of available trust evidence, ranging from a "shortage of evidence", $t_1 + d_1 < 1$ (incomplete information), to an "excess of evidence", viz. $t_1 + d_1 > 1$ (inconsistent or contradictory information). In the extreme cases, there is no information available, denoted as $(0,0)$; or there is evidence that says that the agent is to be trusted fully as well as evidence that states that the agent is completely unreliable, denoted as $(1,1)$.

Considering distrust degree in addition to the traditional trust degree makes it possible to distinguish between absence of trust caused by a lack of knowledge, i.e. $(0,0)$, and absence of trust in the presence of distrust, i.e. $(0,1)$.

Definition 2. *(Trust Network) A trust network is a couple* (U, R) *such that* U *is a set of users and* R *is a* $U \times U \to \mathcal{BL}^\square$ *mapping. For every user* a *and user* b *in* U, *we write* $R(a, b) = (R_T(a, b), R_D(a, b))$,

• $R(a, b)$ is called the trust score of a in b.
• $R_T(a, b)$ is called the trust degree of a in b.
• $R_D(a, b)$ is called the distrust degree of a in b.

R may be thought of as a snapshot taken at a certain time, since the trust learning mechanism involves recalculating trust scores, for instance through trust propagation as discussed next.

3   Development of Efficient Propagation Methods

If only the trust degree, the propagation procedure is quite straightforward. The propagation comes down to the conjunction of the given trust values and adapts to a fuzzy setting by using a triangular t-norm T [5]: an increasing, commutative and associative $[0, 1]^2 \rightarrow [0, 1]$ mapping that satisfies $T(1, x) = x$ for all x in $[0, 1]$. Hence the trust degree of a in c derives from the trust degree of a in b and the trust degree of b in c. Possible choices for T are $T_M(x, y) = \min(x, y)$, $T_P(x, y) = x * y$ and $T_L(x, y) = \max(0, x + y - 1)$. The choice of $T = T_P$ corresponds to a common approach in trust propagation [6].

However if, we consider both the trust and the distrust degree, the problem gets more complicated. User a distrusts c when a trusts b and b distrusts c, or when a distrusts b and b trusts c. Analogously to the t-norm, we use a t-conorm S to model disjunction: an increasing, commutative and associative $[0, 1]^2 \rightarrow [0, 1]$ mapping satisfying $S(x, 0) = x$ for all x in $[0, 1]$. Possible choices are $S_M(x, y) = \max(x, y)$, $S_P(x, y) = x + y - x * y$, and $S_L(x, y) = \min(1, x + y)$. A negator N is a decreasing $[0, 1]^2 \rightarrow [0, 1]$ mapping satisfying $N(0) = 1$ and $N(1) = 0$. The most commonly used one is $N_S(x) = 1 - x$.

In our method, trust propagates through trusted third parties (see Fig. 1). For user a, b and c, $R(a, b)$, $R(b, c)$ and $R(a, c)$ are the trust scores from a in b, b in c and a in c, respectively. $R(a, b) = (R_T(a, b), R_D(a, b))$, where $R_T(a, b)$, $R_D(a, b)$ represent the trust degree and distrust degree of a in b, respectively. A t-norm T, a t-conorm S and a negator N are represented conjunction, disjunction and negation, respectively. We choose $T = T_P$, $S = S_P$ and $N = N_S$, that is, $T(x, y) = x * y$, $S(x, y) = x + y - x * y$, and $N(x) = 1 - x$. We consider six different propagation schemes.

Definition 3. (propagation schemes) $Prop(R(a, b), R(b, c)) = R(a, c)$ with $R(a, b) = (R_T(a, b), R_D(a, b))$, $R(b, c) = (R_T(b, c), R_D(b, c))$ and $R(a, c) = (R_T(a, c), R_D(a, c))$. For all $R(a, b)$ $R(b, c)$ and $R(a, c)$ in $\mathcal{BL}^\square$. The propagation operators Prop1, Prop2, Prop3, Prop4, Prop5 and Prop6 are defined by

Prop 1. $R_T(a, c) = T(R_T(a, b), R_T(b, c))$, $R_D(a, c) = T(R_T(a, b), R_D(b, c))$. In Prop1, user a only takes advice from people whom he trusts: $Prop1((1, 0), (0, 1))$
$= (0, 1)$, and ignores advice from enemies or unknown people: $Prop1((0, 1), (0, 1))$
$= (0, 0)$, $Prop1((0, 0), (0, 1)) = (0, 0)$.

Prop 2. $R_T(a,c) = T(R_T(a,b), R_T(b,c)),$ $R_D(a,c) = T(N(R_D(a,b)), R_D(b,c)).$

Scheme Prop2 is similar to Prop1, but in addition user a takes advice from unknown people, he distrusts the enemy of people whom he does not know: $Prop2((0,0),(0,1)) = (0,1)$.

Prop 3. $R_T(a,c) = S(T(R_T(a,b), R_T(b,c)), T(R_D(a,b), R_D(b,c))),$
$R_D(a,c) = S(T(R_T(a,b), R_D(b,c)), T(R_D(a,b), R_T(b,c))).$

Prop3 corresponds to an interpretation in which the enemy of an enemy is considered to be a friend: $Prop3((0,1),(0,1)) = (1,0)$, and the friend of an enemy is considered to be an enemy: $Prop3((0,1),(1,0)) = (0,1)$. This shows us that useful information can be derived through distrusted user.

Prop 4. $R_T(a,c) = T(R_T(a,b), R_T(b,c)),$
$R_D(a,c) = S(T(R_T(a,b), R_D(b,c)), T(R_D(a,b), R_T(b,c))).$

Prop4 is similar to Prop3, but does not take advice from the enemy of people whom he distrusts: $Prop4((0,1),(0,1)) = (0,0)$. In other words, user c is only trusted by a when a trusts b and b trusts c.

Prop 5. $R_T(a,c) = S(T(R_T(a,b), R_T(b,c)), T(R_D(a,b), R_D(b,c))),$
$R_D(a,c) = T(N(R_D(a,b)), R_D(b,c)).$

In Prop5, user a makes use of information from enemies and unknown people. User a regards the enemy of an enemy as a friend: $Prop5((0,1),(0,1)) = (1,0)$ and the enemy of unknown people as an enemy: $Prop5((0,0),(0,1)) = (0,1)$.

Prop 6. $R_T(a,c) = S(T(R_T(a,b), R_T(b,c)), T(R_D(a,b), R_D(b,c))),$
$R_D(a,c) = T(R_T(a,b), R_D(b,c)).$

Prop6 is similar to Prop4 derive no knowledge through an unknown people, but Prop6 takes advice from people whom he distrusts.

The results of propagation operators for some particular trust scores is shown in table 1, the rows and columns correspond to $(R_T(a,b), R_D(a,b))$ and $(R_T(b,c), R_D(b,c))$, respectively.

4   Evaluation

4.1   Data Sets

We used three major social network datasets to test the propagation methods. These datasets are available at Stanford Large Network Datasets [7]. The networks we use have both positive (trust) and negative (distrust) edges. These Dataset statistics are shown in Table 2.

Epinions — This is a who-trust-whom online social network of a general consumer review site Epinions.com. Members of the site can decide whether to "trust" each other. All the trust relationships interact and form the web of

Table 1. The results of propagation operators for some particular trust scores in different schemes

Prop1	(0,0)	(0,1)	(1,0)	Prop2	(0,0)	(0,1)	(1,0)
(0,0)	(0,0)	(0,0)	(0,0)	(0,0)	(0,0)	(0,1)	(0,0)
(0,1)	(0,0)	(0,0)	(0,0)	(0,1)	(0,0)	(0,0)	(0,0)
(1,0)	(0,0)	(0,1)	(1,0)	(1,0)	(0,0)	(0,1)	(1,0)

Prop3	(0,0)	(0,1)	(1,0)	Prop4	(0,0)	(0,1)	(1,0)
(0,0)	(0,0)	(0,0)	(0,0)	(0,0)	(0,0)	(0,0)	(0,0)
(0,1)	(0,0)	(1,0)	(0,1)	(0,1)	(0,0)	(0,0)	(0,1)
(1,0)	(0,0)	(0,1)	(1,0)	(1,0)	(0,0)	(0,1)	(1,0)

Prop5	(0,0)	(0,1)	(1,0)	Prop6	(0,0)	(0,1)	(1,0)
(0,0)	(0,0)	(0,1)	(0,0)	(0,0)	(0,0)	(0,0)	(0,0)
(0,1)	(0,0)	(1,0)	(0,0)	(0,1)	(0,0)	(1,0)	(0,0)
(1,0)	(0,0)	(0,1)	(1,0)	(1,0)	(0,0)	(0,1)	(1,0)

Table 2. Dataset statistics

Datasets	Epinions	Slashdot	Wikipedia
Nodes	119,217	82,168	7,118
Edges	841,200	948,464	103,747
positive edges	85.0%	77.4%	78.7%
negative edges	15.0%	22.6%	21.2%

trust which is then combined with review ratings to determine which reviews are shown to the user. **Slashdot** — This is a technology-related news website where users can rate each other as friend or foe. We treat those as positive or negative trust ratings, respectively. **Wikipedia** — This is a popular online encyclopedia created by users, and has a set of elected moderators who monitor the site for quality and controversy and who help maintain it. When a user requests admin access, a public discussion page is set up for users to discuss and vote on whether to promote the moderator. Positive and negative votes are counted as positive and negative trust ratings, respectively.

4.2 Evaluation Metrics

we evaluate the performance of these propagation methods in terms of prediction error rate, which is defined as the ratio of incorrect $R(a, b)$ pairs divided by the total $R(a, b)$ pairs to be evaluated. The smaller the metric value, the better is the performance of the method.

$$ErrorRate = \frac{N_{error}}{N_T} \qquad (4)$$

Here, N_{error} is the number of incorrect R(a,b) pairs and N_T is the total number of R(a,b) pairs to be evaluated.

4.3   Experimental Result

To assess the quality of propagation methods, we use them to solve the edge sign prediction problem. For each of our three datasets, we randomly remove a substantial number of edges (3000). The removed edges make up the testing set and the remaining edges make up the training set.

As shown in Table 2, the positive edges outnumber negative edges by a huge margin in three networks: 80 versus 20. Hence, a naive method that always predicts "trust" will incur a prediction error of only 20%. We nevertheless first report our results for prediction on randomly removed edges in the network, as it reflects the underlying problem. However, to ensure that our methods are not benefiting unduly from this bias, we also take the largest balanced subset of the 3000 randomly masked trial edges such that half the edges are trust and the other half are distrust-this is done by taking all the 600 distrust edges in the trial set as well as 600 randomly chosen trust edges from the trial set. Thus, the size of this subset S is 1200. We measure the prediction error in S and call it ErrorRate_S.

Table 3. The prediction error rate on the entire set and subset S

DateSets	Epinions		Slashdot		Wikipedia	
Schemes	ErrorRate	ErrorRate_S	ErrorRate	ErrorRate_S	ErrorRate	ErrorRate_S
Prop1	0.063	0.243	0.099	0.261	0.085	0.254
Prop2	0.053	0.229	0.087	0.236	0.067	0.233
Prop3	0.047	0.167	0.069	0.164	0.055	0.168
Prop4	0.058	0.206	0.071	0.197	0.061	0.229
Prop5	0.041	0.132	0.052	0.129	0.039	0.146
Prop6	0.052	0.196	0.078	0.157	0.079	0.193

From Table 3, we see that the error rates are as low as 0.041 and 0.132 testing on the Epinions on the entire set and the subset S, respectively. The error rates are as low as 0.052 and 0.129 testing on the Slashdot on the entire set and the subset S, respectively. The error rates are as low as 0.039 and 0.146 testing on the Slashdot on the entire set and the subset S, respectively. The prediction error between entire set and subset comes from the fact that the dataset is unbalanced, with more positive edges than negative ones. This seems due to nonuniformity inherent in the dataset.

Fig. 3 shows the error rate on the entire sets, it is clear that Prop5 performs quite well among all others. This is probably because Prop5 makes use of information from distrusted and unknown people. Prop3 is remarkably little decrease in performance regardless of information from unknown people. This suggests that useful information can be derived through distrusted and unknown people.

For each of the three networks we choose several edge removal ratios and perform ten iterations for each of them. Fig.4 shows how prediction error of propagation methods changes when more edges are removed. As shown in Fig.4,

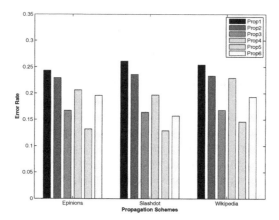

Fig. 3. The error rate on the entire sets

the error rates rise slowly until over 40% of the edges are removed. This implies that these networks have a large amount of redundant information. Not only are edges highly predictable, but they are predictable even without the information from a arge fraction of the edges. Once over 40% of the edges are not provided explicityly, performance falls quickly.

5   Related Work

There are many trust propagation algorithms that take advantage of pair-wise trust values and the structure of a social network. Existing trust models can be classified in several ways, among which probabilistic and gradual algorithms as well as representations of trust and representations of both trust and distrust. Probabilistic algorithms include [8], [9], [10] and [11]. These algorithms treat direct trust as a probability, a higher trust score corresponds to a higher probability that a user can be trusted. Examples of gradual algorithms can be found in [12], [13] and [14]. In this case, instead of trust score is a probability, trust is represented as a matter of degree. The ordering of the trust scores is very important, with "very reliable" representing a higher trust than "reliable", which in turn is higher than "rather unreliable". These algorithms lean themselves better for the computation of trust when the outcome of an action can be positive to some extent, e.g., when provided information can be right or wrong to some degree, as opposed to being either right or wrong. In this paper, we are keeping this kind of application.

Somewhat similar to most existing studies, our method differs in important ways. Firstly, our method is based on a trust score space and preserves trust provenance by simultaneously representing partial trust, partial distrust, partial ignorance and partial inconsistency, and treating them as different, related concepts. Second, trust information is obtained through a trusted third party (TTP).

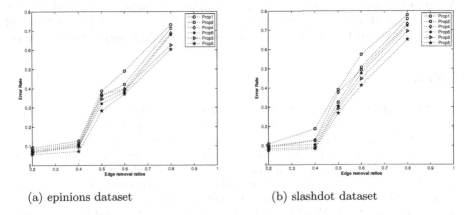

(a) epinions dataset   (b) slashdot dataset

Fig. 4. shows how prediction error of propagation methods changes when more edges are removed

6   Conclusions

In this paper, we are the first to introduce six propagation schemes combing with atomic propagation of trust, distrust and ignorance. The advantages to our methods are as follows: (1) Our methods can alleviate some of the existing problems of trust models, representing trust as a matter of degree and more specifically considering partial trust, partial distrust. (2) Considering ignorance and inconsistency, our methods can differentiate between absence of trust caused by presence of distrust (as towards a malicious user) and versus by lack of knowledge (as towards an unknown user).

Then, we design some experiments to get insight into the performance of propagation schemes testing on three datasets. Experimental results show Prop5 performs quite well because of makes use of information from distrusted and unknown users, which indicates useful information can be derived through distrusted and unknown users. In addition, experimental results predicting error changes of propagation schemes as more edges are removed shows the error rates increase quite slowly until over 40% of the edges are removed. This implies that these networks have a large amount of redundant information. Not only are edges highly predictable, but they are predictable even without the information from a arge fraction of the edges. Once over 40% of the edges are hidden, performance degrades quickly.

Trust and distrust are not static, they can change with the change of time. Therefore, it is also necessary to search for appropriate updating techniques. The future work intend to apply these schemes to other scenarios where time is an issue or influences the development of the scenario. In addition, more recent work on exploiting both the trust and the distrust in the recommendation system to further enhance the quality of recommendations, is a direction of our future work.

Acknowledgements. This work is supported by National Natural Science Foundations of China(No. 61170192), and Natural Science Foundations of CQ(No. CSTC2012JJB40012). L. Li is the corresponding author for the paper.

References

1. Mahinderjit-Singh, M., Li, X.: Trust in RFID-Enabled Supply-Chain Management. International Journal of Security and Networks (IJSN) 5, 96–105 (2010)
2. Riguidel, M., Martinelli, F. (eds.): Security, Dependability and Trust. Thematic Group Report of the European Coordination Action Beyond the Horizon: Anticipating Future and Emerging Information Society Technologies (2006)
3. Kuter, U., Golbeck, J.: Sunny: A new algorithm for trust inference in social networks using probabilistic confidence models, pp. 1377–1382. AAAI (2007)
4. Victor, P., De Cock, M., Cornelis, C., Pinheiro da Silva, P.: Towards a provenance-preserving trust model in agent networks. In: Proceedings of Models of Trust for the Web, WWW 2006 Workshop (2006)
5. Schweizer, B., Sklar, A.: Associative functions and statistical triangle inequalities. Publ. Math.-Debrecen. 8, 169–186 (1961)
6. Golbeck, J.: Computing and applying trust in web-based social networks. PhD thesis (2005)
7. The Stanford Large Network Datasets, http://snap.stanford.edu/data/ (accessed on February 21, 2009)
8. DuBois, T., Golbeck, J., Srinivasan, A.: Rigorous probabilistic trust- inference with applications to clustering. In: The 2009 IEEE/WIC/ACM International Joint Conference on Web Intelligence and Intelligent Agent Technology, vol. 01, pp. 655–658. IEEE Computer Society (2009)
9. Josang, A., Marsh, S., Pope, S.: Exploring different types of trust propagation. Trust Management, 179–192 (2006)
10. Zaihrayeu, I., Pinheiro da Silva, P., McGuinness, D.: IWTrust: Improving user trust in answers from the web. In: The 3rd International Conference on Trust Management, pp. 384–392 (2005)
11. Huang, B., Kimmig, A., Getoor, L., Golbeck, J.: A Flexible Framework for Probabilistic Models of Social Trust. In: Greenberg, A.M., Kennedy, W.G., Bos, N.D. (eds.) SBP 2013. LNCS, vol. 7812, pp. 265–273. Springer, Heidelberg (2013)
12. Victor, P., Cornelis, C., De Cock, M., Herrera-Viedma, E.: Practical aggregation operators for gradual trust and distrust. Fuzzy Sets Systems 184(1), 126–147 (2011)
13. Massa, P., Avesani, P.: Trust-aware collaborative filtering for recommender systems. In: The Federated International Conference on the Move to Meaningful Internet: CoopIS, DOA, ODBASE, pp. 492–508 (2004)
14. Victor, P., Cornelis, C., De Cock, M.: Practical aggregation operators for gradual trust and distrust. Fuzzy Sets and Systems 185(1), 126–147 (2011)

Function Design of Ancient Books Repairing Archives Management System Based on Tacit Knowledge

Fu Jin[1,*], Miaoyan Li[2], and Qingdong Du[2]

[1] Management College, Shenyang Normal University, 110034 Shenyang, China
[2] Software College, Shenyang Normal University, 110034 Shenyang, China
jinfu8655@163.com, {lillian1979,duqingdong}@126.com

Abstract. The ancient books repairing, in which masters teach apprentices to inherit repairing skills in a long time, is a Chinese traditional handicraft industry and has a lot of tacit knowledge. In this paper knowledge management is introduced, and then the ancient books repairing archives database and its management system are analyzed. The content setting of ancient books repairing archives and function design of its management system are tried to solve. The research conclusion has practical significance in promoting the digital construction process of ancient books repairing archives and improving the repairing quality and management efficiency of ancient books.

Keywords: ancient books repairing archives, knowledge management, management system, function design.

1 Significance of Tacit Knowledge Extraction in Ancient Books Repairing And of Management System Development

There are more than thirty million ancient books in China, and more than ten million among them are broken badly and need repairing. There were less than one hundred ancient books repairing personnel before 2007. Since implementing Chinese ancient books protection plan, China has more than three hundred ancient books repairing (ABR) senior experts by cultivating personnel through training courses, which have been held fifteen times. However, being relative to the quantity of ancient books urgently needing repairing, there is obvious shortage in the intellectual resources of ancient books repairing talents. Especially the talents who have sufficient ancient books repairing tacit knowledge, such as Master Du Weisheng who is an ancient books repairing expert working in the national library, seriously lack. [1] Introducing knowledge management into the ancient books repairing industry will help to mine, promote and make use of the skill-category tacit knowledge in the ancient books repairing experts.

Tacit knowledge is in the head of people, unstructured and no-coding and has individual path dependence. And the kind of knowledge is about the individual thought, working knack, skills and experience. Organizational tacit knowledge is accumulated in the organization and not easy to be noticed, such as the organizational

---

* Corresponding author.

M. Wang (Ed.): KSEM 2013, LNAI 8041, pp. 561–569, 2013.
© Springer-Verlag Berlin Heidelberg 2013

process designing and regulations. [2] First, in the ancient books repairing work, the tacit knowledge is the ancient books repairing skills, operational technique and knack. The skills, experience and knowledge of the experts of this industry belong to the knowledge of know how and is the tacit knowledge which is on the level of experts' individual operation. Second, the tacit knowledge is about who can repair ancient books or solve the difficulties in the work of repairing ancient books, which is the knowledge of know who. We should do our best to extract the ancient books repairing experts' individual tacit knowledge, build and manage nation-wide ancient books repairing experts' archives and database in order to share intellectual resources. Third, ancient books repairing experts' cognitive level tacit knowledge includes beliefs, ideas, feelings, values, mental model and so on. This kind of knowledge has profound influence on ancient books repairing work and plays the guiding role in the professionalism and behavior of ancient books repairing personnel. [3] Fourth, some libraries, such as the national library and Zhejiang province library, do well in the ancient books repairing work. Their knowledge of repairing regulation, working process and organizational culture is called the organizational tacit knowledge, which is also the content worth extracting. In short, we can do better in promoting the ancient books repairing quality and improving the ancient books repairing efficiency only by extracting and converting individual, teams', groups' or organizational tacit knowledge through the idea and method of knowledge management.

2   Ancient Books Repairing Management System Process and Its Function Design

The ancient books repairing work includes extracting ancient books needing repairing from the repository of special collection, checking the damaged condition of the ancient books needing repairing, determining the damaged level and value, determining repairing jurisdiction and class, asking for repair, determining repairing solution, implementing repairing operation, dynamically filling archives, auditing repairing result, storing repaired ancient books and entering archive management or reading circulation system. These business processes and tasks are clear. We can make use of the relative techniques of information management and knowledge management and design ancient books repairing management system working process logical model and function modular to promote information technology of ancient books repairing process management, digital repairing archives and intelligent repairing technologies. [4]

The ancient books repairing work is a traditional Chinese handicraft industry. By introducing knowledge management, we can scientifically manage ancient books repairing working process. If we regard the repairing process of one unit of ancient books, which is usually a volume, as a project, we can realize the information management of the business operation process of the entire project. (Fig. 1) The business operation workflow of ancient books needing repairing and the knowledge flow of ancient books repairing archives document information materials which include words, picture data and video or audio information format clear circulation process, and design ancient books repairing management system's basic module function on the basis of them. [5]

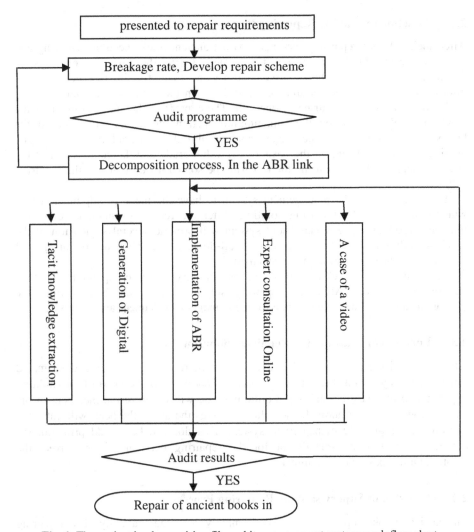

Fig. 1. The ancient books repairing file and its management system work flow chart

2.1   Function of Extracting Ancient Books Needing Repairing

When ancient books repairing personnel extract ancient books needing repairing on the basis of the repairing plan of their own library's ancient books special collection department, the management needs to call data information of ancient books repairing bibliography via the management system and generate ancient books repairing achieves data, in which literature information is generated from ancient books' repairing unit which is usually a volume. After the management calls system, then in the system formats repairing demand and prints out ancient books repairing delivery order, both sides sign on the transfer list and confirm the delivery from the storage of the ancient books needing repairing to make sure the job is taken over from one side to the other.

2.2   Function of Making Repairing Plan

After ancient books repairing personnel extract ancient books needing repairing and receive ancient books repairing task, they login the management system synchronously and get the relative permission, then fill in the before-repair information of ancient books archives record when they regard one volume of the ancient books needing repairing as one unit. The content of the archives includes the single volume's information, the damaging information, the history information of the books used to be repaired, the ancient books' cultural relics grading, the damaging grading and so on. Then, on the basis of the value grade and damaging level of the ancient books needing repairing, the repairing personnel have to choose the repairing solution.

It is necessary for the management system to have the menu prompting function which can suggest the repairing solution. After the repairing personnel enter the relative information, the management system will provide several suggestion of the repairing solution with the mode of the pull-down menu. Then the repairing will choose the repairing solution that meets their need.

The repairing personnel need to choose the estimated date of completion when they choose the solution. After the repairing solution passes the examination, the management system will file the date for the management to query and supervise.

2.3   Function of Auditing the Repairing Solution

After the repairing personnel fill in and submit the repairing solution and estimated finish time, they submit them to the process of audit. After the head of the auditing expert group receives the auditing requirement, he will organize the members of the expert group to do the audit. If the solution passes the audit, the head will enter the auditing result into the management system. If it doesn't, he should print out the detailed suggestion of amendment for the repairing personnel to re-correct the solution until it passes the examination.

2.4   Function of Supervising the Repairing Process

The ancient books repairing business is mainly operated manually. The repairing personnel will make use of the management system and finish the generating process of repairing archives synchronously. The content of the job includes:

- Recording the situation of the repaired material and its sample on the basis of the repairing solution
- Putting down the details and the revealing problems during the repairing process
- Photographing the ancient books' detailed situation before and after the reparation
- Photographing or recording video of the important working process and supplies during the repairing process
- Making material files to the original script removed during the process of reparation and doing the encoding record
- Putting down the repairing experience and extracting the tacit knowledge

2.5   Function of Auditing Repairing Results

After the ancient books repairing operation work is finished, the expert auditing group needs to check the repairing result. If the repairing quality meets the need, the ancient books are allowed to be stored. The detailed standards the Ministry of culture released include:

- The Technical Specification and Quality Requirement of Ancient Books (WH/T23-2006)
- The Damaging Grading Standard of Ancient Books' Special Collection (WH/T22-2006)

On the basis of the above standards, the finished products If they cannot pass the examination, the repairing personnel should print out the detailed reasons and ask the head to make the recovery solution. Then the work enters the circulation, which is from auditing the recovery solution to implementing the operation of recovery and reparation, to checking the repairing results. And the head also should limit the repairing personnel's rights on the basis of their performance.

2.6   Function of Storing Ancient Books Needing Repairing

The ancient books that pass the examination should be stored and circulated. The management should print the storage transfer list where both sides sign and operate on the ancient books management system to make a receiving confirmation.

3   Design of the Ancient Books Reparation Files Database Subsystem

3.1   Definition and Function of Ancient Books Reparation Records

The ancient books repairing files, refers to many kinds of historical forms, such as texts, forms, pictures, audio and video materials, physical materials in the process of repairing the ancient books by those professional repairing personnel with carrying on a comprehensive record and with the value of preservation. The ancient books reparation files database subsystem's function contains: offering the reparation information of reversible process, creative reparation methods, offering the valuable historical information. A lot of experts such as WeiSheng Du and Ping Zhang have emphasized the importance and urgency of creating the ancient books reparation records system. For instance, the reparation of "Zhao Cheng Jin Cang" "Dun Huang Yi Shu" "The Yongle Canon" by the national library does not get the total records and none of the video data. That is a great pity for the study and following reparation. Ancient books reparation files are able to excavate repair engineers' tacit knowledge. By building the ancient books reparation files database, we can select the relevant data and know about the process or details of the ancient books reparation, the description of the appearance before and after reparation, breakage features and causes as well as the locations, breakage extent along with the reparation project, demand and the repair engineers' feelings and experiences and other digital information alike. Ancient books reparation files can leave precious materials and research information for the reparation work in the future.

3.2   Standards and Relative Document Basis of Ancient Books Repairing Archives

Although the ancient books repairing academia and industry start to pay more attention on the building of ancient books repairing archives, the recording format is not unified, the recording degree of complex is different and the recording emphasis is different too. The study shows that it is urgent to build ancient books repairing files and introduce the policy of the unified format and its standard. There are some basis documents that guild recording ancient books repairing files:

- The ancient books repairing technical specifications and quality requirements (WH/T23-2006);
- Standard for distinction of disrepair of ancient   (WH/T22-2006);
- Grading standard of ancient books (WH/T20-2006)

Nowadays, the repairing files content that is in the ancient books repairing technique standard which was published by the Ministry of Culture is too simple, but it lays the frame foundation of building specific standards. In 2008, State Bureau of Cultural Relics published a series of the industry standards of historical relic restoration files recording standards one after another.

- Metal collections of conservation and restoration record specification (WW/T 0010-2008);
- The unearthed bamboo lacquer cultural relics of conservation and restoration record specification (WW/T 0011-2008);
- Stone cultural relics protection and restoration record specification (WW/T0012-2008);
- Collection of silk protection and restoration record specification (WW/T0015-2008).

All of these documents have reference value to the publication of the files recording standard of ancient books which are historical relics.

3.3   Functions Analysis and Design of the Ancient Books Reparation Records Management System

3.3.1   Tacit Knowledge Text Mining Function of the Ancient Books Reparation Records

Ancient books reparation technology values the traditional manual operation characteristic and its inheritance way by the mentorship. Both of them determine the ancient books reparation contained a lot of tacit knowledge such as skills, techniques, knacks and experiences. The ancient books reparation files record the means and methods of ancient books reparation. During the experts' reparation process, there will emerge a lot of tacit knowledge. In order to record the reparation process preferably, video record will be the important means. Video record can be stored and transmitted and different observers have different decoding way, hence, this is a transformation and transmission process through which tacit knowledge becomes explicit knowledge. In fact, by means of the records of the reparation files, individual's introspection to the reparation experience and literal or linguistic expression (including log book, paper writing and discussion exchange and so on) will become the important means to the experts to extract the tacit knowledge. Ancient books reparation files' record and description of the reparation process show the elusory tacit knowledge by words ,pictures, tables and videos, this provide the ideal pattern for the extraction of ancient books reparation tacit knowledge. The procedure of ancient books reparation includes a lot of tacit knowledge such as technologies, skills, knacks, experiences cause of the traditional manual operation and mentorship during the procedure.

3.3.2   Design and Management Process of the Ancient Books Reparation Records Management System Based on Management of Tacit Knowledge

The ancient books reparation records management system may includes: the breakage records and its classification before repairing, making repairing plans, dynamic generation and management of repairing process, the experience record after repairing, repairing identification and evaluation and so on.

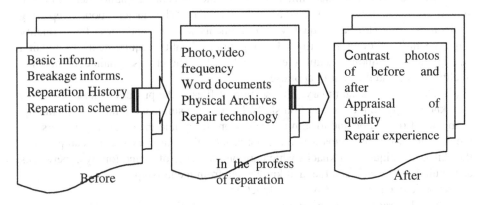

Fig. 2. The flow chart of ancient books repairing archives record set

(1) The basic information of the ancient books. It contains the information of the ancient books, the information of the volume, the information of the breakage and the information of the reparation history.

① The information of the ancient books contains: name, call number, classified index, versions and the year, binding, the number of the volume, cultural value level, the content of book etc.

② The information of the volume means the information of the book which needs reparation : the name of the volume, the sequence of the volume, bookmark, the sequence mark, the cover of the book, the protect page, the number of the pages, materials, PH value, thickness, the style of the book cover, the size of the chase, accessory and the information of the remarks.

③ The information of breakage is a very important part. The scheme of the preparation is according to the information of the breakage. It contains: the location of the breakage, the cause of the breakage, the area of breakage, the level of the breakage, the degree of the breakage, the description of the breakage.

④ The history of the reparation contains: the time of the reparation, the worker of the reparation, the way of the reparation, the note of the reparation history. The history notes contains the valuation of the reparation.

(2) Reparation scheme contains the thinking and methods. The main reparation thinking includes: repair the breakage directly; repair the breakage after deacidification; repair the breakage after dedusting, sterilizing and deacidification; repair the breakage after freeze, repair the breakage after opening the page with the steam.

(3) Reparation process contains the recording of the problems and the method to solve it during the reparation process according to the reparation scheme.

(4) The photograph and video data from beginning to end. The reparation worker record the features of the ancient books and the important picture from beginning to end. These data can serve as the reference of the quality appraised by experts and be saved as the documents.

(5) Reparation quality determination. After the procedure, quality determination should be made by two experts according to the "The ancient books repairing technology standard and quality requirements". That includes the "quality requirements", "Common various binding books quality requirements to repair", "quality degrees" and the detail rules in the "check" .The process' main content is the quality degree and the quality appraise.

(6) The experience summary. This section of files is the process of abstracting the tacit knowledge from the reparation skills. Experience summary is the way to find the problems of reparation and to summarize reasons to remind the others reparateurs and the descendants of problems which are noticeable. The summary section can preserve the skills, techniques and knacks which are on the verge of extinction by experiences accumulation in words and the tacit knowledge from mentorship.

The basic demands of the systemic function:

- The record and generation functions of the ancient books reparation files(with the characteristic of data base)
- Ancient books reparation's regulation of manipulating and the function of intelligently searching the data base
- The data base of showing the typical case videos of ancient books reparation and the function of extracting the experts' tacit knowledge.
- The function of dynamic generation of the data base which is used to search domestic ancient books reparation experts

Acknowledgement. This article is the research achievement of 'The design and realization of tacit knowledge management system of ancient books reparation technology' from the 2012 Natural Science foundation planning program of Liaoning Province Science and Technology Agency (No. 201202199).

References

1. Wu, G.: The investigation and review of Chinese libraries ancient books reparation in recent years. Journal of the National Library of China (1), 30–39 (2002)
2. Nonaka, I., Toyama, R., Konno, N.: SECI, Ba and Leadership: A United Model of Dynamic Knowledge Creation. Long Range Planning (33), 5–34 (2000)
3. Xu, W., Wang, A.: The knowledge management of ancient books reparation. Journal of Academic Libraries (2), 45–49 (2010)
4. Zhang, Z.: The discussion of scientific management of ancient books reparation. Journal of the National Library of China (2), 60–63 (2004)
5. Xu, W.: The design idea of management system of ancient books reparation. Library Development (12), 104–107 (2010)
6. Wang, A., Xu, W.: The analysis of content Settings as well as its importance of ancient books reparation files. Archives Science Bulletin (5), 1–8 (2010)

A Web Tool for Extracting and Viewing the Semantic Markups

Sabin C. Buraga and Andrei Panu

"Alexandru Ioan Cuza" University, Iasi, Romania
Faculty of Computer Science
{busaco,andrei.panu}@info.uaic.ro

Abstract. The paper presents a unified solution for detecting, processing, validating, and exporting various data expressed by different existing semantic markups (microformats, HTML5 microdata, and RDFa) embedded into Web pages. The developed tool is available as a browser extension and a bookmarklet, providing a modular architecture.

Keywords: semantic Web, semantic markups, content extraction, Web interaction, knowledge engineering.

1   Introduction

Currently, the content available on the Web is no longer intended only for human prosumers, but also for machine processing. This gives the possibility to develop intelligent information services, personalized Web sites/applications, and semantically empowered search engines in the large context of the *Semantic Web* or *Web of Data* [1].

When advancing towards this new stage of Web evolution, one of the main obstacles is the effort that the creator of the content (hypermedia information) must put into organizing the (meta)data and knowledge. Also, in the case of various social Web applications (e.g., social network sites, recommendation systems [25], etc.), the users should use vocabularies they must be familiar with for tagging entities and relationships among different resources, in order to make it comprehensible not only for humans, but also for computers. There are several platforms and tools – such as blogs, semantic wikis, content management systems, code editors, etc. – which helps transparently organize data and metadata for machine-comprehensibility, so that the contributors (publishers) do not need to know semantic Web technologies, as the existing systems automatically generate desired constructs to be further managed in an intelligent manner [2].

On the other hand, some patterns in published information and users' requests have emerged [3]. Several existing techniques (microformats, HTML5 microdata, and RDFa) aim to specify – with different syntax, expressivity, and formal background – the proper constructs able to include content semantics directly into Web pages. To identify, collect, and process this valuable knowledge, a different extraction method must be used for each solution of embedding semantics in

M. Wang (Ed.): KSEM 2013, LNAI 8041, pp. 570–579, 2013.

HTML documents. Most important details will be provided in the next section of the paper.

Our aim is to unify, in a single easy to use modular client-side Web tool, the processes of recognizing and viewing multiple semantic markups embedded in various Web documents. Additionally, this piece of software should be able to export such (meta)data in different RDF formats (RDF/XML, Turtle) and non-RDF well-known formats (e.g., CSV – Comma Separated Values).

The modular architecture of our proposed tool, called POSHex, is presented in Section 3. Various implementation details – including availability as browser extension and bookmarklet – are provided by Section 4. After that, the paper gives a list of several common uses (see Section 5) and finishes with conclusions and further directions of research.

2   Enriching the Web Document with Semantic Constructs. Client-Side Extraction Tools

The three main technologies used for enriching the Web content with semantic annotated data are *microformats*, *HTML5 microdata* and *RDFa*. They enable content description by specifying metadata directly into the HTML documents in order to be automatically processed by computers.

2.1   Microformats

Microformats are simple items of semantic markup built upon widely adopted standards that were designed for both humans and machines. According to [4], they are *a clever adaptation of semantic XHTML that makes it easier to publish, index and extract semi-structured information such as tags, calendar entries, contact information and reviews on the Web*. Microformats use existing HTML attributes (like `class` or `rel`) in order to embed into the Web pages various semantic assertions about different aspects of interest: persons (*hCard*), events (*hCalendar*), locations (*geo*), reviews (*hReview*), recipes (*hRecipe*), and many others. Microformats are widely spread – either explicit or through the semantic of content and similar structure of markup.

Some of the most popular browser extensions/plugins that detect microformats are the following:

- *Operator* – a Firefox extension; supports the most important microformats;
- *BlueOrganizer* – for Mozilla Firefox; provides support for *hCard*, *hCal*, *adr*, *rel* via contextual menus, helping users to lookup maps, movies, restaurants, user profiles, and more;
- *Microformats* – available for Google Chrome; it detects and displays *hCard*, *hCalendar*, *hReview*, *hRecipe*, and *geo*;
- *Oomph* – a Microsoft Internet Explorer add-on detecting *hCard*, *hCalendar*, and *hMedia*;
- *SafariMicroformats* – for Apple Safari; has support for hCard and hCalendar.

For other details, including the manner of recommending users certain resources of interest via microformats, consult [3].

2.2   HTML5 Microdata

Microdata is a new HTML5 specification able to annotate the existing content by using specific machine-readable labels – or, more technically, nested groups of ⟨*name, value*⟩ pairs – that are included directly into the documents [5]. The HTML5 standard provides new semantic elements able to add more expressiveness about the data structure, allowing Web publishers to layer rich structured metadata directly into the documents [6].

In order to structure information, a set of standardized vocabularies (schemas) – `schema.org` – was proposed by the major search engine companies in order to denote the most important concepts to be extracted from Web documents. The following important schemas (or conceptual models) are defined and recognized by the main search engines: *Person, Event, Organization, Place, LocalBusiness, Product, Offer, Review, MedicalEntity, CreativeWork, Book, Movie, Recipe,* etc.

From this point of view, HTML5 microdata extends microformats, giving publishers an enhanced flexibility.

The most popular browser extension that detects microdata embedded in Web pages is *Microdata.Reveal* [7] – available only for Google Chrome. For each loaded page, the tool could display a tabular view of the identified microdata. It can also export a JSON (JavaScript Object Notation) representation of the detected HTML5 microdata statements.

2.3   RDFa

RDFa (Resource Description Framework in Attributes) is a W3C recommendation which introduces the idea of embedding RDF assertions into HTML by using several attributes [8]. RDFa 1.0 was specified only for XHTML. Currently, RDFa 1.1 [9] could be used in the context of XHTML and HTML5 formats and, additionally, is working with any XML-based languages – e.g., MathML and SVG (Scalable Vector Graphics). RDFa uses some of the existing HTML attributes, but also introduces new ones, in order to provide support for the linked data paradigm [1].

Because it is just a new serialization alternative to the RDF model, it offers a great expressive power, but with some difficulties regarding the complexity of the implementations. This was a main reason for the development of the microdata specification, being the preferred choice of many search engines [10,11]. Additionally, for practical uses, a subset to RDFa is defined – *RDFa Lite* [12].

Using RDFa constructs, the ⟨*subject, predicate, object*⟩ triples could be easily expressed by using the familiar syntax provided by HTML. The main advantage of RDFa over microformats and microdata is the fact that gives support for specifying any existing vocabularies and conceptual models (taxonomies, thesauri, ontologies). Other related RDF technologies – such as SPARQL queries, GRDDL transformations, consistency checking, etc. – could be successfully used in order to manage the RDF(a) triples. Obviously, the most expressive format is RDFa.

Several browser add-ons able to detect RDFa triples are:

- *Green Turtle* – available for Google Chrome; it is an implementation of the RDFa API [13] and provides a graph-like viewer of the identified RDF triples;
- *RADify* – consists of a simple bookmarklet (in fact, a JavaScript program) able to extract existing RDFa constructs;
- *RDFa Developer* – a Mozilla Firefox extension displaying RDFa triples included into a Web page; additionally, it gives the possibility to execute SPARQL queries on the extracted RDF(a) data;
- *RDFa Triples Lister* – available for Google Chrome; it only lists the RDF triples found into a HTML5 document.

An interesting Web tool is *RDFa Play* providing a real-time editor, data visualizer and debugger for RDFa 1.1 constructs. Also, several RDFa distillers are publicly available.

2.4   Examples

A short comparison between the above mentioned techniques is depicted.

We make use of microformats, HTML5 microdata, and RDFa markups in order to denote some facts about a person:

```
<!-- microformats (using hCard) -->
<p class="vcard">
    <span class="fn">Sabin Buraga</span> holds a
    <span class="title">PhD</span> degree.</p>
<!-- HTML5 microdata, using Person provided by schema.org -->
<p itemscope itemtype="http://schema.org/Person">
    <span itemprop="name">Sabin Buraga</span> holds a
    <span itemprop="title">PhD</span> degree.</p>
<!-- equivalent constructs denoted by RDFa 1.0;
     FOAF (Friend Of A Friend) ontology is used -->
<p xmlns:foaf="http://xmlns.com/foaf/0.1/" typeof="foaf:Person">
    <span property="foaf:name">Sabin Buraga</span> holds a
    <span property="foaf:title">PhD</span> degree.</p>
```

The second example – adapted from [14] – models certain knowledge regarding an academic event by using HTML5 microdata and equivalent RDF triples:

```
<section class="conference" itemscope
    itemid="/event/conference/2013/"
    itemtype="http://schema.org/Event"
    itemclass="SocialEvent BusinessEvent EducationEvent">
  <h1 lang="en">
    <a itemprop="url" href="http://conference.info/2013/">
      <span itemprop="name">Conference 2013</span></a></h1>
  <p itemprop="location" itemscope itemid="/places/Iasi"
        itemtype="http://schema.org/Place">
    <span itemprop="name" lang="en">
```

```
    <a itemprop="url" href="http://places.org/Iasi">Iasi</a>
      (a city in <a href="http://places.org/Europe">Europe</a>)
    </span></p>
  <p class="date"><time itemprop="startDate"
                        datetime="2013-05-15">15</time> --
                  <time itemprop="endDate"
                        datetime="2013-05-17">17 May</time></p>
</section>
```

Using specific mappings (e.g., a XSLT transformation via GRDDL), the equivalent RDF constructions are the following – adopting the Turtle (Terse RDF Triple Language) [15] syntax:

```
@prefix s: <http://schema.org/>
<http://site.edu/event/conference/2013/>
    a s:Event,
      s:SocialEvent, s:BusinessEvent, s:EducationalEvent ;
    s:url <http://conference.info/2013/> ;
    s:name "Conference 2013"@en ;
    s:location <http://places.org/Iasi> ;
    s:startDate <2013-05-15>^^xsd:date ;
    s:endDate <2013-05-17>^^xsd:date .
<http://site.edu/places/Iasi> a s:Place ;
    s:name "Iasi (a city in Europe)"@en ;
    s:url <http://places.org/Iasi> .
```

3 From Conceptual Modeling to System Architecture

Let $S = M \bigcup H \bigcup R$ be a set consisting of the semantic markups – microformats, HTML5 microdata, and RDFa triples – to be detected, processed, displayed, and eventually transformed into other formats (in certain cases, the original semantics might not be preserved).

For each element $s \in S$, the syntax and semantics should be considered – $s = \langle s_{\mathrm{syn}}, s_{\mathrm{sem}} \rangle$ – along with functions regarding detection, validation, transformation, etc. of a given semantic markup. Thus, we could define several relations between microformats and different software components able to perform a certain operation. For example, $\delta(s, d)$ is the relation between a semantic construct and a detector, where $d \in D$ – the set of detectors (i.e. computer programs detecting a semantic construct by using the s_{syn} specification). In a similar way, we can define φ and τ – the relations concerning validation and transformation.

On the basis of this simple formalism, we proposed an ontological model – expressed in OWL (Web Ontology Language) – regarding the tool's architecture. This knowledge is also useful during the analysis and design phases. The main class (*Software*) includes the subclass *ClientSideSoftware* having *Browser-Addon* as a subclass. A set of modules (individuals from *Module* class) will be

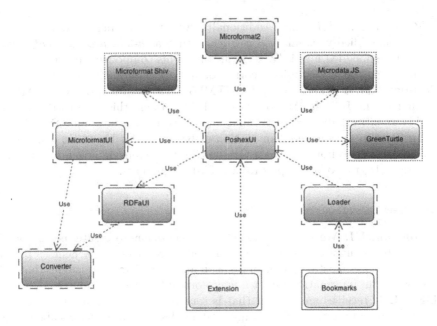

Fig. 1. POSHex system architecture

implemented, each of them providing a specific functionality. Also, the relations between modules or between modules and semantic markups should be defined.

Using this conceptual model, we continue providing details about the current POSHex architecture. Fig. 1 is showing our implemented components (denoted by a dashed line border), third-party library components (depicted by a dotted border) and end user packages (drawn with a continuous line border).

To provide a proper interaction with the users, we have also used the popular jQuery and jQueryUI libraries.

In the next sections, we present each component along with its functionality.

3.1 Parsers

Microformat-shiv. A JavaScript library that implements a parser and provides an API which can be used in various browser extensions/plugins in order to parse semantic content expressed via microformats. This API does not provide a method to extract all microformats at once, so we need to parse each accepted formats (*hCard, hCalendar, hResume, hAtom,* etc.), following a similar approach presented in [3]. The module returns a JSON object containing an array storing detected microformats.

Converter. Offers our implementation of different useful methods to convert JSON and triples retrieved by parsers into various exchange formats: CSV, RDF/XML, and Turtle.

Microformat2. We have developed this parser using the algorithm provided by [16]. The most important accepted traditional microformats are: *adr, geo,* *hCard, hEvent, hResume.* Additionally, we support Microformats 2 which offers simplified vocabulary constructs – such as *h-adr, h-card, h-entry*, etc. – to be embedded into HTML documents. This component returns an array containing the JSON objects corresponding to found microformats.

MicrodataJS. A jQuery plugin for HTML5 microdata, it provides an API similar to the HTML5 Microdata DOM API. Using this plugin [17], we are able to extract all microdata items and convert them into a JSON object.

Green Turtle. It is an implementation of RDFa 1.1 for browsers, distributed as a JavaScript library or as a Google Chrome extension. We are using this library [18] to extract triples from HTML documents.

3.2   User Interface

MicroformatUI. We have developed this component to create menus for all extracted semantic content. This component receives an element as parent and generates a list of submenu items, bound with `onclick` listener that will show useful information about a microformat/microdata element.

RDFaUI. It receives an element that is sent to a RDFa graph processor and creates an interaction grid containing all triples, including details about each subject, predicate, and object of a RDF triple.

4   Installation and Deployment

POSHex [19] is distributed in two ways: as a Web browser extension and as a bookmarklet.

Currently, the extension supports Google Chrome and Mozilla Firefox. The packaging was built by using Crossrider [20], a cloud-based development framework that facilitates the creation of cross-browser extensions. This framework simplifies the installation by automatically detecting user's browser and providing the appropriate package for download.

After installing the *POSHex Extension*, a button will be added in the browser user interface, offering a non-intrusive contextual menu for the extension. The extension functionality is divided in two subcomponents, represented by two source JavaScript files: `background.js` (executed only once, when browser starts; it adds different contextual menus for a proper interaction with the users) and `extension.js` – invoked for each Web page; after the page finishes to load, it runs all the scripts needed for this extension to work in order to detect and extract the semantic constructs.

POSHex Bookmarks is another manner of using this tool without installation. The user should save the four available bookmarks and then manually select the desired functionality by clicking on a bookmark after loading a Web document. These bookmarklets are implemented by several JavaScript methods. The most important bookmark is the one that loads `main.js` from server. It starts retrieving all other resources needed for the POSHex toolbar, creates the toolbar and adds it to the current document.

The current version was tested and works on majority of browsers: Chrome, Firefox, Internet Explorer, Opera, and Safari. The only limitation is that these bookmarks cannot be used on some Websites that do not allow running script outside of domain or do not accept accessing resource via plain HTTP (i.e. for those who use only the HTTPS protocol).

5 Use Cases

When a user visits a page that contains semantic markups (microformats, microdata, or RDFa items), POSHex toolbar will be activated. According to what type of semantic content is found, the corresponding menu will be enabled.

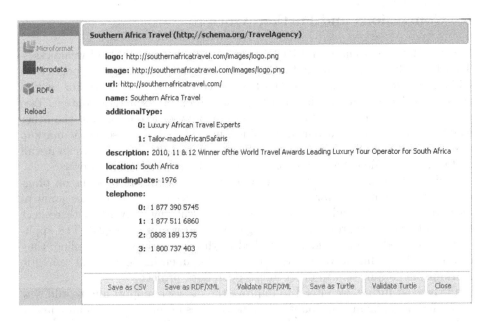

Fig. 2. Several microdata items found on a specific Web page

The user can access the microformats, microdata, or RDFa specific constructs by clicking on an item from menu. A dialog will pop up as in Fig. 2. The user can save extracted data on different formats (CSV, RDF/XML or Turtle) or validate the RDF/XML or Turtle elements by being redirected to the online validators such [21] and [22] – see also Fig. 3.

For documents that are loaded through Ajax requests, the *Reload* button could be used to parse the HTML code again and reload all detected constructs.

Furthermore, the extracted RDF triples could be loaded in a triple-store system (e.g., Apache Marmota, OpenLink Virtuoso) and/or could be transformed according to the JSON-LD specification [23], offering support for the Linked Data initiative [1].

Fig. 3. Viewing, exporting or/and validating the detected RDF triples

6   Conclusions and Further Work

Microformats, microdata, and RDFa have each their strengths and weaknesses. They take different approaches towards the same goal: extending Web pages with additional semantics that are not part of the core HTML language. For each of them, tools have been built in order to parse and extract semantic content. Two main differences and advantages between our browser extension and the other existing solutions are that our proposal is a unified tool for all semantic markup specifications and offers a proper cross-browser capability, so the user can install and invoke a single extension/bookmarklet in all his/her favorite browsers.

POSHex was mainly developed in order to help beginners to focus on pragmatic uses of semantic markups, and not on the tool itself. From this point of view, our proposal could be seen as an useful e-learning software, with respect to several studies regarding methodologies for better education [24]. The application was used by our students enrolled to the Semantic Web discipline; they considered it a valuable Web instrument for experimenting lightweight semantic markups, in conjunction to other linked open data projects.

In the near future, we intend to improve the browser support and to add new useful features like integration with online services of interest (calendars, maps, online communities, recommender systems, etc.), following the ideas exposed in [3] and [25].

Acknowledgements. We are grateful to *Constantin Tiberiu Pasat* for his important support regarding the development and testing.

References

1. Bizer, C., Heath, T., Berners-Lee, T.: Linked Data—The Story So Far. International Journal on Semantic Web and Information Systems 5(3), 1–22 (2009)
2. Dumitriu, S., Girdea, M., Buraga, S.: From Information Wiki to Knowledge Wiki via Semantic Web Technologies. In: Innovations and Andvanced Techniques in Computer and Information Sciences and Engineering, pp. 443–448. Springer (2007)

3. Luca, A.-P., Buraga, S.C.: Enhancing User Experience on the Web via Microformats-based Recommendations. In: Filipe, J., Cordeiro, J. (eds.) Enterprise Information Systems. LNBIP, vol. 19, pp. 321–333. Springer, Heidelberg (2009)
4. Khare, R., Celik, T.: Microformats: A Pragmatic Path to the Semantic Web. In: 15th International Conference on World Wide Web, pp. 865–866 (2006)
5. Hickson, I. (ed.): HTML Microdata, W3C Working Draft (2012), http://www.w3.org/TR/microdata
6. Ronallo, J.: HTML5 Microdata and Schema.org. Code4Lib Journal (16) (2012)
7. Microdata.Reveal Extension, https://github.com/cwa-lml/microdata.reveal
8. Adida, B., et al. (eds.): RDFa 1.1 Primer, W3C Working Group Note (2012), http://www.w3.org/TR/rdfa-primer/
9. Adida, B., et al. (eds.): RDFa Core 1.1, W3C Recommendation (2012), http://www.w3.org/TR/rdfa-core/
10. Steiner, T., Troncy, R., Hausenblas, M.: How Google is using Linked Data Today and Vision For Tomorrow. In: Proceedings of Linked Data in the Future Internet – Future Internet Assembly (2010), http://CEUR-WS.org/Vol-700/Paper5.pdf
11. Finnin, T.: Microdata Chosen over RDFa for Semantics by Google, Bing and Yahoo (2011), http://is.gd/r7S1Cz
12. Sporny, B. (ed.): RDFa Lite 1.1, W3C Recommendation (2012), http://www.w3.org/TR/rdfa-lite
13. Sporny, M., Adrian, B. (eds.): RDFa API, W3C Working Group Note (2012), http://www.w3.org/TR/rdfa-api/
14. Tennison, J.: Microdata and RDFa Living Together in Harmony (2011), http://www.jenitennison.com/blog/node/165
15. Prud'hommeaux, E., Carothers, D. (eds.): Turtle – Terse RDF Triple Language, W3C Candidate Recommendation (2013), http://www.w3.org/TR/turtle/
16. Microformats 2, http://microformats.org/wiki/microformats-2
17. MicrodataJS, https://gitorious.org/microdatajs/
18. Green Turtle, https://code.google.com/p/green-turtle/
19. POSHex Project, http://bit.ly/posh-ex
20. Crossrider Framework, http://crossrider.com/
21. RDF Validator and Converter, http://www.rdfabout.com/demo/validator/
22. W3C RDF Validator, http://www.w3.org/RDF/Validator/
23. Sporny, M., Kellogg, G., Lanthaler, M. (eds.): JSON-LD 1.0. A JSON-based Serialization for Linked Data, W3C Last Call Working Draft (2013), http://www.w3.org/TR/json-ld/
24. Alboaie, L., Vaida, M.-F., Pojar, D.: Alternative Methodologies for Automated Grouping in Education and Research. In: Proceedings of the CUBE International Information Technology Conference, pp. 508–513 (2012)
25. Alboaie, L., Vaida, M.-F.: Trust and Reputation Model for Various Online Communities. Studies in Informatics and Control 20(2), 143–156 (2011)

Using Support Vector Machine for Classification of Baidu Hot Word

Yang Hu and Xijin Tang

Institute of Systems Science, Academy of Mathematics and Systems Science
Chinese Academy of Sciences, 100190 P.R. China
huyang11@mails.gucas.ac.cn, xjtang@amss.ac.cn

Abstract. Support vector machine (SVM) provides embarkation for solving multi-classification problem toward Web content. In this paper, we firstly introduce the workflow of Support Vector Machine. And we utilize SVM to automatically identifying risk category of Baidu hot word. Thirdly, we report the results with some dicsussions. Finally, future research topics are given.

Keywords: SVM, text classification, text extraction, Baidu hot word.

1 Introduction

Text classification is a popular research topic. Methods in text classification include logistic regression, KNN, decision tree, artificial neural network, naive Bayes, etc. As a well-known tool among those, support vector machine is widely utilized [1-3]. In this paper, SVM to text classification is conducted to automatically identify risk category of Baidu hot word.

This paper is organized as follows. Section 2 presents the process of support vector machine applied to Baidu hot word. Section 3 provides the detail of collecting Baidu Hot word and corresponding news. Section 4 discusses the risk classification result of 4 experiments based on SVM. Conclusions and future work are given in Section 5.

2 Process of Support Vector Machine to Text Classification

In the process of SVM to text classification as shown in Figure 1, construction of dictionary is the first step. At first, plain text is segmented into Chinese terms by MMSeg [4], while stop words are eliminated. In this research, stop words form HIT (Harbin Institute of Technology) are obtained. Their stop words contain 767 functional words in Chinese.[1] Finally, the remaining terms constitute the initial dictionary.

Next is feature selection as the way to generate the dictionary of salient terms. Among methods such as information gain, mutual information and chi-square, chi-square out-performed other two methods in test [5]. Here chi-square is tried. Terms within top given ratio on Chi score in each category are filtered into the dictionary.

---

[1] Obtained from http://ir.hit.edu.cn/bbs/viewthread.php?tid=20

M. Wang (Ed.): KSEM 2013, LNAI 8041, pp. 580–590, 2013.

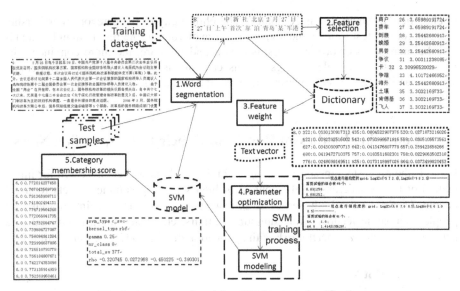

Fig. 1. A process of applying SVM to text classification

According to the dictionary obtained, plain text is transformed into text vectors of terms in the dictionary. Then weights are assigned to feature words in the text vectors. In this research, $tf \cdot idf$ is tried. In addition to $tf \cdot idf$, a pile of methods such as $tf \cdot rf$ (relevance frequency), $tf \cdot \chi^2$, $tf \cdot ig$ (information gain) can be adopted. Here rf measures the unbalance of documents containing the observed term. The formula of rf is $\log(2 + a / b)$, where a is the number of positive documents in bipartition containing the term and b is the number of negative documents containing the term [6].

Afterwards text vectors of feature weights are inputted as samples into SVM training process. In SVM modeling, the unbalance existing among separate categories of samples is irritating. Moreover, categories of words at different time and different context are varied. For example, "Liu Xiang", a proper noun as the name of a well-known Chinese athlete, who withdrew from 2012 London Olympic dramatically because of bruised foot, is labeled as risk category of medical care from August 7, 2012. As more information about operating team exposed, risk category about "Liu Xiang" changes to morals and integrity from August 9, 2012. Therefore, classification precision by SVM is perturbed. So the category membership score is leveraged to enhance the classification precision. The category membership score is computed by Equ.(1).

$$score = \frac{\sum S_i}{2*k} + \frac{k}{2*n} \tag{1}$$

where k is the number of voters supporting a certain category, n is the number of categories, S_i is the score of each supporting voter. As n-category classification problem in SVM can be treated as multiple binary-classification problem, C_n^2 voters as classifiers in bipartition is computed. The philosophy of the category membership score is: for one test sample, the bigger the score of voter as the first item in Equ.(1) is and the more the supporting voters as the second item are, the more convinced the judgment that the sample belongs to this category is. With category membership score, classification result whose score is under the chosen best-fit threshold is ignored.

3   Collecting Baidu Hot Words and Their Corresponding News

Baidu is the biggest Chinese search engine in the world now. People search for information of their concerns and the content of high searching volume reflects focus of people. That's to say, Baidu serves as an instantaneous corpus to maintain a view of people's empathic feedback for community affairs, etc. Thus Baidu is utilized as a perspective to analyze societal risk with application of SVM addressed in Section 2.

3.1   Baidu Hot Word

The portal of Baidu news (http://news.baidu.com) provides routinely 10~20 hot search words which are updated every 5 minutes. Each hot news word corresponds to an individual URL of the word search page which consists of links to news from diverse news portals as shown in Figure 2.

Fig. 2. The portal of Baidu news redirects to word search page consisting of news page URLs

To collect Baidu hot search words, a Web crawler is customized to grab Baidu news page every hour. Then open source package htmlparser is leveraged to extract hot words from these html pages. Afterwards the hot search words are stored as an xml file. Meanwhile, a score is given to hot search word based on its rank in Baidu news page. If there are ten hot words, the 1st one is given 20 points, the 2nd one is 18 points and the last one is 2 points. If there are 20 hot words, the score of 1st hot word is 20, the score of 2nd hot word is 19 and the last one's score is 1. Here score reflects the degree of people's attention. At 23:59 of each day, hot words gathered in the past 24 hours are accumulated and the scores in each hour are added up as day score of each hot word, then hot words are stored in descending order according to their day scores as an xml file. The 1st page among corresponding search pages is also stored. Based on those data, Baidu Hot Word Vision system adopting JSP was developed [7]. Users can get the daily, weekly and monthly rank list of Baidu hot word with frequency distribution, use iView and CorMap to analyze social events and search Baidu hot words containing key words of users' interest.

3.2   News Text Extraction

Based on the 1st page of search results, news text from the news portals whose links are contained in the 1st page is extracted. For big news portal such as Sina (www.sina.com.cn), there exist branch sites such as finance.sina.com.cn, sports.sina.com.cn, etc. Besides, those news portals often adopt new design of Web pages, an obstacle to customization mode of text extraction for each news portal. Hence statistics of news portals is utilized for text extracting. Then target seeds of text extraction are chosen by accumulating counts of hot search word-related news and figuring out affiliated news sites.

3.2.1   Statistics of News Sites

The target seeds for text extraction are determined by combining the set of web sites which account for 80% of hot words-related news. Table 1 lists the top 10 news portals.

Table 1. Top 10 news portals in June and September of 2012

News portal	Rank in June	Rank in September
人民网(www.people.com.cn)	1	5
搜狐(www.sohu.com)	2	1
21CN(www.21cn.com)	3	9
凤凰网(www.ifeng.com)	4	2
第一视频(www.v1.cn)	5	6
财讯(www.caixun.com)	6	10
新浪(www.sina.com.cn)	7	4
和讯网(www.hexun.com)	8	3
山东新闻网(www.sdnews.com.cn)	9	7
网易(www.163.com)	10	8

Figure 3 shows the distribution of frequency that major news portals contribute Baidu hot words in June and September of 2012. The result in Figure 3 shows that top 20% news portals release almost 80% of the news listed at the 1st page of the hot word search results. Furthermore, combination of target seeds in 2 months increases the proportion. Web sites not in the scope of target seeds are also included by employing titles of their news as the origin of corpus for latter processing. Then, 138 news portals are chosen as target seeds for text extraction.

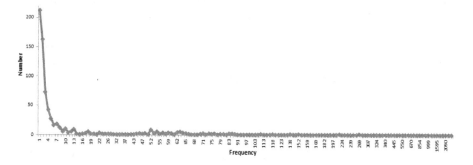

Fig. 3. The distribution of frequency that major news portals contribute Baidu hot words in June and September of 2012

3.2.2   News Text Extraction

By leveraging the computational method called generalized regular expression-based algorithm, the plain texts of diverse Web pages are obtained. The procedure of text extraction is shown below.

```
Input:
    S₁: html files of news pages concerning Baidu hot word
Loop:
    For each html file x in S₁
        Get all div blocks of x
        Abandon div blocks containing less than given threshold Chinese
        characters
        Select div block w of highest share of Chinese character
        Filter HTML tag of w using regular expression
    End for
```

The threshold of the algorithm is set carefully according to the practical application. In this research, the value of threshold is 300 by default.

After the text extraction, the news text is stored as an xml file in directory YYYY-MM-DD/id.xml, where id is the rank of hot words in daily list. Each item consists of 6 sub items including news *title*, *link* to news portal, *site* of news portal, publishing *date*, *id* which is the rank of this news item in word search page and the plain *text* of news page.

```
<item>
    <title>传金正恩下令导弹部队待命攻击美军基地</title>
    <link>http://news.hexun.com/2013-03-29/152626268.html?from=rss</link>
    <site>和讯</site>
    <date>2013-03-29 07:00:00</date>
    <id>13</id>
    <txt>环球外汇3月29日讯在美国连续两次向出动B2隐形轰炸机飞往韩国，向朝
    鲜当局展现军力后，据《路透社》报导，朝鲜领导人金正恩周五(3月29
    日)在紧急会议中命令国内导弹部队随时待命，准备攻击韩国和太平洋
    (601099,股吧)的美军基地。</txt>
</item>
```

3.3   Manual Labeling of Risk for Baidu Hot Word

We map Baidu hot search word into a certain risk category. The risk category adopts the results of a study of risk cognition taken before 2008 Beijing Olympic Games [8, 9]. The risk index compendium sorts out societal risk into 7 categories, national security, economy & finance, public morals, daily life, social stability, government management, resources & environment with 30 sub categories. Figure 4 shows the risk levels of 7 categories between November, 2011 and October, 2012 based on manual labeling of risk category for Baidu hot word. Here the risk level denotes the proportion of total frequency of hot words labeled as one of the 7 risk categories to the total frequency of Baidu hot search words.

As shown in Figure 4, risk level of each category is less than 0.2 and risk level in daily life is higher than those of other 6 risk categories. The risk level is highest in December, 2011 and risk level in daily life takes up one third of the total risk level. We find that rising price and income gap attract most of people's attention at the end of year. On the contrary, the risk level falls down sharply during August, 2012 when London Olympic Games takes place. The majority of hot words are relevant to sports which are risk-free during that period.

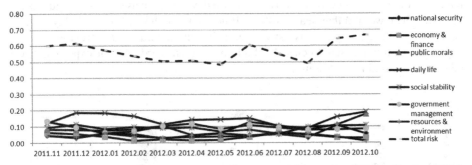

Fig. 4. Risk levels of 7 categories based on Baidu hot word (November, 2011 to October, 2012)

In order to monitor the risk level daily, it is necessary to map each hot word into a risk category and calculate the risk level of each category. Manual labeling of Baidu hot search words is a heavy burden, then machine learning by SVM is explored to automatic labeling.

4   Experiment Results

According to the process addressed in Section 2, we carry out experiments using libSVM [10]. Source data are from January 5, 2013 to February 28, 2013. Based on the manual labeling of each hot search word, we map the corresponding news text into the same risk category as that of the hot search word. The news text with double-repeated news titles and corresponding risk category constitute the sample. Table 2 shows the number of Baidu hot search words and samples of each risk category during that period. From Table 2, we see that the three categories of risk, daily life, social stability and government management, contribute main risks. So the samples among different risk categories are unbalanced.

Table 2. The number of Baidu hot words and samples of each risk category (January 5, 2013 to February 28, 2013)

Risk Category	Num. of Baidu hot words	Num. of samples
national security	124 (8%)	2443 (8%)
economy & finance	57 (4%)	1092 (4%)
public morals	131 (8%)	2492 (8%)
daily life	260 (16%)	4983 (16%)
social stability	137 (9%)	2635 (9%)
government management	356 (22%)	6769 (22%)
resources & environment	81 (5%)	1555 (5%)
risk-free	459 (29%)	8863 (29%)
Total	1605	30832

In our experiments, we classify Baidu hot search words of each day using previous N-day's data (N=1, 2, 3...) as training sets. The design of the experiments is based on the occurrence analysis of Baidu hot search words from January 5, 2013 to February 28, 2013 as shown in Figure 5. Figure 5(a) is the distribution of frequency that Baidu hot search words occur. In total, 261 hot words occur once, 647 hot words occur twice, 8 occur three times, 5 occur four times and 1 occurs six times. Moreover, more than 70% of hot words consecutively occur two days or more as shown in Figure 5(b). In Figure 5(c), we find that nearly 40% of Baidu hot search words of each day already occur consecutively in previous days. Then it is quite natural to use previous several days' data as training sets to classify Baidu hot search words. In our experiment, we take 4 tests to find the fittest model.

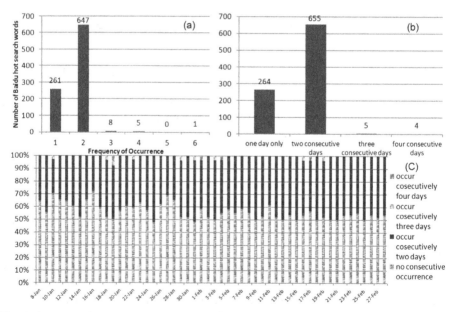

Fig. 5. Statistics of occurrence for Baidu hot search words (January 5, 2013 to February 28, 2013)

Four experiments are designed to test classification in our research. Experiment 1 uses previous day's data for training and today's data for testing, Experiment 2 uses previous 2-day's data as training sets, Experiment 3 uses previous 3-day's data as training sets and Experiment 4 uses previous 4-day's data for training. Figure 6 shows the sample size of each day from January 5, 2013 to February 28, 2013.

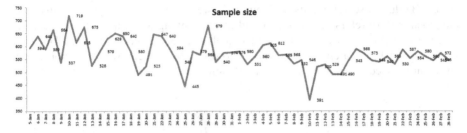

Fig. 6. The number of Baidu hot words corresponding news of each day (January 5, 2013 to February 28, 2013)

In four experiments, RBF (Radial Basis Function) is employed as kernel function of SVM. Terms of top 40% on chi score are chosen to constitute the dictionary. TF-IDF is adopted as the feature weights. To measures the performance of SVM classification, we use the standard definition of precision as shown in Equ.(2) in this research.

$$precision = \frac{\left| S_{L(SVM=L(Manual))} \right|}{\left| S_{sample} \right|} \tag{2}$$

where $S_{L(SVM=L(Manual))}$ is defined as the set of those news that SVM gives the same label as manual labeling. S_{sample} is defined as the set of news in the test samples. For each experiment, the precision values of each day are averaged to obtain a single-number measure of classification performance. The average classification precision of four experiments is given in Table 3. Among four experiments, Experiment 4 gets the highest precision. Figure 7 shows the detail of classification precision for each day of the four experiments.

Table 3. Classification precision for four experiments

	Average classification precision
Experiment 1	68.4%
Experiment 2	69.4%
Experiment 3	70.7%
Experiment 4	71.5%

*Parameter setting: RBF (radial basis function) is employed as kernel function, for feature selection top 40% terms in chi score are chosen, TF*IDF is adopted as feature weights.*

As shown in Figure 7, the classification results on February 4 in Experiment 1, February 9 in Experiment 2, January 24 in Experiment 3 and Experiment 4 get the highest precision in each experiment. On the contrary, January 17 in Experiment 1 and Experiment 4, February 19 in Experiment 1 and Experiment 2 get the lowest precision.

In four experiments, classification precision for one day falls behind the other days when the day contains Baidu hot search words that neither appear before nor are evident in risk classification. Hot words on January 17 in Experiment 1 are one typical example. The classification on January17 gets the worst precision in Experiment 1. Among Baidu hot search words on January17, we find one hot search word that does not appear before refers to one singer and actress talking about a political issue. The machine's label is risk-free, while manual label is national security, instead. Other hot words like GDP and a city mayor who go to office by bicycle, which do not happen before, are indeed ambiguous in risk classification.

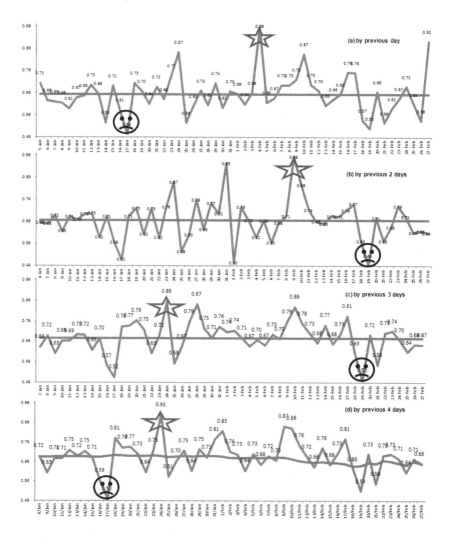

Fig. 7. Classification precision for each day of four experiments, respectively

For four experiments, the best occurs when Baidu hot words happen before. The most typical example is the period of Chinese Spring Festival. Majority of Baidu hot search words concentrate on people's holiday life including traffic problem, price rising and air pollution caused by fireworks, etc.

5   Conclusion

In this paper, we develop one Java program to collect Baidu hot words and their corresponding news, then we leverage the process of SVM to text classification to automatically identify risk category of Baidu hot words. Four experiments using different previous day's data are tested to find the fittest model to label the risk of today's hot words. The results show that classification for today by previous 4 days' data gets the highest precision. Based on the risk classification, we may have a vision of societal risk through Baidu hot words.

A lot of work needs to be done in the future. As a classification problem, empirical comparison with other methods is needed in our paper. And the effect of different removal percentage of feature words and different kernel types on classification precision is promising [11]. We also need to apply SVM to Baidu hot word using longer longitudinal data as more data are accumulated.

Acknowledgements. This research is supported by National Basic Research Program of China under Grant No. 2010CB731405 and National Natural Science Foundation of China under Grant No.71171187.

References

1. Yang, Y.M., Liu, X.: A re-examination of text categorization methods. In: Proceedings of the 22nd Annual International ACM SIGIR Conference on Research and Development Information Retrieval, pp. 42–49 (1999)
2. Tong, S., Koller, D.: Support Vector Machine Active Learning with Applications to Text Classification. Journal of Machine Learning Research 2, 45–66 (2001)
3. Sebastiani, F.: Machine learning in automated text categorization. ACM Computing Surveys (CSUR) 34, 1–47 (2002)
4. Tsai, C.H.: MMSEG: A word identification system for Mandarin Chinese text based on two variants of the maximum matching algorithm,
   http://www.geocities.com/hao510/mmseg/
5. Yang, Y.M., Pedersen, J.O.: A comparative study on feature selection in text categorization. In: Proceeding of the 14th International Learning Conference on Machine Learning, pp. 412–420. Morgan Kaufmann Publishers (1997)
6. Lan, M., Tan, C.L., Su, J., Lu, Y.: Supervised and traditional term weighting methods for automatic text categorization. IEEE Transactions on Pattern Analysis and Machine Intelligence 31, 721–735 (2009)

7. Wu, D., Tang, X.J.: Preliminary analysis of Baidu hot words. In: Proceedings of the 11th Workshop of Systems Science and Management Science of Youth and 7th Conference of Logistic Systems Technology, pp. 478–483. Wuhan University of Science and Engineering Press (2011) (in Chinese)
8. Tang, X.: Qualitative meta-synthesis techniques for analysis of public opinions for in-depth study. In: Zhou, J. (ed.) Complex 2009. LNICST, vol. 5, pp. 2338–2353. Springer, Heidelberg (2009)
9. Zheng, R., Shi, K., Li, S.: The Influence Factors and Mechanism of Societal Risk Perception. In: Zhou, J. (ed.) Complex 2009. LNICST, vol. 5, pp. 2266–2275. Springer, Heidelberg (2009)
10. Chang, C.C., Lin, C.J.: libsvm2.8.3.,
    http://www.csie.ntu.edu.tw/~cjlin/libsvm/
11. Zhang, W., Yoshida, T., Tang, X.: A Study on Multi-word Extraction from Chinese Documents. In: Ishikawa, Y., He, J., Xu, G., Shi, Y., Huang, G., Pang, C., Zhang, Q., Wang, G. (eds.) APWeb 2008 Workshops. LNCS, vol. 4977, pp. 42–53. Springer, Heidelberg (2008)

Context-Free and Context-Dependent Service Models Based on "Role Model" Concept for Utilizing Cultural Aspects

Hisashi Masuda[1,2], Wilfrid Utz[3], and Yoshinori Hara[4,2]

[1] Japan Advanced Institute of Science and Technology, Japan
masuda@jaist.ac.jp
[2] Japan Science and Technology Agency, RISTEX
[3] Research Group Knowledge Engineering, University of Vienna, Austria
wilfrid@dke.univie.ac.at
[4] Graduate School of Management, Kyoto University, Japan
hara@gsm.kyoto-u.ac.jp

Abstract. Today's service economy experiences a bi-polarization: a few large organizations are capable to offer services that scale and are applicable on a global market (as context-free services) while a plethora of small organizations provide their services locally due to the contextual dependencies of their offerings (as context-dependent services). As a future challenge and next generation of service offerings we investigate in this paper how context-dependent services can scale-up to a global market while maintaining the domestic and cultural aspects they embody. We propose the "role-model" concept to overcome the limitation in integrating aspects from both types. The concept is evaluated by analysing the actual service workflows of two representative cases using a common modelling approach realized on the ADOxx meta-modelling platform.

Keywords: Cultural computing, Service economy, Meta-modelling.

1 Introduction

1.1 Background

Today's service economy experiences bi-polarization in how organisations provide their offerings on the market. While a few companies manage to provide services on a global market and leverage on effects of globalisation, a multitude of small and medium-sized companies act in their local market and domain due to the dependencies on cultural aspects and local background of services offered. For this paper we define the former - global services as "context-free services" whereas the later - local services as "context-dependent services".

Context-free services are widely available in today's economy around the world. Examples range from the food industry as provided by McDonald's or Burger King to IT-services and products as offered by Google, Apple or Microsoft. The common denominator for all these services lies within the fact that

M. Wang (Ed.): KSEM 2013, LNAI 8041, pp. 591–601, 2013.

these services are are "readable" and understood by anyone who is willing to consume the service.

On the other hand, context-dependent services exist in the local surrounding and domain where the service is offered. Examples for the Japanese environment include Edomae-sushi (high class sushi restaurant), Kyo-kaiseki(high-class restaurant in Kyoto). These services target a specific audience and depend on domestic culture. For the Japanese cases as mentioned above, the Japanese high maturity level consumers in their cultural surrounding are in focus, synonymously applicable for other services in other countries. For the context-dependent services, the actual service value as perceived by the consumer is to a high extent related to the contextual environment of the service rather than service functionality/related product itself. For the sushi case, the actual service functionality - raw fish on rice, is enhanced by the environment and cultural setting provided by the restaurant. Concepts like atmosphere, communication, decoration are regarded more important in this example. For context-dependent services the scaling from the local domain to a global market is difficult to achieve - aspects from "understanding" of services has to be integrated in the deep and cultural-founded quality of services. This integrated, new style of services offered is understood as a fundamental differentiation from the current situation.

In the field of Human Computer Interaction (HCI) different approaches have been investigated to analyse the cultural aspects in information systems. The ZENetic approach as introduced in [8] and [9] as well as Alice's adventure as described in [4] and [7] introduce new physical interface to represent the cultural aspects of human interaction in a culture aware setting. Both examples from the cultural computing domain are promising candidates on how local cultural aspects can be integrated in the personal HCI experience/expectation of the user. With respect to the service domain, the envisioned integration approach - mainly in relation to scalability of cultural foundation of context-dependent services - is currently not sufficiently analysed and covered.

1.2   Research Question and Objectives

The research issue as outlined above lies within the analysis procedure for both service types using a common approach and platform independent of a service's type. As a result, a comparison and differentiation of aspect of the two type services is feasible, the "bottleneck" factors for context-dependent services are externalized and allow an evaluation why certain services cannot scale in other countries. Based upon the results of the analysis, we can consecutively propose a new business model for service providers.

1.3   Approach

The approach followed in the research work conducted and presented in this paper consists of two elements: as a first element we propose the concept of "role model" as a representation and conceptual model of the behaviour of the service provider and service consumer and their interaction. Secondly, meta-modelling concepts are applied to implement a prototypical analysis environment for the approach introduced. As a development and realisation environment the ADOxx[1] meta-modelling platform is used; the development process is driven by the Open Models Initiative (OMI) approach. Several research and industrial projects have adopted the OMI approach and developed modelling tool for conceptually modelling enterprise information systems[3] on ADOxx[2].

To analyse the context-dependent service model it is essential to understand the consumers' and providers' aspects simultaneously and concurrently, also considering dynamic aspects of service offering [5], [6] in different stages of the service life-cycle. Service provider and consumers act based on specific role models or mindsets, a service provider in a context-dependent setting aims to simulate the target/consumer role model to increase the service value in the encounter stage. Important aspects to be considered in this adaptation process are cultural/local background (e.g. Japanese, German, English, Italian cultural mindset) and the maturity level of the consumer among others. Maturity level is defined as the experience the consumer has built up in "using" the service (number of re-visits, frequency of usage). For context-free services, a general-type role-model is assumed for the provider and consumer hence the service consumer has no problem to consume the service since only minor adaptations and modifications are needed by the service provider in the encounter stage. Such minor adaptations include language adaptation or implementing technology for certain non-functional aspects of the service.

Using ADOxx as a development environment, a graphical modelling approaches for the role-model as introduced is implemented supporting different cultural settings. Different viewpoints of cultural aspects are supported by a hybrid approach to combine and adapt meta-models to include/exclude specific aspects in a domain of interest, still maintaining model value functionality implemented on meta-level (e.g. multi-level analysis functionalities, dynamic evaluation options, visualisation mechanisms).

Within the next section we introduce cultural computing aspects as related work to the research conducted for the service economy, mapped to the research questions of the previous subsection. In section 3 we introduce the role model concept in detail before evaluating the concept through a case study in section 4 for McDonald's as a context-free service and Edomae-sushi as a context-dependent service. The results of the case study are interpreted in section 5 before concluding the paper in section 6.

2 Related Work

In the field of Human Computer Interaction , the interaction with information systems has continuously evolved from personal computing, cooperative computing, social computing to cultural computing. Two related experiments in the

---

[1] see http://www.adoxx.org for further details
[2] see http://www.openmodels.at/web/omi/omp for current projects

field of cultural computing have been executed to investigate Eastern and Western culture in the computing domain: a) ZENetic computing represent eastern culture by enabling new physical computing interfaces [1] and b) Alice's adventure. An implementation example for a) is "Sekitei" where the user operates on physical objects. The operation of matter is related to the graphical effect in the computer. Additional elements included in the interaction are movie and sound devices since it is important in ZEN culture to feel "five sense".

An example for b) is based on Alice's adventures in the wonderland. The approach inherited the ZENetic computing paradigm in HCI providing object base interaction based on the adventure metaphor.

For both experiments, the objective lies within detailed Eastern/Western cultural aspects, still both did not discuss how to converge from one aspect to the other and reach the objective to scale one approach in another cultural surrounding/setting. In this paper, we aim to establish a common concept to allow in a first step analysis and comparison of approaches which could potentially lead to a convergence and realisation of synergistic situations in a second step.

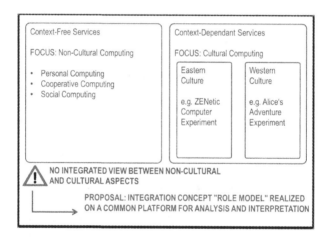

Fig. 1. Research goals mapped to related work

Fig. 1 shows the target outcome of the research work to compare cultural and non-cultural aspects in the service industry by applying the same scheme for analysis.

3   Constructing Context-Free and Context-Dependent Service Models

The guiding concept to construct service models is the "role model". In this section we firstly introduce the role model structure followed by the actual construction of a generic model for both types of services.

3.1 "Role Model" Concept

The assumption of the research work is that actions of service providers and consumers are based on their role models. For the service providers' case, the role model is a simulation of consumers' behaviour during service encounter, specific needs and demands. The consumers' role model is a simulation of the providers' behaviour: what type of service is available, how to order and act in the encounter stage, what services types are available and the like.

The role model is constructed upon a common knowledge base for both the service provider and the service consumer. The common knowledge base acts as a persistence mechanism on service operations to support the restricted memory of human actors - it is assumed that all information needed for service provision and consumption is provided through this layer. The operational knowledge for service provision is provided through different adaptation mechanisms for both service types: in the context-free model, the service provider has a constant setting from the background knowledge base; the dynamic aspect of the role model becomes clear for context-dependent services where a continues adaptation to the expectation and cultural background on both ends occurs. From a consumers' perspective the own low level maturity and expected high level/experience of service provider is matched against each other to reach a common interaction level. Hence the service provider has to design the service according to the expected heterogeneous levels of consumers and "fill-up" the knowledge-base accordingly with cultural-rich information.

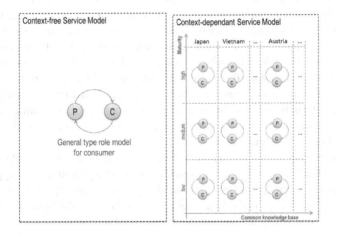

Fig. 2. Role model concept for 2 dimensions

Fig. 2 conceptually shows the role model approach exemplifying on 2 dimensions for service-dependent models namely "maturity level of consumer/provider" and "service location". As a neutral concept, the role model approach could potentially be applied for other dimensions. In the discussion of the case study we follow the same example dimensions as a) an adaptation means for consumer behaviour and b) localisation aspects to be considered. The characteristics of the role model for both service types are defined in the following.

3.2   Role Model for Context-Free Services

The context-free service as offered by a few large and widely operating organisation - wide in a sense of target audience and market - is designed to fit to the common knowledge base of a large set of consumers. The designed service uses a general type of role model, there is no need to adapt and fit the providers' role model to the observed model by consumers. The service offering is similar independent of location, cultural background, experience and maturity level, with the exception of minor non-functional aspects in service design (language, compliance to national legislation and the like). Because of this characteristic, the workflow for providing the service can be designed a priori and allows for a continuous scaling up. From an evaluation perspective services designed using this type do no aim to get the full-score of all consumers, but aim to get an average score from a lot of consumers.

3.3   Role Model for Context-Dependent Services

Context-dependent services as offered by a multitude of organisations that act on local scale and include domestic/local culture are designed for a sub-set of consumers' common knowledge base. For example, in the case of Edomae-sushi (Japanese high class sushi service), the service provider aims to provide his service only to high maturity level sushi consumers[1]. For these high-quality Japanese services, providers aim "to sense what their customers want from subtle cues and deliver a customized service without explicitly advertising the effort"[1] for doing so. The provider observes consumer behaviour and gradually modifies the role model to fit to the behaviour recognized. As a dynamic interaction process the consumer might adapt the role model as well and therefore the service provision is evolving ad-hoc during the encounter stage. For the context-dependent model, the provider aim at a full score from specific type of consumers, but do not attract and score from consumers of all segments. The applicability of the role-model concept is tested using the means of an explorative case study analysis using an IT prototype implementation on the ADOxx platform. The results of the case study are presented in the next section.

4   Case Study Results

For the evaluation of the concept as presented above a exploration approach was selected using a prototype developed following the OMI approach. Further details on the prototype and the project on OMI are provided in [10].

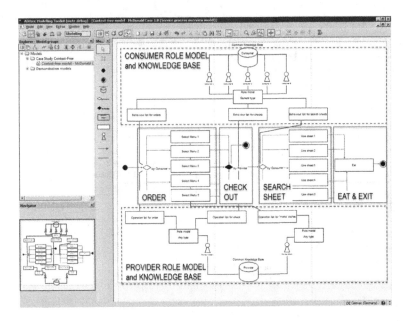

Fig. 3. Workflow Representation and Role Model Interaction for McDonald's Case

4.1 Context-Free Model: The McDonald's Case

The results of the case-study developed using the Multi-Context Modelling tool for the McDonald's case are shown in Fig.3.

For the context-free case it is possible to represent the service design process in a single workflow structure, the process phases are clearly defined and the consumer act in accordance with a general type role model. From a providers' perspective pre-defined role-models can be designed and determined in advanced for different stages of the process.

4.2 Context-Dependent Model: The Edomae-Sushi Case

The results of the case-study developed using the Multi-Context Modelling tool for the Edomae-Sushi case are shown in Fig.4.

For the context-dependent case, a single, static structure can not be determined in advance. During the execution of the process, the provider continuously and on an ad-hoc basis adapts the interaction schema with the consumer. The initial mind-set is the high-maturity consumer, in case the actual consumer is different, the role model is exchanged accordingly, resulting in a new definition of the process on instance level. For the example given in Fig.4, the different paths of modifications are explicitly modelled on an abstract level with the possibility to drill-down and refine the view. An observation from the sample using the current approach relates two two aspects: a) how could the actual modification stage be defined and described to explicate it in a readable/meaningful way and b) how are ad-hoc process/workflow defined and analysed.

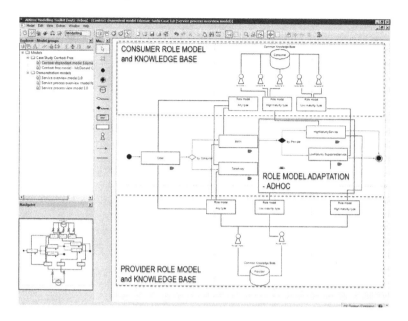

Fig. 4. Workflow Representation and Role Model Interaction for Edomae-Sushi Case

From an implementation perspective, the importance of information technology and cultural aspects in the service execution are discussed in the following section.

5   Implementation

For the application and implementation of context-free and context-dependent models in operation, two main characteristics are analysed: the IT factor and the recognition of the importance of culture in service delivery.

5.1   Distinction IT and Human Part

A main distinction point between the 2 types of models lies within the execution aspect of the design to support the mentioned scale-up. For context-free services, the main target is to rationalize the execution, ideally to support machine interpretation. Support of information technology and machine interpretation is feasible since the processes and workflows are create a priori with well-defined interfaces. In contrary, context-dependent services lack the a priori definition, for the interpretation of ad-hoc processes and adaptation mechanism the human factor play an important role.

5.2   Essential of Cultural Aspects

The Japanese Sushi case exemplifies on how the transformation from a highly-contextualized creative service to a context-free service has occurred over the last years. On a rather context-free basis, Japanese Sushi is available world-wide, where the original service as consumable mainly in Japan experienced a transformation from one type to the other. By comparing these two service offerings, the cultural aspect becomes clear. This approach is not limited to the Japanese culture, similar examples can be found around the globe (e.g. Viennese coffee house culture, France fashion culture, etc.) The concept of "role model" is positioned as an approach to capture these cultural aspects initially and provide means for integration in a human-driven workflow.

6   Conclusion

The research work presented in this paper constructs the concept of the "role model" on a common integration platform for service design. Using a common platform, the analysis of service from any type becomes feasible and is supported by mechanism and algorithms of a meta-modelling platform using conceptual models as a foundation. The constructed role model is regarded as an essential building block to explicate service interaction but needs to be embedded into a coherent and comprehensive environment that also considers the imposed business model from a strategic viewpoint, enactment support for defined work-flows exchanging flexibly the modelling approach (graph-based, natural language based, rule-based, logic-based), and dynamic evaluation aspects as a feedback loop to gather audit rails. Following this roadmap, the authors expect to be able to more precisely understand cultural aspects and scalability consideration in service economies of today.

Limitations and Further Research Work

The major limitation of the work presented relates to the approach followed when constructing the role model, namely a bottom-up methodology that focuses on two service enterprises to study the specifics of the role model concept of both types introduced earlier. Our work did elicit on the differentiation of Japanese traditional/"cultural" service and standard ones, to understand the Japanese context using a multi-context model framework. The concept is supposed to applicable to analyse other countries' context-dependent, non-standard service, the level of adaptation needed to make the model fit and suitable for different cultural settings needs to be studied further.

Another limitation of the work presented relates to the focus on the service encounter stage only. For a comprehensive and precise study, a wider perspective is needed, also considering the service design phase, the evaluation and feedback mechanisms and continuous adaptation/improvement (learning) as applicable. The future work direction is defined in Fig.5 to enable a platform for multi-modelling of the service process in context-free and context-dependent settings.

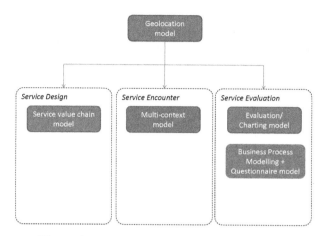

Fig. 5. Overview for a multi-modelling platform

Acknowledgments. The research project that led to the results presented in this paper is funded by JST-RISTEX. The authors wish to express their appreciation to Dr. Yamauchi as a team-member of the research project responsible for analysing the Sushi case as a core building block of developing the context-dependent model for Japanese Creative Services.

References

1. Hara, Y., Yamauch, Y., Yamakawa, Y., Fujisawa, J., Ohshima, H., Tanaka, K.: How Japanese traditional "Omonpakari" services are delivered - a multidisciplinary approach. In: SRII 2012 (2012)
2. Karagiannis, D., Fill, H.G., Hoefferer, P., Nemetz, M.: Metamodeling: Some application areas in information systems. In: Kaschek, R., et al. (eds.) UNISCON, vol. 5, pp. 175–188. Springer, Heidelberg (2008)
3. Karagiannis, D., Grossmann, W., Hoefferer, P.: Open model initiative - a feasibility study (2008), http://cms.dke.univie.ac.at/uploads/media/OpenModelsFeasibilityStudySEPT2008.pdf (April 04, 2010)
4. Kooijmans, T., Rauterberg, M.: Cultural Computing and the Self Concept: Towards Unconscious Metamorphosis. In: Ma, L., Rauterberg, M., Nakatsu, R. (eds.) ICEC 2007. LNCS, vol. 4740, pp. 171–181. Springer, Heidelberg (2007)
5. Masuda, H., Hara, Y.: A Dynamic Evaluation Model based on Customer Expectation and Satisfaction. In: SRII 2011, pp. 401–408 (2011)
6. Masuda, H., Hara, Y.: Using Value-in-use: A Dynamic Model for Value-in-Exchange and Value-in-Use. In: HSSE 2012, vol. 2012, pp. 5972–5980 (2012)
7. Rauterberg, M.: From Personal to Cultural Computing: how to assess a cultural experience, Information nutzbar machen. In: Kemper, G., von Hellberg, P. (eds.) Pabst Science Publ., pp. 13–21 (2006)

8. Tosa, N., Matsuoka, S.: Cultural Computing: ZENetic Computer. In: Proceedings of ICAT 2004, Korea, pp. 75–78 (2004)
9. Tosa, N., Matsuoka, S., Ellis, B., Ueda, H., Nakatsu, R.: Cultural Computing with Context-Aware Application: ZENetic Computer. In: Kishino, F., Kitamura, Y., Kato, H., Nagata, N. (eds.) ICEC 2005. LNCS, vol. 3711, pp. 13–23. Springer, Heidelberg (2005)
10. JCS Project on Open Models Platform,
    http://www.openmodels.at/web/jcs/home

Author Index